T0342192

SMART SOLAR PV INVERTERS
WITH ADVANCED
GRID SUPPORT FUNCTIONALITIES

SMART SOLAR PV INVERTERS
WITH ADVANCED
GRID SUPPORT FUNCTIONALITIES

Rajiv K. Varma
Electrical and Computer Engineering Department
The University of Western Ontario
London, ON
Canada

IEEE PRESS

WILEY

Published by John Wiley & Sons, Inc., Hoboken, New Jersey.
Published simultaneously in Canada.

For general information on our other products and services or for technical support, please contact our Customer Care Department within the United States at (800) 762-2974, outside the United States at (317) 572-3993 or fax (317) 572-4002.

Wiley also publishes its books in a variety of electronic formats. Some content that appears in print may not be available in electronic formats. For more information about Wiley products, visit our web site at www.wiley.com.

Library of Congress Cataloging-in-Publication Data
Names: Varma, Rajiv K., author.
Title: Smart solar PV inverters with advanced grid support functionalities
 / by Rajiv K. Varma.
Description: Hoboken, New Jersey : Wiley-IEEE Press, [2022] | Includes
 bibliographical references and index.
Identifiers: LCCN 2021032004 (print) | LCCN 2021032005 (ebook) | ISBN
 9781119214182 (cloth) | ISBN 9781119214205 (adobe pdf) | ISBN
 9781119214212 (epub)
Subjects: LCSH: Photovoltaic power systems. | Smart power grids.
Classification: LCC TK1087 .V37 2022 (print) | LCC TK1087 (ebook) | DDC
 621.31–dc23
LC record available at https://lccn.loc.gov/2021032004
LC ebook record available at https://lccn.loc.gov/2021032005

Cover Design: Wiley
Cover Images: "Morning sun over solar panels" courtesy of ABB Asea Brown Boveri Ltd, Zurich, Switzerland; "Nighttime PV-STATCOM demonstration site" courtesy of Bluewater Power Group of Companies, Sarnia, ON, Canada.

This book is dedicated to

Our Gurudev

CONTENTS

ABOUT THE AUTHOR

RAJIV K. VARMA is a *Professor* in the Electrical and Computer Engineering Department at the University of Western Ontario (UWO). He is a *Fellow of the Canadian Academy of Engineering.* He was also the *Hydro One Chair in Power Systems Engineering* from 2012 to 2015.

He has a rich experience of 32 years of teaching and research in academia. He is an internationally renowned researcher in Flexible AC Transmission Systems (FACTS), and grid integration of solar and wind power systems. He has received 13 *Teaching Excellence* awards at UWO.

He was the principal coauthor of the IEEE Press/Wiley book on *"Thyristor-Based FACTS Controllers for Electrical Transmission Systems,"* published in 2002. This book has been translated into Chinese and also has a Southeast Asian edition.

He is presently the Chair of the IEEE Power & Energy Society (PES) *"HVDC and FACTS Subcommittee."* He was also the Chair of the *"IEEE Working Group on HVDC and FACTS Bibliography"* from 2004 to 2019 and the editor of *"IEEE Transactions on Power Delivery"* during 2003–2008. He has actively contributed to the development of IEEE Standard 1547-2018 and IEEE Standard P2800.

He was the team lead for the first-ever IEEE Tutorial on *"Smart Inverters for Distributed Generators"* in IEEE PES T&D Conference in 2016 and in PES General Meetings in 2017, 2018, and 2019. He co-delivered the IEEE Substations Committee Tutorial on *"Static Var Compensator (SVC)"* six times at different IEEE PES T&D Conferences and PES General Meetings during 2005–2012.

He has also delivered several tutorials, courses, workshops, and webinars on smart inverters, FACTS, SVC, HVDC, and solar/wind integration in different countries including the United States, Canada, Colombia, Nepal, and India for utility engineers, system planners, and researchers.

He has led several research grants, including multiuniversity multiutility projects, on grid integration of solar PV systems and FACTS, totaling over $11 million. He has also published more than 180 papers in international journals and conferences.

He has developed a set of innovative technologies of utilizing PV solar farms in the *night* and day as a dynamic reactive power compensator – STATCOM (a FACTS Controller), which he named as *PV-STATCOM*. These novel PV-STATCOM technologies on existing solar farms can provide a 24/7 functionality of a STATCOM at a significantly lower cost for the same benefits. The PV-STATCOM technology both for *night* and day applications was successfully installed and demonstrated for the *first time in Canada, and perhaps in the world*, on 13th December 2016 in the utility network of Bluewater Power Distribution Corporation, in Sarnia, Ontario.

Dr. Varma holds 23 granted patents and 9 pending patents on this technology in the United States, Canada, Europe, China, and India. For this research, he received the *Prize Paper Award* from IEEE Power & Energy Society (PES) in 2012 and the *First Place Poster Award* in the 7th International IRED Conference in 2016.

He received the prestigious *IEEE PES Nari Hingorani FACTS Award* in 2021 "for advancing FACTS controllers application in education, research, and professional society and for developing an innovative STATCOM technology utilizing PV solar farms." He became a *Fellow of the Canadian Academy of Engineering* in 2021 with the citation, "...Among his pioneering contributions has been a major ground-breaking utility-implemented award-winning technology, PV-STATCOM, that enables solar PV plants to provide FACTS functionalities at one-tenth cost of FACTS themselves..."

He obtained B.Tech. and Ph.D. degrees in Electrical Engineering from the Indian Institute of Technology (IIT) Kanpur, India, in 1980 and 1988, respectively. He started his academic career as an Assistant Professor at the Indian Institute of Technology Kanpur, in 1989. He was awarded the Government of India *BOYSCAST Young Scientist Fellowship* in 1992–1993 to conduct research on FACTS at the University of Western Ontario, London, Canada. He continued as a Visiting Assistant Professor at UWO until December 1994. He returned to IIT Kanpur and was promoted to Associate Professor in 1997. He was awarded the *Fulbright Travel Grant* of the U.S. Educational Foundation in India to travel to the United States in 1998 and do research in High Voltage DC (HVDC) transmission and FACTS at Bonneville Power Administration, U.S. Dept. of Energy, Portland, Oregon. He became a Professor at the Indian Institute of Technology Kanpur, in 2001 prior to joining the University of Western Ontario in December 2001.

Dr. Varma has held *Adjunct Professor* positions at the University of Waterloo and Ryerson University, Toronto. He is *Senior Member* of IEEE, a *Member* of CIGRE, and also a licensed *Professional Engineer* in the province of Ontario.

FOREWORD

During the course of my 33-year career to date, I have had the privilege to contribute to a wide variety of activities associated with the application of Flexible AC Transmission Systems (FACTS) Controllers to improve power system dynamic behavior. This interest has taken me on a career journey with two different manufacturers, including direct experience on numerous utility-scale FACTS installations.

Through our mutual interest with the application of power electronics-based equipment to improve power system dynamic behavior, it was inevitable that Rajiv and I would meet by way of our common activities and volunteer work within the IEEE Power & Energy Society (PES) Transmission and Distribution (T&D) Committee and its HVDC and FACTS Subcommittee and participate in a number of its working groups including Performance and Modeling, Economics and Operating Strategies, and Education. Rajiv and I also interacted over the years on the subject of FACTS Controllers in the IEEE PES Substations Committee in multiple subcommittees and working groups. In addition, when I was an instructor in Bill Long's University of Wisconsin Engineering Professional Development program on the Dynamic Reactive Power Control Short Course series, Rajiv's book titled "Thyristor-Based FACTS Controllers for Electrical Transmission Systems," was the text for several editions of that course. Through these experiences, I became quite familiar with Rajiv's work and contributions to FACTS and smart solar photovoltaic (PV) inverters.

With the advent of more cost-effective equipment, along with the growing interest in decarbonization via renewable generation, the utilization of solar PV installations has grown significantly over the past decade and more. The application of this technology will continue to grow in the coming years as government mandates for Renewables Portfolio Standards (RPS) (or equivalent) increase and expand, while the total costs of PV installations continue to decrease. Beyond standard PV installations, which primarily focus on the control of the active power generated by the solar panels, lie opportunities for the application of smart solar PV inverters. Through advanced controls, smart solar PV inverters utilize the full range of capability for both active and reactive power, which in turn allows for a variety of benefits to improve power system dynamic behavior. These concepts are highlighted in detail in the various pages of this book.

Around 2009, Rajiv began to develop ground-breaking techniques to transform solar PV inverters into the performance of a STATCOM, which provides various functionalities both during the night and day. Rajiv termed this advancement as PV-STATCOM, which is essentially a new FACTS Controller. The PV-STATCOM controls can provide several advanced grid support functions such as dynamic voltage control, power oscillation damping, mitigation of subsynchronous control interactions and torsional oscillations, improved Fault-Induced Delayed Voltage Recovery (FIDVR), stabilization of remote critical motors, and fast frequency response, all of which can lead to improved dynamic performance, increased power transfer and load serving, and enhanced connectivity of neighboring wind plants and solar plants. It is worth noting that the PV-STATCOM technology is about 10 times more economical than an equivalent-sized STATCOM. This advancement by Rajiv, along with a field demonstration for the first time in 2016, is described in the various pages of this book in several chapters.

Rajiv, by providing this book to the industry, captures his career-long dedication to, and knowledge of, power electronics-based systems to improve dynamic performance. This book is a timely addition on the growing topic of smart solar PV installations. The topics cover a wide range of interest from smart PV inverter functions, modeling and control, applications (both distribution and transmission), hosting capacity, coordinated control, and emerging trends. The treatment further supports the characterization that the smart solar PV inverter, designated as PV-STATCOM, is a new FACTS Controller. This book will be of great interest from beginners to experts in academia, research, and industry of all competencies including utilities, system operators, developers, integrators, regulators, manufacturers, and beyond.

John Paserba, Fellow IEEE
Vice President, Power Systems Group
Mitsubishi Electric Power Products, Inc. (MEPPI)
Warrendale, PA, USA

PREFACE

Solar photovoltaic (PV) systems are the fastest growing renewable energy systems, worldwide. It is expected that by 2050 about 35% of global electricity will be provided by solar PV systems. While this technology is helping reduce greenhouse gas emissions and meeting the climate targets for the planet, researchers worldwide have been engaged in developing technologies for additional and novel usages of solar systems.

Solar PV systems are based on inverters which have traditionally provided only active power generation from solar energy. Power electronics however allows several additional capabilities to be realized from the same inverters which can be of tremendous benefit in enhancing the stability and reliability of power systems. Efforts have been ongoing worldwide to develop such "advanced" or "smart" functionalities on solar PV inverters. Such inverters have been termed as "advanced inverters" or more commonly as "smart inverters." Smart inverter functions have been shown to not only mitigate the problems of integration of solar systems themselves but also to alleviate challenges in power systems caused by other sources, such as disturbance events. The developments of smart inverter functionalities have outpaced the Standards responsible for integrating solar PV systems in the grid. The interest and engagement of academics, utilities, system planners, regulators, operators, and manufacturers in the development of the smart inverter technologies is very high.

A wealth of literature has been published over the last two decades describing the controls, simulation studies, laboratory implementations and operating experiences of various functionalities of smart inverters, which continues to grow every day at a very rapid pace. Unfortunately, this invaluable literature is scattered and not available in a comprehensive form.

In 2008, the province of Ontario in Canada undertook a major initiative to develop innovative technologies for integrating solar PV systems at a large scale in transmission and distribution systems. The author of this book was privileged to be selected to lead three highly funded ($8.2 Million) multi-university multi-disciplinary multi-utility research grants in Ontario to achieve this objective. During this period, the author developed a new patented technology of utilizing solar PV systems in the night and day as STATic Synchronous COMpensator (STATCOM), naming it PV-STATCOM, for providing various grid support functionalities which are typically provided by Flexible AC Transmission System (FACTS).

The author was also fortunate to be the team-lead for the first-time delivery of the IEEE Tutorial on "Smart Inverters" with leading experts from EPRI, NREL, National Grid, Southern California Edison, Enphase Energy, and First Solar. This Tutorial was delivered with great success for four consecutive years in IEEE Power & Energy Society (PES) Conferences. This also provided the author a great learning experience of different perspectives of smart inverters. The author also presented several panels sessions in IEEE PES General Meetings, and invited lectures, courses and workshops on smart inverters in different countries.

All the above provided the motivation to compile and organize, even though on a minute scale, the enormously rich and vastly distributed literature on smart inverters in the form of a book. The author is very grateful to the immense knowledge contributed by researchers worldwide, and leading organizations such as IEEE, EPRI, NREL, NERC, WECC, LBNL, CAISO, CIGRE, IEA PVPS Task 14 Group, to name only a few, from whose knowledge and contributions this book has greatly benefited.

This book is organized into nine chapters.

Chapter 1 presents the concepts of reactive power and active power control, which form the basis of smart inverter functions. The impact of such controls on system voltage and frequency are explained. Different challenges of high solar PV penetration in transmission and distribution systems are briefly described. The evolution of smart inverter technology is then presented.

Chapter 2 presents different smart inverter functions for both reactive power and active power based voltage control. The voltage and frequency ride through functions are explained and their implementation in different Standards such as IEEE Standard 1547-2018 and NERC's Standard PRC 024-3 are described. Smart inverter functions for battery energy storage systems are further elucidated. The prioritization of different smart inverter functions are discussed. Emerging smart inverter functions are then introduced.

Chapter 3 presents the basic concepts of active and reactive power flow in a smart inverter system. The operating principles and models of different subsystems in the power circuit and control circuit of a smart PV inverter system are described. The implementation methodology of different smart inverter controls is explained with smart inverter voltage controller as an example. The principle of achieving a decoupled control of active power and reactive power is presented. The modeling needs of different smart inverter controllers are discussed.

Chapter 4 presents the basic concepts of FACTS technology and two of its main-shunt connected member Controllers – the Static Var Compensator (SVC) and STATCOM. The focus of this Chapter is to present a new technology developed by this book's author, of utilizing PV solar farms both during nighttime when solar farms are typically idle and during any time of system need during daytime as a STATCOM, named PV-STATCOM. The different nighttime and daytime operating modes of the PV-STATCOM are illustrated. The cost of transforming an existing solar PV system into PV-STATCOM as well as its operating costs are analyzed. Subsequently, the potential of PV-STATCOM technology in providing various benefits in transmission and distribution systems, is elucidated.

Chapter 5 describes different night and day applications of PV-STATCOM technology for providing various grid support functions related to distribution systems, with case studies. These include dynamic voltage control, enhancing connectivity of PV solar farms, increasing connectivity of neighboring wind farms, and stabilization of critical motors. These are the functions for which typically SVCs or STATCOMs are employed, which are quite expensive.

Chapter 6 presents different night and day grid support functions provided by PV-STATCOM in transmission systems. These comprise improving power transfer capacity in transmission lines, damping of power oscillations and alleviation of Fault Induced Delayed Voltage Recovery (FIDVR). These functionalities are provided by reactive power modulation at night and by a combination of active and reactive power modulation during daytime. PV-STATCOM applications are also presented for mitigation of subsynchronous oscillations in synchronous generators and induction generator based wind farms connected to series compensated transmission lines.

A unique PV-STATCOM functionality of simultaneously providing fast frequency response and power oscillation damping is also described with a case study.

Chapter 7 explains the concept of hosting capacity for solar PV systems and its enhancement in distribution networks. Different non smart inverter based methods for increasing hosting capacity are presented. The characteristics of different smart inverter functions and their effectiveness in improving hosting capacity are discussed. The methodologies and guidelines for selecting the settings of different smart inverter functions are explained. Several simulation studies of increasing hosting capacity in utility networks are described. Finally, different worldwide field implementations of smart inverters in enhancing hosting capacity are presented and their key takeaways highlighted.

Chapter 8 presents the concepts of control coordination and discusses the lessons learned from control coordination of FACTS Controllers, which would be helpful in resolving control interaction issues in smart inverters. Control coordination issues of smart PV inverters with conventional voltage control equipment are presented. Case studies of control interactions between same and different smart inverter functions among neighboring smart inverters are described. A detailed small signal study of the various factors causing control interaction between two smart inverters in a distribution feeder, validated by electromagnetic transients simulations, is presented. A comprehensive control coordination study of 100 MW PV-STATCOM and 100 MW Doubly Fed Induction Generator (DFIG) based wind farm connected to series compensated line in mitigating subsynchronous oscillations is also described.

Chapter 9 deals with some of the fast-emerging trends with smart PV inverters. Some application examples are presented of enhanced grid support capabilities enabled by integrating the smart inverter functionalities of solar PV inverters, Battery Energy Storage Systems and Electric Vehicle Chargers. A new technology of "grid forming inverters" that is presently being widely researched across the world, is introduced.

The main focus of this Chapter is to describe the field demonstrations of novel smart PV inverter functions which can provide significant cost savings and benefits to power transmission and distribution systems. These advanced grid support functions are presently not mandated in any Standard worldwide for grid interconnection of solar PV systems. These functionalities include fast frequency response, flexible solar operation, reactive power at night, and night and day PV-STATCOM technology for providing several FACTS functionalities. This Chapter presents some thoughts on potential financial compensation mechanisms to smart PV inverters for providing grid support functionalities that go beyond being just "good citizens" on the power transmission and distribution systems.

This book is intended for academics, graduate students, utility engineers, system planners, system regulators, system operators, and inverter manufacturers. It starts by providing fundamental understanding of various aspects of smart inverter controls and their functionalities. It then presents advanced controls and novel functionalities of smart inverters for enhancing power system stability and reliability through detailed small signal and electromagnetic transient simulation studies. The book however does not cover protection systems and communications systems for smart inverters.

This book is written both to provide intuitive understanding of smart inverter concepts for beginners as well as advanced knowledge for adepts. The book therefore treads a middle path of presenting mathematical formulations with only a moderate level of complexity. Since the available knowledge on smart inverters is extremely vast the approach in the book is to explain the essential aspects and provide an exhaustive list of references for subsequent reading.

It is hoped that this book will inspire readers into the realm of smart inverters. This is the first book exclusively devoted to smart inverters, to the best of author's knowledge, and is very likely to have inadvertent errors and omissions. The author sincerely apologizes to all the readers for the same, and requests that these errors may kindly be communicated to him so that they may be rectified later.

Rajiv K. Varma

ACKNOWLEDGMENTS

I consider myself extremely fortunate to have learned and to be inspired by some of the extraordinarily distinguished and selfless teachers, researchers, and individuals, to whom I shall forever be grateful beyond words. Their immense wisdom has tremendously helped shape my career and which has eventually led to this book.

I first express my profound gratitude to Dr. K.R. Padiyar, my teacher and PhD thesis supervisor at IIT Kanpur, who initiated me into power systems and FACTS, and taught me the fundamentals of how to do research and write it. He had so much to teach me, but I could learn only little due to my own limitations. It is indeed a blessing in my life to be his student.

My sincere thanks to Dr. Narain Hingorani, the inventor of FACTS technology, who has been an enormous inspiration in my career.

My heartfelt gratitude to Dr. M.A. Pai, who has continuously supported and encouraged me throughout my academic career and during the process of this book writing. I am immensely grateful to Late Dr. Prabha Kundur, who incessantly inspired and motivated me all along in my career, and especially so, while writing this book. My true gratitude is also due to Late Michael Henderson for his constant encouragement in my research and in this book writing.

John Paserba has played a very important role in my career and for this book, for which I can never thank him enough. I am indebted to him for firstly reviewing the book's manuscript and providing very meticulous comments which greatly helped in improving the book. He then very kindly agreed to write the Foreword which is indeed an enormous honor for this book.

I also greatly appreciate Dr. Ram Adapa, Dr. Benjamin Jeyasurya, and Oleg Popovsky for reviewing the initial book proposal and graciously recommending that this book be published by Wiley/ IEEE Press.

I sincerely thank IEEE, CIGRE, Electric Power Research Institute (EPRI), National Renewable Energy Laboratory (NREL), North American Electric Reliability Corporation (NERC), CEATI International Inc., Western Electricity Coordinating Council (WECC), Lawrence Berkeley National Laboratory (LBNL), California ISO (CAISO), and EU-PVSEC for providing copyright permissions to reproduce some of their material in this book.

I am immensely grateful to Janice McMichael Dennis, President and CEO, Bluewater Power Group of Companies, and Tim Vanderheide, former Vice President, Bluewater Power, Sarnia, for their constant research support. They were extremely generous to provide their solar farm site for demonstrating my PV-STATCOM technology for the first time in Canada, and perhaps in the world. I am further very appreciative of Bluewater Power for providing the nighttime picture of their solar farm site for PV-STATCOM field demonstration, and ABB Asea Brown Boveri Ltd., Switzerland, for the morning picture of their solar farm, which adorn the cover of this book.

I sincerely thank Ben Mehraban, Stephen Williams, and Dr. Hemant Barot for providing valuable insights on the practical aspects of solar PV systems and FACTS, which I have included in this book.

I convey my indebtedness to The University of Western Ontario for having me, and providing all the support and facilities for performing my research in solar PV systems and FACTS. My profound thanks to Dr. Ken McIsaac, Chair of Electrical and Computer Engineering Department, for his support for my PV solar systems research and in writing this book.

I greatly appreciate my former MESc students Sridhar Bala Subramaniam, Mahendra A.C., Byomakesh Das, and Vishwajitsinh Atodaria; former PhD students Shah Arifur Rahman, Ehsan M. Siavashi, Hesamaldin Maleki, Reza Salehi, Mohammad Akbari, and Sibin Mohan; and former Post Doctoral Fellows Dr. Vinod Khadkikar and Dr. Iurie Axente for their research on different aspects of PV-STATCOM technology. Each one has also helped me in a unique way in this book writing by performing some system studies included in this book, doing literature search, and drawing figures. To each one of them, individually, I express my sincere thanks.

I wish to express my gratitude to the entire team of John Wiley & Sons led by Mary Hatcher for kindly agreeing to publish this book and for providing all the support throughout the publication process. I also wish to sincerely thank Victoria Bradshaw for developing the book cover, Teresa Netzler for supervising the book's production, and Viniprammia Premkumar for preparing the final version of the book in such an excellent manner for printing.

Finally, I owe this book to my wife Malini and children Sarvesh and Ratna without whose immense sacrifices, immeasurable patience, and limitless support this book would have never happened.

Rajiv K. Varma

LIST OF ABBREVIATIONS

AGC	automatic generation control
ANSI	American National Standards Institute
APC	active power curtailment
APS	Arizona Public Service
AVR	automatic voltage regulator
BESS	battery energy storage system
BOS	balance of system
BPS	bulk power system
CAISO	California Independent System Operator
CIGRE	Conseil International des Grands Réseaux Electriques, translated as, International Council on Large Electric Systems
CPUC	California Public Utilities Commission
DER	distributed energy resource
DERMS	distributed energy resource management system
DFIG	doubly fed induction generator
DG	distributed generator/generation
DMS	distribution management system
DVAR	dynamic VAR
EHV	extra high voltage
EPC	engineering, procurement, and construction
EPRI	Electric Power Research Institute
EPS	electric power system
ERCOT	Electric Reliability Council of Texas
ESS	energy storage system
FACTS	Flexible AC Transmission System
FERC	Federal Energy Regulatory Commission
FFR	fast frequency response
FRO	frequency response obligation
HC	hosting capacity
HFRT	high frequency ride through
HV	high voltage
HVDC	high voltage direct current
HVRT	high voltage ride through
IBR	inverter based resource

IEA	International Energy Agency
IEEE	Institute of Electrical and Electronics Engineers
IGBT	insulated gate bipolar transistor
ILC	inverter level controller
IM	induction motor
LBNL	Lawrence Berkeley National Laboratory
LFRT	low frequency ride through
LTC	load tap changer
LV	low voltage
LVRT	low voltage ride through
MPPT	maximum power point tracking
MV	medium voltage
MVA	mega volt ampere
MVAR	mega volt ampere reactive
MW	megawatt
NERC	North American Electric Reliability Corporation
NREL	National Renewable Energy Laboratory
OLTC	on load tap changer
OPF	optimal power flow
p.u.	per unit
PCC	point of common coupling
PF	power factor
PFR	primary frequency response
PI	proportional–integral
PII	permitting, inspection, and interconnection
PLL	phase locked loop
PMU	phasor measurement unit
PoC	point of connection
POI	point of interconnection
PPC	power plant controller
PQ	power quality
PSDC	power swing damping controller
PSS	power system stabilizer
PV	photovoltaic
PVPS	photovoltaic power systems
PV-STATCOM	photovoltaic static synchronous compensator
PWM	pulse width modulation
QSTS	quasi static time series
RMS	root mean square
ROCOF	rate of change of frequency
RPC	reactive power control
RTDS	real time digital simulator
SCADA	supervisory control and data acquisition
SCE	Southern California Edison
SEIG	self-excited induction generator
SF	solar PV farm

SG	smart grid
SI	smart inverter
SIL	surge impedance loading
SIR	synchronous inertial response
SIWG	Smart Inverter Working Group
SLG	single line to ground fault
SOC	state of charge
SPWM	sinusoidal pulse width modulation
SRP	Salt River Project
SSDC	subsynchronous damping controller
SSO	subsynchronous oscillations
SSR	subsynchronous resonance
STATCOM	static synchronous compensator
SVC	static var compensator
TCR	thyristor-controlled reactor
THD	total harmonic distortion
TOV	transient overvoltage
TSC	thyristor-switched capacitor
UFLS	under-frequency load shedding
UPF	unity power factor
VAR	volt amp reactive
VCO	voltage controlled oscillator
VSC	voltage source converter
VSI	voltage source inverter
VVC	volt–var control
WAMS	wide area measurement system
WECC	Western Electricity Coordinating Council
WF	wind farm
WTG	wind turbine generator

1

IMPACTS OF HIGH PENETRATION OF SOLAR PV SYSTEMS AND SMART INVERTER DEVELOPMENTS

Solar Photovoltaic (PV) power systems are being integrated at an unprecedented rate in both bulk power systems and distribution systems worldwide. It is expected that by 2050, solar PV systems will provide about 35% of global electricity generation [1]. Different countries, and their provinces and states, are setting up ambitious targets for PV system installations up to 100% renewables with substantial share of solar PV systems. Several grid impact studies with 100% Inverter Based Resources (IBRs) and Distributed Energy Resources (DERs) with a major component of solar PV systems have already been performed [2, 3]. While these systems significantly help in reducing overall greenhouse gas emissions, they present unique integration challenges which need to be understood and mitigated to derive full benefits from their applications. The solar PV systems are based on inverters. Power electronics technology provides new "smart" capabilities to the inverters in addition to their primary function of active power generation. These capabilities not only help solar PV systems mitigate different adverse impacts of their integration but also provide several valuable grid support functions.

This chapter presents the concepts of reactive power and active power control, which form the basis of smart inverter operation. The impact of such controls on system voltage and frequency is explained. The different challenges of integrating solar PV systems on a large scale in transmission and distribution systems are briefly described [4]. The evolution of smart inverter technology is then presented.

1.1 Concepts of Reactive and Active Power Control

1.1.1 Reactive Power Control

1.1.1.1 Voltage Control

Injection of reactive power at a bus causes the voltage to rise whereas absorption of reactive power causes the bus voltage to decline. Figure 1.1 illustrates a simple power system having an equivalent voltage E and equivalent network short circuit impedance with reactance X and resistance R. An inductor X_L is connected as load at a bus termed Point of Common Coupling (PCC) to show the effect of reactive power absorption. The PCC voltage and inductor current are denoted by V and I, respectively. The impact of reactive power absorption by the inductor on the PCC voltage is examined through phasor diagrams for three cases of network impedance. The phasor diagrams for cases

Smart Solar PV Inverters with Advanced Grid Support Functionalities, First Edition. Rajiv K. Varma.
© 2022 The Institute of Electrical and Electronics Engineers, Inc. Published 2022 by John Wiley & Sons, Inc.

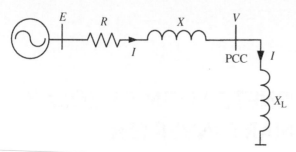

Figure 1.1 A simple power system with an inductor connected at PCC.

(a)

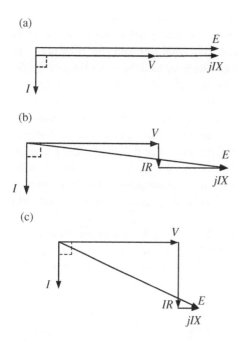

(b)

(c)

Figure 1.2 Phasor diagrams for network with inductive load; (a) network with $R = 0$; (b) network with $X/R = 3$; (c) network with $X/R = 1/3$.

(a) $R = 0$ (purely inductive network), (b) $X/R = 3$ (substantially reactive network), and (c) $X/R = 1/3$ (substantially resistive network) are depicted in Figure 1.2a–c, respectively. The phasor diagrams are drawn with the phasor $\boldsymbol{V}$ as reference, which has same magnitude in all the three cases. The phasor diagrams can also be drawn with equivalent voltage $\boldsymbol{E}$ as reference phasor having the same magnitude, although the conclusions will be the same in both cases.

In the absence of inductor X_L, the PCC voltage is $\boldsymbol{E}$. The lagging inductor current causes a voltage drop $IR + jIX$ across the network impedance, thereby reducing the PCC voltage to $\boldsymbol{V}$. Stated alternately, the reactive power absorption by the inductor reduces PCC voltage by an amount $|\boldsymbol{E}| - |\boldsymbol{V}|$.

For case (a) $R = 0$, it is evident from Figure 1.2a that the change in voltage is directly proportional to network reactance and the magnitude of inductive current I (which in turn is dependent on the size of the bus inductor X_L). Hence for same inductive current, the larger the network reactance, larger is the change in bus voltage. This also implies that higher reactive power absorption (corresponding to higher I) will cause a larger reduction in voltage in weak systems.

The impact of system X/R ratio is seen from Figure 1.2b corresponding to $X/R = 3$, and from Figure 1.2c relating to $X/R = 1/3$. The same amount of reactive current and reactive power absorption in inductor X_L causes a larger voltage drop in the network with higher X/R ratio.

Figure 1.3 A simple power system with a capacitor connected at PCC.

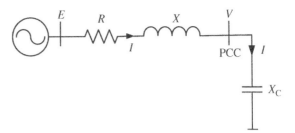

Consider a capacitor X_C being connected as load at the PCC as depicted in Figure 1.3. The impact of reactive power injection by the capacitor on the PCC voltage is investigated through phasor diagrams for three cases of network impedance. The phasor diagrams for cases (i) $R = 0$ (purely inductive network), (ii) $X/R = 3$ (substantially reactive network), and (iii) $X/R = 1/3$ (substantially resistive network) are displayed in Figure 1.4a–c, respectively. In the absence of capacitor, the PCC voltage is E. The leading capacitor current causes a voltage drop $IR + jIX$ across the impedance of the network, thereby increasing the PCC voltage to V. Stated alternately, the reactive power injection by capacitor increases the PCC voltage by an amount $|V| - |E|$.

The change in voltage due to capacitive load is thus directly proportional to network reactance and the magnitude of capacitive current I (which in turn is dependent on the size of the bus capacitor X_C), as seen from Figure 1.4a. Hence for same capacitive current, the larger the network reactance, higher is the change in voltage. This also demonstrates that higher reactive power injection (corresponding to higher I) will cause a larger increase in voltage in weak systems.

The impact of system X/R ratio is observed from Figure 1.4b corresponding to $X/R = 3$, and from Figure 1.4c relating to $X/R = 1/3$. The same amount of reactive current and reactive power injection by capacitor X_C will cause a larger voltage rise in networks with higher X/R ratio.

(a)

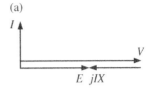

(b)

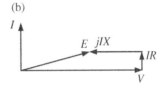

(c)

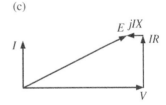

Figure 1.4 Phasor diagrams for network with capacitive load; (a) network with $R = 0$; (b) network with $X/R = 3$; (c) network with $X/R = 1/3$.

The above analysis demonstrates that a voltage control strategy based on reactive power exchange at a bus will be more effective in weak systems and in systems with higher X/R ratio, i.e. in largely inductive networks. Conversely, reactive power exchange will be less effective in strong systems and also in substantially resistive networks.

Low-voltage distribution systems are typically characterized by low X/R ratios. Hence a purely reactive power based voltage control strategy will be less effective. This may necessitate the use of voltage regulation strategies based on active power control.

1.1.1.2 Frequency Control

Reactive power exchange has no direct impact on the system frequency. However, several studies [5, 6] have reported that reactive power exchange indirectly influences the system frequency in systems having a large number of voltage-dependent loads and in microgrids. In this case, change in bus voltages caused by reactive power exchange can vary the active power absorbed by the loads leading to a mismatch in the generation and loads and consequent frequency deviation. Reasonably effective strategies of frequency regulation through reactive power control can, therefore, be developed especially in power systems having substantial component of voltage-dependent loads.

1.1.2 Active Power Control

1.1.2.1 Voltage Control

i) ***Exchange of Active Power:*** The influence of active power exchange on voltage regulation is discussed in this section. A simple power system is considered with a solar PV system connected at the PCC as illustrated in Figure 1.5. The system has an equivalent source voltage E, whereas V is the voltage at PCC. The network has equivalent reactance and resistance X and R, respectively. The current I injected by the PV system is considered to be purely resistive I_R, i.e., at unity power factor. Impact of this active current injection on systems with different X/R ratios is now investigated. Figure 1.6a,b depicts the phasor diagram for systems with $X/R = 3$, and $1/3$, respectively. Active power injection is seen to increase the voltage in both cases. However, the rise in voltage is more in the network having $X/R = 1/3$ than in network having $X/R = 3$. This implies that active power injection will result in a greater voltage rise in primarily resistive networks. Similarly, active power absorption, as in energy storage systems, will be more effective in decreasing the voltage in resistive networks.

ii) ***Exchange of Active and Reactive Power:*** In this case, the solar PV system is considered to perform both active power injection and reactive power exchange. A simple explanation of the impact of active and reactive power is described here. The complex current injected by the solar PV system is denoted by I, and the complex power injected by the solar PV system is expressed by $P + jQ$. Assuming the open-circuit voltage E remains constant, the change in PCC voltage due to this current injection is given by:

$$\Delta V = V - E = I(R + jX) \tag{1.1}$$

$$I = \frac{(P - jQ)}{V^*} = \frac{(P - jQ)}{V} \tag{1.2}$$

since V is chosen to be the reference phasor with angle $= 0$.

Substituting Eq. (1.2) in Eq. (1.1), the change in voltage is obtained as

$$\Delta V = \frac{(P - jQ)}{V}(R + jX) \tag{1.3}$$

$$\Delta V = \left[\frac{(RP + XQ)}{V}\right] + j\left[\frac{(XP - RQ)}{V}\right] \tag{1.4}$$

It is noted that Q can be either inductive or capacitive. Also, P can be either positive, if injected as in case of PV system, or negative if absorbed by an Energy Storage System.

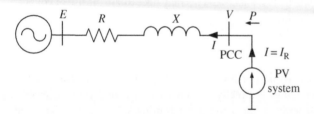

Figure 1.5 A simple power system with a PV solar system connected at PCC.

Equation (1.4) demonstrates that the PCC voltage is influenced by both active and reactive power exchange from the DER [7]. The magnitude of voltage change is dependent on the system short circuit impedance and its X/R ratio.

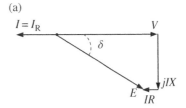

1.1.2.2 Frequency Control
A balance between active power generation and the sum of loads and line losses results in a stable system frequency. Hence, active power exchange (injection/absorption) by IBRs and DERs with the power system directly impacts system frequency. In general, the system frequency increases with active power injection and decreases with active power absorption. The magnitude of frequency change is dependent upon the relative value of active power exchange in comparison with power generation from the remaining generators in

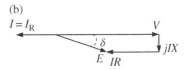

Figure 1.6 Phasor diagrams for network with active power injection from the solar PV system; (a) network with $X/R = 3$; (b) network with $X/R = 1/3$.

the power system. Greater the active power exchange higher is the variation in system frequency. Consequently, the impact of DER active power exchange on microgrid system frequency is much larger than in grid connected environments. The impact of active power control on system frequency is briefly described in the next section.

1.1.3 Frequency Response with Synchronous Machines

A steady-state system frequency results when synchronous power generation matches the system load and the losses in power system supplying those loads. In case of a system disturbance, such as sudden loss of a major generation unit, the typical variation in system frequency and its subsequent restoration to the pre-disturbance level is depicted in Figure 1.7 [8]. The frequency restoration is enabled through three sequential stages of frequency control – primary frequency control, secondary frequency control, and tertiary frequency control [8–13].

Assume that the power system is operating at steady state at $t = 0^-$ and a large generation loss occurs at $t = 0^+$. The kinetic energy of all the synchronous machines (generators, condensers, motors) is autonomously extracted to supply the load (inertial response), leading to a decline in the speed of generators and consequently the system frequency. The decline in frequency continues till additional power injection from synchronous generators balance out the load. The rate at which the frequency decreases is termed "Rate of Change of Frequency (ROCOF)." The lowest level at which the frequency is eventually arrested is known as "frequency nadir." The time period from the onset of disturbance to reaching the frequency nadir is known as "arresting period."

Primary frequency control (also referred as Frequency Containment Reserve [FCR]) is provided by synchronous generator turbine governors by injecting power from the generators during the arresting period and continuing thereafter. This causes the frequency to stabilize at the "settling frequency," which is higher than the nadir but still lower than the steady-state frequency before the disturbance. This time period until settling frequency is reached is termed "rebound period."

Secondary frequency control through Automatic Generation Control (AGC) is then exercised to restore system frequency to its pre-disturbance scheduled level. The period over which this secondary frequency control is provided is known as "recovery period," which extends over 5–10 minutes (or more).

Tertiary frequency control is provided subsequently which involves restoration of the synchronous generator and other reserves which provided primary frequency control, to their preset levels, so that they can respond to any future loss-of-generation events. The tertiary control involves

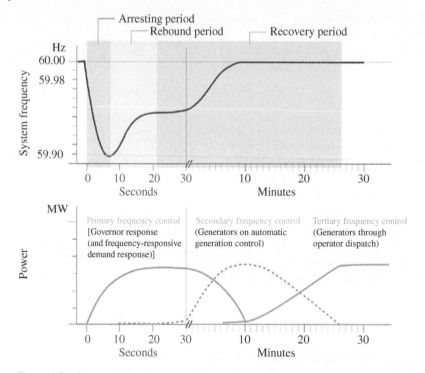

Figure 1.7 Sequential frequency controls after a sudden loss of generation and their impact on system frequency. *Source:* Eto et al. [8]. Reprinted with permission from Lawrence Berkeley National Laboratory, Berkeley, CA, USA.

coordinated changes in the dispatch levels (outputs) of different generators. In this control, some generators are dispatched down to restore their reserve capability while some other generators are dispatched up by a corresponding amount, while maintaining the scheduled system frequency. Deployment of tertiary frequency control is the final stage of frequency restoration in the recovery period.

The focus of this book is on the arresting period, frequency nadir, and initial parts of the recovery period with and without the high penetration of inertia-less solar PV systems. The key performance indicators involved in frequency response are explained below.

1.1.3.1 Rate of Change of Frequency

The ROCOF is a measure of how quickly the frequency changes following a sudden imbalance between generation and load [9]. ROCOF at any point on the frequency response curve is the tangential line at that point. However, it is usually calculated as the change in frequency over a short period of time, immediately after the sudden loss of generation.

The initial ROCOF after generation loss at $t = 0$ s is typically calculated as:

$$\text{ROCOF}_{0.5} = \frac{f_{0.5} - f_0}{0.5\,\text{s}} \tag{1.5}$$

A high ROCOF indicates that the system frequency can potentially fall to a level where Under Frequency Load Shedding (UFLS) may be initiated. ROCOF is used to determine the approximate time required to reach the UFLS threshold. This helps systems planners and operators to provide adequate amount of frequency response within this period to prevent load shedding. It is noted that as soon as the frequency starts to decline, several factors come in to play such as inertial response, load response, and other nonlinear actions which may alter the actual time to reach UFLS [9].

1.1.3.2 Factors Impacting ROCOF

ROCOF is influenced by the following factors [9]:

1) Size of the contingency, i.e. amount of the lost generation or magnitude of power over a major tie line (e.g. HVDC line) from neighboring interconnections or load (for overfrequency events).
2) Total system inertia contributed by synchronous machines (synchronous generators, synchronous condensers, and synchronous motors).
3) Magnitude and the speed of energy injected from generating resources subsequent to contingency.
4) Sensitivity of loads to change in system frequency.
5) Incremental losses in the bulk power system due to modification in power flows after the contingency.

Factors (1), (2), and (3) have a more significant contribution to ROCOF among all the above factors.

1.1.3.3 System Inertia

Synchronous inertia is defined as "the ability of a power system to oppose changes in system frequency due to resistance provided by rotating masses" [14]. Inertia of a synchronous machine is defined as the ratio of stored kinetic energy to the MVA rating of the machine and is expressed as MW-second/MVA, i.e. in seconds. The inertia constant effectively indicates the time duration over which the kinetic energy stored in the rotating mass will allow the production of rated output of the synchronous machine.

The system inertia, however, represents the aggregation of kinetic energy stored in the rotating masses of all the synchronous machines (generators, condensers, and motors) in the system. This system inertia provides the Synchronous Inertial Response (SIR) to arrest the system frequency as soon as the frequency starts to decline following the loss of a major generation [13].

The system inertia dictates the initial ROCOF as below [9]:

$$\text{ROCOF} = \frac{\Delta P_{\text{loss}}}{2 \times \left(\text{KE}_{\text{sys}} - \text{KE}_{\text{loss}}\right)} \times 60 \tag{1.6}$$

$$\text{KE}_{\text{sys}} = \sum_{i \in I} H_i \times \text{MVA}_i \tag{1.7}$$

where ΔP_{loss}, loss in generation; KE_{loss}, inertia of the lost generating system; KE_{sys}, total system kinetic energy of the online synchronous machines; H_i, inertia constant of the online synchronous machine; MVA_i, MVA rating of the ith online synchronous machine; I, set of all the online synchronous machines.

The above formulation demonstrates that ROCOF is directly proportional to the size of generation loss and inversely proportional to twice the SIR of the system.

In isolated microgrid environments, the relative value of active power exchanged from the DERs in comparison with the remaining generators is high. Hence, the impact of DER generation loss on microgrid system frequency is much larger than in grid-connected environments.

It is noted that as the penetration of static IBRs (solar PV, wind generators, battery energy storage systems [BESSs]) increases in power systems, the system kinetic energy will decline.

1.1.3.4 Critical Inertia

Critical inertia is defined as the minimum level of system inertia necessary to ensure that frequency responsive reserves have sufficient time to be deployed and prevent the operation of the first stage of UFLS after the largest credible contingency [9]. Critical inertia in an interconnected system is also referred as "inertia floor" for that system.

Critical inertia is dependent on various factors, such as the amount of fast frequency response (FFR) and primary frequency response (PFR) available in the power system, size of critical contingency, the thresholds for initiating UFLS, etc. The critical inertia can be reduced if (i) FFR from loads or other mechanisms can be provided faster, (ii) the thresholds for UFLS are lowered, or (iii) the size of the largest contingency is made smaller.

1.1.3.5 Size of Largest Contingency

The largest credible contingency is described by NERC Reliability Standard BAL-003 (Resource Loss Protection Criteria) [9]. It is typically the loss of the largest generating unit or a combination of generating units or a large transmission tie line. This largest contingency results in the maximum imbalance between generation and loads.

Larger the size of the contingency, larger will be the reduction in system inertia and consequently, a steeper and more negative ROCOF will be experienced. On the other hand, if the inertia of the lost generator is small relative to the total inertia in the power system, the ROCOF will not be as steep. This will be beneficial for the power system.

1.1.4 Fast Frequency Response

FFR is the power injected into (or absorbed from) the grid in response to changes in measured or observed frequency during the arresting phase of a frequency excursion event to improve the frequency nadir or initial ROCOF [9].

In systems dominated by synchronous machines, FFR is provided by the inertial response of synchronous machines and conventional turbine generator response. The different types of frequency response including inertial response, PFR and FFR, which can act in coordination, are illustrated in Figure 1.8 [9].

The beneficial impacts are imparted by both nonsustained inertial response and the sustained FFR and PFR. The sustained frequency response methods provide continuous injection of power until secondary frequency controls bring the frequency back to the scheduled level. It should be noted that the nonsustained response such as inertial response which ceases after a short time period does not adversely impact the frequency during the arresting period or the recovery period. Hence there is a need for inertial response to be coordinated with other mechanisms of FFR and PFR.

FFR can be initiated based on a knowledge of magnitude of frequency deviation, ROCOF, or other factors. There is, however, a delay involved in estimating these quantities which needs to be accounted for. FFR is provided by the following methods, either individually or in combination [9, 15]:

i) Active power injection which is proportional to the measured frequency deviation (proportional response)
ii) Injection of fixed magnitude of active power as soon as the frequency reaches a prespecified trigger point (step response)
iii) Active power injection that is proportional to the computed ROCOF (derivative response)

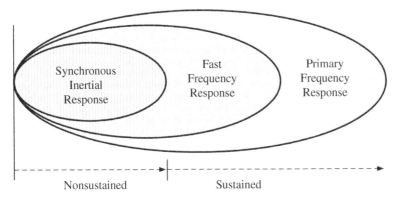

Figure 1.8 Simultaneous contributions of inertial response, primary frequency response, and fast frequency response. *Source:* Reprinted with permission from North American Electric Reliability Corporation [9]. This information from the North American Electric Reliability Corporation's website is the property of the NERC and available on the Standards page (https://www.nerc.com/pa/Stand/Pages/default.aspx). This content may not be reproduced in whole or any part without the prior express written permission of the North American Electric Reliability Corporation.

iv) Injection of fixed amount of active power as soon as a prespecified ROCOF is attained (step response)
v) Controlled decrease in load in proportion to measured frequency deviation or ROCOF (proportional or derivative response)
vi) Controlled decrease of a fixed amount of load once a prespecified frequency or ROCOF is reached (step response)

It should be ensured that any of the step responses described above does not adversely impact system stability or system frequency.

It is also desirable that the FFR must be timely and sustained rather than injected for a short period and then withdrawn [16].

FFR and inertia are two distinct entities. Inertia is a characteristic of synchronous generators which provides autonomous response to slow down the ROCOF but cannot restore power system frequency. Inertial response, in fact, provides time for active power injection to correct the supply–demand imbalance. On the other hand, FFR results from control action which is tunable based on system needs. FFR injects active power into the system, helps correct the generation–load imbalance, and subsequently restores the power system frequency. Inertia and FFR are, therefore, two distinct services which cannot be interchanged. They are characterized by the following features [17]:

- Large power systems presently require a minimum level of inertia, below which no amount of FFR can be utilized to ensure a stable power system.
- The magnitude and type of FFR required to maintain grid reliability are related to the amount of existing system inertia.
- The relationship between needed FFR and system inertia is nonlinear and can be evaluated through detailed system modeling. It also depends upon several operational considerations.
- FFR through IBRs is controllable and can be designed to provide responses which are much wider than that provided by synchronous generators.

1.2 Challenges of High Penetration of Solar PV Systems

Solar PV systems are typically classified as IBRs or DERs when connected to bulk power systems or distribution systems, respectively. The impacts of solar PV systems occur both at a local level, i.e. on the interconnecting distribution feeder or substation, or at the interconnected bulk power system level, or both. A summary of the impact of high penetration of solar PV systems in different countries is presented in the IEA PVPS Program report [18, 19]. A detailed coverage of technical challenges and experiences with high levels of PV integration is presented in [3, 4, 20–27].

A summary of the adverse impacts of high penetration of solar PV systems is presented below. However, a detailed coverage of these impacts and their potential solutions utilizing smart inverters are provided in subsequent chapters. Although the challenges are described with respect to solar PV systems, they also apply to other IBRs such as wind generators.

1.2.1 Steady-state Overvoltage

Solar PV systems tend to raise the voltages due to active power injection in distribution feeders to which they are connected [27, 28]. Voltage rise is prominent especially during low load conditions or when the PV power output exceeds the load during high PV penetration scenarios. The voltage rise depends upon the power generated by the DERs, location of the DERs, and the electrical characteristics, i.e. system short circuit level and X/R ratio of the feeder. Higher voltage excursions are experienced in weak systems having relatively high short circuit impedances. The voltage rise is more when the solar PVs are connected at the end of long feeders, and gets aggravated in presence of shunt capacitors. The steady-state voltage rise often goes beyond the acceptable voltage limits of the utility. This leads to the utility putting a restriction on the number of solar PVs that can be connected in their grid.

A typical example of voltage rise caused by a solar PV-based DER located at the far end of a distribution feeder connected to a substation is shown in Figure 1.9 [29]. In the absence of DER, the voltage is much below the ANSI voltage limit of 1.05 pu [30]. The power injection by DER causes the voltage to rise beyond the ANSI limit. If the DER is made to absorb reactive power (through any smart inverter function), the voltage goes below the stipulated limit.

The solutions typically employed by utilities to alleviate such overvoltage conditions are [27]:

i) reducing the use of fixed shunt capacitor banks and instead using switchable capacitor banks which can be switched during high power production from the DERs. It must be understood that while changing capacitor settings is helpful for voltage control, varying capacitor values impacts network resonant conditions which may potentially cause harmonic amplification in the grid. This possibility must be examined beforehand, and solutions implemented, if required.

ii) lowering the voltage setpoints on the Load Tap Change (LTC) transformers and line voltage regulators, or altering the control modes of operation of such voltage control equipment.

Conventional transformer taps, voltage regulators, switched capacitors, etc., are used to regulate voltages but may not be adequate under all scenarios. In cases where very rapid voltage control in time frames of 2–3 cycles is required, Flexible AC Transmission system (FACTS) Controllers such as Static VAR systems (SVCs) and STATCOMs or Dynamic VARs (DVARs) are employed for this purpose [31, 32]. These controllers although very effective and are quite expensive.

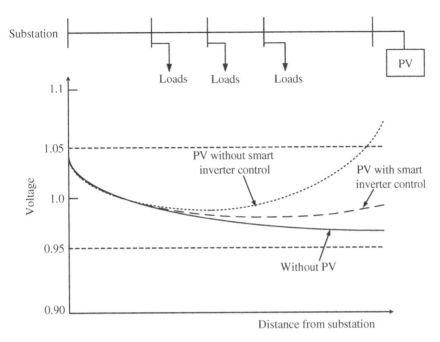

Figure 1.9 Voltage rise due to active power injected by solar PV based DER in a feeder. *Source:* Based on Zandt [29]. Used with permission from EPRI.

1.2.2 Voltage Fluctuations

Events of cloud passing cause frequent and large changes in the output of solar PV farms [22]. EPRI has made available two YouTube videos depicting the impacts of solar variability on PV power production and voltages along a distribution feeder [33, 34]. Varying wind conditions cause similar changes in power output from wind farms. This leads to voltage fluctuations on the distribution feeder, the magnitude of which is influenced by system strength and the X/R ratio of the interconnecting feeder at the PCC. Such variations in voltage profile adversely influence the operation of voltage-sensitive loads such as induction motors and process control equipment.

1.2.3 Reverse Power Flow

Large penetration levels of solar PV may offset the feeder loads and cause excess power to flow in the reverse direction in the grid. This can potentially occur during daytime with solar PV-based DERs, and typically in the night (during high wind conditions) with wind-based DERs [27]. Power systems are traditionally designed for unidirectional power flow. However, with the growing installations of DERs, power can flow potentially in a bidirectional manner.

Wind typically blows during nighttime, as in Ontario, Canada. In feeders where wind farms are connected, reverse power flow usually occurs during night since loads are much lower during nighttime as compared to daytime. Such reverse flow of power causes voltages to rise in the feeder that may exceed acceptable limits during the prevalent light-loading conditions.

A case study is presented for a realistic distribution feeder in Ontario which has a solar PV farm and a wind farm connected on the same feeder [35]. The study system shown in Figure 1.10 consists of a 27.6 kV radial distribution network 45 km long connected to a supply substation through 118/27.6 kV transformer. A solar farm of 8.5 MW is connected 5 km away from a self-excited induction

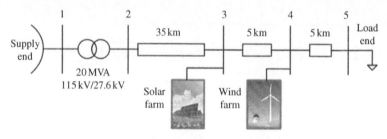

Figure 1.10 A realistic distribution feeder in Ontario.

generator-based wind farm. The load is considered to be lumped at the feeder end. The daytime peak load is considered to be 4.82 MW and 2.19 Mvar whereas the nighttime load is considered to be 2.1 MW, 0.73 Mvar (1.76 MVA at 0.91 lagging power factor).

The objective of this study is to determine the maximum amount of wind power that can be connected during daytime and nighttime without violating the utility steady-state voltage limit of 1.06 pu. The wind farm PCC voltage (obtained through load flow studies) is plotted with increasing amount of wind power for nighttime and three cases for daytime: Case 1 with 0% solar PV power output (0.0 MW), Case 2 with 50% PV output (4.25 MW), and Case 3 with 100% PV output (8.5 MW). Load flow studies are performed for all three cases with increasing wind farm output. The maximum wind power that can be integrated into the grid without violating voltage limits in the grid is 13.1 MW for Case 1, 8.8 MW for Case 2, and 4.36 MW for Case 3. During nighttime, when the load is lowest, the maximum wind power output permissible is only 4.04 MW. This demonstrates that in the night when wind availability is the highest and feeder load is minimum, only a very small amount of wind power can be connected to the grid due to steady-state voltage rise limitations. Such voltage limit violations have led utilities to restrict the amount of wind power systems to be connected on specific feeders. On another note, it is seen that a high level of wind power generation (within the thermal limit of the feeder) tends to decrease the voltage profile due to reactive power absorption in the feeder and the transformer connecting the wind farm to the feeder (Figure 1.11).

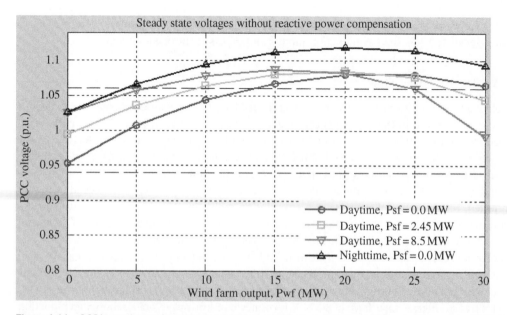

Figure 1.11 PCC bus voltages for various solar farm output cases with increasing wind farm output (daytime and nighttime).

Reduction of active power production is one means to address this steady-state overvoltage issue [36]. An adaptive voltage control strategy utilizing an LTC and automatic voltage control relay was proposed in [37] to increase the active power output from the DERs while still keeping the voltage regulated within limits. In certain jurisdictions, FACTS Controllers, such as Static VAR Compensators, STATCOMs, or DVARs, have been installed to stabilize the dynamic voltage variations caused by varying wind power generation [38–41].

Protection and relaying systems, voltage regulators, LTCs with line drop compensation, are typically designed for one way flow of power. Reverse power flow, therefore, adversely impacts the operation of such systems. High amount of reverse power flows may cause interruption rating of the circuit protection systems to be exceeded as well as cause sympathetic tripping of adjacent circuits.

Sectionalizers and fault indicators that are commonly used in distribution systems are based on radial flow of power. They can misoperate in scenarios where power flow can change direction. These radial devices need to be replaced or reconfigured, else they may provide erroneous information about fault locations making system restoration difficult [26].

Voltage regulators, even if they are reconfigured for bidirectional power flow, can only control the voltage in close vicinity. They are unable to regulate the voltage between the regulator and the DER when voltage increases occur due to reverse power flow [26]. The customers near the DER may still experience high voltages in cases of large reverse flow due to high penetration of DERs.

To circumvent these issues due to reverse power flow, the protection system settings, voltage regulators, and tap changer operations must be reconfigured to respond correctly during bidirectional power flows. In case of reverse power flow in single-phase systems, phase balancing must be pursued as a mitigation measure [24].

However, there are many distribution systems having large PV plants where reverse power flow routinely occurs and the protection systems are well-designed to handle such normal operations.

1.2.4 Transient Overvoltage

Unsymmetrical faults, such as single-line-to-ground faults, on DER connected distribution lines, can cause Transient Overvoltages (TOVs) [27]. TOVs may also be caused by sudden connection or tripping of DERs, tripping of downstream feeders or loads, large and sudden fluctuations of DER power output due to cloud passage/wind flow changes [27]. The magnitude of TOV is, of course, dependent upon the connection scheme of the coupling transformers (start-star, star-delta, etc.) [42]. These TOVs can potentially have a damaging impact on other equipment connected to the distribution line.

The magnitude of TOV is seen to increase with the number of DER connections or increasing power output from the DERs. For this reason, utilities often impose a restriction on the number of DERs to be connected to their distribution feeders based on this criterion.

For the same study system in Section 1.2.3, Figure 1.12 displays the TOV in phase voltages when a single line to ground fault of six cycles occurs at the load end, during nighttime. This represents a low loading condition. This TOV exceeds the utility specified limit of 1.3 pu in the worst case [42].

Transient overvoltages are evaluated for nighttime and three solar farm output cases (0, 50, and 98% solar farm output) during daytime. The maximum wind power that can be integrated into the grid without violating TOV limit of 1.3 pu in the grid is 11.1 MW for Case 1, 5.4 MW for Case 2, and 2.1 MW for Case 3. During nighttime, when the load is lowest, the maximum wind power output permissible is only 3.96 MW. It is noted that the TOV criterion restricts the DER connectivity even more than the steady-state voltage criterion.

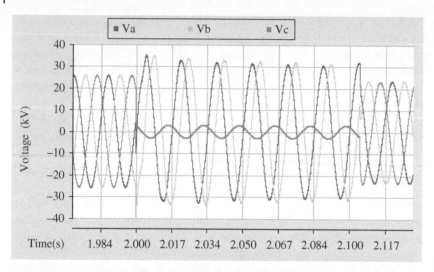

Figure 1.12 TOV in PCC phase voltages for fault at load end.

In addition, there are other overvoltage issues such as rapid voltage change (transformer energization), load rejection overvoltage [43], and ground fault overvoltage [44].

The above TOV situation can be avoided by using transformers of different configurations which provide an effective grounding on the feeder side. One of the mechanisms of limiting TOVs is the use of grounding transformers or a fast ground overvoltage protection strategy [21] to prevent supply of power to the loads on an ungrounded system after a fault or unplanned switching event. Recently, SVCs have been employed to mitigate TOVs in wind farms [45].

1.2.5 Voltage Unbalance

Unbalance in power supply voltages results in unsymmetrical phase currents which can be potentially damaging to voltage-sensitive equipment, transformers, and motors [46]. Unbalanced currents can cause torque pulsations, mechanical stresses, higher losses, and overheating of windings that can reduce their lifecycle. Grid codes typically specify that voltage unbalances [47] must be constrained within 2% of the PCC voltage.

Voltage unbalance issues are experienced with PV systems connected to single-phase distribution systems [48, 49]. Small PV and wind DERs (up to 10 kW) are typically connected to single-phase low-voltage networks. Many a times the combination of PV system and interconnected loads do not appear as balanced systems across the three phases at the substation. This unbalance causes neutral voltage to rise and phase voltage to increase in one or two phases. Ideally, phase balancing should be performed at the time of installation; however, it may not be possible to do so due to the extensive effort involved. Phase balancing is usually undertaken if a power quality complaint is received [18]. In Ontario, several cases of PV inverter shutdowns have been reported due to voltage unbalance in the feeders.

1.2.6 Decrease in Voltage Support Capability of Power Systems

High proliferation of utility-scale solar PV systems and rooftop solar PV systems cause displacement of conventional fossil-fuel-based synchronous generators which have voltage control capabilities through their automatic voltage regulators. This results in a decrease of reactive power support for voltage control in the grid. This requires installation of additional voltage support equipment such as switched capacitors, SVC, STATCOM, or synchronous condenser on transmission and distribution systems, as needed.

1.2.7   Interaction with Conventional Voltage Regulation Equipment

Intermittency and variability in power generated by the DERs result in continuous voltage variations in the voltage at the PCC, as described above. The utilities typically install voltage-regulating equipment in their networks such as line voltage regulators, load tap changers, and capacitor banks. These equipment constantly monitor the voltages in the lines to which they are connected and take appropriate control action, e.g. capacitor switching to keep the voltages within acceptable limits. Unpredictable and rapid variations caused by power generation from DERs cause more frequent operation of such installed voltage control equipment than what they are designed for [19, 27]. This leads to a potential reduction in their operating life. The operating time of these voltage control equipment is typically 30–90 seconds [21], hence step-voltage control operations may occur in intervals of one to two minutes on the feeder. The line voltage regulators and LTCs often employ line drop compensation for voltage control. Such a voltage regulation technique is dependent upon the actual line current. DERs modify this line current due to their power generation and may adversely impact the voltage regulation process [27]. Frequent switching of capacitor banks also causes changes in the reactive power flow in the lines leading to unintended voltage variations and increased line losses.

1.2.8   Variability of Power Output

A significant adverse impact of solar PV systems is that their outputs are variable, and hence they are not dispatchable as conventional generators. Their power output can be frequently and substantially impacted by environmental conditions such as cloud coverage. In addition, an increasing amount of solar PV generation (e.g. rooftop) is not observable by the system operator, which creates challenges in maintaining system reliability with such large intermittent PV generation resources.

Some techniques for predicting and addressing the variability are presented below.

Techniques are continuously being developed to accurately predict the environmental conditions for both wind and PV-based IBRs and DERs in very short intervals of time [50]. Solar PV power forecast reduces the uncertainty of the intermittency and variability of such power generation. This helps system operators to commit and decommit synchronous generators to handle situations of high and low power output from these renewable energy systems. The forecasts are also of great value in lowering the amount of operating reserves required for the system, thereby reducing the overall system operating cost.

The variability from solar power arises due to cloud passage. Short-term cloud forecasting can be done by sky imaging which can predict approaching clouds. Forecasting in the time frame of next few hours can be based on satellite imagery of clouds. Longer-term forecasting can be done on the basis of weather models to determine the formation of clouds [50].

Despite the gradual emergence of accurate forecasting, the power system must still respond and adapt to the changing power output from such IBRs and DERs so as not to impact system reliability and continuity of the power supply to customers.

1.2.9   Balancing Supply and Demand

Increasing solar generation reduces the net load especially during periods of relatively low loads and high generation such as in Spring. CAISO has produced a plot of "net load" which is the difference between forecasted load and expected electricity production from variable generation

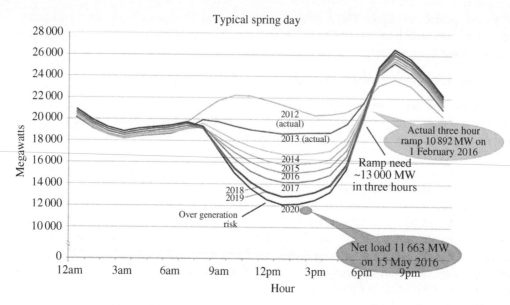

Figure 1.13 California ISO (CAISO) Duck Chart. *Source:* CAISO [52]. Licensed with permission from the California ISO. Any statements, conclusions, summaries, or other commentaries expressed herein do not reflect the opinions or endorsement of the California ISO.

sources such as solar PV systems. In certain times of the year, these curves produce a "belly" like appearance in the mid-afternoon that quickly ramps up to produce an "arch" similar to the neck of a duck – hence the popular industry given name "The Duck Chart" shown in Figure 1.13 [51]. The more solar generation increases, the more the curve looks like the belly of a duck. This problem mainly occurs during springtime when the weather is sunny but still cool. At this time, solar generation is high but electricity demand is less as air-conditioning loads are low [52].

This behavior of net load creates special problems for system operator to balance generation and load. As the sun sets and the solar generation reduces to zero, a significant net demand ramp (the arch of duck's neck) is experienced by the system operator. It must, therefore, dispatch significant generation resource to meet the demand ramp. Further, as the net demand decreases in late evening, when a downward ramp is experienced the system operator must rapidly reduce or shut down the generation to offset the downward ramp.

Another major problem is the oversupply during midday which may lead to the system operator curtailing solar power generation thus reducing its economic and environmental benefits. Several mitigative measures which minimize the risk of overgeneration are proposed in [51].

1.2.10 Changes in Active Power Flow in Feeders

The active power generated by solar PV systems can modify the loading levels on the connected feeders [21]. They can partially or completely offset the existing loads on distribution feeders, which results in a reduced power import on the feeder from the grid. They can also increase power flow in certain segments of a distribution feeder causing overload conditions. Presently employed voltage control devices which have been optimized for their size, location, and performance based on the existing loading patterns will no longer provide the same voltage regulation performance as originally planned [26]. These need to be re-optimized.

It is further important to examine if the feeder sections connecting the substation to DERs have enough power carrying capacity and also whether the intervening switchgear have adequate ratings to accommodate this additional DER power.

Reconductoring may be needed to allow the power flow from solar PV systems. In some cases, when large solar PV systems are to be connected, a dedicated feeder may be needed to avoid the adverse impacts of a conventional solar PV connection [21]. Such a dedicated feeder without any loads can be directly connected to the substation. It may be noted that such reconductoring or network upgrades are quite expensive and time consuming to implement.

1.2.11 Change in Reactive Power Flow in Feeders

Voltage variations due to active power generation by solar PV systems coupled with operation of voltage regulators and switching of capacitor banks leads to change in reactive power flow in the lines [53]. For instance, if a capacitor is switched off, reactive power earlier being supplied by the capacitor is no longer available and needs to be imported from neighboring lines or utilities. This may also have financial impacts on the distribution utilities that have a high import component of power from neighboring utilities.

1.2.12 Line Losses

Line I^2R heating losses are proportional to the current flow in the feeders. If the solar PV systems partially offset the line loading the losses will decrease [54, 55]. It is shown on an AEP network that low to moderate level of solar PV penetration tends to lower the line losses since the power produced by the solar PV systems tends to offset the load, and consequently, the import of current from the grid source reduces [56]. However, if the solar PV systems cause a net increase in line current the losses will increase. In some cases, the line losses with solar PV systems may become higher than the case with no solar PV systems installed [21]. These losses require increased power import over the lines to supply the same load, increased heating in lines resulting in faster aging, and decrease in lifespan of line infrastructure.

1.2.13 Harmonic Injections

Solar PV systems inject harmonics into the grid which may increase the THD of the bus voltages [57]. Although individual solar PV inverters are certified to be harmonic compliant with existing Standards, such as [47, 58] harmonic injections from several inverters within a solar PV plant or multiple solar PV plants connected to a feeder system can get combined. The harmonics, however, do not add up in an arithmetic manner. Certain amount of harmonic cancellation occurs due to phase differences amongst the harmonics from different solar PV systems [57]. Solar PV systems connected to LV feeder systems tend to create more harmonic distortion than at higher voltage systems [59].

Two types of harmonic sources exist in a power distribution system. One is the background harmonics in the grid while the other is the harmonic injections from the power electronic sources such as a wind generator plant, adjustable speed drive, PV inverters, and dynamic reactive power compensators (SVC or STATCOM), etc.

Both distribution and transmission networks exhibit network resonances due to the interaction of inductive and capacitive elements present therein. The inductive elements are contributed by system short circuit reactance, reactance of generators, and the inductances of transformers, transmission lines, cables, bus reactors, etc. The capacitive component is provided by bus capacitors, power factor correction capacitors, line charging capacitance, cable capacitance, capacitors connected at wind plants, filter capacitors at solar PV and wind plants, etc. Interaction amongst the above inductive and capacitive elements results in several network resonant modes [59, 60]. These resonances can be both series resonance and parallel resonance [60]. The network resonances get aggravated in (i) systems with low short circuit level, i.e. weak systems and (ii) lightly loaded systems. It is noted that excitation of network resonances can restrict the connectivity of PV solar farms and wind farms in transmission and distribution networks [61].

In distribution systems, series resonance is characterized by minimum impedance values at the resonant frequencies. The ambient harmonic voltages in the grid drive high harmonic currents in the downstream network at these low impedances. This is the harmonic distortion problem caused by series resonance [60]. On the other hand, parallel resonance is exhibited due to high impedances at the resonant frequencies. The currents from the harmonic current sources (PV inverters, wind turbine generators, Adjustable Speed Drives, etc.) interact with these high impedances resulting in significantly high harmonic voltages in the network. This is the harmonic distortion problem caused by parallel resonance [60].

Series and parallel resonance can either exist independently or coexist at the same location. These resonances are a characteristic of a specific power system. By themselves, they do not cause any adverse effect on the system. Just as it takes two hands to clap, i.e. only when there is an alignment of a network resonant frequency and a harmonic injection at the same frequency, a potential of adverse harmonic distortion can occur.

Consider a distribution line with series inductance connected to a wind plant with a power factor correction capacitor. For the background harmonic voltages in the grid, the above inductor–capacitor combination presents a situation of series resonance. The background harmonic voltages see a low impedance path and inject high amount of harmonic currents resulting in substantial voltage distortion at the terminals of the wind plant. Meanwhile, the same inductor–capacitor combination appears as a parallel resonant circuit for the harmonic currents injected by the inverters in the wind plant. If the parallel resonance frequencies align with the harmonics injected by the wind plant, the harmonic currents from the wind plant see large impedance resulting in substantially high harmonic voltages at the wind plant terminals [60].

Figure 1.14 depicts the frequency scan of a power distribution network, i.e. a plot of network impedance Z (pu) as a function of frequency. Series resonance is indicated by the dips (minimum impedance points) in the plot of the frequency-dependent impedance shown in Figure 1.14. On the other hand, parallel resonance is characterized by peaks (high impedance points) in the plot of the frequency-dependent impedance shown in Figure 1.14. A case of harmonic amplification due to parallel resonance is illustrated below.

The vertical bold arrow qualitatively depicts harmonic current injection I_h at the 11th harmonic (660 Hz). The bus voltage (pu) is also illustrated on the y-axis. Three cases of frequency scan are depicted. Figure 1.14a,b illustrates the cases when the network resonant peak does not coincide with the injected 11th harmonic at 660 Hz. For both of these cases, the harmonic voltage ($V_h = I_h \times Z_h$) corresponding to the 11th harmonic current injection is not high. Figure 1.14c portrays the frequency scan when the network resonant frequency aligns with 11th harmonic (660 Hz) current injection. In this case, the injected harmonic sees very high impedance Z_h. The corresponding harmonic voltage ($V_h = I_h \times Z_h$) becomes very high and may exceed the utility acceptable voltage limit (indicated by the horizontal dashed line) and potentially damage customer equipment.

The above example is for one injected harmonic for one network resonance mode. However, in active power systems, there may be several network resonant frequencies and several harmonic current injections.

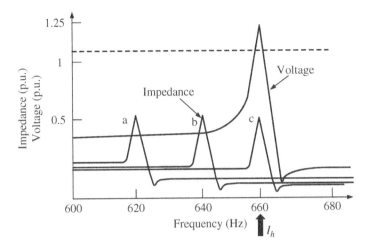

Figure 1.14 (a, b) Network resonant frequency not coincident with harmonic frequency. (c) Network resonant frequency coincident with 11th harmonic frequency (660 Hz).

Inverter-based generators such as PV systems produce a spectrum of odd harmonics that are injected into the network [37]. If for any network condition(s), these injected harmonics align with any one of the resonance peaks, the voltage at that harmonic frequency will be amplified [26, 27], and excessive voltage harmonic distortions may occur.

These harmonic distortions can potentially cause overheating and failure of equipment such as transformers, motors, and other voltage sensitive equipment connected in the vicinity. Cases have been reported where harmonic amplification due to network resonances has led to shutdown of wind turbines.

In a practical scenario of high voltage distortion due to network resonance, it may be difficult to identify if the harmonic resonance is caused by series or parallel resonance. Background harmonics need to be measured by harmonic monitoring equipment over a prolonged period of time, especially before the installation of the wind plant (in this case). The harmonic injections are typically provided by PV inverter manufacturers and wind plant inverter under laboratory conditions, where ambient harmonics are not present. The harmonic injections may be different when connected to the grid having ambient harmonics.

Still, harmonic measurements at the PV plant and wind plant during no active power output conditions and operation at different power levels under different grid conditions can help in identifying the cause of harmonic distortion and developing mitigation strategies.

In a utility study [62], it is shown that bus capacitors interact with system short circuit impedance (inductance) and cause network impedance resonances. For realistic short circuit levels, the network impedance resonant frequencies get aligned with the harmonics injected by the PV inverters and result in harmonic amplification and voltage THD in excess of the stipulated limits. For the study system, even though 20 MW of PV generation does not cause any violation of limits related to steady-state voltage and TOV, only 6.9 MW of PV generation can cause the utility THD limit to be violated. Thus the limit of PV connectivity was dictated by total harmonic distortion considerations than from overvoltage criterion.

Another example of harmonic resonance in a 54 MW solar PV plant caused by capacitor switching is described in [63].

The various harmonic mitigation measures adopted in utilities include: (i) changing the size and sequence of energization of capacitor banks within a plant, (ii) detuning the network resonances by network reconfiguration, (iii) reduction of harmonic generation, (iv) converting existing power factor correction capacitors into passive filters, (v) utilizing separate passive filters, and (vi) employing active filters, etc. [60].

1.2.14 Low Short Circuit Levels

As solar PV systems increasingly displace conventional synchronous generators, the amount of short circuit current capability in the power system decreases resulting in weak grids [16]. The stability of inner current control loops and the phase-locked loop declines when inverters are connected to networks which are weak, i.e. where the network voltage is susceptible to substantial variability. Low short circuit levels may also cause challenges in disconnecting solar PV systems during unintentional islanding scenarios.

1.2.15 Protection and Control Issues

Protection systems of power distribution systems comprise overcurrent relays, fuses, and reclosers that are designed to operate for only one direction of power flow, i.e. from the grid to the loads or to the possible fault location [20, 64]. With the high proliferation of solar PV systems, the following challenges are experienced [16, 20, 64–69, 70]:

i) The fault current may become bidirectional as it will be fed both from the grid and the different solar PV systems in the distribution system.
ii) It is not necessary that the fault current will always increase with the introduction of solar PV systems: it can both increase and decrease depending upon the location of the concerned PV systems with respect to the fault location [71]. Solar PV systems installed upstream of the fault location would increase the fault current while locating them downstream could reduce the fault current level. Hence the protection coordination done earlier without the solar PV systems, i.e. with only unidirectional flow of power will not apply when multiple solar PV systems are connected.
iii) The power output from solar PV systems is variable. It can be either zero as in the night or maximum with full solar irradiance. Furthermore, the power output is variable due to varying cloud coverage. These operating conditions lead to changing fault current levels. Hence, variability in solar and wind DER power production can adversely impact the operation of overvoltage and overcurrent protection systems.
iv) The conventional power systems were not designed for power flow in reverse direction. The relays and protection systems have to be modified to adapt to the backflow of power. Presence of high amount of solar PV systems results in current flows from multiple sources, which makes protection coordination even more complex and difficult. Protection device settings need to be modified or relays may need to be replaced in distribution systems where reverse power flow is expected [26].

1.2.16 Short Circuit Current Issues

Solar PV systems contribute fault currents in a different manner than synchronous generators and hence impact the operation of protection systems differently [27].

The fault currents from PV inverters are much smaller than those from synchronous generators and are for much shorter duration (typically less than a cycle to few cycles) [72]. The magnitude and duration of fault currents are determined by inverter control and not so much by the impedance between the PV system and the fault location. This unique behavior of DER inverters during system faults may cause different issues such as enhanced short circuit currents, relay desensitization, sympathetic tripping, breaker reduction of reach, protection coordination issues, etc. [16].

The short circuit current contribution from solar PV systems is dependent on the controls employed in inverter systems. This is quite different from the behavior of synchronous generators during faults, which depends upon its different effective reactances during the fault, i.e. subtransient, transient, and steady-state reactances. Moreover, the inverter fault current does not include zero sequence component and the negative sequence current is typically partially or fully suppressed depending on the inverter control [73].

The short circuit current impact of solar PV systems is a function of the following factors:

i) size of solar PV system
ii) location of solar PV system
iii) nature of controlled short circuit current from the solar PV system, which is dependent upon the inverter control employed
iv) trip time of solar PV systems with decreased terminal voltages subsequent to the fault occurrence
v) constitution of the short circuit current, i.e. the relative magnitudes of active and reactive currents therein. It is noted that solar PV systems operating on unity power factor control will inject primarily real short circuit current, whereas solar PV systems operating with LVRT characteristics will emanate a higher component of reactive currents
vi) configuration of interconnection transformer
vii) grounding techniques employed, and the resulting flow of zero sequence currents.

All the above factors impact conventional protection and relaying schemes. Studies have been reported [74] that even if the flow of power is unidirectional, increasing penetration levels of solar PV systems can cause miscoordination between the different protection devices.

Proliferation of solar PV systems in the network increases the short circuit level due to their short circuit current contribution during faults [21, 75–77]. The short circuit current contribution from a PV system inverter is typically in the range of 1.2 times rated current for the large size inverter (1 MW), 1.5 times (500 kW) for medium size inverter and between 2 and 3 times for smaller inverters [72, 78]. While the short circuit current contribution from an individual solar PV system may be small, the total amount of short circuit current contribution may become appreciably large for high penetration of PV DERs [79].

It is a reasonable expectation that short circuit current contributions from a large number of solar systems in the distribution networks may add up to levels that could damage circuit breakers. Hence, circuit breakers will need to be upgraded and substations will need to be modified at a significant cost to the concerned utility. This apprehension actually resulted in the denial of about 45% applications for solar farm connections in Ontario, Canada, during 2011–2013. Consequently, a major effort was launched by CanSIA, the national trade association representing the solar energy industry in Canada, to investigate the actual impact of fault currents from PV inverters, and their remedial measures [72], so that more PV systems can be connected in Ontario.

The problem of short circuit currents can be solved by substation upgrades, installing series reactors [80], or by employing fault current limiters [81].

1.2.17 Unintentional Islanding

An "island" is any segment of a power system which has its own generation and loads, and hence can operate independent of the power system for at least some time period. While intentionally formed islands can be helpful in supporting system reliability, unintentionally created islands can present several technical and safety challenges, and risks.

Distribution feeders with a high penetration level of solar PV power systems can potentially cause the problem of unintentional islanding [82, 83]. The solar PV system may continue to feed a load downstream of its location in an islanded mode, if due to a fault/disturbance the utility has isolated the feeder upstream of the DER. The PV system may not be designed to maintain voltage and frequency for customers in that island when the source of power from utility side is disconnected. If the islanded system is not provided with effective grounding systems, Transient Overvoltages will result due to the power being fed by ungrounded generators, which can have a damaging impact on the customer equipment connected in that island [67]. Moreover, the continued energization of the island by the solar PV systems may pose a safety threat to utility personnel working in that area.

In some cases, an island may be planned ahead of time by the utility to provide continued service to customers. However, when such islanded condition is not preplanned by the utility, the distribution system isolated from the utility source is referred as an "unintentional island". In case of an unintended island, the solar PV system shall detect the island, cease to energize the area electric power system, and trip within a very short time (typically two seconds) of the formation of the island [47].

PV-based DERs have a low fault current and especially with variable power output may not be able to correctly detect fault conditions on the grid. They are, therefore, critically dependent on anti-islanding control signals issued by the grid to determine when to trip.

Anti-islanding methods in grid-connected solar PV systems can generally be classified into two major groups, which include: (i) communication-based methods and (ii) noncommunication-based methods [84]. The noncommunication-based methods include passive methods, active methods, and hybrid methods. Communication-based strategies include (i) transfer trip and (ii) power line signaling. Direct Transfer Trip is one of the communication-based techniques for tripping solar PV systems [84]. However, with the unprecedented growth of solar PV systems in distribution systems, anti-islanding techniques based on traditional transfer trip have become quite expensive to implement on each member of the large fleet of solar PV systems.

Alternate transfer trip strategies for disconnecting solar PV systems during events of upstream disruptions of power supply are being implemented recently. A power line carrier-based anti-islanding scheme is successfully operating in the network of National Grid in Massachusetts, USA. A signal indicating the upstream breaker "ON" status is continuously transmitted on the power line. This signal is received by all the PV inverters connected to that line or network, which continue to perform with their smart inverter functions, as long as they read this ON signal. If the upstream breaker opens due to a system fault or disturbance this signal is no longer available on the disconnected feeder and all DER inverters connected downstream, simply trip. This cost-effective technique has been operating very successfully.

A study on the variation of frequency during an islanding condition with different ratios of rotating and inverter-based generators is reported in [23]. During a high power mismatch between the generation and the load, it is shown that:

i) in a rotating generator dominated case, frequency declines if the load exceeds generation, however, frequency increases if generation exceeds the load. In other words, the frequency is impacted by the active power mismatch.

ii) in an inverter dominated case, the frequency can even increase if the load exceeds generation when there is a surplus of reactive power. Alternatively stated, frequency variations may also be governed by reactive power mismatches.

It is further reported [23] that anti-islanding protection systems will trip much faster if the island is dominated by inverter-based generation and there is a surplus of reactive power. It is, therefore, recommended that to have an improved anti-islanded protection in systems having a dominance of inverter-based generation, excess reactive power must be made available in the islanded portion. This can be achieved through activation of shunt capacitors.

1.2.18 Frequency Regulation Issues due to Reduced Inertia

Increasing penetration of IBRs such as solar PV, wind, BESSs, etc., and the retirement of large conventional thermal synchronous generators is leading to a substantial decline in inertia in power systems. Reduced inertia can have the following consequences [11, 13]:

1) Larger or steeper ROCOF resulting in shorter time to reach UFLS thresholds.
2) Load shedding thresholds are reached even before the PFR or FCR becomes available.
3) Lower nadirs and activation of UFLS will result in potential disconnection of domestic customers.
4) Faster frequency variations may adversely impact system protection.
5) Low-frequency conditions may cause generators to trip. Loss of generation can potentially lead to cascading effects and a partial or even full system blackout.
6) Risk of islanding segments of distribution networks with large number of solar PV systems.
7) Lack of rotating inertia also presents problems in microgrid environments, where there are no large conventional generators to stabilize the system frequency.

Some other important considerations regarding low inertia systems are as follows [13]:

1) In large interconnected systems, the level of electromechanical damping in different regions may also have an impact. This can result in a higher ROCOF close to the lost generation than at locations farther in the system. Hence locational impacts of system inertia and disturbance need to be examined.
2) The loss of generation may result in system islanding where a small part of the system gets separated and operates autonomously. Since the inertia of the islanded system is even lower than the main low inertia system, different control challenges such as system oscillations may be experienced. Such islands may form even in distribution networks.

1.2.18.1 Under Frequency Response

A study of the US Western Interconnection depicting the impact of extra high penetration of solar PV systems on overall system frequency response is described below [2]. This study considers a total renewable penetration of 80% (65% solar and 15% wind) at the interconnection level and 100% penetration (all PV) at a regional level.

A realistic model of the Western Interconnection is utilized. The 2022 Light Spring (LSP) planning case is considered due to the relatively low level of online synchronous generation during a light load condition. This represents an ideal (most severe) case to benchmark the frequency response performance.

The 2022 LSP case is implemented in GE's Positive Sequence Load Flow analysis simulation software. The LSP case is modeled in detail with over 4000 generators and more than 19 000 buses. Each power plant has its own specific detailed dynamic models, including a generator, exciter, and turbine governor. HVDC systems and protective systems such as UFLS and line protection are also included in the Western Interconnection dynamic model. As PV systems are entirely based on inverters, they are represented by GE Type 4 wind power plant model. This includes the converter model, active and reactive power control model, and low-voltage power logic [2].

The total generation in the 2022 LSP case is 117 GW; with approximately 13.9% instantaneous penetration of wind and 1.1% instantaneous penetration of solar PV. To increase PV penetration the existing synchronous generators are replaced by solar PV systems in each area, so no transmission upgrades are required. The generators are replaced in the sequence of coal power plants, gas generators, hydro generators, and then nuclear generators. As synchronous generators gradually get replaced by PV systems, the net system inertia decreases almost linearly, as shown in Figure 1.15 [2]. The system equivalent inertia level for the base case is 3.4 seconds. The inertia levels for 20, 40, 60, and 80% renewable penetration levels are noted to be 3.17, 2.31, 1.68, and 0.92 seconds, respectively.

The system frequency response is examined in terms of three metrics: (i) ROCOF which reflects the inertia response; (ii) frequency nadir which is determined by both inertia and PFR; and (iii) settling frequency which represents the PFR of the system.

Three severe contingencies are considered as follows:

1) Largest N-2 contingency involving loss of two largest generators in the Palo Verde nuclear plant, equivalent to a loss of 2525 MW capacity.
2) Typical N-2 contingency involving loss of two largest generators in the Colstrip coal power plant, representing a loss of 1514 MW capacity.
3) Typical N-1 contingency involving loss of one large generator in the Comanche power plant, indicative of a loss of 904 MW capacity.

To demonstrate the severity of frequency decline as a result of different renewable penetration levels, models of all the protection systems including UFLS are disabled in this simulation. The frequency response for the largest N-2 and typical N-2 contingencies are portrayed in Figure 1.16 [2]. For both contingencies, increasing PV penetration worsens the overall frequency response. It increases ROCOF, lowers frequency nadir, and decreases settling frequency. The 2.6 GW generation loss with 80% PV penetration, causes the frequency to drop below 59.4 Hz triggering the first stage of UFLS. This implies that the Western Interconnection is not capable of hosting 80% solar PV due to its low inertia and diminished frequency response.

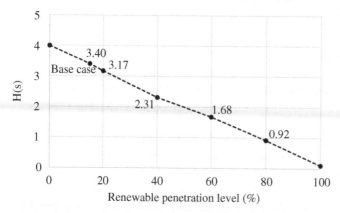

Figure 1.15   System equivalent inertia at different PV penetration levels. *Source:* Modified from Tan et al. [2].

(a)

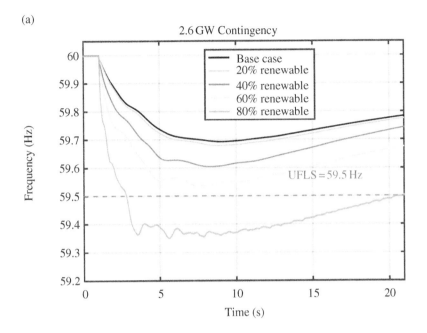

(b)

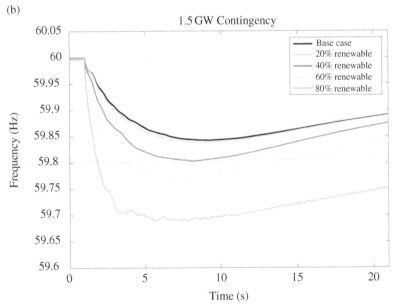

Figure 1.16 WECC frequency response under high PV penetration scenarios: (a) 2625 MW generation loss, (b) 1514 MW generation loss. *Source:* Tan et al. [2].

Frequency response obligation (FRO) for the Western Interconnection is described as the power that the system must provide during the first couple of seconds after an event to prevent the decline of frequency and avoid activation of first stage of UFLS. The FRO is set at 906 MW/0.1 Hz which implies that the power from all the generators should increase by 906 MW for a frequency decline of 0.1 Hz. This FRO is updated by NERC every year, based on system measurements.

Studies on the Western Interconnection reveal that the prescribed FRO is met with 20, 40, and 60% renewables; however, it is not satisfied in case of 80% renewable penetration. The FRO is noted to be 599, 692, and 724 for the largest N-2 contingency, typical N-2 contingency, and typical N-1 contingency, respectively. These values are much lower than the NERC specified FRO of 906 MW/0.1 Hz. This deficit of FRO will endanger the reliability and safety of the interconnected system.

1.2.18.2   Over Frequency Response

An overfrequency event is caused by loss of a large load or an exporting tie line (e.g. HVDC line). Displacement of conventional fossil-fuel-based conventional generators by IBRs such solar PV systems decreases system inertia. This exacerbates over frequency response of power systems. The typical behavior of power systems with different levels of stored kinetic energy during an overfrequency event is depicted in Figure 1.17 [11].

Systems with low inertia (e.g. 100 GWs) experience a higher "zenith" (peak frequency) subsequent to loss of a large load, than systems with higher inertia. The ROCOF is positive for all systems but higher for low inertia systems.

Overfrequency events have the following features [11]:

1) Lesser risk of system collapse as compared to underfrequency events.
2) Synchronous generators may provide an unexpected response due to sudden rise in frequency and may disconnect in some cases. Still, there is a low risk of cascaded loss of generation.
3) Less likelihood of disconnection of domestic loads.

Since there are no substantial adverse impacts on the system by overfrequency events, system operators are generally less concerned about these events. Overfrequency events are typically alleviated by reduction in power output from synchronous generators or IBRs.

1.2.19   Angular Stability Issues due to Reduced Inertia

Low-frequency electromechanical power oscillations (typically 0.1–2 Hz) are recognized as one of the major limiting factors in power transfer over long transmission lines [85]. Conventionally, these oscillations are damped by Power System Stabilizers (PSSs) integrated with synchronous

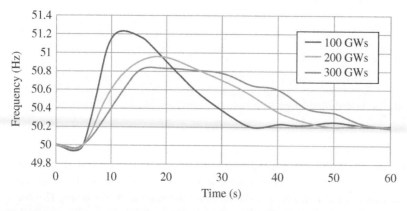

Figure 1.17 Typical behavior of power systems with different levels of stored kinetic energy during an overfrequency event. *Source:* Reprinted with permission from EPRI [11].

generators. Displacement of fossil fuel-based steam turbine generators by inertialess IBRs such as solar PV systems results in overall reduction in damping of electromechanical and inter-area oscillations, as conventional generators [86–89]. This especially becomes a concern during system transients such as faults or large equipment/line outages.

An example study of reduced damping due to high penetration of solar PV systems is presented in [87]. Eigenvalue analysis and transient stability studies are performed on a test system representing the entire Western Electricity Coordinating Council (WECC) network ranging from 34.5 to 500 kV. The synchronous generators are modeled with excitation systems, PSS, and governors. Solar PV systems comprising both rooftop systems and utility-scale plants are added in a region with a high potential of their growth. The utility-scale PV systems are fixed at 600 MW while the amount of rooftop PV systems are varied to achieve different PV penetration scenarios.

The percent PV penetration is considered to be the ratio of total PV generation to total system generation. The solar PV penetration is increased by displacing conventional generators while still keeping critical generators providing reactive power support in service. To maintain the generation–load balance, the outputs of the critical generators are reduced to accommodate the increased penetration of solar PV systems. The rooftop PV systems are modeled with unity power factor while the utility-scale solar PV systems are represented by full converter model and having reactive power based voltage control capability. A scenario of high solar PV output and low load is considered. The DSA Tools software package is used for performing both small-signal studies and transient stability studies.

Critical electromechanical modes having a damping less than 10% are obtained from Eigen analysis. The variation of damping ratio of this system mode with increasing solar PV penetration is illustrated in Table 1.1 [87]. It is seen that increasing PV penetration has a detrimental impact on the damping of the critical electromechanical mode.

The generators having highest participation in this critical mode are identified through participation factor analysis. A single-line diagram of the region close to these generators which are connected at buses 2102 and 2103 is depicted in Figure 1.18 [87].

A three-phase fault at 345 kV bus numbered 2104 is initiated for four cycles. The speed of generator located at bus 2103 is portrayed in Figure 1.19 [87].

Clearly, the speed of this generator becomes more oscillatory with increasing solar PV penetration, i.e. with increasing displacement of conventional synchronous generators. It is, therefore, recommended that when solar PV penetration levels are likely to increase in a system, critical generators must be identified and still kept in service to maintain adequate damping levels of the critical electromechanical modes.

Table 1.1 Impact of PV penetration on damping of critical mode.

PV penetration (%)	Real part of critical eigenvalue	Imaginary part of critical eigenvalue	Frequency (Hz)	Damping ratio (%)
0	−1.0926	11.2221	1.786	9.69
20	−0.9135	11.1828	1.7798	8.14
30	−0.9423	11.1346	1.7721	8.43
40	−0.5173	11.0029	1.7512	4.7
50	−0.4876	10.9947	1.7499	4.43

Source: Eftekharnejad et al. [87].

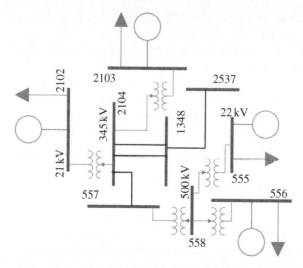

Figure 1.18 Single-line diagram of the study system near the participating generators. *Source:* Eftekharnejad et al. [87].

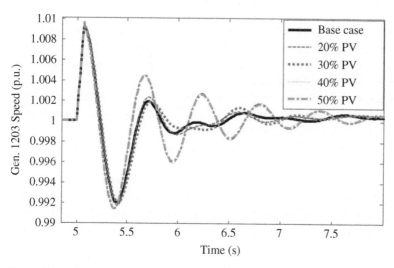

Figure 1.19 Speed of generator 2103 after a three-phase fault at bus 2104. *Source:* Eftekharnejad et al. [87].

1.3 Development of Smart Inverters

"Smart inverters" or "advanced inverters" for providing various grid support functions have been developed over a period of time by exploiting various potential capabilities of power electronic inverters. These active and reactive power-based control functions on solar PV inverters can alleviate several adverse impacts of high penetration of solar PV and other IBRs, as discussed in previous section. Smart inverter technology is a living technology which is growing at a very rapid rate worldwide. However, some early-stage initiatives in the development of smart inverter technology, to the best of author's knowledge, including [90–97], are described below.

1.3.1 Developments in Germany

The 2006 German E.ON grid code prescribed minimum requirements for grid connection of generating plants including solar PV systems [90]. These included certain provisions of reactive power, power factor and voltage control, frequency droop, dynamic reactive current during faults, etc. The 2008 German technical guideline for generating plants including solar PV systems connecting to medium voltage (MV) networks also laid down specifications for steady-state voltage control and dynamic network support (reactive current injection during faults, etc.) [91]. Reactive power-based functionalities such as fixed power factor, fixed reactive power, watt–power factor, and volt–var were specified. These were in addition to frequency-based active power controls.

Germany experienced a major challenge known as the "50.2 Hz problem." According to the German Grid Code DIN V VDE V 0126-1-1, all low-voltage generation plants were required to switch off immediately if the system frequency increased to 50.2 Hz [98]. At the time of creation of these standards, the solar installed capacity was quite small to cause any appreciable impact on the grid. However, by 2010, solar connections grew at a phenomenal rate and reached a cumulative capacity of 12.7 GW at the low-voltage levels. A detailed study revealed that in a worst-case scenario, about 9000 MW of solar power systems would get disconnected immediately if the system frequency increased to 50.2 Hz. This would lead to potential frequency instability problems for the German grid.

It was recommended on the basis of extensive studies, that PV inverters should be retrofitted with an updated control strategy by which they will reconnect if the frequency stays below 50.05 Hz for more than a specified time (30 or 60 seconds). In this manner, the power generation from the PV systems will once again become available to the grid after the disturbance has passed, instead of being lost completely. This would prevent the possibility of frequency instability in the grid.

This retrofitting in software was implemented during 2012–2014 on about 315 000 PV systems of rating greater than 10 kW. The cost of this retrofit exercise was estimated to be between €65 and €175 million.

1.3.2 Developments in the USA

In 2010, EPRI performed a smart inverter functionality survey of inverter manufacturers [93]. It subsequently published a pioneering report in 2013 (updated in 2016), on Smart Inverters describing a range of smart inverter functions including reactive power control and active power control of solar PV inverters for accomplishing several grid support functions [36].

In 2013, with lessons learned from the German experience, California decided to proactively update its interconnection Rule 21 requiring implementation of several smart inverter functions in the DER inverters seeking connection in California [99]. The Smart Inverter Working Group (SIWG) was formed out of a collaboration between the California Public Utilities Commission (CPUC) and California Energy Commission (CEC), that identified the development of advanced inverter functionality as an important strategy to mitigate the impact of high penetrations of DERs [99].

1.3.3 Development in Canada of Night and Day Control of Solar PV Farms as STATCOM (PV-STATCOM)

In 2009, a novel patented control paradigm of solar PV farms was developed by this book's author, whereby solar PV farms can be operated in nighttime as a STATCOM with full inverter capacity and during daytime with inverter capacity remaining after active power generation, for providing various grid support functions [35, 92, 97]. This new control of solar PV system as STATCOM was named PV-STATCOM [100]. In 2013, an enhanced patented control was developed whereby solar PV farms can operate as a STATCOM with full inverter capacity at any time in the day during periods of system need. Different applications of PV-STATCOM for providing benefits to transmission and distribution systems, including its first-in Canada field demonstration in 2016, are described in subsequent chapters.

1.4 Conclusions

This chapter illustrates the concepts of control of reactive power and active power and their corresponding impacts on the grid. This forms the basis of the different smart inverter functions from solar PV inverters, which will be described in the book. The various issues and challenges of high penetration of solar PV systems both in distribution and bulk power systems, are described. Finally, the early-stage evolution of smart inverter technology is presented.

References

1 International Renewable Energy Agency (IRENA) (2019). Global Energy Transformation – A Roadmap to 2050. *International Renewable Energy Agency (IRENA) Report*.

2 Tan, J., Zhang, Y., You, S., Liu, Y., and Liu, Y. (2018). Frequency response study of U.S. Western interconnection under extra-high photovoltaic generation penetrations. In *Proc. 2018 IEEE Power & Energy Society General Meeting*, 1–5.

3 TenneT TSO GmbH (2020). The Massive InteGRATion of Power Electronic Devices (MIGRATE). Bayreuth, Germany, *Technical Brief*.

4 CEATI International Inc. (2017). Mitigation of the Negative Impacts of Solar and Wind DG Connections in Distribution Systems. Montreal, QC: CEATI *Rep. No. T164700 #50/134*.

5 Moeini, A. and Kamwa, I. (2016). Analytical concepts for reactive power based primary frequency control in power systems. *IEEE Transactions on Power Systems* 31: 4217–4230.

6 Farrokhabadi, M., Cañizares, C.A., and Bhattacharya, K. (2017). Frequency control in isolated/ islanded microgrids through voltage regulation. *IEEE Transactions on Smart Grid* 8: 1185–1194.

7 Carvalho, P.M.S., Correia, P.F., and Ferreira, L.A.F.M. (2008). Distributed reactive power generation control for voltage rise mitigation in distribution networks. *IEEE Transactions on Power Systems* 23: 766–772.

8 Eto, J., Undrill, J., Mackin, P. et al. (2010). Use of a Frequency Response Metric to Assess the Planning and Operating Requirements for Reliable Integration of Variable Renewable Generation. Berkeley, CA: Lawrence Berkeley National Laboratory *Rep. No. LBNL-4142E*.

9 NERC (2020). Fast Frequency Response Concepts and Bulk Power System Reliability Needs. Atlanta, GA: NERC NERC Inverter-Based Resource Performance Task Force (IRPTF) *White Paper*.

10 NREL (2013). Variable Renewable Generation can Provide Balancing Control to the Electric Power System. Denver, CO: NREL *Rep. NREL/FS-5500-57820*.

11 EPRI (2019). Implications of Reduced Inertia Levels on the Electricity System. EPRI, Palo Alto, CA *Rep. No. 3002015132*.

12 EPRI (2019). Implications of Reduced Inertia Levels on the Electricity System. EPRI, Palo Alto, CA, USA, *Techn. Update Rep. 3002014970*.

13 EPRI (2019). Meeting the Challenges of Declining System Inertia. EPRI, Palo Alto, CA, USA, *White Paper*.

14 Eto, J.H., Undrill, J., Mackin, P., and Ellis, J. (2018). Frequency Control Requirements for Reliable Interconnection Frequency Response. Berkeley, CA: Lawrence Berkeley National Laboratory *Rep. No. LBNL-2001103*.

15 Miller, N., Lew, D., Piwko, R. et al. (2017). Technology capabilities for fast frequency response. *Report prepared by GE Energy Consulting for Australian Energy Market Operator*, Schenectady, NY, USA.

16 IEEE (2018). Impact of Inverter Based Generation on Bulk Power System Dynamics and Short-Circuit Performance. IEEE/NERC Task Force on Short-Circuit and System Performance Impact of Inverter Based Generation, New York, NY, USA, *IEEE PES Techn. Report. PES-TR68.*

17 AEMO (2017). Fast frequency response in the NEM-Working paper. Australian Energy Market Operator, Australia. *Report.*

18 International Energy Agency (2014). *High Penetration of PV in Local Distribution Grids: Subtask 2: Case Study Collection*. International Energy Agency PVPS Program, *Rep. IEA PVPS T14-02.*

19 Stetz, T., Marten, F., and Braun, M. (2013). Improved low voltage grid-integration of photovoltaic systems in Germany. *IEEE Transactions on Sustainable Energy* 4: 534–542.

20 Walling, R.A., Saint, R., Dugan, R.C. et al. (2008). Summary of distributed resources impact on power delivery systems. *IEEE Transactions on Power Delivery* 23: 1636–1644.

21 Katiraei, F. and Aguero, J.R. (2011). Solar PV integration challenges. *IEEE Power and Energy Magazine* 9: 62–71.

22 Coster, E.J., Myrzik, J.M.A., Kruimer, B., and Kling, W.L. (2011). Integration issues of distributed generation in distribution grids. *Proceedings of the IEEE* 99: 28–39.

23 Katiraei, F., Sun, C., and Enayati, B. (2015). No inverter left behind: protection, controls, and testing for high penetrations of PV inverters on distribution systems. *IEEE Power and Energy Magazine* 13: 43–49.

24 Cheng, D., Mather, B.A., Seguin, R. et al. (2016). Photovoltaic (PV) impact assessment for very high penetration levels. *IEEE Journal of Photovoltaics* 6: 295–300.

25 Obi, M. and Bass, R. (2016). Trends and challenges of grid-connected photovoltaic systems – a review. *Renewable and Sustainable Energy Reviews* 58: 1082–1094.

26 Bravo, R.J., Salas, R., Bialek, T., and Sun, C. (2015). Distributed energy resources challenges for utilities. In *Proc. 2015 IEEE 42nd Photovoltaic Specialist Conference (PVSC)*, 1–5.

27 Seguin, R., Woyak, J., Costyk, D., Hambrick, J., and Mather, B. (2016). High-Penetration PV Integration Handbook for Distribution Engineers. NREL, Golden, CO, USA, *Techn. Rep. NREL/ TP-5D00-63114.*

28 Masters, C.L. (2002). Voltage rise: the big issue when connecting embedded generation to long 11 kV overhead lines. *Power Engineering Journal* 16: 5–12.

29 Zandt, D.V. (2019). Applications of DER advanced functions and settings. In *Proc. ITWG Meeting*, 26 June.

30 (2016). *Electric power systems and equipment—voltage ratings (60 Hz)*. ANSI C84.1-2016 Standard.

31 Hingorani, N.G. and Gyugyi, L. (1999). *Understanding FACTS*. Piscataway, NJ: IEEE Press.

32 Mathur, R.M. and Varma, R.K. (2002). *Thyristor-Based FACTS Controllers for Electrical Transmission Systems*. New York: Wiley-IEEE Press.

33 EPRI (2015). *YouTube Video – EPRI High Penetration Solar Impacts*. EPRI.

34 EPRI (2015). *YouTube Video – Solar PV Impacts to Distribution Feeder*. EPRI.

35 Varma, R.K., Rahman, S.A., Mahendra, A.C., Seethapathy, R., and Vanderheide, T. (2012). Novel nighttime application of PV solar farms as STATCOM (PV-STATCOM). In *Proc. 2012 IEEE Power & Energy Society General Meeting*, 1–8.

36 EPRI (2016). Common Functions for Smart Inverters, 4e. EPRI, Palo Alto, CA. *Techn. Rep. 3002008217.*

37 Chen, L., Qi, S., and Li, H. (2012). Improved adaptive voltage controller for active distribution network operation with distributed generation. In *Proc. 47th International Universities Power Engineering Conference (UPEC '12).*

38 Foster, S., Xu, L., and Fox, B. (2006). Grid integration of wind farms using SVC and STATCOM. In *Proc. 41st International Universities Power Engineering Conference*, 157–161.

39 Ronner, B., Maibach, P., and Thurnherr, T. (2009). Operational experiences of STATCOMs for wind parks. *IET Renewable Power Generation* 3: 349–357.

40 Han, C., Huang, A.Q., Baran, M.E. et al. (2008). STATCOM impact study on the integration of a large wind farm into a weak loop power system. *IEEE Transactions on Energy Conversion* 23: 226–233.

41 Lahaçani, N.A., Aouzellag, D., and Mendil, B. (2010). Contribution to the improvement of voltage profile in electrical network with wind generator using SVC device. *Renewable Energy* 35: 243–248.

42 (2013). Distributed Generation Technical Interconnection Requirements. Hydro One Networks Inc. *Rep. DT-10-015 R3.*

43 Nelson, A., Hoke, A., Chakraborty, S. et al., (2015). Inverter Load Rejection Over-Voltage Testing – Solar City CRADA Task 1a Final Report. NREL, Golden, CO, USA, *Techn. Rep. NREL/TP-5D00-63510.*

44 Hoke, A., Nelson, A., Chakraborty, S. et al. (2015). Inverter Ground Fault Overvoltage Testing. NREL, Golden, CO, USA, *Techn. Rep. NREL/TP-5D00-64173.*

45 Boynuegri, A.R., Vural, B., and Tascikaraoglu, A. (2012). Voltage regulation capability of a prototype Static VAr Compensator for wind applications. *Applied Energy* 93: 422–431.

46 Ching-Yin, L. (1999). Effects of unbalanced voltage on the operation performance of a three-phase induction motor. *IEEE Transactions on Energy Conversion* 14: 202–208.

47 (2018). *IEEE standard for interconnection and interoperability of distributed energy resources with associated electric power systems interfaces.* IEEE Std 1547-2018 (Revision of IEEE Std 1547-2003).

48 Shahnia, F., Majumder, R., Ghosh, A.1, Ledwich, G., and Zare, F. (2010). Sensitivity analysis of voltage imbalance in distribution networks with rooftop PVs. In *Proc. 2010 IEEE PES General Meeting*, 1–8.

49 Reiman, A.P., McDermott, T.E., Reed, G.F., and Enayati, B. (2015). Guidelines for high penetration of single-phase PV on power distribution systems. In *Proc. 2015 IEEE Power & Energy Society General Meeting*, 1–5.

50 Bird, L., Milligan, M., and Lew, D. (2013). Integrating Variable Renewable Energy: Challenges and Solutions. NREL, Golden, CO, USA, *Techn. Rep. NREL/TP-6A20-60451.*

51 (2016). What the duck curve tells us about managing a green grid. https://www.caiso.com/Documents/FlexibleResourcesHelpRenewables_FastFacts.pdf (accessed 08 February 2020).

52 (2017). Confronting the duck curve: how to address over-generation of solar energy. https://www.energy.gov/eere/articles/confronting-duck-curve-how-address-over-generation-solar-energy (accessed 08 February 2020).

53 Kraiczy, M., Wang, H., Schmidt, S. et al. (2018). Reactive power management at the transmission–distribution interface with the support of distributed generators – a grid planning approach. *IET Generation, Transmission & Distribution* 12: 5949–5955.

54 Hoff, T. and Shugar, D.S. (1995). The value of grid-support photovoltaics in reducing distribution system losses. *IEEE Transactions on Energy Conversion* 10: 569–576.

55 Hung, D.Q. and Mithulananthan, N. (2013). Multiple distributed generator placement in primary distribution networks for loss reduction. *IEEE Transactions on Industrial Electronics* 60: 1700–1708.

56 S. K. Solanki, V. Ramachandran, and J. Solanki (2012). Steady state analysis of high penetration PV on utility distribution feeder. In *Proc. IEEE PES T&D Conference and Exposition*, 1–6.

57 Enslin, J.H.R. and Heskes, P.J.M. (2004). Harmonic interaction between a large number of distributed power inverters and the distribution network. *IEEE Transactions on Power Electronics* 19: 1586–1593.

58 (2014). *IEEE recommended practice and requirements for harmonic control in electric power systems.* IEEE Std 519-2014 (Revision of IEEE Std 519-1992).

59 Arrillaga, J. and Watson, N.R. (2003). *Power System Harmonics.* New York: Wiley.

60 Bradt, M., Badrzadeh, B., Camm, E. et al. (2011). Harmonics and resonance issues in wind power plants. In *Proc. 2011 IEEE Power & Energy Society General Meeting*, 1–8.

61 CEATI International Inc. (2016). Analysis of Parallel and Series Resonance on the Electrical Distribution System. CEATI, Montreal, QC, Canada, *Rep. No. T154700 #5171.*

62 Varma, R.K., Berge, J., Axente, I., Sharma, V., and Walsh, K. (2012). Determination of maximum PV solar system connectivity in a utility distribution feeder. In *Proc. 2012 IEEE PES T&D Conference and Exposition*, 1–8.

63 NERC (2019). Improvements to Interconnection Requirements for BPS-Connected Inverter-Based Resources. NERC, Atlanta, GA, USA, *Reliability Guideline*.

64 Pan, Y., Ren, W., Ray, S., Walling, R., and Reichard, M. (2011). Impact of inverter interfaced distributed generation on overcurrent protection in distribution systems. In *Proc. 2011 IEEE Power Engineering and Automation Conference*, 371–376.

65 Margossian, H., Capitanescu, F., and Sachau, J. (2013). Feeder protection challenges with high penetration of inverter based distributed generation. In *Proc. 2013 EUROCON*, 1369–1374.

66 Ko, A.D., Burt, G.M., Galloway, S. et al. (2007). UK distribution system protection issues. *IET Generation, Transmission & Distribution* 1: 679–687.

67 Short, T.A. (2003). *Electric Power Distribution Handbook*. Boca Raton, FL: CRC Press.

68 Sa'ed, J.A., Favuzza, S., Ippolito, M.G., and Massaro, F. (2013). Investigating the effect of distributed generators on traditional protection in radial distribution systems. In *Proc. 2013 IEEE Grenoble Conference*, 1–6.

69 Mahadanaarachchi, V.P. and Ramakumar, R. (2008). Impact of distributed generation on distance protection performance – a review. In *Proc. 2008 IEEE Power & Energy Society General Meeting*, 1–7.

70 NERC (2017). Integrating Inverter-based Resources into Low Short Circuit Strength Systems. NERC, Atlanta, GA, USA, *Reliability Guideline*.

71 Jafari, M., Olowu, T.O., Sarwat, A.I., and Rahman, M.A. (2019). Study of smart grid protection challenges with high photovoltaic penetration. In *Proc. 2019 North American Power Symposium*, 1–6.

72 Johnston, W. and Katiraei, F. (2012). Impact and sensitivity studies of PV inverters contribution to faults based on generic PV inverter models – Ontario Grid Connection Study, Canadian Solar Industries Association *Report*.

73 Kou, G., Chen, L., VanSant, P. et al. (2020). Fault characteristics of distributed solar generation. *IEEE Transactions on Power Delivery* 35: 1062–1064.

74 Marchesoni, M., Marinopoulos, A., Massucco, S., and Picco, V. (2012). High penetration of very large scale PV Systems into the European electric network. In *Proc. 2012 IEEE International Energy Conference and Exhibition (ENERGYCON)*, 658–662.

75 Nimpitiwan, N., Heydt, G.T., Ayyanar, R., and Suryanarayanan, S. (2007). Fault current contribution from synchronous machine and inverter based distributed generators. *IEEE Transactions on Power Delivery* 22: 634–641.

76 Baran, M.E. and El-Markaby, I. (2005). Fault analysis on distribution feeders with distributed generators. *IEEE Transactions on Power Systems* 20: 1757–1764.

77 Brahma, S.M. and Girgis, A.A. (2004). Development of adaptive protection scheme for distribution systems with high penetration of distributed generation. *IEEE Transactions on Power Delivery* 19: 56–63.

78 Bravo, R.J., Yinger, R., and Robles, S. (2013). Three phase solar photovoltaic inverter testing. In *Proc. 2013 IEEE Power & Energy Society General Meeting*, 1–5.

79 Hooshyar, H. and Baran, M.E. (2013). Fault analysis on distribution feeders with high penetration of PV systems. *IEEE Transactions on Power Systems* 28: 2890–2896.

80 Dalal, S.B., Knuth, W., Gaun, A., and Grisenti, A. (2018). Fault current mitigation using 550 kV air core reactors. In *Proc. 2018 IEEE PES T&D Conference and Exposition*, 1–9.

81 Ohrstrom, M. and Soder, L. (2011). Fast protection of strong power systems with fault current limiters and PLL-aided fault detection. *IEEE Transactions on Power Delivery* 26: 1538–1544.

82 McGranaghan, M., Ortmeyer, T., Crudele, D. (2008). Renewable Systems Interconnection Study: Advanced Grid Planning and Operations. Sandia National Laboratorics, Albuquerque, NM, USA, *Rep. SAND2008-0944 P*.

83 Ye, Z., Walling, R., Garces, L., et al. (2004). Study and Development of Anti-Islanding Control for Grid-Connected Inverters. NREL, Golden, CO, USA, *Techn. Rep. NREL/SR-560-36243*.

84 John Sundar, D. and Kumaran, M.S. (2015). A comparative review of islanding detection schemes in distributed generation systems. *International Journal of Renewable Energy Research* 5: 1016–1023.

85 Kundur, P. (1994). *Power System Stability and Control*. New York: McGraw-Hill.

86 Achilles, S., Schramm, S., and Bebic, J. (2008). Transmission System Performance Analysis for High-Penetration Photovoltaics. NREL, Golden, CO, USA, *Rep. NREL/SR-581-42300*.

87 Eftekharnejad, S., Vittal, V., Heydt, G.T. et al. (2013, 2013). Small signal stability assessment of power systems with increased penetration of photovoltaic generation: a case study. *IEEE Transactions on Sustainable Energy* 4: 960–967.

88 Tamimi, B., Cañizares, C., and Bhattacharya, K. (2013). System stability impact of large-scale and distributed solar photovoltaic generation: the case of Ontario, Canada. *IEEE Transactions on Sustainable Energy* 4: 680–688.

89 Miller, N.W., Shao, M., Pajic, S., and D'Aquila, R. (2014). Western Wind and Solar Integration Study Phase 3 – Frequency Response and Transient Stability: Executive Summary. NREL, Golden, CO, USA, *Rep. NREL/SR-5D00-62906-ES*.

90 E.ON Netz GmbH (2006). *Grid Code for High and Extra High Voltage*. E.ON Netz GmbH, Bayreuth, Germany.

91 (2013). *Technical Guideline: Generating Plants Connected to the Medium-Voltage Network. (Guideline for generating plants' connection to and parallel operation with the medium-voltage network)*, BDEW (Bundesverband der Energie- und Wasserwirtschaft e.V.), Berlin, Germany, June 2008, revised January 2013.

92 Varma, R.K., Khadkikar, V., and Seethapathy, R. (2009). Nighttime application of PV solar farm as STATCOM to regulate grid voltage. *IEEE Transactions on Energy Conversion (Letters)* 24: 983–985.

93 EPRI (2010). Smart Inverter Functionality Survey: A State-of-the-Industry Assessment. EPRI, Palo Alto, CA, USA, *Techn. Update, Rep. No. 1022239*.

94 Casey, L.F., Schauder, C., Cleary, J., and Ropp, M. (2010). Advanced inverters facilitate high penetration of renewable generation on medium voltage feeders – impact and benefits for the utility. In *Proc. 2010 IEEE Conference on Innovative Technologies for an Efficient and Reliable Electricity Supply*, 86–93.

95 Walling, R.A. and Clark, K. (2010). Grid support functions implemented in utility-scale PV systems. In *Proc. 2010 IEEE PES T&D Conference and Exposition*, 1–5.

96 Albuquerque, F.L., Moraes, A.J., Guimaraes, G.C. et al. (2010). Photovoltaic solar system connected to the electric power grid operating as active power generator and reactive power compensator. *Solar Energy* 84: 1310–1317.

97 Varma, R.K., Rahman, S.A., and Seethapathy, R. (2010). Novel control of grid connected photovoltaic (PV) solar farm for improving transient stability and transmission limits both during night and day. In *Proc. 2010 World Energy Conference*, Montreal, Canada.

98 Boemer, J.C., Burges, K., Zolotarev, P. et al. (2011). Overview of German grid issues and retrofit of photovoltaic power plants in Germany for the prevention of frequency stability problems in abnormal system conditions of the ENTSO-E region continental Europe. Presented at the First International Workshop on Integration of Solar Power into Power Systems, Aarhus, Denmark, October.

99 Smart Inverter Working Group (2019). https://www.cpuc.ca.gov/General.aspx?id=4154#:~:text=The%20Smart%20Inverter%20Working%20Group,distributed%20energy%20resources%20 (DERs) (accessed 08 February 2020).

100 Varma, R.K., Das, B., Axente, I., and Vanderheide, T. (2011). Optimal 24-hr utilization of a PV solar system as STATCOM (PV-STATCOM) in a distribution network. In *Proc. 2011 IEEE Power & Energy Society General Meeting*, 1–8.

2

SMART INVERTER FUNCTIONS

Conventional inverters in renewable energy systems are intended to convert DC power from the energy source into AC power for the grid at unity power factor [1, 2]. The DC power may be obtained from photovoltaic (PV) solar panels, battery energy storage systems, etc. No reactive power exchange is envisaged.

Power electronics has, however, imparted new capabilities to inverters to not only generate active power but modulate it in addition to performing controlled reactive power exchange (injection/absorption) with the grid [1, 2]. Use of Voltage Source Converter (VSC) also provides both inversion and rectification capabilities which can be utilized for respectively discharging and charging battery energy storage systems that may be operating either independently or coupled with renewable energy systems. Integration of communication systems further allows such inverters to be controlled in response to commands issued from substations or utility control centers.

Such inverters that allow exchange of both active power and reactive power in a controlled manner either autonomously or in response to communicated signals have been termed "smart inverters" or "advanced inverters."

The concepts and functions of smart inverters were initially presented in [3–15]. These functions are systematically described in the publications from E.ON Netz GmbH [3], BDEW [4], IEC [11], EPRI [16, 17], Smart Inverter Working Group in California [18], IEEE Standard 1547-2018 [19], UL Standard 1741 SA [20], Hawaii Electric's Rule 14H [21], NERC Standard PRC-024-3 [22], and NERC Reliability Guidelines [23, 24].

The technology of smart inverters with advances in communication systems is continuing to outpace the existing grid codes. Standards and grid codes in various countries are being revised to adopt and implement different evolving functions of smart inverters through extensive deliberations among various stakeholders. The principles of some of the key smart inverter functions are described in this chapter [25]. The implementation details of each of these functions are incorporated in various operating standards and grid codes. Smart inverter functions for battery energy storage systems are also briefly presented in this chapter. The prioritization of different smart inverter functions are discussed. Emerging smart inverter functions are then introduced. This list of smart inverter functions is expected to only grow with time.

2.1 Capability Characteristics of Distributed Energy Resource (DER)

Distributed Energy Resource (DER) is a general terminology for distributed generators, i.e. solar PV, wind generators, and energy storage systems which have the capability of exporting active power to the area electric power system (EPS) to which they are connected.

Smart Solar PV Inverters with Advanced Grid Support Functionalities, First Edition. Rajiv K. Varma.

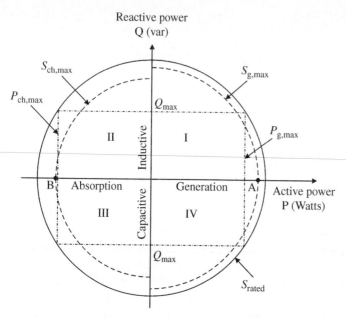

Figure 2.1 P–Q capability curve of a distributed energy resource. *Source:* Based on [17]. Used with permission from EPRI.

The operating *P–Q* capability of a DER is depicted in Figure 2.1 [17]. The total apparent power MVA nameplate capacity "S_{rated}" of the smart inverter is depicted as a circle having active power along the *x-axis* and reactive power along the *y-axis*. It is understood that,

$$Q = \sqrt{(S^2 - P^2)} \tag{2.1}$$

where *P* is active power; *Q* is reactive power; and *S* is the total apparent power.

The right half plane represents active power generation for distributed generators or active power discharge from energy storage systems. The left half plane portrays active power absorption or charging of energy storage systems. The DER can operate in quadrants I and IV with no energy storage but can function in all four quadrants with energy storage.

The maximum apparent power generation capability is illustrated through the semicircle S_{gmax} on the right half plane, whereas the maximum apparent power absorption (charging) capability is illustrated by the semicircle S_{chmax} on the left half plane. The nameplate apparent power rating S_{max} of the DER may be more than either of the two capabilities S_{gmax} or S_{chmax}. Furthermore, it may be noted that S_{gmax} and S_{chmax} may or may not be identical. The maximum active power generation and absorption (charging) limits are indicated by P_{gmax} and P_{chmax}, respectively, whereas the magnitude of maximum reactive power that the DER can produce or absorb is given by Q_{max}.

Some manufacturers can provide P_{gmax} and P_{chmax} equal to S_{gmax} at unity power factor, while some manufacturers can provide Q_{max} equal to S_{gmax} at zero active power production.

Reactive Power Capability Characteristic of a Synchronous Generator

The reactive power capability of a typical synchronous generator is shown in Figure 2.2 [26]. The reactive power is limited from several considerations such as heating constraints associated with flow of field and stator currents, stability limits, etc. In general, synchronous generators may not be able to exchange reactive power at zero loading conditions, except synchronous condensers. On the other hand, DER inverters can potentially exchange rated reactive current with the grid

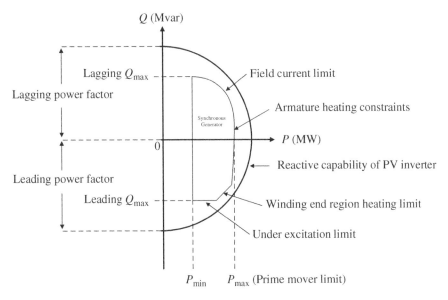

Figure 2.2 Reactive power capability of a typical synchronous generator compared to a PV Inverter. *Source:* Reprinted with permission from the National Renewable Energy Laboratory [26].

even at zero active power outputs. Hence the reactive power exchange capability of a PV inverter is much superior to that of synchronous generators. This enhanced capability can provide several grid support functions as described later in the book.

2.2 General Considerations in Implementation of Smart Inverter Functions

The connection point of an individual DER unit with a local EPS is denoted as Point of Connection (PoC) [19]. This indeed corresponds to the electrical terminals of the DER unit. The node where the local EPS connects with the area EPS is termed as Point of Common Coupling (PCC) as shown in Figure 2.3. There may exist intervening lines, cables, local loads, or an interface transformer between the PoC of an individual DER unit and the PCC. For voltage control functions, a distinction is made between voltage control at the PoC or voltage control at the PCC. Typically, the voltage control at PCC is considered relevant from the viewpoint of system operator. An offset voltage, therefore, has to be added to the voltage at PoC to obtain the voltage at the PCC.

In some cases, a reference PCC may be more relevant. For instance, if an energy storage system is required to smoothen the voltage fluctuations at the terminals of a remote PV solar farm, the terminals of the remote PV solar farm will be considered as reference PCC [19]. In an extended context, synchrophasor measurements from remote PCC locations may be utilized for the operation of a DER.

Extensive monitoring of data from the DER and bidirectional communication with the utility energy control center is envisaged in the implementation of smart inverter functions.

The parameters and settings of different smart inverter functions (as described in the following sections) may be adjusted locally. They may also be remotely controlled through communication links by the area EPS operator with the understanding that the DER will perform these functions in the best possible manner without compromising its own safety.

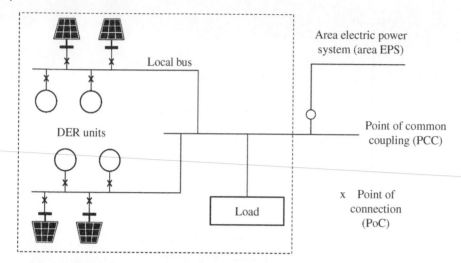

Local electric power system (local EPS)/ DER facility

Figure 2.3 Depiction of different interconnecting buses.

The response of different smart inverter functions is described in terms of open-loop response time, which refers to the time duration between the initiation of the step change in control signal input to the instant at which the output changes by 90% of the desired change, before the occurrence of any overshoot.

It is important to understand the sign convention of active and reactive power. Standards such as IEEE Standard 1547-2018 [19] define the operation of DERs with the generator sign convention which is opposite of load convention. The active power generated by the DER is considered positive and that absorbed by the DER is treated as negative. If the DER current lags the bus voltage, it generates (or injects) reactive power which is considered positive and has the effect of increasing bus voltage. This is the over-excitation mode of DER operation. However, if the DER current leads the bus voltage, it absorbs (or consumes) reactive power which is treated as negative and has the influence of decreasing the bus voltage. This is the under-excitation mode of DER operation. Reactive power at a DER bus is generally expressed in terms of leading or lagging power factor. Utility experience has shown that to avoid any confusion it is preferable to use terms such as reactive power injected by the DER or reactive power absorbed by the DER at a bus, instead of leading reactive power or lagging reactive power.

Smart inverter functions are described in great detail in [17, 19–23, 27]. The concepts of some major smart inverter functions as summarized from the above references are presented below. It may be noted that in any given application different smart inverter functions may be prioritized, for instance, if they are impacting the same variable – active power or reactive power. Furthermore, the settings of different smart inverter functions may be adjusted locally or remotely by the area EPS operator through communication systems.

2.2.1 Performance Categories

Different standards may specify different requirements during normal and abnormal performance of smart inverters. For instance, IEEE Standard 1547-2018 [19] has classified DER performance in different categories, as below. Each performance category specifies the minimum equipment capability requirements.

2.2.1.1 Normal Performance
Category A:

- covers minimum performance capabilities for reactive power control and voltage regulation, which are achievable by generally all types of commonly used DER technologies.
- relates to applications where DER penetration is low and the power output of DERs does not undergo frequent and substantial variations.

Category B:

- covers all requirements specified under Category A.
- specifies additional capabilities for applications where DER penetration is high or where the power output of DERs undergo frequent and substantial variations.

2.2.1.2 Abnormal Performance
Category I:

- covers minimum bulk power system (BPS) needs, which are achievable by generally all types of commonly employed DER technologies.
- ride-through requirements are derived from the German BDEW Standard [4];
- applicable for low penetration levels of DER.
- not consistent with reliability standards specified for bulk power generation resources.

Category II:

- covers all the BPS reliability needs, in accordance with the NERC's reliability standard PRC-024-2 [22].
- specifies additional requirements than in [22] to address the potential of fault-induced delayed voltage recovery on distribution systems.

Category III:

- specifies the highest level of disturbance ride-through capabilities and provides enhanced BPS security.
- applies to EPSs with very high levels of DER penetration.
- requirements in accordance with California Rule 21 [28] Smart Inverter requirements.
- specifies additional requirements than in [22] to address the potential of fault-induced delayed voltage recovery on distribution systems.

2.2.2 Reactive Power Capability of DERs

The minimum reactive power capability of DERs is specified by different Standards. For instance, the minimum reactive power capability when the DER is providing active power in excess of 20% of its rated active power, as specified by IEEE Standard 1547-2018 [19] is presented in Table 2.1 [19] and depicted in Figure 2.4 [19].

Table 2.1 Minimum reactive power injection and absorption capability.

Category	Injection capability as % of nameplate apparent power (kVA) rating	Absorption capability as % of nameplate apparent power (kVA) rating
A (at DER rated voltage)	44	25
B (over the full extent of ANSI C84.1 range A [29])	44	44

Source: IEEE Std 1547-2018 [19].

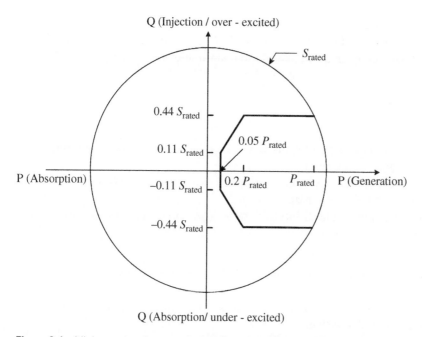

Figure 2.4 Minimum reactive power capability of Category B DERs. *Source:* IEEE Std 1547-2018 [19].

It is noted that to meet the above minimum requirement at rated active power output, the DER may need to: (i) curtail active power or (ii) be oversized, i.e. the nameplate MVA rating of the DER should be more than its active power rating, as indicated in Figure 2.1.

Different Standards may specify different minimum reactive power capabilities, e.g. in India [30]. There are two priority modes for reactive power exchange by DERs.

2.2.2.1 Active Power (Watt) Precedence Mode

In this watt precedence or watt priority mode, preference is given to active power generation, and reactive power is exchanged with the power system with inverter capability remaining after active power injection. The active power (watts) is not curtailed to provide the needed reactive power. This implies that the reactive power exchanged with the power system may become less than the amount needed to achieve a certain objective, especially during periods of high active power generation.

2.2.2.2 Reactive Power (Var) Precedence Mode

In this var precedence or var priority mode, preference is given to reactive power exchange. This implies that active power will need to be curtailed (or the inverter be oversized) to release enough inverter capacity to exchange the desired amount of reactive power with the power system, especially during periods of high active power generation.

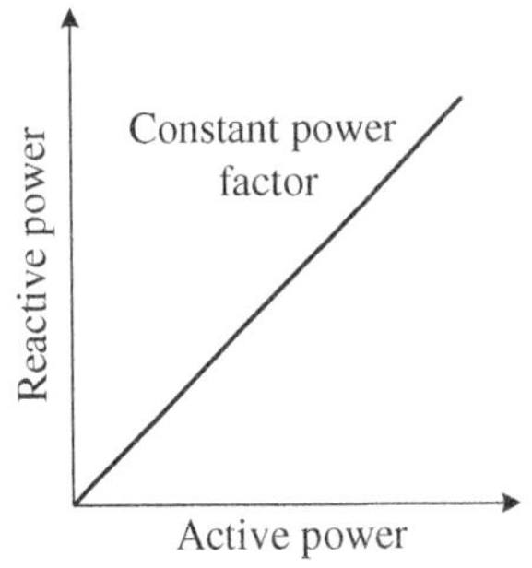

Figure 2.5 Constant power factor function.

2.3 Smart Inverter Functions for Reactive Power and Voltage Control

2.3.1 Constant Power Factor Function

This function allows the DER to operate at a constant (nonunity) value of power factor as specified by the area EPS operator. This power factor may be set at any value between +1.00 and −1.00, typically, between 0.9 inductive and 0.9 capacitive. The inverter exchanges (injects or absorbs) appropriate reactive power at a given value of active power to ensure that the power factor is maintained at the specified value as depicted in Figure 2.5. The reactive power for the specified power factor shall not exceed the reactive power capability specified in the applicable standard, e.g. Table 2.1 [19].

The maximum response time of the DER to maintain the desired power factor is specified by the prevalent standard and is typically 10 seconds or less [19]. The constant power factor settings can be adjusted locally and/or remotely as specified by the area EPS operator.

2.3.2 Constant Reactive Power Function

In this mode of operation, the DER shall maintain a constant reactive power – inductive or capacitive as specified by the area EPS operator. The reactive power for the specified power factor shall not exceed the reactive power capability specified in the applicable standard, e.g. Table 2.1 [19]. The response time of DER to maintain the required constant reactive power is typically less than 10 seconds [19] or specified by the relevant standard. The constant reactive power settings can be adjusted locally and/or remotely as specified by the area EPS operator.

2.3.3 Voltage–Reactive Power (Volt–Var) Function

This function enables the DER to control its reactive power exchange (production/absorption) in response to the PCC voltage. If the system voltage at PCC is high, the DER absorbs reactive power (to reduce the voltage). However, if the bus voltage is low, the DER injects reactive power (to increase the voltage).

The piecewise-linear operating characteristic of the volt–var function is depicted in Figure 2.6. The DER produces capacitive reactive power Q_1 if the bus voltage is below V_1, but ramps down to zero reactive power as the voltage varies between V_1 and V_2. No reactive power is exchanged when the bus voltage is between V_2 and V_3, i.e. when the voltage is within the utility acceptable range around V_{ref}. The region between V_2 and V_3 is termed as "deadband" which is typically kept between 0.02 and 0.03 pu on either side of V_{Ref}. If the voltage increases above V_3, the DER absorbs increasing amount of inductive reactive power in a ramped manner until the voltage becomes V_4. The DER absorbs its maximum reactive power Q_4 at V_4 and continues to stay at that level even if the system voltage rises beyond V_4. V_L and V_H are the lower and upper voltage limits for DER

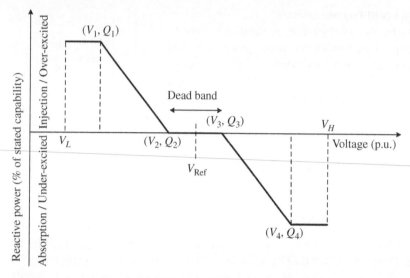

Figure 2.6 Volt–var curve of a DER. *Source:* IEEE Std 1547-2018 [19].

continuous operation. The open-loop response time, i.e. the time taken to reach 90% of the desired reactive power change (before any overshoot) from the initiation of the step change in voltage is specified to be in the range 1–90 seconds (typically 5–10 seconds) [19]. The volt–var settings can be adjusted locally and/or remotely as specified by the area EPS operator.

A hysteresis function may be introduced to prevent chattering or fluttering of reactive power due to small changes in system voltage around each voltage breakpoint. The voltage may be individual phase voltage (for single-phase DER) or the average of the three-phase RMS voltage or positive sequence voltages over one cycle, or as specified in the relevant Standard or Grid Code.

Different standards across the globe specify different default settings for volt–var curves [31]. It is noted that international standards are based on different statutory voltage limits, i.e. 0.9–1.1 Vpu in Europe, 0.94–1.1 Vpu in Australia, and 0.95–1.05 Vpu in the United States. The parameters of these international default settings as normalized to ANSI voltage limits [29] are compiled in Table 2.2 [31] and plotted in Figure 2.7 [31]. This is just to indicate that different settings (slopes, deadbands, reactive power limits, voltage limits, etc.) are specified as default by different standards in different countries or jurisdictions.

The volt–var function is implemented with either watt priority or var priority, depending on the application. It is important to note that the PV inverter will be operating at variable power factor while exercising volt–var control.

The volt–var control function may also be utilized for maintaining voltages at the levels specified under Conservation Voltage Reduction (CVR) [32].

2.3.4 Active Power–Reactive Power (Watt–Var or P–Q) Function

This control function allows the DER to actively control its reactive power exchange (injection or absorption) in response to active power output in accordance with a specified active power–reactive power (watt-var or P–Q) characteristic. A typical watt-var characteristic is illustrated in Figure 2.8. The different corner points of this characteristic for active power injection and absorption are indicated as (P_i, Q_i) and (P'_i, Q'_i), respectively, where $i = 1, 2,$ and 3. The left half characteristic pertains

Table 2.2 Default volt–var settings for different international standards.

Volt–Var settings	California	Hawaii	IEEE – Cat. A	IEEE – Cat. B	Australia	Europe
V_1	0.92	0.97	0.9	0.92	0.91	0.95
Q_1	+30%	+44%	+25%	+44%	+30%	+43.6%
V_2	0.967	0.97	1	0.98	0.966	0.97
Q_2	0	0	0	0	0	0
V_3	1.033	1.03	1	1.02	1.038	1.031
Q_3	0	0	0	0	0	0
V_4	1.07	1.06	1.1	1.08	1.1	1.05
Q_4	−30%	−44%	−25%	−44%	−30%	−43.6%

Source: Based on [31]. Used with permission from EPRI.

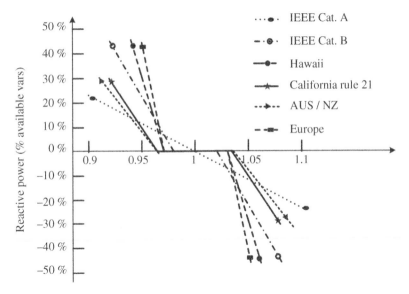

Figure 2.7 Default volt–var settings for different international standards. *Source:* Based on [31]. Used with permission from EPRI.

only to DERs with active power absorption capability such as energy storage systems. As active power output increases, the DER starts absorbing reactive power to counter the voltage rise associated with high active power injection. The response time of the watt–var control is typically less than 10 seconds [19]. The volt–watt settings can be adjusted locally and/or remotely as specified by the area EPS operator.

This function is particularly relevant for controlling the variability of voltage resulting from intermittent power generation from renewable energy systems such as solar PV and wind power systems.

A related smart inverter function is the Active Power – Power Factor (watt-power factor) function according to which the power factor changes from capacitive at very low values of active power output to inductive at high values of active power output [16, 17].

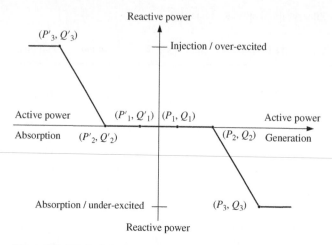

Figure 2.8 Typical watt–var characteristic. *Source:* IEEE Std 1547-2018 [19].

2.3.5 Dynamic Voltage Support Function

Dynamic voltage support function enables a DER or an inverter-based resource (IBR) to provide rapid exchange of reactive power with the EPS to stabilize voltage variations in the power system. Dynamic voltage support enhances voltage stability when voltage excursions enter into the low voltage ride through or high voltage ride-through regions. Under agreement with system operator, dynamic voltage support may be provided during the ride-through regions (i.e. mandatory or permissive operating regions [19]) and extend for a short period after the voltage has returned to within the normal continuous operating region of theDER/IBR, as specified by the relevant standard.

DERs/IBRs providing dynamic voltage support and continuing current injection to the grid during disturbance events provide useful support in mitigating potential transient and voltage instability issues in thpower system [33].

Dynamic voltage support may be provided by different techniques as described below.

2.3.5.1 Dynamic Network Support Function

German Grid Codes require that generating plants (i.e. solar PV) connected to power networks must be able to support the network voltage during a network fault by feeding a reactive current into the network [3, 4]. The dynamic reactive current injection characteristic as adapted from [3] is depicted in Figure 2.9. ΔI_B and I_N represent the additional injected reactive current and the rated current of the inverter, respectively. ΔV and V_N denote the change in voltage during system disturbance (e.g. fault) and the rated voltage, respectively.

The PV inverters need to inject minimum 2% of the rated current per 1% voltage drop below 0.9 pu [3]. A higher reactive current up to the rated inverter current must also be possible in some cases. This dynamic network support is provided with reactive power precedence. Hence, the PV inverter is required to reduce its active power to provide this reactive current. The voltage support is continued for a short period of time after the voltage becomes higher than 0.9 pu. Similarly, during overvoltage scenarios, the PV inverters need to absorb minimum 2% of the rated current per 1% voltage rise above 1.1 pu [3]. The rise time of dynamic current injection is <20 ms.

It is noted that after fault clearance, the generating plant must not extract more inductive reactive power from the network than prior to the occurrence of the fault [4].

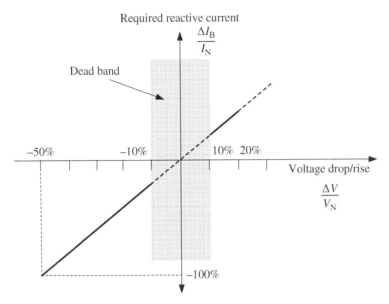

Figure 2.9 Dynamic reactive current injection. *Source:* Modified from E.ON Netz GmbH [3].

2.3.5.2 Dynamic Reactive Current Support Function

This grid support function is intended for dynamic voltage stabilization. It provides reactive current support during dynamic variations of voltage [17, 34]. This function is in contrast to voltage control provided by volt–var control which is intended to stabilize slowly varying voltages and is also much slower in response. While the volt–var function provides reactive power in response to the absolute value of voltage, this function provides appropriate reactive current in response to voltage deviation. For instance, if the absolute voltage is low needing capacitive reactive power, and the voltage rises rapidly in the meanwhile, this function produces an inductive current to restrain the rise in voltage. This function is conceptually similar to momentum or inertia, in which any quick change is opposed. The reactive current contributed by this function may be considered as an additional current, i.e. a current that is added to the preexisting reactive current provided by the volt–var function of constant power factor function, if in operation.

The dynamic reactive current function as adapted from [17] is depicted in Figure 2.10. If the voltage deviation ΔV increases during a voltage swell condition, an inductive current is produced in a ramped manner. Similarly, if the voltage deviation ΔV decreases during a voltage sag condition, a capacitive current is injected in a ramped manner. This generated current is an additional current over the steady-state current produced by the DER which may be inductive or capacitive due to the already active volt–var function or constant power factor function. The voltage deviation may be obtained as the difference between the PCC RMS voltage and a moving average of the PCC voltage over a prespecified previous time window and is expressed as percentage of the reference voltage V_{ref}. A dead band is provided to avoid any additional current generation if the voltage deviation is between ΔV_{max} and ΔV_{min}. The response time of this function is typically a few cycles.

A prioritization function may be provided by which the dynamic reactive current function is given a higher priority over active power production. In this case, active power may need to be reduced to release DER inverter capacity for reactive current exchange, as in var-precedence mode.

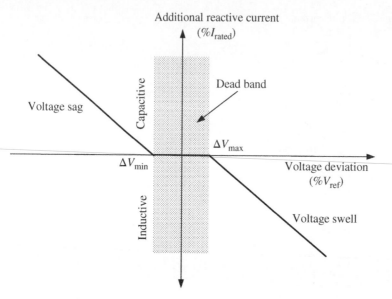

Figure 2.10 Dynamic reactive current support function. *Source:* Based on [17]. Used with permission from EPRI.

Due to the rapidity of response, this function allows time for other slower voltage control equipment to respond. This function is advantageous if the DER is connected in a feeder which is experiencing a problem of voltage flicker. This function is also useful for mitigating Fault Induced Delayed Voltage Recovery (FIDVR) [35, 36] and preventing the stall of induction motors.

Implementation of this function requires careful evaluation of the following aspects [19]:

i) coordination with distribution protection systems due to the reactive current injection during faults on the distribution feeders
ii) potential overvoltages during fault-clearing
iii) possible interaction with DER's anti-islanding detection process
iv) potential of overvoltage in healthy phase during unbalanced faults if this function is implemented symmetrically in all the three phases. This can, however, be obviated by utilizing negative sequence control of the DER inverter.

2.4 Smart Inverter Function for Voltage and Active Power Control

2.4.1 Voltage–Active Power (Volt–Watt) Function

In this mode, the DER shall actively control its active power output in response to voltage (typically, at the PCC) in accordance with a specified piecewise-linear characteristic. This function allows the DER to reduce its active power output if the system voltage becomes high and goes beyond utility acceptable limits. A high system voltage usually occurs if there is a large amount of power generation from the PV solar systems connected on the same distribution feeder during conditions of light load on the feeder.

The typical operating characteristics of the volt–watt function for a DER without and with active power absorption capability (i.e. energy storage) are depicted in Figure 2.11a,b, respectively [19]. The DER produces active power P_1 as long as the PCC voltage remains within utility-acceptable

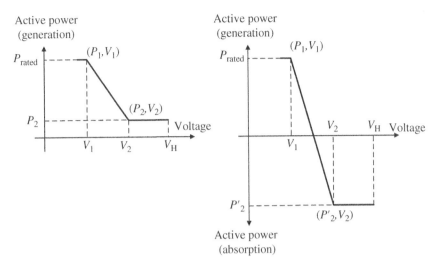

Figure 2.11 Volt–watt characteristic for DER (a) without energy storage; (b) with energy storage. *Source:* IEEE Std 1547-2018 [19].

range. If the voltage exceeds V_1, the active power is reduced in a ramped manner to P_2 or if this is outside the continuous operating capability of the DER, to the minimum DER capability instead of shutting down.

If the DER is equipped with energy storage device, in order to prevent unacceptable voltage rise the DER first decreases its power output and subsequently starts absorbing power to a level P_2' to limit the voltage rise to V_2. If P_2' is outside the continuous operating capability of DER, the DER absorbs power to its maximum absorption capability instead of shutting down. For this function, the response time can be in the range 0.5–60 seconds (typically 10 seconds) [19]. The volt–watt settings can be adjusted locally and/or remotely as specified by the area EPS operator.

It may be noted that the capability of energy storage device to vary its active power output may be limited by the following considerations:

i) its existing state of charge
ii) any requirement of maintaining a specific power output, e.g. for electricity market operations, providing peak shaving functions, etc.

Different international standards use different default volt–watt settings. Some international settings normalized to the ANSI voltage limits are compiled in Table 2.3 and displayed in Figure 2.12 [31].

Table 2.3 International default volt–watt settings normalized to ANSI voltage limits.

Volt–Watt settings	IEEE 1547-2018, California Rule 21, Hawaii Rule 14H	Australia	Europe
V_1	1.06	1.038	1.05
P_1	100%	100%	100%
V_2	1.1	1.1	1.07
P_2	0%	20%	0%

Source: Based on [17]. Used with permission from EPRI.

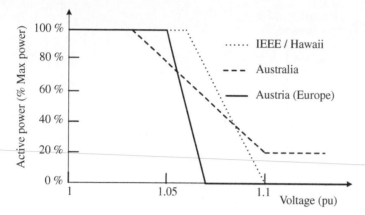

Figure 2.12 International default volt–watt settings normalized to ANSI voltage limits. *Source:* Based on [17]. Used with permission from EPRI.

2.4.2 Coordination with Volt–Var Function

Active power is generally given preference over reactive power, i.e. the capability of the reactive power control is exhausted before deciding to curtail active power. This is because active power brings revenue for the DER.

Some considerations in the simultaneous implementation of volt–var and volt–watt functions are

- Volt–var control is more effective in networks having a higher reactance, i.e. those with higher X/R ratios
- Volt–watt control is more effective in networks having a higher resistance, i.e. those with lower X/R ratios
- Volt–var control is activated first since reactive power capacity on inverters can be immediately utilized
- If the volt–var control is unable to regulate the voltage within acceptable limits, volt–watt control is activated.

An example of combined operation of volt–var and the volt–watt control functions is illustrated in Figure 2.13 [19]. At first, active power generation is given priority and the volt–var mode controls voltage. As the voltage continues to increase the DER changes its reactive power output from capacitive to inductive, reaching its maximum available inductive reactive power limit. If the voltage continues to rise further, the active power is gradually reduced, eventually to zero power output [16, 17]. If the DER is equipped with energy storage system, the DER can further start absorbing active power (commence charging) to help restrict the rise in voltage.

If active power needs to be decreased to lower the system voltage, the reduction or curtailment in active power output releases inverter capacity which can be further utilized to absorb reactive power in volt–var mode to help decrease the system voltage. This aspect is not demonstrated in Figure 2.13.

2.4.3 Dynamic Volt–Watt Function

This function allows a DER, especially with energy storage systems, to dynamically generate or absorb additional active power in response to a deviation in the voltage level at the PCC. This DER function helps alleviate fast voltage deviations through dynamic active power control within

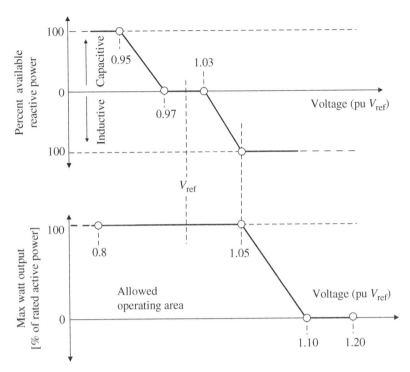

Figure 2.13 Combined operation of volt–var mode and volt–watt mode of operation. *Source:* Based on [17]. Used with permission from EPRI.

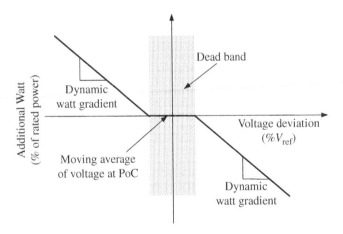

Figure 2.14 Dynamic volt–watt function. *Source:* Based on [17]. Used with permission from EPRI.

the DER's own operating capability and safety. This function is different from the volt–watt function. While the volt–watt function varies active power in response to absolute value of voltage, this function provides additional active power exchange in response to voltage deviation.

The dynamic volt–watt function as adapted from [17] is depicted in Figure 2.14. If the voltage deviation ΔV increases during a voltage swell condition, additional watts are reduced in a ramped manner. Similarly, if the voltage deviation ΔV decreases during a voltage sag condition, additional watts are injected in a ramped manner. These additional watts correspond to additional active power over the steady-state active power exchanged by the DER. This function essentially enables "dynamic watt gradient" which represents the additional watts expressed in percent maximum power rating P_{max} of the DER relative to the deviation in voltage expressed as percent of rated voltage V_{ref}. For instance, a value of "−1.0" indicates that the DER will absorb/reduce additional 1% P_{max} for each 1% V_{ref} increase in voltage deviation. The voltage deviation is obtained as the difference between the PCC RMS voltage and a moving average of the PCC voltage over a pre-specified previous time window. A dead band is provided to avoid any additional active power exchange if the voltage deviation is between ΔV_{max} and ΔV_{min}.

This function will operate with certain limitations, such as DERs with energy storage can provide both active power generation and absorption, whereas DERs without energy storage such as solar PV systems can only reduce their active power.

2.5 Low/High Voltage Ride-Through (L/H VRT) Function

This function describes the response of a DER during abnormal voltage disturbances in accordance with the performance requirements of the area EPS as well as the BPS to which the area EPS is connected. The performance requirements relate to system stability, reliability, and safety of both personnel as well as equipment connected to the system including the DER itself [19]. At a conceptual level, this function describes the voltage levels and time intervals during which the DER shall remain connected or must get disconnected. This function essentially allows the DER to stay connected for longer durations so as not to cause any adverse effects on grid performance, while maintaining its own safety constraints.

Voltages during transient disturbances may violate specified limits temporarily but return to levels within specified limits in a short period of time. During large penetration of DERs in a grid when a significant number of loads are dependent on power supply from the DERs, it is important that any transient voltage disturbance does not lead to the tripping of the DERs. This is because after the transient has passed, the absence of DERs may lead to a delay in recovery of voltage and possibly undesirable power outages. This function allows the DERs to ride-through temporary voltage transients, preventing them from undesirable disconnections and allowing time for voltage to recover while the DERs stay connected.

Even when the DER proliferation is not high in systems, such as small systems or microgrids where loads mainly derive power from the DERs, it is preferred that DERs remain connected for longer periods even during voltage disturbances.

The importance of ride-through capabilities was brought out during two major events in the United States [37]. A normally cleared 500 kV fault caused by the Blue Cut Fire in the Southern California area caused the tripping of approximately 1200 MW of BPS – connected solar PV systems due to frequency and voltage related protective functions in August 2016 [37]. In another event, normally cleared faults on the 220 kV and 500 kV systems caused the tripping of approximately 900 MW of BPS-connected solar PV systems due to action of voltage protective relays, on 9 October 2017 [38]. In both disturbance events, the generating resources did not get disconnected directly due to occurrence of fault. Instead, the PV inverters tripped due to their response to the detected terminal voltage and frequency conditions [27]. This led to NERC issuing new guidelines on the ride-through functions.

Different jurisdictions around the world may have different L/H VRT curves relevant for their systems. Some specific L/H VRT curves described in different standards are presented below.

2.5.1 IEEE Standard 1547-2018

IEEE Standard 1547-2018 [19] describes the criteria and requirements as applicable to all DER technologies interconnected to EPSs at typical primary or secondary distribution voltage levels. This standard does not apply to energy resources connected to transmission or networked sub-transmission systems.

IEEE Standard 1547-2018 [19] has defined three categories related to response to area EPS abnormal conditions. Over- and under-voltage ride-through criteria are specified distinctly for the three categories as voltage ride-through is dependent on the penetration of DER, with varying reliability needs [27, 33]. The three Categories I, II, and III are described in Section. 2.2.1.

As examples, the L/H VRT characteristics for Categories I and III are illustrated in Figure 2.15 and Figure 2.16, respectively [19]. These characteristics make reference to German MV Code Synchronous Generators [4], NERC Standard PRC-024-2 [22], and Hawaii Rule 14H [21]. The different operating regions in these characteristics are defined as below [19]:

Mandatory operation: Required continuance of active current and reactive current exchange of DER with area EPS as prescribed, notwithstanding disturbances of the area EPS voltage or frequency having magnitude and duration severity within defined limits.

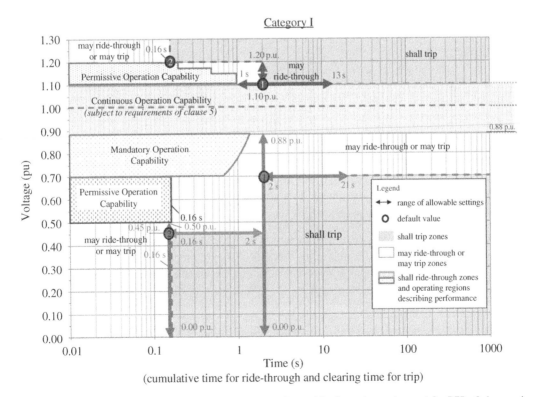

Figure 2.15 DER response to abnormal voltages and voltage ride-through requirement for DER of abnormal operating performance Category I. *Source:* IEEE Std 1547-2018 [19].

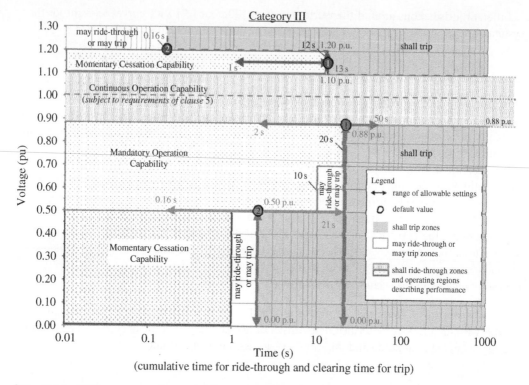

Figure 2.16 DER response to abnormal voltages and voltage ride-through requirements for DER of abnormal operating performance Category III. *Source:* IEEE Standard 1547a-2020 (Amendment to IEEE Standard 1547-2018) [19].

Cease to energize: Cessation of active power delivery under steady-state and transient conditions and limitation of reactive power exchange.

There are, however, a few clarifications to this definition:

i) This may lead to momentary cessation or trip
ii) This does not necessarily imply, nor exclude galvanic separation (disconnection) or a trip
iii) Limited reactive power exchange may continue as specified, e.g. through filter banks, etc.
iv) Energy storage systems are allowed to continue charging but not beyond their maximum state of charge.

Momentary cessation: Temporarily *cease to energize* an EPS, while connected to the area EPS, in response to a disturbance of the applicable voltages or the system frequency, with the capability of immediately restoring output of operation when the applicable voltages and the system frequency return to within defined ranges.

Permissive operation: Operating mode where the DER performs ride-through either in *mandatory operation* or in *momentary cessation*, in response to a disturbance of the applicable voltages or the system frequency.

Trip: Cease to energize or disconnect from the area EPS within a specified clearing time but without immediate return to service.

For low voltage ride through, the relevant voltage at any time is considered to be the least magnitude of the individual phase-to-neutral, phase-to-ground, or phase-to-phase voltage relative to the corresponding nominal system voltage. For high voltage ride-through, the relevant voltage at any given time shall be the greatest magnitude of the individual phase-to-neutral, phase-to-ground, or phase-to-phase voltage relative to the corresponding nominal system voltage [19]. Undervoltages and overvoltages generally occur independently and hence, HVRT and LVRT requirements are based on specified cumulative durations and analyzed independently [19].

The specified voltage disturbance ride-through requirements do not apply when frequency is outside of the ride-through range specified under L/H FRT criteria [19]. The voltage and time set points can be adjusted locally or remotely by the system operator.

The DERs of Categories II and III should by default not reduce their total apparent current during the disturbance period in mandatory operation mode below 80% of the pre-disturbance value or of the corresponding active current level subject to the available active power, whichever is less, subject to certain conditions [19].

The Standard also specifies the number of consecutive temporary voltage disturbances which a DER should ride through and their cumulative durations. For each performance category, if the cumulative duration of voltage within any ride-through operating region exceeds the maximum allowed duration for the specified severity of voltage disturbance in that region, the DER may trip. Also, if several consecutive disturbances lead to initiation of electromechanical oscillations which may cause potential system instability concerns or loss of DER synchronism, or damage to DER, the DER is allowed to trip [19].

IEEE Std. 1547-2018 has made dynamic voltage support operation *optional* during L/H VRT for all categories of DERs, even though it recognizes that such support of applicable voltage can provide benefit to area EPS and the BPS. It is stated that under agreement with the area EPS operator the DER may provide dynamic voltage support to the area EPS during Mandatory Operation and Permissive Operation both during LVRT and HVRT operation. The dynamic voltage support provides rapid exchange of reactive power with the power system during voltage excursions. If such dynamic voltage support is provided, this shall continue for a short period (typically up to five seconds) after the applicable voltage has returned within the relevant lower or upper threshold of continuous operation region. This dynamic voltage support should not cause DER to cease to energize under scenarios where the DER would not cease to energize without dynamic voltage support [19]. The area operator may examine and ensure that dynamic voltage support does not cause any adverse impact on protection systems.

Another form of dynamic voltage support can be provided through dynamic reactive current support function (Section 2.3.5.2) while exercising the L/HVRT function. This can help alleviate low or high voltage conditions during the system disturbance. The dynamic reactive current function is required under some Standards or Grid codes such as [3, 4] but may be optional in some others [19].

It is important that techniques to detect islands should be coordinated with L/H VRT settings so as not to adversely impact anti-islanding requirements.

2.5.2 North American Electric Reliability Corporation (NERC) Standard PRC-024

In North America, while IEEE 1547-2018 applies only to DERs connected to the distribution systems, IBRs connected to BPSs are required to follow the recommendations laid down by NERC [22–24]. These IBRs will also be required to meet the future requirements of IEEE P2800 [39] which is the transmission counterpart of IEEE Standard 1547-2018, presently under development in 2020.

The 60 Hz Voltage Ride Through Curve for NERC is illustrated in Figure 2.17 [22]. This curve indicates the no trip and trip zones both for low voltage and high voltage conditions. For low voltage conditions, the voltage considered is the minimum fundamental frequency phase-to-ground or phase-to-phase voltage, whereas for high voltage conditions, the higher of maximum RMS or peak phase-to-phase voltage is utilized. The per unit voltage is the nominal operating voltage specified by

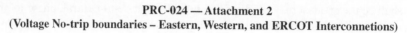

PRC-024 — Attachment 2
(Voltage No-trip boundaries – Eastern, Western, and ERCOT Interconnetions)

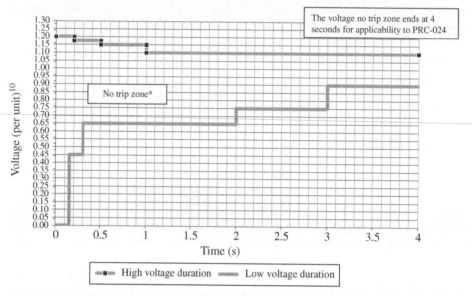

Figure 2.17 Voltage ride-through time duration curve from NERC. *The area outside the "No Trip Zone" is not a "Must Trip Zone." [10]: Voltage at the high side of the Bulk Electric System (BES) generator step up (GSU) transformer or main power transformer (MPT); *Source:* Reprinted with permission from North American Electric Reliability Corporation [22]. This information from the North American Electric Reliability Corporation's website is the property of the NERC and available on the Standards page (https://www.nerc.com/pa/Stand/Pages/default.aspx). This content may not be reproduced in whole or any part without the prior express written permission of the North American Electric Reliability Corporation.

the Transmission Planner for analyzing the reliability of the Interconnected Transmission Systems at the point of interconnection to the Bulk Electric System (BES). The VRT curve is developed based on three-phase transmission system zone 1 faults with normal clearing in less than or equal to nine cycles [22].

BES-connected IBRs are expected to continue current injection inside the "No Trip" zone of the voltage ride-through curves.

The differences in ride-through features of NERC PRC-024-3 [22] and IEEE 1547-2018 [19] are only because IEEE 1547-2018 applies only to distribution-connected resources while PRC-024-3 prescribes performance for BPS connected resources [27]. This serves as an example of that the ride-through characteristics may be different for distribution connected and BPS connected inverter resources.

2.6 Frequency–Watt Function

This function is used in situations when transient disturbances in power system result in short-term frequency variations which need to be corrected in a rapid manner. Also, in small systems such as microgrids, system disturbances or generation-load mismatches may result in longer-term frequency changes that need to be minimized. This function is based on the concept that excess power generation results in frequency rise and shortfall of generation lead to frequency decline.

Two types of frequency–watt functions are proposed in [17]. The Frequency Watt Function 1 is simpler and does not incorporate energy storage system. On the other hand, the Frequency Watt Function 2 is more flexible for use in different situations and incorporates two-way flow of power, i.e. both generation and absorption as with energy storage systems. Both functions are mutually exclusive and only one can be implemented at a time.

2.6.1 Frequency–Watt Function 1

The frequency–watt function 1 is demonstrated in Figure 2.18 [17], which illustrates a plot of deviation in generated power versus deviation in grid frequency. If the grid frequency increases beyond a maximum value Δf_m, the generated power output P_{gmax} at that frequency is made fixed. If the frequency increases further, the power is decreased in a ramped manner until the power either reduces to zero or to a limit specified by the grid code. A typical ramp rate may be 40% P_{gmax} per Hz.

A hysteresis function is provided such that when the deviation in grid frequency reduces to Δf_1, the power is unfixed (released) and is allowed to return to its normal value. The power is not allowed to return to its nominal value abruptly, but with a rate of rise which is typically 10% $P_{gmax}/$min.

The frequency–watt control is now explained through an example. Let the nominal grid frequency be 60 Hz. Due to excess generation in the grid, the frequency increases to 60.2 Hz. At this frequency, let us assume that the DER is producing 2000 W. This power is chosen as the reference value P_{gmax}. As the frequency continues to rise to 61.8 Hz the DER power output is reduced with a ramp rate of 40% $P_{gmax}/$Hz. The DER power output at 61.8 Hz becomes P_{g2}, where,

$$P_{g2} = P_{gmax} - (\Delta f_2 - \Delta f_m) \times 40\% \times P_{gmax}/\text{Hz}$$
$$= 2000 - (1.8 - 0.2)\,\text{Hz} \times 40\% \times 2000\ \text{W/Hz}$$
$$= 720\ \text{W}$$

After some time, the grid frequency reduces to 60.05 Hz and the corresponding frequency deviation Δf_1 becomes 0.05 Hz. At this time, the DER power output is unfixed and allowed to increase to pre-disturbance value of 2000 W with a ramp of 10% $P_{gmax}/$minute.

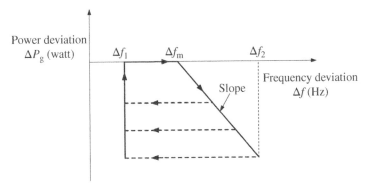

Figure 2.18 Frequency watt function 1. *Source:* Based on [17]. Used with permission from EPRI.

2.6.2 Frequency–Watt Function 2

The frequency–watt function 2 is demonstrated in Figure 2.19 [17], which illustrates a plot of generated power (expressed as % of maximum power rating of DER) versus actual grid frequency f. If the system frequency increases beyond the nominal frequency f_{nom} to a specified value f_1, the generated power P_g is reduced until it reduces to zero at frequency f_2 and continues to be zero until frequency rises to f_3 and beyond. If the system frequency starts to reduce, a hysteresis function is invoked, which allows the power to start returning to its nominal value of P_g only when the system frequency reduces to f_4. The horizontal bidirectional dashed lines within the hysteresis curve indicate the path if the frequency starts decreasing while P_g is getting reduced, or if the frequency starts increasing as P_g is getting increased at any intermediate point. The generated power returns to its original nominal value of P_g when the system frequency decreases to f_5.

While P_g is being reduced (after frequency exceeds f_1), if the system frequency starts to reduce, the power will return to its nominal value P_g following the dashed lines within the hysteresis band. It is noted that the slopes (ramp rates) of power decrease and increase may not be the same. The ramp times both during increase and decrease of power output are also specified in units of %P_{gmax}/s.

2.6.3 Frequency Droop Function

Frequency-droop functions are also described in IEEE Std. 1547-2018 [19] in terms of active power output in percent of nameplate rating versus frequency for different DER loading conditions. These curves incorporate tripping conditions (described in Section 2.7). An example of three frequency-droop function curves corresponding to different pre-disturbance power output levels of a DER with a 5% droop and 36 mHz deadband is illustrated in Figure 2.20 [19]. For small frequency deviations resulting in power change of less than 5% rated active power, the default open-loop response time is typically five seconds. For large frequency deviations, the DER response rate is typically no larger than 20% nameplate rating per minute.

2.6.4 Frequency–Watt Function with Energy Storage

The DERs with battery energy storage system allow a two-way flow of power through their VSC. The frequency–watt curve of a DER equipped with battery energy storage system is demonstrated in Figure 2.21 [17]. The battery is normally in the charging mode absorbing power P_{ch} at the nominal system frequency f_{nom} and beyond. If the system frequency decreases below a specified value f_1, the absorbed power P_{ch} is reduced until it reduces to zero at frequency f_2. If the frequency continues to decrease, the DER starts discharging (injecting power P_g into the grid) until the frequency decreases to f_3 and even below. A hysteresis function is integrated such that if the frequency starts to rise

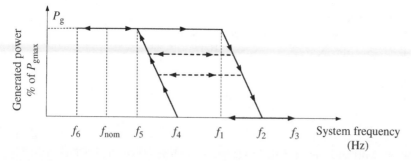

Figure 2.19 Frequency watt function 2. *Source:* Based on [17]. Used with permission from EPRI.

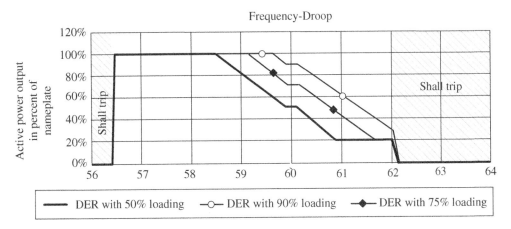

Figure 2.20 Frequency-droop function curves. *Source:* Modified from IEEE Std 1547-2018 [19].

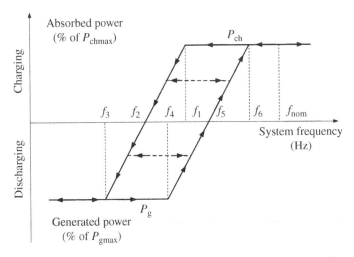

Figure 2.21 Frequency watt function with battery energy storage. *Source:* Based on [17]. Used with permission from EPRI.

beyond f_4 the DER decreases its discharge (power injection P_g) into the grid until the power exchange becomes zero at frequency f_5. The DER subsequently starts absorbing power till the frequency increases to f_6 (close to the nominal value f_{nom}) when its power absorption level returns to P_{ch}. The horizontal bidirectional dashed lines within the hysteresis curve indicate the path if the frequency starts increasing while P_{ch} is getting reduced, or if the frequency starts decreasing as P_{ch} is getting increased at any intermediate point.

2.7 Low/High Frequency Ride-Through (L/H FRT) Function

Frequency ride through is essential for ensuring that generation and load remain balanced across the interconnected BPS and distribution systems. All generating resources need to ride together during abnormal frequency conditions, with sufficient ride-through capability, to prevent any over-frequency and under-frequency load shedding conditions [27].

During large penetration of DERs in a grid when a significant number of loads are dependent on power supply from the DERs, it is important that any transient frequency deviation does not lead to the disconnection of the DERs as this may lead to a delay in recovery of frequency and more importantly lead to undesirable power outages. It was indeed for this reason that very expensive retrofits of frequency ride-through functions were done on hundreds of thousands of solar PV systems larger than 10 kW in Germany during 2012–2014 due to the 50.2 Hz problem described in Chapter 1 [13].

Even when the DER proliferation is not high, as in microgrids, where loads mainly derive power from the DERs, it is desirable that DERs remain connected for longer periods even during frequency disturbances.

It is argued that the DERs should not disconnect even when the other conventional generators remain connected during a frequency dip. Further, during frequency rise, instead of disconnection, the output from the DERs may be lowered to help support the power system.

The L/H FRT function essentially describes conditions under which a DER must remain connected or get disconnected from the grid during frequency disturbances. Different jurisdictions around the world may have different L/H FRT curves relevant for their systems. Some specific L/H FRT curves described in different standards are presented below.

2.7.1 IEEE Standard 1547-2018

In this Standard, the frequency ride-through requirements are harmonized for the different categories as frequency is a system-wide phenomenon and not dependent on the penetration levels of DER. The harmonized L/H FRT characteristic for all the three Categories I–III of DERs, as incorporated in the IEEE Std. 1547, is presented in Figure 2.22 [19]. The specified frequency ride-through requirements do not apply when voltage is outside of the ride-through range specified under L/H VRT criteria [19].

In one of the specifications for L/H FRT [19], it is stated that during temporary frequency disturbances, for which the system frequency is less than 58.8 Hz and greater than or equal to 57.0 Hz, and having a cumulative duration below 58.8 Hz of less than 299 seconds in any ten-minute period, the DER shall be capable to Ride-Through and:

a) shall maintain synchronism with the area EPS
b) shall not reduce its active power output lower than the values specified for different categories as below:

Category I: 80% nameplate active power rating or pre-disturbance active power output whichever is less

Categories II and III: pre-disturbance active power output

In general, during both LFRT and HFRT in the Mandatory Operating Mode, the DER of all categories shall stay connected and modulate active power to mitigate under-frequency and over-frequency conditions, respectively. The IEEE Standard 1547-2018 does not require DERs to be equipped with energy storage systems or operate at a power level lower than the available power to create a headroom (reserve capacity) for increasing the power output during underfrequency events [19].

It is specified that the DER shall have the capability of mandatory operation with frequency-droop (frequency–watt) characteristic during both LFRT and HFRT conditions.

The Standard [19] also states that the DERs shall continue to ride through and not trip if the magnitudes of rate of change of frequency (ROCOF) are equal or below specified values, during

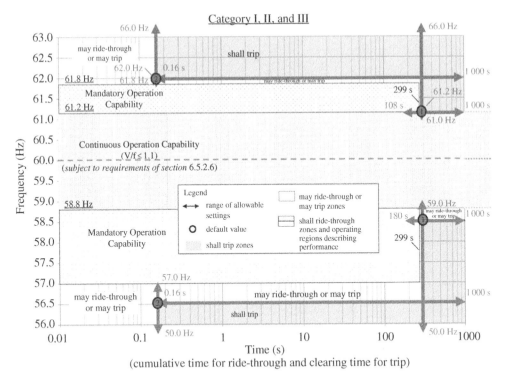

Figure 2.22 DER default response to abnormal frequencies and frequency ride-through requirements for DER of abnormal operating performance Category I, Category II, and Category III. *Source:* IEEE Std 1547-2018 [19].

the Continuous Operating Region and the L/H FRT Operating Regions. The ROCOF is computed as the average ROCOF over an averaging window of at least 0.1 seconds.

2.7.2 North American Electric Reliability Corporation (NERC) Standard PRC-024

The L/H FRT or off-nominal frequency capability curves for a select few jurisdictions are depicted in Figure 2.23 [22]. Figure 2.23a–c illustrates the setpoints for Eastern Interconnection of USA; Western Interconnection of USA; and Quebec, Canada [22]. It is noted that the area outside the "No Trip Zone" is not a "Must Trip Zone." It is important that frequency measurements may be done appropriately and not be based on individual phase measurements [37].

BPS-connected IBRs are expected to continue current injection inside the "No Trip" zone of the frequency and voltage ride-through curves [23]. It is noted that the frequency ride-through capability requirements in IEEE Standard 1547-2018 [19] meet and exceed the PRC-024 specifications [22, 27].

2.8 Ramp Rate

The rate of change of power output of DER from one level to another level is termed ramp rate. A sudden transition of DER power output from one level to another with uncontrolled ramp rate may result in undesirable voltage spikes or dips, harmonics, and/or electromechanical oscillations.

(a)

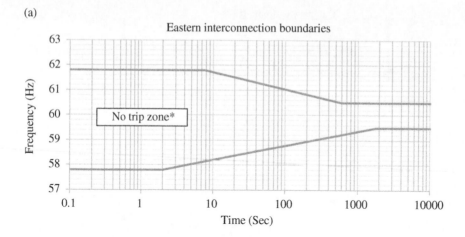

(b)

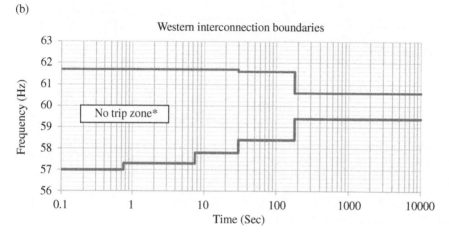

(c)

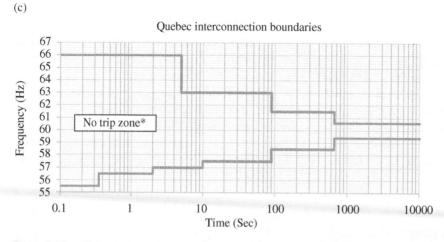

Figure 2.23 Off nominal frequency capability curve from NERC (a) Eastern interconnection boundaries, (b) Western interconnection boundaries, (c) Quebec interconnection boundaries. *The area outside the "No Trip Zone" is not a "Must Trip Zone." *Source:* Reprinted with permission from North American Electric Reliability Corporation [22]. This information from the North American Electric Reliability Corporation's website is the property of the NERC and available on the Standards page (https://www.nerc.com/pa/Stand/Pages/default. aspx). This content may not be reproduced in whole or any part without the prior express written permission of the North American Electric Reliability Corporation.

This is especially true if the DER power output is large or there are aggregated DERs operating together. For this reason, the power output of DER must be changed in a ramped manner.

Different ramp rates may be specified by manufacturers for different conditions such as "normal ramp up rate," "emergency ramp-up rate" and "soft-start connect ramp-up rate." A default ramp rate can be 2% of maximum current output per second, although it can be varied. Soft start implies turn-on of different PV systems connected on a single feeder system to full production in a staggered manner to prevent sudden voltage spikes, or rises.

IEEE Std. 1547 [19] specifies that during enter service period, the DER shall increase output of active power or exchange of active power for DER with energy storage systems, linearly or in a step-wise linear ramp. The enter service is the process of commencing DER operation with an energized power system. The average rate of change shall not exceed the DER nameplate active power rating divided by the enter service period. The duration of enter service period ranges over 1–1000 seconds with a default time of 300 seconds (i.e. five minutes). The above is the maximum ramp rate and the DER may restore power slower than the above-specified values.

2.9 Fast Frequency Response

This function allows DERs to vary their active power in proportion to the ROCOF. This function is termed "inertial response" in [19], although "fast frequency response" is the preferred name according to NERC [33, 40]. This function is not required but is permitted [19].

This capability of DERs or IBRs to respond to rapidly changing frequency is extremely helpful in enhancing system reliability especially in systems with low system inertia that experience high ROCOF conditions [33].

2.10 Smart Inverter Functions Related to DERs Based on Energy Storage Systems

2.10.1 Direct Charge/Discharge Function

This function allows an entity external to the energy storage system (e.g. battery) control the charging and discharging of the storage system at specified rates based on its capabilities [17, 19, 21]. This function utilizes the following general energy storage related settings:

Maximum intermittency ramp rate: This determines the maximum rate at which the active power output of the DER may be ramped up or down at the PCC. It is assumed that the DER has information about the actual (active-time) power level at the PCC. Based on this information, the DER is expected to control the rate of ramp-up or ramp-down of power at the PCC utilizing its charging and discharging function. This setting is especially useful in managing the variability of power output from renewable energy sources such as solar PV or wind-based systems.

Minimum reserve for storage: This indicates the minimum level to which the storage system may be discharged as a percent of the total storage capacity. This is different from the minimum battery State of Charge (SOC) or the equivalent quantity for the relevant energy storage system as specified by the manufacturer. This setting allows the battery to keep a certain amount of charge as reserve for use under special circumstances such as customer backup or ancillary services.

Maximum storage charge rate: This is the maximum power or rate at which the storage system may be charged, in watts.

Maximum storage discharge rate: This is the maximum power or rate at which the storage system may be discharged, in watts.

The external entity may specify the charging rates and discharging rates as a percent of the above-described maximum rates for charging or discharging, as applicable. Furthermore, the charging and discharging rates may not be the same.

The DER waits for a time duration specified by the "randomized time window" to allow the communicate systems to access the different DER units within the group of DER before implementing this requested function. The DER changes the present charge/discharge rate to the requested charge/discharge rate in a linear ramped manner over a specified "ramp time." The function stays enabled over a "reversion timeout" period after which the DER returns to its default charge/discharge rate.

A charge or discharge schedule may also be implemented by specifying the charge/discharge rate on a daily, weekly, or seasonal basis.

2.10.2   Price-Based Charge/Discharge Function

This function allows DER with an energy storage system (e.g. battery) to autonomously control its charging and discharging based on an externally communicated "price of energy" signal [17]. This function is also termed as "Price-Based Active Power Function." Based on this price signal, the energy storage system decides by itself, about when to charge or discharge and to what levels. In determining the above, the storage system takes into account several factors: (i) preferences of its owner; (ii) capabilities, limitations, and lifecycle of storage system; (iii) estimations about variations in energy pricing; (iv) optimization of its asset based on past learning and future estimations, etc.

This function utilizes the same general energy storage-related settings as in the Direct Charge/Discharge function described in previous subsection. The DER receives the communicated "set price" signal together with a reading of the "present price."

The DER waits for a time duration specified by the "randomized time window" to allow the communication systems to access the different DER units within the group of DER before implementing the requested energy price setting. The DER then changes the present charge/discharge level to achieve the level required by the "set price" signal in a linear ramped manner over a specified "ramp time." The function stays enabled over a "reversion timeout" period after which the DER returns to its default charge/discharge level.

A schedule of price may also be incorporated in addition to a specific price setting. The schedule may be based on Time of Use (TOU) or follow some daily, weekly, or seasonal pattern. Adaptive learning mechanisms may also be incorporated for determining the price based on historical price data.

2.10.3   Coordinated Charge/Discharge Management Function

This function is intended for DERs with energy storage system that perform dual functions of meeting local owner requirements and participation in grid support activities [17, 34]. As an example, the energy storage system in an electric vehicle (EV) for such dual purposes (i) needs to be fully charged to the user-specified level by the time the EV is required by the user and (ii) operate as a distributed energy resource for the grid in the interim period. Similarly, a community energy storage (CES) system may continue serving local needs but must be fully charged to provide support to the utility by the time a weather emergency such as a storm is likely to hit the region.

This function thus enables the coordination of local needs of the energy storage system with the grid support functions in terms of the target charge level and the time schedule for its availability. The parameters specified in the previous section under Direct and Price-Based Charge/Discharge functions are applicable for this function as well. This function is illustrated through two characteristics, as below.

2.10.3.1 Time-Based Charging Model

The time-based charging of this function is depicted in Figure 2.24 [17]. The external entity, i.e. system operator, makes a request to the DER to absorb an amount of energy equal to "Energy Request" from the grid to vary the SOC of the DER from its value at the "Time of Reference t_0" to a "Target SOC." The DER commences charging at the "Maximum Charge Rate" for the "Duration at Maximum Charge Rate." This duration is determined by the ESS based on its available capacity to absorb energy including losses. This duration lasts until time t_1 when the ESS reaches a SOC beyond which charging at maximum rate cannot be sustained. The charging rate decreases for further charging until the target SOC is reached, or until 100% SOC is reached at time t_2.

The "Minimum Charging Duration" indicates the minimum time required to increase the SOC from its value at the time of reference t_0 to the target SOC. This parameter is computed by the DER and is continually updated as SOC keeps getting changed during the charging (or discharging) process. The "Time Charge Needed" represents the time by which the ESS must reach the target SOC. The "Target State of Charge" is the desired state of charge that the ESS is required to achieve as a percent of the usable capacity. This target SOC may be determined locally as in case of an EV which needs to acquire a specific charge by a specified time. This target SOC may also be determined by an external managing entity. For instance, a utility may communicate to a community energy storage system to be charged to a specific level in case of impending storm conditions at a specific time.

2.10.3.2 Duration at Maximum Charging and Discharging Rates

The reference charging and discharging power limit curves for an ESS based DER are depicted in Figure 2.25 [17]. The plot illustrates that the ESS can charge at the maximum charging rate until a SOC level S_2 is reached beyond which the charging rate has to decrease for achieving 100% SOC. Similarly, the ESS can discharge at a maximum discharging rate until a SOC level S_1 is reached below which the discharge has to reduce for further discharge until a minimum state of charge S_{min} is reached. These limiting parameters are known internally to the ESS and generally not available to

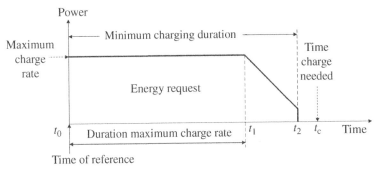

Figure 2.24 Charging function of the ESS-based DER. *Source:* Based on [17]. Used with permission from EPRI.

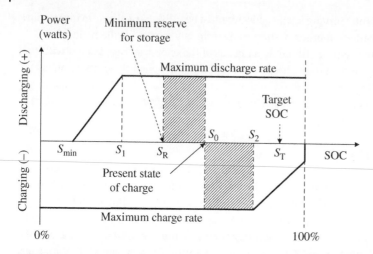

Figure 2.25 State of charge (SOC)-based model of ESS. *Source:* Based on [17]. Used with permission from EPRI.

the external entity. The shaded charging area represents the capability of the DER to absorb energy from the grid at the maximum charge rate from its present state of charge S_0. The shaded discharging area indicates the energy available in the ESS for discharging into the grid at the maximum discharge rate from its present state of charge S_0 until the state of charge S_R corresponding to minimum reserve for storage is reached by the ESS.

The duration for which energy can be delivered at the Maximum Discharge Rate is termed "Duration Maximum Discharge Rate." This is determined by the ESS based on its available capacity for production until the S_R corresponding to "Minimum Reserve for Storage" or the SOC level S_1 below which maximum discharge rate cannot be sustained, is reached. The curves further indicate that additional energy may be available for charging/discharging but at rates lower than the maximum rates for charging/discharging, respectively.

The "Duration at Maximum Charge Rate" and "Duration at Maximum Discharge Rate" are two major parameters determined by the ESS and made available to the external controlling entity based on which the controlling entity can plan the management of ESS based DER. On the other hand, the DER may itself override the other settings based on the "Target State of Charge" and "Time Charge Needed" parameters. Further details of implementation of this function are available in [17, 34].

2.11 Limit Maximum Active Power Function

2.11.1 Without Energy Storage

This function enables the maximum active power from a DER, or from an aggregation of DERs with associated loads to be limited to a specified value. This function is invoked typically during localized feeder overvoltage conditions, overloading of grid assets (e.g. transformers), or overvoltages caused by high penetration of DERs during light load conditions, etc. This may also be necessary to avoid reverse power flows into network during abnormal or emergency situations.

The limit, e.g. P_{glim} is set as a percentage of the maximum active power generation capability P_{gmax} of the DER. The DER waits for a period termed "Time Window" after a command to initiate this function is received. This time window is to accommodate the communication systems to access the different set of DER units (within this group of DER) which need to respond to this command. The DER limits its active power to the specified limit in typically 30 seconds.

The DER then linearly ramps down its upper power output limit, e.g. 100% P_{gmax} to the specified limit P_{glim} over a specified "ramp time." After a "reversion timeout" period over which the reduced power output limit is in place, the DER returns to its upper power output limit, e.g. 100% P_{gmax}.

2.11.2 With Energy Storage System

This function ensures that the power output from the Energy Storage System is limited in order to maintain a specific power flow in the distribution feeder or a transformer station. Similar parameters as specified above are also implemented in this case.

2.12 Set Active Power Mode

This function allows the active power injection or absorption from a DER to be set at a specified value. This function is different from the Limit Maximum Active Power function.

2.13 Active Power Smoothing Mode

This function allows a DER to dynamically generate or absorb additional active power in response to a deviation in the power level of a reference generation or load. This function performs in a similar manner as the dynamic reactive current injection which dynamically exchanges reactive current in response to deviations in the voltage.

This DER function essentially compensates for intermittent renewable energy systems and transiently variable loads within the DER's own operating capability and safety. Application of this function may involve energy storage systems to be coupled with the DER.

2.14 Active Power Following Function

This function requires the DER active power production/absorption to track a reference signal provided by the system operator. This may involve both load following and generation following.

In case of load following, the power output from the DER is varied to meet the reference load demand. Hence the DER output increases as the load increases. However, for generation following, the DER power absorption increases as the power output from the reference generation increases, to keep the total generation at a specified value. This function is only applicable to DERs with energy storage.

2.15 Prioritization of Different Functions

It is important to prioritize different smart inverter functions as several of them may be operating at the same time [17, 34]. For instance, different reactive power control functions may be operational simultaneously, or different active power control functions may be active at the same time. Furthermore, different functions may be impacting the same variable such as voltage or frequency.

2.15.1 Active Power-related Functions

The smart inverter functions related to active power in the order of their priority are described in [17, 19]. Some of these priorities are enunciated below.

2.15.1.1 Functions Affecting Operating Boundaries

Volt–Watt and Frequency–Watt functions belong to this category. In case both functions are simultaneously active, since both functions lead to reduction in active power output, the function which results in lower active power output (close to zero) is given priority [17, 19].

2.15.1.2 Dynamic Functions

"Active Power Smoothing Function" and "Dynamic Volt–Watt" function come under this classification. Since both functions provide additional watts that may be added or subtracted from the existing power output of the DER, they will not conflict with each other and are therefore given equal priority.

2.15.1.3 Steady-State Functions Managing Watt Input/Output

The "Direct Charge/Discharge Function," "Price-Based Charge/Discharge Function," "Peak Power Limiting Function," "Load/Generation Following Function" fall under this category as all of them manage the active power exchange (generation/absorption) from the DER. All these functions are mutually exclusive and cannot operate at the same time.

2.15.2 Reactive Power-Related Functions

The smart inverter functions related to reactive power in order of their priority are as follows [17, 19]:

2.15.2.1 Dynamic Functions

The "Dynamic Reactive Current Support Function" comes under this category, which exchanges (injects/absorbs) additional reactive current to the current being presently produced by the DER.

2.15.2.2 Steady-State Functions

The "Constant Power Factor," "Volt–Var" "Watt–Var" functions belong to this category. These functions are mutually exclusive and do not conflict with each other as only one may be effective at any time. Therefore, they have equal priority.

In case commands for different reactive power control functions are received by the DER, the most recent function to be made active by either direct command from system operator or by operating schedule will receive priority.

2.15.3 Smart Inverter Functions Under Abnormal Conditions

The DER Ride Through requirements shall take precedence over all other smart inverter functions related to voltage and frequency control [19].

2.16 Emerging Functions

There are other smart inverter functions which are under different stages of development [23, 33, 34, 39]. It is expected that these functions and other capabilities being developed by researchers around the globe, may become the basis of new smart inverter functions in the near future.

2.16.1 PV-STATCOM: Control of PV inverters as STATCOM during Night and Day

During 2009–2011, a new smart solar PV inverter technology, termed PV-STATCOM, was developed by this book's author, which transforms a solar PV system into a STATCOM and provides various STATCOM functionalities both during night and day [5, 14, 41]. The PV-STATCOM controls can provide several advanced grid support functions such as dynamic voltage control [5, 42, 43], increasing power transfer capacity by power oscillation damping [14, 43–45], mitigation of subsynchronous control interactions, and subsynchronous torsional oscillations [46–48], alleviation of FIDVR [49], stabilization of remote critical motors [50], simultaneous fast frequency response and power oscillation damping [51, 52], enhancement of connectivity of neighboring wind plants [5] and solar plants [53], etc. The PV-STATCOM control has also been field demonstrated for the first time in Canada in 2016 for stabilizing critical motors both during night and day [54]. The PV-STATCOM technology is about 10 times more economical than an equivalent sized STATCOM. The PV-STATCOM functionalities are described in detail in Chapters 4–6 and 10.

2.16.2 Reactive Power at No Active Power Output

The 2017 EPRI Report [55] has stated under guidance to utilities that joint industry efforts should discuss the merits of advanced inverters performing functions (such as VAR support) continuously (not just during the day), i.e., during nighttime as well.

The 2018 IEEE PES Report [27] recommends that smart inverters may be enhanced in future to help mitigate grid issues that may be relevant. The report states that some futuristic features may include the following: "Reactive power support during night times: future inverters could be designed to provide voltage support even when there is no input energy."

The 2018 NERC Report [23] has also described different benefits of dynamic reactive power support from PV inverters during periods of no active power output, i.e. during nighttime for PV inverters and during periods of zero wind for wind farm inverters. These benefits include the following:

- Voltage support from BPS-connected solar plants in evenings after sunset can assist in controlling BPS voltages particularly on hot summer evenings to meet the demand of HVAC loads.
- Voltage support in the morning before sunrise and in the evening after sunset can help stabilize BPS voltages during sharp load ramp periods caused by heating loads especially during winter time.
- Voltage regulation at night to mitigate overvoltage issues caused due to light loads and line charging of lightly loaded overhead lines and underground cables.
- Improved voltage regulation, reduction in voltage variability, and enhancement of voltage instability in the BPS.

The report states that reactive power support during these times may be able to lower the transmission investments that would otherwise be needed in absence of such dynamic reactive power support capabilities [23].

The 2019 NERC Reliability Guideline [24] has recommended improvements to interconnection requirements related to IBR performance. One of the recommendations is that Transmission Operators may require IBRs to exchange reactive power with the BPS (to provide voltage control) when no active power is generated, i.e., during night for PV inverters.

2.17 Summary

This chapter presented the features of some of the major smart inverter functions that have been included in different Standards. Suggestions from leading North American organizations about potential new smart inverter functions at zero active power output of IBRs are described. A new smart PV inverter PV-STATCOM which provides various STATCOM functionalities both during night and day is introduced.

References

1 Rashid, M.H. (2007). *Power Electronics Handbook: Devices, Circuits, and Applications.* Elsevier.

2 Mohan, N., Undeland, T.M., and Robbins, W.P. (2003). *Power Electronics: Converters, Applications, and Design.* New York: Wiley.

3 E.ON Netz GmbH (2006). *Grid Code for High and Extra High Voltage.* Bayreuth, Germany: E.ON Netz GmbH.

4 BDEW (Bundesverband der Energie- und Wasserwirtschaft e.V.) (2008). *Technical Guideline: Generating Plants Connected to the Medium-Voltage Network. (Guideline for Generating Plants' Connection to and Parallel Operation with the Medium-Voltage Network).* BDEW, Berlin, Germany. Revised January 2013.

5 Varma, R.K., Khadkikar, V., and Seethapathy, R. (2009). Nighttime application of PV solar farm as STATCOM to regulate grid voltage. *IEEE Transactions on Energy Conversion (Letters)* 24: 983–985.

6 Albuquerque, F.L., Moraes, A.J., Guimaraes, G.C. et al. (2010). Photovoltaic solar system connected to the electric power grid operating as active power generator and reactive power compensator. *Solar Energy* 84: 1310–1317.

7 Casey, L.F., Schauder, C., Cleary, J., and Ropp, M. (2010). Advanced inverters facilitate high penetration of renewable generation on medium voltage feeders - impact and benefits for the utility. In *Proc. 2010 IEEE Conference on Innovative Technologies for an Efficient and Reliable Electricity Supply,* 86–93.

8 Walling, R.A. and Clark, K. (2010). Grid support functions implemented in utility-scale PV systems. In *Proc. 2010 IEEE PES T&D Conference and Exposition,* 1–5.

9 Smith, J.W., Sunderman, W., Dugan, R., and Seal, B. (2011). Smart inverter volt/var control functions for high penetration of PV on distribution systems. In *Proc. 2011 IEEE/PES Power Systems Conference and Exposition,* 1–6.

10 Gonzalez, S., Hoffmann, F., Mills-Price, M., Ralph, M., and Ellis, A., Implementation of advanced inverter interoperability and functionality. In *Proc. 2012 38th IEEE Photovoltaic Specialists Conference,* 1362–1367.

11 IEC 61850-90-7 (2012). *Distributed energy management (DER): advanced power system management functions and information exchanges for inverter-based DER devices modelled in IEC 61850-90-7,* Version 27 (June 2012).

12 Fawzy, Y.T., Allert, C., Paetzold, A. et al. (2012). A summary report of innovative small and large PV inverters. Deliverable D2.1, MetaPV, European Commission, *Report*.

13 Boemer, J.C., Burges, K., Zolotarev, P. et al. (2011). Overview of German grid issues and retrofit of photovoltaic power plants in Germany for the prevention of frequency stability problems in abnormal system conditions of the ENTSO-E region continental Europe. Presented at the First International Workshop on Integration of Solar Power into Power Systems, Aarhus, Denmark (October 2011).

14 Varma, R.K., Rahman, S.A., and Seethapathy, R. (2010). Novel control of grid connected photovoltaic (PV) solar farm for improving transient stability and transmission limits both during night and day. In *Proc. 2010 World Energy Conference*, Montreal, Canada.

15 Reno, M.J., Broderick, R.J., and Grijalva, S. (2013). Smart inverter capabilities for mitigating over-voltage on distribution systems with high penetrations of PV. In *Proc. 2013 IEEE 39th Photovoltaic Specialists Conference (PVSC)*, 3153–3158.

16 EPRI (2014). Common Functions for Smart Inverters, Version 3. EPRI, Palo Alto, CA, USA, *Technical Report 3002002233*.

17 EPRI (2016). Common Functions for Smart Inverters, 4e. EPRI, Palo Alto, CA, USA, *Technical Report 3002008217*.

18 Smart Inverter Working Group (2014). *Recommendations for Updating the Technical Requirements for Inverters in Distributed Energy Resources*. California: Smart Inverter Working Group (January 2014).

19 IEEE (2018). *IEEE standard for interconnection and interoperability of distributed energy resources with associated electric power systems interfaces*, and IEEE Standard 1547a-2020 (Amendment to IEEE Std. 1547-2018).

20 Underwriters Laboratories (2016). UL standard 1741 Supplement SA: *Grid support utility interactive inverters and converters*, Underwriters Laboratories, USA.

21 Hawaii Electric Company, Inc. (2018). *Rule 14H Interconnection of distributed generating facilities with the company's distribution system*, Hawaii Electric Company, Inc.

22 NERC (2020). NERC standard PRC-024-3 *Frequency and voltage protection settings for generating resources*.

23 NERC (2018). BPS-Connected Inverter-Based Resource Performance. NERC, Atlanta, GA, USA, *Reliability Guideline*.

24 NERC (2019). Improvements to Interconnection Requirements for BPS-Connected Inverter-Based Resources. NERC, Atlanta, GA, USA, *Reliability Guideline*.

25 CEATI International Inc. (2016). Investigation of smart inverters. CEATI International Inc., Montreal, QC, Canada. *Report No. T154700 #50/128*.

26 Loutan, C., Morjaria, M., Gevorgian, V. et al. (2017). Demonstration of essential reliability services by a 300-MW solar photovoltaic power plant. Technical Report NREL/TP-5D00-67799, National Renewable Energy Laboratory, Golden, CO, USA https://www.nrel.gov/docs/fy17osti/67799.pdf, Accessed November 16th, 2020.

27 IEEE (2018). Impact of IEEE 1547 standard on smart inverters. *Technical Report PES-TR67*, IEEE Power & Energy Society, New York, NY, USA.

28 CPUC (2019). California electric tariff Rule 21. https://www.cpuc.ca.gov/Rule21 (accessed 08 February 2020).

29 ANSI (2016). *Electric power systems and equipment—voltage ratings (60 Hz)*, ANSI Standard C84.1-2016, USA.

30 Ministry of New and Renewable Energy (2020). Technical requirements for Photovoltaic Grid Tie Inverters to be connected to the Utility Grid in India. Ministry of New and Renewable Energy, New Delhi, India, *Draft Standard* (April 2020).

31 EPRI (2018). Mitigation methods to increase feeder hosting capacity. EPRI, Palo Alto, CA, *Report 3002013382*.

32 US Department of Energy (2010). Evaluation of conservation voltge reduction (CVR) on a national level. Pacific Northwest National Laboratory, US Department of Energy, *Report PNNL-19596*.

33 NERC (2020). Bulk Power System Reliability Perspectives on the Adoption of IEEE 1547-2018. NERC, Atlanta, GA, USA, *Reliability Guideline*.

34 Smart Inverter Working Group (2017). SIWG Phase 3 DER Functions: Recommendations to the CPUC for Rule 21, Phase 3 Function Key Requirements, and Additional Discussion Issues. Smart Inverter Working Group, California

35 NERC (2009). A Technical Reference Paper Fault Induced Delayed Voltage Recovery, Version 1.2, NERC, Princeton, NJ, USA. *Technical Paper*

36 EPRI (2015). Recommended settings for voltage and frequency ride-through of distributed energy resources. EPRI, Palo Alto, CA, *White Paper*.

37 NERC (2017). 1,200 MW Fault induced solar photovoltaic resource interruption disturbance report – Southern California 8/16/2016 Event, NERC, Atlanta, GA, USA, *Report*.

38 NERC (2018). 900 MW Fault induced solar photovoltaic resource interruption disturbance report – Southern California Event: October 9, 2017: Joint NERC and WECC Staff Report, NERC, Atlanta, GA, USA, *Report*.

39 IEEE Standards Association. *P2800 – Draft Standard for Interconnection and Interoperability of Inverter-Based Resources Interconnecting with Associated Transmission Electric Power Systems*. New York, NY, USA.

40 NERC (2020). Fast frequency response concepts and bulk power system reliability needs. NERC, Atlanta, GA, USA, Inverter-Based Resource Performance Task Force (IRPTF) *White Paper*.

41 Varma, R.K., Das, B., Axente, I., and Vanderheide, T. (2011). Optimal 24-hr utilization of a PV solar system as STATCOM (PV-STATCOM) in a distribution network. In *Proc. 2011 IEEE Power & Energy Society General Meeting*, 1–8.

42 Varma, R.K. and Siavashi, E.M. (2018). PV-STATCOM: A new smart inverter for voltage control in distribution systems. *IEEE Transactions on Sustainable Energy* 9: 1681–1691.

43 Varma, R.K., Rahman, S.A., Mahendra, A.C., Seethapathy, R., and Vanderheide, T. (2012). Novel nighttime application of PV solar farms as STATCOM (PV-STATCOM). In *Proc. 2012 IEEE Power & Energy Society General Meeting*, 1–8.

44 Varma, R.K., Rahman, S.A., and Vanderheide, T. (2015). New control of PV solar farm as STATCOM (PV-STATCOM) for increasing grid power transmission limits during night and day. *IEEE Transactions on Power Delivery* 30: 755–763.

45 Varma, R.K. and Maleki, H. (2019). PV solar system control as STATCOM (PV-STATCOM) for power oscillation damping. *IEEE Transactions on Sustainable Energy* 10-4: 1793–1803.

46 Varma, R.K. and Salehi, R. (2017). SSR mitigation with a new control of PV solar farm as STATCOM (PV-STATCOM). *IEEE Transactions on Sustainable Energy* 8: 1473–1483.

47 Salehi, R. and Varma, R.K. (2019). PV solar farm control as STATCOM (PV-STATCOM) for alleviating subsynchronous oscillations. In *Proc. 2019 CIGRE Canada Conference*, Montreal, QC, Canada.

48 Varma, R.K. and Salehi, R. (2020). PV-STATCOM for mitigating subsynchronous resonance associated with inverter-based resources. Presented at the Panel Session, Voltage Control and Low and High Voltage Ride-Through of Wind and Solar Photovoltaic Systems. In *Proc. 2020 IEEE PES General Meeting*.

49 Varma, R.K. and Mohan, S. (2020). Mitigation of fault induced delayed voltage recovery (FIDVR) by PV-STATCOM. *IEEE Transactions on Power Systems* 35-6: 4251–4262.

50 Varma, R.K., Mohan, S., and McMichael-Dennis, J. (2020). Multi-mode control of PV-STATCOM for stabilization of remote critical induction motor. *IEEE Journal of Photovoltaics*, 10-6: 1872–1881.

51 Varma, R.K. and Kelishadi, M.A. (2020). Simultaneous fast frequency control and power oscillation damping by utilizing PV solar system as PV-STATCOM. *IEEE Transactions on Sustainable Energy* 11-1: 415–425.

52 Varma, R.K. and Akbari, M. (2018). A novel reactive power based frequency control by PV-STATCOMs during day and night. In *Proc. 2018 IEEE Power & Energy Society General Meeting*, 1–5.

53 Varma, R.K. and Siavashi, E.M. (2019). Enhancement of solar farm connectivity with smart PV inverter PV-STATCOM. *IEEE Transactions on Sustainable Energy* 10-3: 1161–1171.

54 Varma, R.K., Siavashi, E.M., Mohan, S., and Vanderheide, T. (2019). First in Canada, night and day field demonstration of a new photovoltaic solar based Flexible AC transmission System (FACTS) device PV-STATCOM for stabilizing critical induction motor. *IEEE Access* 7: 149479–149492.

55 EPRI (2017). Arizona Public Service Solar Partner Program - Advanced Inverter Demonstration Results, EPRI, Palo Alto, CA *Report 3002011316*.

3

MODELING AND CONTROL OF THREE-PHASE SMART PV INVERTERS

This chapter describes the basic concepts of active and reactive power flow in a smart inverter system. These are subsequently related to the power flows in a smart solar photovoltaic (PV) inverter system. The operating principles and models of different subsystems in the power circuit and control circuit of a smart PV inverter system are described. The principle of *abc-dq* transformation is elucidated and the application of phase-locked loop (PLL) in achieving a decoupled control of active power and reactive power is presented. The modeling of different smart inverter controllers is discussed and two variants of smart inverter voltage controllers are presented as examples. Subsequently, the essential modeling features of PV plant control are enunciated.

3.1 Power Flow from a Smart Inverter System

The basic principles of active and reactive power flow from a smart inverter system are described in this section. The smart inverter system here refers to the entire plant of a distributed energy resource (DER), which may be a solar PV system, a wind power plant, an energy storage system (ESS), electric vehicle charging station, etc., or any combination of the above. The control system design of the smart inverter system will be based on the understanding of these concepts.

Figure 3.1 illustrates a smart inverter system, e.g. a solar PV system connected to a power system. The power system comprises several buses, however, only the reference infinite bus and the point of common coupling (PCC) bus are depicted.

The voltages at the infinite bus and PCC bus are indicated by $V_{inf} \angle 0°$ and $V_1 \angle \theta_1$, respectively. The infinite bus provides the reference angle for the entire power system. The voltage magnitude and angles of different buses in the power system are obtained from load flow studies. The smart inverter system is connected to the power system through a reactance, which primarily corresponds to the reactance of the coupling transformer that connects the smart inverter system to the power system. This reactance may also include the series reactance of the filter of the smart inverter. Strictly speaking, an impedance should be considered between the smart inverter system and power system, however, to simplify the analysis and to understand the basic concepts, only reactance is considered. The terminal voltage of the smart inverter system is given by $V_2 \angle \theta_2$. The current of the smart inverter system I is assumed to lag the smart inverter system voltage by an angle φ (representing the power factor). Figure 3.2 displays the phasor diagram of the above system.

The current I flowing from the smart inverter system into the power system PCC bus is given as

$$I = \frac{V_2 \angle \theta_2 - V_1 \angle \theta_1}{X \angle 90°} = \frac{V_2}{X} \angle \left(\theta_2 - 90°\right) - \frac{V_1}{X} \angle \left(\theta_1 - 90°\right) \tag{3.1}$$

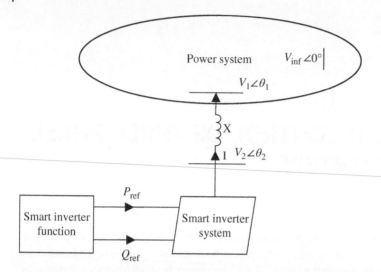

Figure 3.1 A smart inverter system connected to the power system.

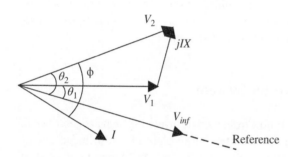

Figure 3.2 Active and reactive power flow from the smart inverter system toward the power system.

The complex power flow from the smart inverter system into the power system PCC bus is expressed as

$$P_1 + jQ_1 = V_1 \angle \theta_1 * I^*$$

$$= V_1 \angle \theta_1 \left[\frac{V_2}{X} \angle (90° - \theta_2) - \frac{V_1}{X} \angle (90° - \theta_1) \right]$$

$$= \frac{V_1 V_2}{X} \angle (90° + \theta_1 - \theta_2) - \frac{V_1^2}{X} \angle 90° \tag{3.2}$$

The active and reactive components of the above equation are separated to obtain the individual active and reactive power flows as below

$$P_1 = \frac{V_1 V_2}{X} \cos \left[90° + (\theta_1 - \theta_2) \right] - \frac{V_1^2}{X} \cos 90° \tag{3.3}$$

$$Q_1 = \frac{V_1 V_2}{X} \sin \left[90° + (\theta_1 - \theta_2) \right] - \frac{V_1^2}{X} \sin 90° \tag{3.4}$$

The active power flow P_1 from the smart inverter system into the power system is simplified as

$$P_1 = \frac{V_1 V_2}{X} \sin (\theta_2 - \theta_1) \tag{3.5}$$

3.1.1 Active Power Flow

3.1.1.1 Magnitude of Active Power Flow

The magnitude of power flow is governed by the magnitude of smart inverter system voltage magnitude V_2 and more importantly by its angular difference with respect to the PCC voltage given by $\theta_2 - \theta_1$. The power output from the smart inverter system is varied in response to input power (e.g. solar irradiance) and is also modulated to accomplish various smart inverter functions as described in Chapter 2. The smart inverter control system varies V_2 and θ_2, accordingly.

3.1.1.2 Direction of Active Power Flow

The direction of active power flow is from the bus with higher voltage angle to the bus with lower voltage angle. The smart inverter system should, therefore, be controlled to have a terminal voltage angle θ_2 higher than the PCC bus voltage angle θ_1 for active power to flow from the smart inverter system to the power system. In situations where active power needs to flow into the smart inverter system for instance to charge the DC capacitor of the solar PV system, i.e. during nighttime for STATCOM operation, or charge the batteries in the BESS, the smart inverter system bus voltage angle θ_2 is made lower than the PCC bus voltage angle θ_1. This is implemented by the control system of the smart inverter system.

3.1.2 Reactive Power Flow

The reactive power flow Q_1 from the smart inverter system into the power system is expressed as

$$Q_1 = \frac{V_1 V_2}{X} \cos(\theta_2 - \theta_1) - \frac{V_1^2}{X} = \frac{V_1}{X}\left[V_2 \cos(\theta_2 - \theta_1) - \frac{V_1}{X}\right] \tag{3.6}$$

3.1.2.1 Magnitude of Reactive Power Flow

The magnitude of reactive power is a function of voltage magnitude of the smart inverter system. The smart inverter control system thus needs to control the magnitude of smart inverter voltage magnitude V_2 for exercising reactive power control.

For any active power to flow in either direction from the smart inverter system, the angular difference $\theta_2 - \theta_1$ must be a finite quantity, i.e. it cannot be zero. This implies that $\cos(\theta_2 - \theta_1)$ will be less than unity. Therefore, with increasing active power output from the smart inverter system, if some amount of reactive power injection is desired, V_2 needs to be increased by a larger amount to offset the decreasing value of $\cos(\theta_2 - \theta_1)$. This implies that for reactive power injection, V_2 must be higher than V_1.

The control system of the smart inverter thus needs to appropriately vary the magnitude of the smart inverter bus voltage to achieve reactive power exchange (injection or absorption) with the power system. This control enables the smart inverter to perform various reactive power-based smart inverter functions described in this book.

Reactive power exchange (flow in either direction) can occur even if the active power flow is zero, i.e. when $\theta_2 - \theta_1 = 0$. This provides smart inverters such as solar PV systems and inverter-based wind farms the capability to exchange reactive power during nighttime and periods of no wind, respectively.

3.1.2.2 Direction of Reactive Power Flow

The convention of reactive power flow from the smart inverter system (similar to convention for synchronous generators) is as below:

i) If $V_2 > V_1$, reactive power is injected by the smart inverter system into power system. This is capacitive reactive power. The smart inverter system current lags the smart inverter bus voltage in this case.

ii) If $V_2 < V_1$, reactive power is absorbed by the smart inverter system from the power system. This is inductive reactive power. The smart inverter system current leads to the smart inverter bus voltage in this case.

For the special case of zero active power flow between the smart inverter system and the power system, i.e. when $\theta_2 - \theta_1 = 0$, the phasor diagrams for cases (i) and (ii) are shown in Figures 3.3 and 3.4, respectively.

3.1.3 Implementation of Smart Inverter Functions

Figure 3.1 illustrates that the smart inverter system receives two inputs P_{ref} and Q_{ref}, which are provided by the various smart inverter functions. It is seen from above analysis that by appropriately varying the magnitude and angle of its terminal voltage the smart inverter can perform desired smart inverter function as below:

i) Active power generation, absorption, and modulation based on the active power reference P_{ref}, and

ii) Reactive power generation, absorption, and modulation based on the reactive power reference Q_{ref}

A decoupled control is described later which allows smart inverters to control both active power and reactive power simultaneously in an independent manner.

Legacy (nonsmart) inverters provide only active power generation or absorption based on active power reference P_{ref}, without any modulation. Furthermore, they operate at unity power factor, hence Q_{ref} is set to zero.

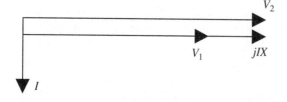

Figure 3.3 Reactive power injection by the smart inverter system.

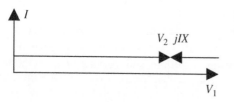

Figure 3.4 Reactive power absorption by the smart inverter system.

3.2 Smart PV Inverter System

Figure 3.5 depicts a typical smart PV inverter system connected to the power system. The PV solar array generates a DC current I_{pv} in response to the voltage V_{DC} appearing across the array. The DC power is converted to AC power by the voltage source converter (VSC) also referred to as the inverter. The three-phase sinusoidal voltage and current at the output of VSC are denoted by V_{tabc} and i_{tabc}, respectively. The VSC is comprised of IGBT-based switches which are turned on and off by gate pulses generated by the sinusoidal pulse width modulation (SPWM) unit. The PWM-based switching of VSC produces several high-frequency components which need to be filtered out to allow only the fundamental voltage to appear at the VSC output. A low pass LC filter is utilized for this purpose. The series inductance and resistance of the filter are represented by L_f and R_f, respectively, whereas the shunt filter capacitor and the damping resistance are denoted by C_f and R_c, respectively. The output terminal of the filter is connected to the power system grid through a coupling transformer T. In reality, there are two transformers – an isolation transformer followed by the interconnection transformer. The isolation transformer is an integral part of the PV system. The interconnection transformer is used to connect a single or multiple PV inverter systems within a solar PV plant with the grid system, as described in Section 3.6. However, both of these are considered as a single transformer for ease of analysis. V_{PCC} represents the voltage at the output (grid-side) of the interconnection transformer which may be considered as the voltage at PCC. A nighttime charging circuit is included to facilitate charging of the DC-link capacitor at night, for performing PV-STATCOM functions. The different constituents of the smart PV inverter system are explained later in this chapter.

For analyzing the performance of an individual smart inverter system, the capacitor voltage $V_c \angle \rho$ is considered to be the grid voltage $V_g \angle \rho$ (also equivalent to $V_1 \angle \theta_1$ shown in Figure 3.1). This is because in an actual solar PV power plant, several such individual smart PV inverters are connected to transformer T, as explained above. The three-phase sinusoidal voltage components of grid voltage are represented by V_{gabc} or V_{cabc}, depending on the context. The phase vector of the VSC terminal voltage is represented by $V_t \angle \rho + \delta$ (equivalent to $V_2 \angle \theta_2$ shown in Figure 3.1). Its three-phase sinusoidal voltage components are represented by V_{tabc}. The operation of VSC control system is based on rotating d-q reference frame quantities. This enables independent or decoupled control of active and reactive power from the VSC. A PLL is utilized to generate the phase angle ρ of the grid voltage, and the grid frequency ω_0. These are utilized in the abc-dq transformation block to transform the sinusoidal quantities, e.g. grid voltage V_{gabc}, VSC current i_{tabc}, and PCC voltage V_{PCCabc} (if used for PCC voltage control), into DC quantities V_{gdq}, i_{tdq}, and V_{PCCdq}, respectively. The control system generates the modulation index m_{dq} in the d-q reference which is subsequently inverse transformed into abc reference frame to generate the modulation index m_{abc}. This modulation index is used with the triangular carrier frequency to produce the gate pulses for switching the IGBTs in the VSC.

Different smart inverter functions related to DC-link voltage control ($V_{DC\text{-}SI}$), active power control ($P_{SI1}...P_{SIn}$), and reactive power control ($Q_{SI1}...Q_{SIn}$) are incorporated in the smart PV inverter control system. Each smart inverter function has a corresponding input signal, such as IN_{VDC}, $IN_{PSI1}...IN_{PSIn}$, $IN_{QSI1}...IN_{QSIn}$, as the case may be. The active power reference P_{ref} and reactive power reference Q_{ref} values are generated by different smart inverter controllers. The VSC is controlled to provide active power and reactive power according to these reference values.

A straightforward option as explained in Section 3.1 is to modulate both the magnitude and phase angle of the fundamental component of the VSC terminal voltage, with respect to the voltage magnitude and phase angle of the grid voltage. This results in generation of VSC active power output

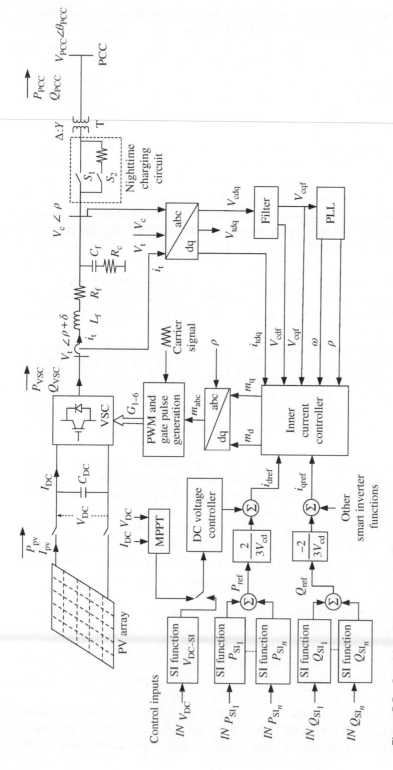

Figure 3.5 Control system of a smart PV inverter system.

P_{PCC} equal to active power reference P_{ref} and VSC reactive power output Q_{PCC} equal to reactive power reference Q_{ref}. This is known as "voltage-mode control."

The voltage-mode control, although simple, suffers from a discrepancy. Since only the inverter voltage is regulated and not the inverter current, the VSC may experience large current flows during grid faults or transients. This may potentially damage the IGBT switches. An alternate "current-mode control" is, therefore, employed, which directly regulates the VSC current. The active power and reactive power of VSC are, therefore, controlled by the d-axis component i_{td} and q-axis component i_{tq} of the VSC current, respectively. The d-q axis current references corresponding to P_{ref} and Q_{ref} are computed as i_{tdref} and i_{tqref}, and the individual currents i_{td} and i_{tq} are then controlled to these reference values, respectively. This current control strategy thus indirectly controls the magnitude and phase angle of the fundamental component of the VSC terminal voltage.

It is noted that this smart PV inverter obtains the voltage angle and frequency reference from the grid and are, therefore, known as "grid following" inverters.

The smart solar PV system is constituted by three subsystems: (i) power circuit, (ii) VSC control circuit, and (iii) smart inverter controllers. Each of these constituents is described in the following sections.

3.3 Power Circuit Constituents of Smart Inverter System

The power circuit consists of PV panels and the associated power point tracking system, VSC, AC filter at the output of VSC, and the isolation transformer.

3.3.1 PV Panels

The PV array comprises a number of PV panels in series and parallel to provide the needed DC voltage and DC current output capacity of the solar PV system. Each PV panel consists of a number of PV cells in series and parallel. The typical I–V characteristic of a PV panel for varying levels of solar irradiance and varying temperatures are depicted in Figure 3.6a,b, respectively. The solar irradiance level is expressed in the units of Sun, where one Sun is equivalent to an irradiance level of 1000 watt/m^2.

An equivalent single diode circuit model for PV panels, which has been validated with manufacturer's data from several commercially available PV panels, has been developed in PSCAD software [1]. The PV panel output current according to this model is given as:

$$
\begin{aligned}
I_{pvpanel} = {} & \frac{\left[n_s I_{SC(STC)}\{1+K_i(T-T_s)\}\{R_{sh}+R_s\}-n_p V_{oc}\right]\left[e^{\frac{qV_{oc(STC)}\{1+K_v(T-T_s)\}}{nkTn_s}}-1\right]G}{n_s R_{sh}\left[e^{\frac{qV_{oc(STC)}\{1+K_v(T-T_s)\}}{nkTn_s}}-e^{\frac{qI_{sc(STC)}\{1+K_i(T-T_s)\}R_s}{nkTn_p}}\right]G_{nom}} \\[2ex]
& -\frac{\left[n_s I_{SC(STC)}\{1+K_i(T-T_s)\{R_{sh}+R_s\}-n_p V_{oc}\right]\left[e^{\frac{q\left(n_p V_{pvpanel}+n_s I_{pvpanel}R_s\right)}{nkTn_s n_p}}-1\right]}{n_s R_{sh}\left[e^{\frac{qV_{oc(STC)}\{1+K_v(T-T_s)\}}{nkTn_s}}-e^{\frac{qI_{sc(STC)}\{1+K_i(T-T_s)\}R_s}{nkTn_p}}\right]} \\[2ex]
& +\frac{Gn_p V_{oc(STC)}\{1+K_v(T-T_s)\}}{n_s R_{sh}G_{nom}}-\frac{\left(n_p V_{pvpanel}+n_s I_{pvpanel}R_s\right)}{n_s R_{sh}}
\end{aligned}
$$

$$(3.7)$$

(a)

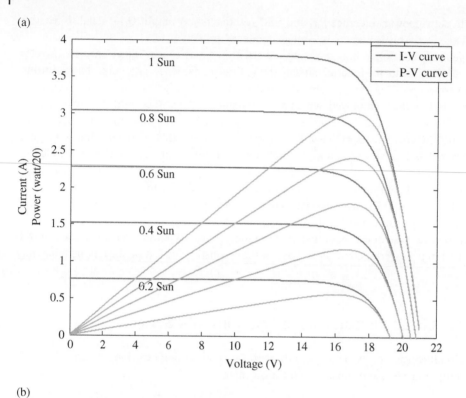

(b)

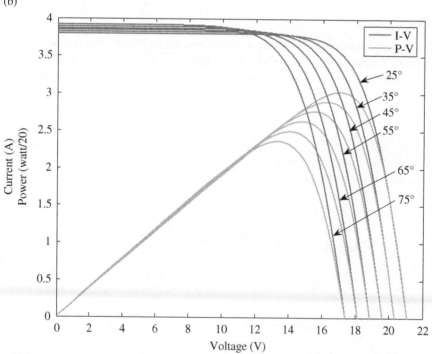

Figure 3.6 Typical *I–V* characteristic of a PV module at (a) different solar irradiation levels at 25 °C and (b) different temperatures at 1 Sun irradiation.

where, $I_{pvpanel}$ is the output current of a PV panel, $I_{SC(STC)}$ is the panel short circuit current at standard test conditions (STC), K_i is the temperature coefficient of PV short circuit current, T is the operating temperature of the PV panel, T_s is the standard temperature at STC, n_s is the number of cells in series per panel, n_p is the number of cells in parallel per panel, R_{sh} is the equivalent shunt resistance of each cell, R_s is the equivalent series resistance of each cell, V_{oc} is the open-circuit voltage of a PV panel, q is the charge of an electron, $V_{oc(STC)}$ is the open-circuit voltage of a PV panel at STC, K_v is the temperature coefficient of PV open-circuit voltage, G is the solar radiation, n is the diode ideality factor, k is the Boltzmann's constant, G_{nom} is the solar radiation at STC, $V_{pvpanel}$ is the output voltage of a PV panel.

The output voltage and current of the PV array is given by

$$I_{pv} = I_{pvpanel} \times N_p \tag{3.8}$$

$$V_{pv} = V_{pvpanel} \times N_s \tag{3.9}$$

where, I_{pv} is the output current of the PV array, V_{pv} is the output voltage of the PV array, N_s is the number of panels in series, N_p is the number of panel series strings in parallel.

The expression for a PV array output current is found by substituting Eqs. (3.8) and (3.9) in Eq. (3.7). The expression for PV array output current is given by

$$I_{pv} = a_1 G - a_2 \left\{ e^{a_3 V_{pv} + a_4 I_{pv}} - 1 \right\} - a_5 V_{pv} \tag{3.10}$$

where,

$$a_1 = \frac{N_p I_{phx}}{K_1 \times 1000} \tag{3.11}$$

$$a_2 = \frac{N_p I_{ox}}{K_1 \times 1000} \tag{3.12}$$

$$a_3 = \frac{1000 \times q}{nkTn_sN_s} \tag{3.13}$$

$$a_4 = \frac{1000 \times qR_s}{nkTn_pN_p} \tag{3.14}$$

$$a_5 = \frac{n_pN_p}{1000 \times N_sn_sR_{sh}K_1} \tag{3.15}$$

$$K_1 = 1 + \frac{R_s}{R_{sh}} \tag{3.16}$$

$$I_{phx} = \frac{\left[n_s I_{SC(STC)}\{1 + K_i(T - T_s)\}\{R_{sh} + R_s\} - n_p V_{oc} \right] \left[e^{\frac{qV_{oc(STC)}\{1 + K_v(T - T_s)\}}{nkTn_s}} - 1 \right]}{n_s R_{sh} \left[e^{\frac{qV_{oc(STC)}\{1 + K_v(T - T_s)\}}{nkTn_s}} - e^{\frac{qI_{sc(STC)}\{1 + K_i(T - T_s)\}R_s}{nkTn_p}} \right] G_{nom}} \tag{3.17}$$

$$+ \frac{n_p V_{oc(STC)}\{1 + K_v(T - T_s)\}}{n_s R_{sh} G_{nom}}$$

$$I_{ox} = \frac{\left[n_s I_{SC(STC)}\{1 + K_i(T - T_s)\{R_{sh} + R_s\} - n_p V_{oc} \right]}{n_s R_{sh} \left[e^{\frac{qV_{oc(STC)}\{1 + K_v(T - T_s)\}}{nkTn_s}} - e^{\frac{qI_{sc(STC)}\{1 + K_i(T - T_s)\}R_s}{nkTn_p}} \right]} \tag{3.18}$$

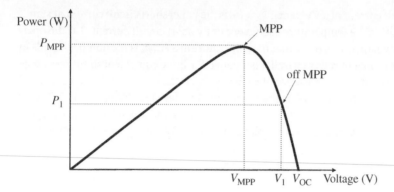

Figure 3.7 Variation of power output from a PV solar panel.

3.3.2 Maximum Power Point Tracking (MPPT) Scheme

The power generated by the solar PV panels is a nonlinear function of voltage across the PV panel as shown in Figure 3.7. This shows that for a given solar irradiance and temperature, the solar panel produces a maximum amount of DC power at a specific value of DC voltage. This operating point of the PV panel is known as Maximum Power Point (MPP). A DC voltage controller is employed in solar PV VSC system which regulates the PV panel voltage at this MPP voltage to allow maximum amount of power to be extracted from the PV panels. While this voltage is specifically defined for a single solar panel, it also applies for a solar array with multiple panels in series and parallel, as long as all of them are subjected to same solar irradiance and temperature. Several manufacturers provide more than one maximum power point tracking (MPPT) systems for panels with different orientations within a same solar PV plant.

In a single-stage solar PV system, the voltage across PV array is provided by the DC-link capacitor of VSC. An MPPT technique is used to determine the optimum voltage (MPP voltage) to result in maximum PV power output. There are several MPPT techniques proposed in literature [2, 3] such as Perturb and Observe (P&O) technique, Incremental Conductance method, Artificial Neural Network (ANN) method, Fuzzy Logic method, etc. Among them, the P&O and IC method, and their variants are most popularly utilized in practice. The MPPT algorithm determines the MPP voltage and provides that as DC voltage reference V_{ref} to the DC-link voltage controller, which is tasked to regulate the DC capacitor voltage at this voltage level.

3.3.3 Non-MPP Voltage Control

Smart inverters allow modulation of active power from VSCs of solar PV plants, in contrast to legacy inverters which provide a constant (maximum) active power output based on solar irradiance. The modulation of active power is achieved by varying the DC voltage across the PV array in accordance with a desired smart inverter active power control strategy. It is noted from Figure 3.7 that the DC power output from a PV array varies from maximum to zero, as the DC voltage across the PV array is increased from the MPP voltage to open-circuit voltage, respectively.

For such non-MPP (or, off-MPP) voltage control, the DC voltage reference is provided by the desired smart inverter active power control strategy. This voltage reference is then implemented by the DC voltage controller and made to appear across the DC-link capacitor of the VSC.

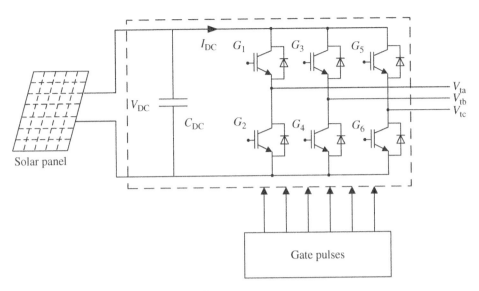

Figure 3.8 A two-level Voltage Source Converter.

3.3.4 Voltage Source Converter (VSC)

The conversion of the harvested solar PV power into AC power for the grid is performed by a VSC [4, 5]. Different topologies have been developed for VSCs such as two-level, multi-level, H-bridge, etc., each offering different benefits [6]. A simple and commonly employed topology is the two-level six pulse VSC with a capacitor on its DC side, as shown in Figure 3.8. Each branch has two IGBT switches with antiparallel diodes which, on being appropriately switched by gate pulses, generate one phase of the AC voltage v_{tabc} using the input DC voltage V_{DC} across the DC-link capacitor C_{DC}. The gate pulses for the VSC are provided by the Pulse Width Modulation (PWM) unit described later.

The DC power to the VSC comes from PV arrays during daytime. The DC-link voltage V_{DC} for a two-level VSC is determined from the following relation

$$V_{DC} \geq 2 \times \sqrt{\frac{2}{3}} V_{t-LL} \tag{3.19}$$

where, V_{DC} is the DC bus voltage and V_{LL} is the fundamental frequency component of line – line voltage V_t at the AC terminal of the inverter.

In the context of DC voltage, a term DC-to-AC ratio is often used, which is defined as the ratio of power rating of PV arrays to the rating of the VSC (or inverter). Since the prices of PV modules have declined much more rapidly than that of the inverters, many utility-scale PV developers are over-sizing the PV arrays with respect to the inverters. This results in rated power output from the inverters even when solar irradiance is not at its maximum. It implies that (i) some solar power is being spilled (i.e. not being fully utilized) and (ii) the inverter utilization has increased. It is seen that the financial gain from increased AC power output outweighs the loss from spilling DC power during sunny periods [7, 8]. Some typical values of DC : AC ratios for residential (<20 kW), commercial and industrial (20–5 MW), and utility scale (> 5 MW) are 1.13, 1.18, and 1.23, respectively [9], although DC : AC ratios of 1.3 have also been reported for utility-scale central inverters.

During nighttime for STATCOM operation, the DC power for charging the DC capacitor and to compensate for the losses in VSC semiconductor switches is obtained from the AC grid. A nighttime charging circuit is provided as shown in Figure 3.5. This circuit utilizes the diodes across the IGBT switches of the inverter to charge the DC-link capacitor. The charging circuit consists of a resistor which is used to limit the inrush current during charging. Switch S_2 is kept ON while switch S_1 is turned OFF for nighttime charging. Once the DC-link capacitor is charged, the charging circuit is disabled to prevent losses in the charging circuit by turning the switch S_1 ON and S_2 OFF. Thereafter, the DC voltage reference is switched to the constant value $V_{DCrated}$ for STATCOM operation.

3.3.4.1 DC-Link Capacitor

The DC-link capacitance C_{DC} plays an important role in the stability of DC-link voltage control loop. The DC-link capacitor provides a low impedance path for the ripple current caused by the VSC switching frequency. It is shown that the DC-link voltage control loop suffers from the effect of right half plane (RHP) pole if the PV array operates at a voltage less than the MPP voltage [10]. The location of this pole is affected by the value of C_{DC}. A stability criteria is established in [10] to decide the minimum value of C_{DC} relating the cross-over frequency of DC-link voltage control loop ω_c, the current, and voltage at the critical operating point of the PV array. The value of DC-link capacitor is obtained as

$$C_{DCmin} = 3 \times \frac{I_{sc}}{V_{pvmin}\omega_c} \tag{3.20}$$

where, V_{pvmin} is the minimum voltage of operation of PV array and I_{sc} is the short-circuit current of PV array. ω_c is generally chosen to be 2–10 times slower than the inner current control loop [11]. A safety factor of 25% is included in the minimum value of C_{DC} to arrive at the final value of DC-link capacitance.

3.3.4.2 AC Filter

The output of VSC contains multiple high-frequency components caused by the carrier frequency (typically 2–15 kHz) employed in the PWM process. These high-frequency components are minimized or reduced to meet the requirements of different Standards, i.e. [12] by different types of filters such as second-order LC filter or third-order LCL filter. A commonly employed LCL filter is shown in Figure 3.9. A filter inductance L_f is inserted in series with the output of the VSC. The resistance of this inductor together with the ON-state resistance of the IGBTs are represented by resistance R_f. A filter capacitance C_f and a damping resistor R_c are connected in shunt. L_t and R_t represent the inductance and resistance, respectively, of the isolation transformer utilized to connect the VSC to the area electric power system at PCC.

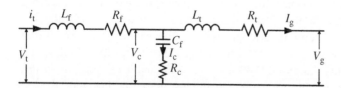

Figure 3.9 AC filter.

The different filter components are designed based on the following criteria. The filter inductance L_f and filter capacitance C_f are chosen to fulfill the following criteria [3, 13–15]:

Design of inductance L_f:

i) The inductance L_f is selected to limit the inverter current ripple within 10–15% and is given by

$$L_f = \frac{V_{DC}}{8 \times \Delta i_{max} \times f_s} \tag{3.21}$$

where, V_{DC} is the DC-link voltage, Δi_{max} is the maximum allowed ripple current, and f_s is the switching frequency.

ii) The voltage drop across L_f should be less than 0.3 pu on inverter rating base. A higher voltage drop across L_f will require a larger DC-link voltage to provide the desired AC voltage at the VSC output. L_f is typically selected between 0.1 and 0.25 pu [3].

Design of capacitor C_f and resistance R_c:

i) The value of capacitor C_f is chosen so that the reactive power exchanged by the capacitor is less than 0.05 pu of the VSC rating [13] and is given by

$$C_f = \frac{x \times S_n}{\omega_0 \times V_{LL}^2} \tag{3.22}$$

where, x is the amount of reactive power in pu consumed by the capacitor C_f, S_n is the apparent power rating of VSC, ω_0 is the nominal angular frequency, and V_{LL} is the rated line-line voltage at VSC terminal.

ii) The resonant frequency f_r of the designed filter should lie within the range

$$10 f_0 < f_r < 0.5 f_s \tag{3.23}$$

where,

$$f_r = \frac{1}{2\pi} \sqrt{\frac{L_t + L_f}{L_f C_f L_t}} \tag{3.24}$$

f_0 is the nominal system frequency and f_s is the switching frequency of VSC.

iii) The damping resistance R_c is designed to provide adequate damping at the resonant frequency [16], such that there is enough damping at the resonant frequency. The resistance R_c is designed to be one-third of the admittance of the filter capacitor C_f at resonant frequency [16]

$$R_c = \frac{1}{3 \times 2 \times \pi \times f_r \times C_f} \tag{3.25}$$

The dynamics of LCL filter are described by the following equations:

$$L_f \frac{di_t}{dt} = V_t - R_f i_t - V_g$$

$$C_f \frac{dV_{cf}}{dt} = i_t - i_{PCC}$$

$$L_t \frac{di_g}{dt} = V_g - R_t i_g - V_{PCC}$$

$$V_g = -V_{cf} + R_c(i_t - i_{PCC}) \tag{3.26}$$

where, V_t and i_t are the VSC AC terminal voltage and current, respectively. V_{cf} and V_g (or V_c) are the voltage across capacitor and grid voltage, respectively. V_{PCC} and i_{PCC} are the voltage and current at the PCC, respectively.

A detailed design of the LCL filter is given in [13].

3.3.4.3 Isolation Transformer

Utility networks in North America are generally operated with a solidly grounded neutral conductor, while PV systems are operated with a grounded DC-link [3], although this practice varies in different jurisdictions. Hence an isolation transformer is used to provide galvanic isolation between the PV system and the grid. It makes the PV system impervious to the zero sequence fault currents on the grid side. This transformer typically has a delta-wye configuration with delta winding on PV side and the wye grounded winding on the grid side. Such a configuration does not allow *triple-n* harmonics to percolate into the grid and instead circulates them within the transformer winding. As mentioned earlier, a separate interconnection transformer is installed to connect a single or a group of several PV systems within a solar PV plant to the grid. This is typically of a wye-grounded –wye-grounded configuration [3]. In this book, both the isolation transformer and interconnection transformer are considered as a single coupling transformer unit for ease of modeling.

3.4 Control Circuit Constituents of Smart Inverter System

The VSC control circuit of a solar PV system comprises measurement filters, PLL, current controller, and PWM unit. These are described below.

3.4.1 Measurement Filters

The smart PV inverter utilizes several electrical quantities for performing various smart inverter functions, i.e. the VSC terminal voltage V_t, VSC current i_t, grid voltage V_g at the electrical point of connection (PoC), active power P_{PPC}, reactive power Q_{PPC}, PCC voltage V_{PCC}, etc.

These signals contain some PWM switching side-band harmonics and also incur modulation by 60 Hz through the *abc* to *dq* frame transformation. The high-frequency signal components need to be removed using low pass filters. Measurement systems are used to measure these electrical quantities after appropriate filtering. These measurement systems usually incur delays. For instance, the voltage measurement process may incur the following delays:

i) delay due to voltage RMS or average computation, typically half a cycle
ii) delay due to filtering of the transduced voltage signal, signal processing, and moving average computation, etc.
iii) delays due to communication of the voltage signal from the measurement location to the input of the specific inverter by solar PV plant controller, as well as any communication delay between the controllers of the specific inverter [17].

The combined delays are modeled by an appropriate time constant τ corresponding to the specific measurement system.

The *dq* axis components of the grid voltage V_g or V_c are known as grid voltage feed-forward signals. They help the closed-loop current controller in providing an improved response during start-up transient, grid voltage disturbance, and unbalances [3]. Filters associated with the measurement of these voltage signals are referred to as grid voltage feed-forward filters.

3.4.2 abc-dq Transformation

3.4.2.1 Concept

Transforms make engineering simple. They transport intricate problems from the "real world" which is the "world of harsh and complicated reality" to the "world of transforms" which is a "world of simplicity." The problem is solved using simple techniques in the "world of simplicity" or the "world of transforms" and the results are transported back to the "real world" with ease.

For instance, complicated multiplication and division problems in the "real world" are transformed to the "simple world of logarithms" and solved as simple addition and subtractions, respectively. Intricate differential equations in the "real world" are transformed to the "simple world of Laplace" and solved as simple algebraic equations.

Similarly, in power systems and in smart inverter systems, sinusoidal time-varying signals are transformed from the "real world of *abc* reference frame" to the "simple world of *d-q* (direct-quadrature) reference frame" and solved as DC signals. Even more, they can be controlled in a decoupled manner.

The concept of *abc-dq* transformation can be explained with a simple example of merry-go-round which has toy-horses circling around at a constant speed with kids seated on them. The merry-go-round is near a stationary pole (i.e. lamp post). Consider two kids – Kid F in the front horse and the other Kid B in the horse behind. If Kid B asks Kid F, "Where are you?" Kid F states his position as, "Right now, I am ten and three-quarter rounds with respect to that stationary pole near the merry go round, and this position is going to change continuously." Kid F then asks Kid B, "Where are you"? Kid B describes his position as, "At this time, I am ten and five-eighth rounds with respect to that same stationary pole near the merry go round, and also my position is going to change continuously." Kid F thinks for a moment and then says to Kid B, "Let us think simple. Why don't we say that you are 45° behind me and your position will always be constant with respect to me?"

One can easily see from this very familiar example, that just by changing the reference frame from a stationary axis (with respect to the pole) to a rotating axis (with respect to one of the kids), the continuously changing position with time (which was difficult to remember) transforms into a constant value (which is easy to grasp).

In case of power systems and smart inverters, all the voltages and currents are time-varying sinusoidal quantities in the stationary *abc* reference frame. The advantages of the *abc-dq* transformation are as follows:

i) time-varying sinusoidal quantities in the stationary *abc* reference frame are transformed into constant DC quantities by choosing a reference frame that rotates at the same (synchronous) speed as the sinusoidal quantities.

ii) simpler controllers (e.g. Proportional-Integral) with lower-order dynamics can be designed with DC quantities. This is not possible with sinusoidal time-varying quantities.

3.4.2.2 Theoretical Basis

The mathematical concepts of *abc* to *dq* transformation are described in [3, 6], which are summarized here:

A general balanced three-phase sinusoidal function is described in terms of its individual phases in the *abc* reference frame as below:

$$f_a(t) = \hat{f} \cos(\omega t + \theta_0)$$

$$f_b(t) = \hat{f} \cos\left(\omega t + \theta_0 - \frac{2\pi}{3}\right)$$

$$f_c(t) = \hat{f} \cos\left(\omega t + \theta_0 - \frac{4\pi}{3}\right) \tag{3.27}$$

where, $\hat{f}$, ω, and θ_0 are the amplitude, angular frequency, and initial phase angle of the function, respectively.

The space phasor corresponding to this sinusoidal function is defined as

$$\vec{f}(t) = \frac{2}{3}\left[f_a(t) e^{j0} + f_b(t) e^{j\frac{2\pi}{3}} + f_c(t) e^{j\frac{4\pi}{3}} \right] \tag{3.28}$$

Substituting Eq. (3.27) in Eq. (3.28), and using appropriate trigonometric identities, we get

$$\vec{f}(t) = \left(\hat{f} e^{j\theta_0} \right) e^{j\omega t} \tag{3.29}$$

$$= f e^{j\omega t}$$

where, $f = \hat{f} e^{j\theta_0}$ denotes a vector in the *Real-Imaginary* complex plane, having initial angle θ_0.

Figure 3.10 illustrates the phasor diagram of the space phasor $\vec{f}(t)$ in the stationary *abc* reference frame and *dq* reference frame rotating anti-clockwise at synchronous speed ω. The stationary Cartesian coordinate system, referred as $\propto - \beta$ reference frame is also depicted in Figure 3.10.

The angles that synchronously rotating *d*-axis and space phasor $\vec{f}(t)$ make with respect to the stationary *a*-axis are represented by $\rho(t)$ and $\theta(t)$, respectively, where,

$$\theta(t) = \omega_0 t + \theta_0 \tag{3.30}$$

The space phasor $\vec{f}(t)$ is transformed into stationary $\propto - \beta$ reference frame, for controller design purposes, as

$$\vec{f}(t) = f_\propto(t) + jf_\beta(t) \tag{3.31}$$

where, $f_\propto(t)$ and $f_\beta(t)$ are the $\propto - \beta$ axis components of the space phasor $\vec{f}(t)$ as shown in Figure 3.10. The $\propto, \beta$ axis correspond to the real and imaginary axis, respectively.

Equating the real and imaginary components of Eq. (3.29) and Eq. (3.31) provides the *abc* -$\propto\beta$ axis transformation as,

$$\begin{bmatrix} f_\propto(t) \\ f_\beta(t) \end{bmatrix} = \frac{2}{3}[C] \begin{bmatrix} f_a(t) \\ f_b(t) \\ f_c(t) \end{bmatrix} \tag{3.32}$$

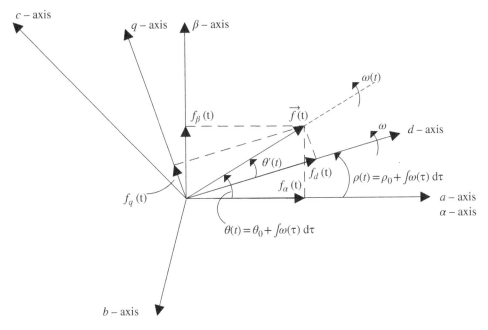

Figure 3.10 Phasor diagram of space phasor $\vec{f}(t)$ in *abc* and *dq* reference frames.

where,

$$[C] = \begin{bmatrix} 1 & \dfrac{-1}{2} & \dfrac{-1}{2} \\ 0 & \dfrac{\sqrt{3}}{2} & \dfrac{-\sqrt{3}}{2} \end{bmatrix} \tag{3.33}$$

Conversely,

$$\begin{bmatrix} f_a(t) \\ f_b(t) \\ f_c(t) \end{bmatrix} = [C]^T \begin{bmatrix} f_\propto(t) \\ f_\beta(t) \end{bmatrix} \tag{3.34}$$

Furthermore,

$$f_\propto(t) = \hat{f}(t)\cos[\theta(t)] \tag{3.35}$$
$$f_\beta(t) = \hat{f}(t)\sin[\theta(t)] \tag{3.36}$$

where, both $f_\propto(t)$ and $f_\beta(t)$ are sinusoidal functions of time having amplitude $\hat{f}$ and angular frequency $\omega = \frac{d\theta}{dt}$.

The controller design in stationary $\propto - \beta$ reference frame presents the following issues:

1) Controllers need to be designed for sinusoidal signals which are more complex than those needed for DC signals.
2) Controllers are of higher-order and also difficult to optimize.
3) Bandwidth of the controllers must be larger than the frequency of reference signals.

These problems are easily addressed using the rotating *dq* frame of reference. This can be done through two approaches:

i) *abc – αβ* transformation followed by *αβ – dq* transformation.
ii) Direct *abc – dq* transformation.

The *abc – αβ* reference frame transformation is already described above. The transformation from stationary *αβ* reference frame to rotating *dq* reference frame is defined as:

$$f_d + jf_q = \left(f_\alpha + jf_\beta \right) e^{-j\rho(t)} \tag{3.37}$$

where, f_d and f_q are the *d*-axis and *q*-axis components of the space vector $\vec{f}$.

The transformation amounts to phase-shifting the space vector $\vec{f}(t)$ by an angle $-\rho(t)$, as shown in Figure 3.10. Multiplying both sides of Eq. (3.37) by $e^{j\rho(t)}$ results in the following expression for the space vector,

$$\vec{f}(t) = f_\alpha + jf_\beta = \left(f_d + jf_q \right) e^{j\rho(t)} \tag{3.38}$$

This also represents the inverse transformation from *dq* reference frame to *αβ* reference frame. We have from Eqs. (3.29) and (3.31),

$$\vec{f}(t) = f_\alpha + jf_\beta = \hat{f}(t) e^{j\left[\theta_o + \int \omega(\tau)d\tau\right]} \tag{3.39}$$

where, $\omega(t)$ in general is the time-varying frequency and θ_o is the initial angle of the sinusoidal function described in Eq. (3.27).

Let the angle $\rho(t)$ be defined as

$$\rho(t) = \rho_o + \int \omega(\tau)d\tau \tag{3.40}$$

The above implies that the *dq* reference frame is rotating with angular frequency ω. Substituting Eqs. (3.39) and (3.40) in Eq. (3.37) gives

$$f_d + jf_q = \hat{f}(t) e^{j\left[\theta_o + \int \omega(\tau)d\tau\right]} \cdot e^{-j\left[\rho_o + \int \omega(\tau)d\tau\right]} = \hat{f}(t) e^{j[\theta_o - \rho_o]} \tag{3.41}$$

This clearly shows that the space phasor $f_d + jf_q$ is time invariant or stationary. Also, that the components f_d and f_q are DC quantities.

It is noted that:

i) In general, $\theta(t)$ is not the same as $\rho(t)$, although,

$$\frac{d\theta(t)}{dt} = \frac{d\rho(t)}{dt} \tag{3.42}$$

ii) Both the space vector $\hat{f}(t)$ and the *dq* reference frame are rotating anticlockwise with the same frequency $\omega(t)$.
iii) In case of constant frequency-based VSC system such as smart inverters, $\omega(t)$ is set to a constant value ω_o which is the nominal frequency of power system.
iv) If ρ is selected to be equal to grid voltage angle, the active and reactive powers of the smart inverter VSC can be independently controlled by the *d*-axis and *q*-axis components of VSC current, respectively. This is achieved with the help of a PLL described later.

The $\alpha\beta - dq$ reference frame transformation given in Eq. (3.37) is expanded using the identity

$$e^{j\rho t} = \cos \rho t + j \sin \rho t \tag{3.43}$$

as below

$$\begin{bmatrix} f_d(t) \\ f_q(t) \end{bmatrix} = \begin{bmatrix} \cos \rho(t) & \sin \rho(t) \\ -\sin \rho(t) & \cos \rho(t) \end{bmatrix} \begin{bmatrix} f_\alpha(t) \\ f_\beta(t) \end{bmatrix} \tag{3.44}$$

The inverse transformation from $dq - \alpha\beta$ reference frame is expressed as

$$\begin{bmatrix} f_\alpha(t) \\ f_\beta(t) \end{bmatrix} = \begin{bmatrix} \cos \rho(t) & -\sin \rho(t) \\ \sin \rho(t) & \cos \rho(t) \end{bmatrix} \begin{bmatrix} f_d(t) \\ f_q(t) \end{bmatrix} \tag{3.45}$$

The direct transformation from abc frame to dq frame is, therefore, obtained by substituting Eq. (3.32) in Eq. (3.44)

$$\begin{bmatrix} f_d(t) \\ f_q(t) \end{bmatrix} = \frac{2}{3} \begin{bmatrix} \cos \rho(t) & \cos \left[\rho(t) - \dfrac{2\pi}{3} \right] & \cos \left[\rho(t) - \dfrac{4\pi}{3} \right] \\ \sin \rho(t) & \sin \left[\rho(t) - \dfrac{2\pi}{3} \right] & \sin \left[\rho(t) - \dfrac{4\pi}{3} \right] \end{bmatrix} \begin{bmatrix} f_a(t) \\ f_b(t) \\ f_c(t) \end{bmatrix} \tag{3.46}$$

The inverse transformation from $dq - abc$ reference frame is given by

$$\begin{bmatrix} f_a(t) \\ f_b(t) \\ f_c(t) \end{bmatrix} = \frac{2}{3} \begin{bmatrix} \cos \rho(t) & \sin \rho(t) \\ \cos \left[\rho(t) - \dfrac{2\pi}{3} \right] & \sin \left[\rho(t) - \dfrac{2\pi}{3} \right] \\ \cos \left[\rho(t) - \dfrac{4\pi}{3} \right] & \sin \left[\rho(t) - \dfrac{4\pi}{3} \right] \end{bmatrix} \begin{bmatrix} f_d(t) \\ f_q(t) \end{bmatrix} \tag{3.47}$$

Referring to the phasor diagram shown in Figure 3.10, we have

$$\hat{f}(t) = \sqrt{f_d^2(t) + f_q^2(t)} \tag{3.48}$$

$$\cos [\theta'(t)] = \frac{f_d(t)}{\hat{f}(t)} = \frac{f_d(t)}{\sqrt{f_d^2(t) + f_q^2(t)}} \tag{3.49}$$

$$\sin [\theta'(t)] = \frac{f_q(t)}{\hat{f}(t)} = \frac{f_q(t)}{\sqrt{f_d^2(t) + f_q^2(t)}} \tag{3.50}$$

where, $\theta'(t) = \theta(t) - \rho(t)$

3.4.2.3 Power in abc and dq Reference Frame

Consider a balanced three-phase network having terminal voltages and currents represented by $v_{abc}(t)$ and $i_{abc}(t)$, wherein v_{abc} and i_{abc} may not be balanced but $i_a + i_b + i_c = 0$.

The instantaneous power is given as [3]

$$P(t) = \mathrm{Re} \left\{ \frac{3}{2} \vec{v}(t) \vec{i}^*(t) \right\} \tag{3.51}$$

$$Q(t) = \text{Im}\left\{ \frac{3}{2} \vec{v}(t) \vec{i}^{*}(t) \right\} \tag{3.52}$$

$$S(t) = P(t) + jQ(t) = \frac{3}{2} \vec{v}(t) \vec{i}^{*}(t) \tag{3.53}$$

where, $\vec{v}(t)$ and $\vec{i}(t)$ are the space phasors corresponding to the sinusoidal voltages v_{abc} and current i_{abc} as defined in Eqs. (3.27) and (3.28).

In dq reference frame, $\vec{v}(t)$ and $\vec{i}^{*}(t)$ are expressed as

$$\vec{v}(t) = \left(v_d + jv_q \right) e^{j\rho(t)} \tag{3.54}$$

$$\vec{i}^{*}(t) = \left(i_d - ji_q \right) e^{-j\rho(t)} \tag{3.55}$$

Substituting Eqs. (3.54) and (3.55) in Eqs. (3.51) and (3.52) respectively, we get

$$P(t) = \frac{3}{2} \left[v_d(t)i_d(t) + v_q(t)i_q(t) \right] \tag{3.56}$$

$$Q(t) = \frac{3}{2} \left[-v_d(t)i_q(t) + v_q(t)i_d(t) \right] \tag{3.57}$$

The above equations show that active power $P(t)$ is controlled by both d and q axis components of $v(t)$ and $i(t)$. Same is the case with reactive power $Q(t)$. In other words, both $P(t)$ and $Q(t)$ are coupled with both $i_d(t)$ and $i_q(t)$.

If $v_q(t)$ is made equal to zero, the active and reactive power will become, respectively

$$P(t) = \frac{3}{2} v_d(t)i_d(t) \tag{3.58}$$

$$Q(t) = -\frac{3}{2} v_d(t)i_q(t) \tag{3.59}$$

Thus, active power $P(t)$ is controlled only by $i_d(t)$ and reactive power is controlled only by $i_q(t)$. Stated alternatively, the controls of active power and reactive power become decoupled. This is the basis of control of smart PV inverter. The condition $v_q(t) = 0$ is achieved by the help of a PLL described later.

3.4.3 Pulse Width Modulation (PWM)

The concept of switching is described in a simple manner as below. In the VSC shown in Figure 3.8, if a single switching pulse is applied to IGBT switch G_1 to turn it on, a positive DC voltage will appear at the AC output. Subsequently, if a switching pulse is applied to G_2 a negative DC voltage will be produced at the AC output terminal. This process will result in a square voltage wave which is far from the sine voltage wave desired at the output. A PWM strategy is, therefore, employed, in which a series of switching pulses are applied with each pulse having a different pulse width. The variation of pulse width is so performed that the fundamental component of the train of voltage pulses appearing at the corresponding AC terminal will be a sinusoidal wave.

Various VSC switching strategies are described in power electronics literature for two-level and multi-level VSCs [6]. These include PWM, SPWM, space vector modulation [6], etc. A commonly used PWM strategy is the SPWM technique. In this strategy, the firing pulses are generated by comparing a periodic triangular wave of high frequency known as carrier frequency wave with a sinusoidal signal known as the modulating signal. The modulating signal represents the sinusoidal

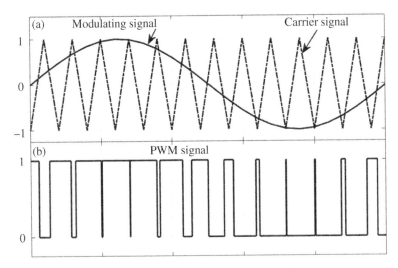

Figure 3.11 Sinusoidal pulse width modulation. (a) Modulating signal and carrier signal (b) pulse width modulated signal.

voltage desired at the output of the VSC. Switching pulses are generated at the intersection of the triangular wave voltage with the modulating sinusoidal signal. The generation of PWM switching pulses for a three-phase VSC is illustrated in Figure 3.11. A detailed description of the PWM process is presented in [4].

The frequency of the carrier triangular wave is a tradeoff between generated harmonics and switching losses. Higher frequency results in reduced harmonics and a cleaner sinusoidal voltage but incurs increased switching losses in the converter. Typical switching frequencies are selected between 2 and 15 kHz.

For a two-level VSC with SPWM modulation, the relationship between the DC voltage and the AC voltage at the VSC output is given by Eq. (3.19).

The relationship between the fundamental components of the AC terminal voltages V_{tabc} of the VSC and its DC terminal voltage V_{DC} is expressed as:

$$V_{ta}(t) = \frac{V_{DC}}{2} m_a(t)$$

$$V_{tb}(t) = \frac{V_{DC}}{2} m_b(t)$$

$$V_{tc}(t) = \frac{V_{DC}}{2} m_c(t) \tag{3.60}$$

In space vector notation, the above relationship is defined by

$$\overrightarrow{V_t} = \frac{V_{DC}}{2} \overrightarrow{m} \tag{3.61}$$

where, $\overrightarrow{V_t}$ is the output AC voltage of VSC in space phasor domain, V_{DC} is the input DC voltage of VSC, and $\overrightarrow{m}$ is the modulation index of VSC in space phasor domain. The sinusoidal modulation indexes m_{abc} are related to the modulation indexes m_{dq} in the dq reference frame by the transformation described below,

$$
\begin{bmatrix} m_a(t) \\ m_b(t) \\ m_c(t) \end{bmatrix} = \frac{2}{3} \begin{bmatrix} \cos \rho(t) & \sin \rho(t) \\ \cos \left[\rho(t) - \dfrac{2\pi}{3} \right] & \sin \left[\rho(t) - \dfrac{2\pi}{3} \right] \\ \cos \left[\rho(t) - \dfrac{4\pi}{3} \right] & \sin \left[\rho(t) - \dfrac{4\pi}{3} \right] \end{bmatrix} \begin{bmatrix} m_d(t) \\ m_q(t) \end{bmatrix}
\tag{3.62}
$$

The VSC can be modeled by: (i) switched model or (ii) averaged model. The switched model incorporates detailed transient switching process of the IGBTs together with switching losses. Such accurate models are typically used in electromagnetic transients simulation software such as PSCAD and EMTP, etc. However, to achieve higher simulation speeds especially in large power systems, an equivalent average-value model is utilized [3] known as "averaged model." In this model, the switching dynamics are ignored and the VSC terminal variables such as voltages and currents are approximated by their per-switching cycle moving average values. It is noted that the averaged model cannot be used to study voltage or current harmonics generated by VSC. Moreover, this model is valid only for frequencies up to one-third of the switching frequency of VSC and also for the operation of VSC in the linear modulation region, i.e. $m_{abc} \leq 1$, or $\sqrt{(m_d^2 + m_q^2)} \leq 1$ [3].

The dynamics of the DC-link voltage of VSC in the "averaged model" is given by [3]

$$
C_{DC} \frac{dV_{DC}}{dt} = I_{pv} - I_{DC}
\tag{3.63}
$$

where, C_{DC} is the DC-link capacitance of VSC and I_{DC} is the input DC current of the VSC.

The input DC current is given by [3]

$$
I_{DC} = \frac{3\left(m_d i_{td} + m_q i_{tq}\right)}{4}
\tag{3.64}
$$

where, m_d and m_q are the modulation index in dq reference frame, i_{td} and i_{tq} are the inverter output currents in dq reference frame.

The modulation process thus controls the amplitude and the phase angle of the generated sinusoidal voltage on the AC side of the VSC. This in turn controls the active power and reactive power flowing out of the VSC as explained in previous section.

3.4.4 Phase Locked Loop (PLL)

The VSC is connected to the AC power system typically at the grid bus (e.g. PoC, PCC). The voltage at the grid bus is continuously subject to changes in frequency, even if small; unbalance in different phases; and distortions at harmonic and other frequencies either in steady state or during grid faults. It is important that despite the above nonideal conditions the VSC stays synchronized with the frequency and phase of the fundamental frequency positive sequence voltage component of interconnected AC power system. The PLL system is utilized to ensure this synchronization. It detects the frequency and phase of the grid voltage rapidly and accurately, and outputs the phase angle of the grid voltage phasor.

Majority of the VSC control systems are based on the synchronously rotating d-q reference frame. In such systems, the PLL extracts the d- axis component of the bus voltage for control purposes, and further provides the phase angle of the bus voltage phasor required in the VSC PWM process.

The concept of PLL operation can be understood by the phasor diagram in Figure 3.12, which depicts the conditions before synchronization. This phasor diagram depicts the stationary three-phase *abc* reference frame axes, the stationary α-β reference frame axes, and the rotating *d-q*

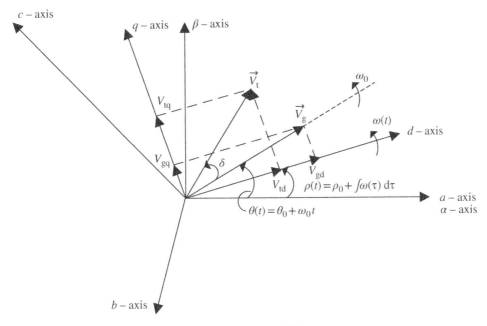

Figure 3.12 Phasor diagram prior to synchronization by PLL.

reference frame axes. The phasor diagram further illustrates the space vectors corresponding to the grid voltage $\overrightarrow{V_g}$ and VSC terminal voltage $\overrightarrow{V_t}$, and their components along the axes of different reference frames. The grid voltage vector is rotating anticlockwise with the system nominal angular frequency ω_0 whereas the dq reference frame is rotating anticlockwise with a frequency of $\omega(t)$. At a given instant of time, the grid voltage vector $\overrightarrow{V_g}$ makes an angle of $\theta(t) = \theta_0 + \omega_0 t$, whereas the d-axis makes an angle of $\rho(t) = \rho_0 + \int \omega(\tau)d\tau$ with the stationary a-axis. The PLL varies the rotational speed of dq axis to ω_0 and further aligns the grid voltage vector $\overrightarrow{V_g}$ with the d-axis component of the VSC voltage vector V_{td}, thereby making the q-axis component of grid voltage V_{gq} zero. Once this is achieved the angle $\theta(t) = \rho(t)$. This angle ρ is then utilized for PWM process in the VSC. The phasor diagram corresponding to this condition, i.e. after synchronization is achieved is shown in Figure 3.13.

Different types of PLLs each having different features are employed in power electronics industry [18–20]. However, a commonly used PLL known as synchronous reference frame PLL is shown in Figure 3.14 and described below.

The sinusoidal voltage components of the grid voltage space vector $\overrightarrow{V_g}$ hereon referred as the capacitor voltage $\overrightarrow{V_c}$ before synchronization are decomposed into the synchronously rotating d-q axis components V_{cd} and V_{cq} using the transformation (3.46).

$$V_{cd} = \hat{V}_c \cos(\omega_0 t + \theta_0 - \rho) \tag{3.65}$$

$$V_{cq} = \hat{V}_c \sin(\omega_0 t + \theta_0 - \rho) \tag{3.66}$$

$$\frac{d\rho}{dt} = \omega(t) \tag{3.67}$$

Where, $\hat{V}_c$ denotes the magnitude of the capacitor voltage $\overrightarrow{V_c}$

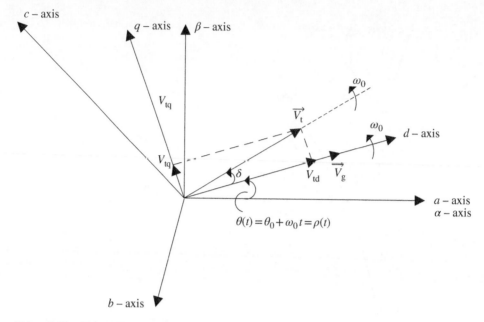

Figure 3.13 Phasor diagram after synchronization by PLL.

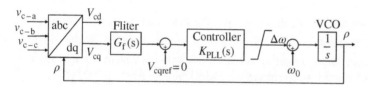

Figure 3.14 Block diagram of a Phase Locked Loop.

As explained above, the q-axis component of the voltage space vector V_{cq} needs to be made zero to achieve a decoupled control of active and reactive power. This implies that $\rho(t)$ must be regulated to $(\omega_0 t + \theta_0)$. At first, V_{cq} is passed through a low pass filter $G_f(s)$ to eliminate any distortions and harmonic components. The filtered component $V_{cqf}(s)$ is compared with the reference signal $V_{cq,ref}$ which is set to zero. The error signal is processed through a controller $K_{PLL}(s)$ which produces an error frequency signal $\Delta\omega(s)$, where $\Delta\omega(s)$ is the deviation in frequency from system nominal frequency ω_o during transients.

$$\Delta\omega(s) = K_{PLL}(s)V_{cqf}(s) \tag{3.68}$$

A frequency bias of ω_o is added to $\Delta\omega$ to ensure that the frequency excursions are around ω_o [21]. The output is also limited by a saturation block with limits ω_{min} and ω_{max} in order to restrict the variations of ω. The frequency signal ω is provided as an input to a voltage controlled oscillator (VCO) which essentially increases or decreases the speed of the dq reference frame with respect to ω_o. In steady state:

$$\omega = \omega_o \tag{3.69}$$

$$V_{cd} = \hat{V}_c \tag{3.70}$$

$$V_{cq} = 0 \tag{3.71}$$

The VCO is basically a resettable integrator which converts frequency to angle. It outputs an angle ρ which is the angle of the grid voltage vector. This angle reference ρ is utilized in abc to dq reference frame conversion.

The compensator $K_{\text{PLL}}(s)$ is a Proportional Integral (PI) controller and is given by

$$K_{\text{PLL}}(s) = b_1 + \frac{b_2}{s} \tag{3.72}$$

where, b_1 is the proportional gain and b_2 is the integral gain. The design of PLL controller for a required bandwidth and phase margin is described in [6, 22].

It may appear that PLL plays only the role of synchronizing VSC voltage with the grid voltage under different grid operating conditions, while the major function is performed by the VSC control system in exchange of active and reactive power with the grid. However, PLL dynamics can impact power system stability especially in weak grids. The design of PLL controller parameters must, therefore, be performed carefully.

Effect of PLL on Active and Reactive Power Output of VSC

The active power P_{VSC} and reactive power Q_{VSC} of VSC leaving the low pass filter inductor (but before the shunt filter capacitor) are given by

$$P_{\text{VSC}} = \frac{3}{2} \left[V_{\text{cd}} i_{\text{td}} + V_{\text{cq}} i_{\text{tq}} \right] \tag{3.73}$$

$$Q_{\text{VSC}} = \frac{3}{2} \left[-V_{\text{cd}} i_{\text{tq}} + V_{\text{cq}} i_{\text{td}} \right] \tag{3.74}$$

Since PLL ensures that the phase angle is locked to the angle of grid voltage V_{c}, V_{cq} becomes zero at steady state. This simplifies the above equations as

$$P_{\text{VSC}} = \frac{3}{2} \left[V_{\text{cd}} i_{\text{td}} \right] \tag{3.75}$$

$$Q_{\text{VSC}} = \frac{3}{2} \left[-V_{\text{cd}} i_{\text{tq}} \right] \tag{3.76}$$

The above equations clearly demonstrate that the control of active power and reactive power of VSC have become decoupled. The active power of VSC can be controlled only by controlling i_{td} whereas the reactive power of VSC can be controlled by controlling only i_{tq}.

It is however noted that in the VSC controller, Q_{VSC} is calculated using the filtered component of V_{cd} i.e. V_{cdf} and i_{tq}. Q_{VSC} is, therefore, expressed as

$$Q_{\text{VSC}} = \frac{3}{2} \left[-V_{\text{cdf}} i_{\text{tq}} \right] \tag{3.77}$$

The total reactive power output of the smart inverter is the sum of Q_{VSC} and the reactive power generated by the filter capacitor at the terminal of VSC.

3.4.5 Current Controller

As described earlier, the active and reactive power of the VSC are not directly controlled by the magnitude and phase angle of the voltage of VSC, as there is potential of inverter current becoming unacceptably large during faults and network transients. Instead, the active and reactive power of the VSC are controlled by controlling the inverter current. Specifically, the d-axis component of

VSC current i_{td} and the q-axis component of VSC current i_{tq} are regulated to their reference values $i_{td, \text{ref}}$ and $i_{tq, \text{ref}}$, to generate the desired active power and reactive power output, respectively.

The AC side dynamics of VSC is described as below

$$\vec{V}_t = R_f \vec{i_t} + L_f \frac{d\vec{i_t}}{dt} + \vec{V}_c \tag{3.78}$$

Transforming the above into dq reference frame using Eqs. (3.27) and (3.46), the AC side dynamics of VSC are rewritten as

$$L_f \frac{di_{td}}{dt} = L_f \omega(t) i_{tq} - R_f i_{td} + V_{td} - V_{cdf} \tag{3.79}$$

$$L_f \frac{di_{tq}}{dt} = L_f \omega(t) i_{td} - R_f i_{tq} + V_{tq} - V_{cqf} \tag{3.80}$$

where, the subscripts d, q denote the d and q axis component of the respective variable. V_{cdf} and V_{cqf} represent the filtered d, q axis components of V_c.

In steady-state, the PLL locks to the phase angle of grid voltage making $V_{cd} = \hat{V}_g$ and $V_{cq} = 0$. Hence, $\omega(t) = \omega_0$.

For the two-level VSC with SPWM the VSC AC-side terminal voltage $V_{t, abc}$ is controlled as below

$$V_{t,abc} = \frac{V_{DC}}{2} m_{abc} \tag{3.81}$$

where, V_{DC} and m_{abc} are DC bus voltage and modulation indexes in abc-frame, respectively.

Transforming Eq. (3.81) into dq frame of reference results in

$$V_{t,dq} = \frac{V_{DC}}{2} m_{dq} \tag{3.82}$$

Substituting Eq. (3.82) in Eqs. (3.79) and (3.80),

$$L_f \frac{di_{td}}{dt} + R_f i_{td} = L_f \omega_0 i_{tq} + \frac{V_{DC}}{2} m_d - V_{cdf} \tag{3.83}$$

$$L_f \frac{di_{tq}}{dt} + R_f i_{tq} = -L_f \omega_0 i_{td} + \frac{V_{DC}}{2} m_q - V_{cqf} \tag{3.84}$$

It is noted that the presence of $L_f \omega_0$ makes the dynamics of i_{td} and i_{tq} coupled and nonlinear [6]. In order to decouple and linearize the dynamics, m_d and m_q are determined based on the following control laws using u_d and u_q as two new control inputs [6]:

$$m_d = \frac{2}{V_{DC}} \left[u_d - L_f \omega_0 i_{tq} + V_{cdf} \right] \tag{3.85}$$

$$m_q = \frac{2}{V_{DC}} \left[u_q + L_f \omega_0 i_{td} + V_{cqf} \right] \tag{3.86}$$

m_d and m_q are substituted into Eqs. (3.83) and (3.84) to result in

$$L_f \frac{di_{td}}{dt} = -R_f i_{td} + u_d + V_{cdf} \tag{3.87}$$

$$L_f \frac{di_{tq}}{dt} = -R_f i_{tq} + u_q + V_{cqf} \tag{3.88}$$

Equations (3.87) and (3.88) show that the dynamics of i_{td} and i_{tq} are decoupled and linear. Now, i_{td} and i_{tq} can be controlled by controlling u_d and u_q, respectively, while V_{cdf} and V_{cqf} can be

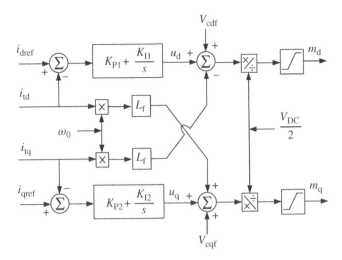

Figure 3.15 Block diagram of the current controller.

considered as disturbance inputs. The control block diagram for the current controller is shown in Figure 3.15. The saturation block is employed in current controller to protect the VSC from instantaneous overcurrents during network disturbances, e.g. faults [3].

It is seen that u_d is the output of compensator $K_d(s)$ which processes the error $i_{tdref} - i_{td}$. Similarly, u_q is the output of compensator $K_q(s)$ which processes the error $i_{tqref} - i_{tq}$.

The compensators $K_d(s)$ and $K_q(s)$ are PI controllers and are given by

$$K_d(s) = K_{P1} + \frac{K_{I1}}{s} \tag{3.89}$$

$$K_q(s) = K_{P2} + \frac{K_{I2}}{s} \tag{3.90}$$

where, K_{P1}, K_{P2} are proportional gains and K_{I1}, K_{I2} are integral gains.

The parameters of PI controllers can be designed for a particular bandwidth and phase margin of the current controllers using symmetrical optimum technique [6, 22]. The controllers are designed so that their closed-loop bandwidth is at least 10 times smaller than the switching frequency of VSC [6].

Smart inverter controls can directly generate reference signals i_{tdref} and i_{tqref}. Some smart inverter controls generate reference power signals P_{ref} and Q_{ref} for the VSC, from which i_{tdref} and i_{tdref} are obtained using Eqs. (3.75) and (3.76), respectively

$$i_{tdref} = \frac{2}{3V_{cd}} P_{ref} \tag{3.91}$$

$$i_{qdref} = -\frac{2}{3V_{cd}} Q_{ref} \tag{3.92}$$

3.4.6 DC-Link Voltage Controller

The DC-link voltage controller regulates the voltage across the DC-link capacitor V_{DC} at a voltage reference value $V_{DC,ref}$ provided by the MPPT algorithm or the non-MPPT technique, whichever is applicable. This allows the corresponding DC power from the PV array to be delivered to the AC grid after DC-AC conversion.

The charge and consequently the voltage of the DC-link capacitor continues to decline due to the losses in the VSC switches. The DC-link capacitor, therefore, needs to absorb a slight amount of active power to maintain the DC voltage at the reference value $V_{DC,ref}$. During daytime, the DC-link voltage control utilizes a small amount of DC power from the solar panels to replenish the lost charge and transfers the remaining solar power into the grid. During nighttime, while performing a smart inverter function such as PV-STATCOM (described in Chapter 4), the VSC control absorbs a small amount of active power from the grid and charges the capacitor through VSC antiparallel diodes connected across the IGBT switches.

The power balance equation at DC side of VSC is expressed as

$$\frac{d}{dt}\left[\frac{C_{DC}V_{DC}^2}{2}\right] = P_{pv} - P_{VSC,in} \tag{3.93}$$

$$P_{pv} = V_{pv}I_{pv} \tag{3.94}$$

where, V_{pv}, I_{pv}, and P_{pv} are the voltage, current, and power output of the PV panel, respectively. $P_{VSC,\,in}$ is the input DC power of the VSC. Also, $V_{DC} = V_{pv}$ when PV panels are connected to the VSC.

It is seen from above that for a given PV panel power P_{pv}, the voltage V_{DC} can be controlled by controlling $P_{VSC,\,in}$. However, as shown before, $P_{VSC,\,in}$ can be controlled by i_{td} in steady state. In other words, V_{DC} can be controlled by controlling i_{td}.

The error between V_{DCref}^2 and V_{DC}^2 is processed by a compensator $K_{DC}(s)$, which generates the current reference i_{tdref} for current controller. The compensator $K_{DC}(s)$ is a PI controller in series with a low pass filter [23] and is given by

$$K_{DC}(s) = \left(K_{PDC} + \frac{K_{IDC}}{s}\right)\left(\frac{1}{\tau_{DC}s + 1}\right) \tag{3.95}$$

where, K_{PDC} is the proportional gain and K_{IDC} is the integral gain of the PI controller. τ_{DC} is the time constant of the low pass filter.

The control block diagram for DC-Link voltage controller is shown in Figure 3.16.

The compensator $K_{DC}(s)$ is designed based on bandwidth and phase margin requirements [23].

3.5 Smart Inverter Voltage Controllers

Various smart inverter functions for controlling reactive power and active power of the solar PV inverter for providing different grid support functionalities are described in Chapters 2. Different PV-STATCOM smart inverter functions corresponding to the control of PV systems as STATCOM

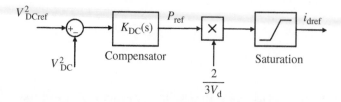

Figure 3.16 Block diagram of DC-link voltage controller.

for different grid benefits are presented in Chapters 5, 6, and 9. Figure 3.5 depicts the general block diagram of these smart inverter functions. Each smart inverter function eventually controls the VSC active current reference i_{dref} and reactive current reference i_{qref}, either individually or in an appropriate combination.

Two types of smart inverter function for voltage control are described below. One relates to the volt–var voltage control function, while the other corresponds to closed-loop voltage control function, as in PV-STATCOM.

3.5.1 Volt–Var Control

The implementation of a volt–var control smart inverter function is depicted in Figure 3.17 [17]. The smart inverter function is modeled by the volt–var curve, as described in Chapter 2. The PCC instantaneous voltage V_{PCC} is processed through an averaging window to generate the average voltage. A look-up table representing the volt–var function is used to produce the reactive power reference Q_{ref} for the VSC current controller. The open-loop response time of the volt–var control is in the range 1–90 seconds, although the typical values for DERs of Category A and B are 10 seconds and 5 seconds, respectively [12].

3.5.2 Closed-Loop Voltage Controller

This smart inverter function controls the PCC voltage V_{PCC} at a specified reference voltage level V_{PCCref} in a closed-loop manner. As described in Chapter 2, PCC voltage is influenced by both active power exchange and reactive power exchange with the grid. Reactive power exchange is more effective in grids with high X/R ratio whereas active power exchange is more effective in grids with low X/R ratio. This section describes a PCC voltage controller based on control of reactive power Q_{VSC}, which can be controlled in turn by i_{qref}.

In dq reference frame, V_{PCC} is given by

$$V_{PCC} = \sqrt{V^2_{PCCdf} + V^2_{PCCqf}} \qquad (3.96)$$

where, the subscripts d, q denote the d, q axis components of V_{PCC}, respectively. The subscript f represents the filtered component of the respective voltage.

The PCC voltage controller is depicted in Figure 3.18. The error between V_{PCCref} and V_{PCC} is processed by a compensator $K_{VAC}(s)$. A reactive power reference Q_{ref} is generated, which is subsequently expressed in terms of the current reference i_{qref} for current controller. The compensator $K_{VAC}(s)$ is typically a PI controller as below,

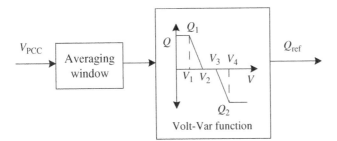

Figure 3.17 Implementation of volt–var smart inverter function.

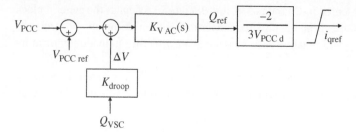

Figure 3.18 Block diagram of PCC voltage controller.

$$K_{\text{VAC}}(s) = \left(K_{\text{PV}} + \frac{K_{\text{IV}}}{s}\right) \tag{3.97}$$

where, K_{PV} is the proportional gain and K_{IV} is the integral gain of the controller.

A slope or droop K_{droop} is introduced in the smart inverter voltage controller similar to the slope in the *V-I* characteristic of SVCs and STATCOMs [3, 24, 25]. This droop in the *V-I* characteristic of the smart PV inverter provides the following advantages:

i) reduces the rating of the VSC required for controlling the voltage in the utility acceptable range
ii) prevents the smart inverter from utilizing its entire reactive power capacity (inductive to capacitive) for small changes in PCC voltage, especially in strong systems
iii) facilitates reactive power sharing among multiple smart PV inverters performing voltage control

The VSC reactive power Q_{VSC} is multiplied by K_{droop} to provide a small voltage deviation ΔV which is added to the voltage reference V_{PCCref}. Instead of Q_{VSC}, the VSC current i_t or smart inverter current i_{PCC} may also be utilized for providing the droop function. This voltage deviation ΔV is positive for reactive power absorption and negative for reactive power injection. The typical value of droop is 1–3% [24].

Eigenvalue sensitivity studies are performed to choose the best gains for the voltage controller $K_{\text{VAC}}(s)$ to provide a fast step response time (<1 cycle), overshoot less than 10%, and a rapid settling time (3–5 cycles). The controller parameters are selected such that the bandwidth of the PCC voltage controller is at least three times lower than the bandwidth of the q axis current controller [6].

3.6 PV Plant Control

A large utility-scale solar PV plant consists of several clusters of PV inverters, each fed by solar arrays as portrayed in Figure 3.19 [25]. Each cluster of inverters is connected to a common PV inverter step-up transformer. The outputs of these PV inverter transformers are connected through several medium voltage (MV) radial feeders or cables. These feeders are subsequently connected at the input of the substation transformer which ties to the high voltage grid at the point of interconnection (POI). Some PV plants have reactive power compensation equipment such as fixed capacitors and switched capacitors. These operate in conjunction with the PV inverters to meet the reactive power capability and control requirement at the POI.

The PV plant level controller measures different power system parameters such as voltage, frequency, active, and reactive power flows at the POI. It also receives commands issued by the PV fleet remote operations center or directly from the transmission system operator. Based on all this information, the plant controller communicates the required smart inverter function set points to

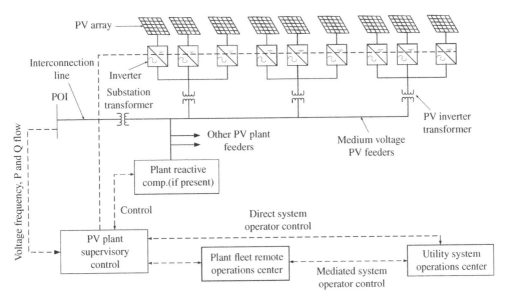

Figure 3.19 Typical topology of a solar PV plant. *Source:* Modified from [25]. Used with permission from WECC.

individual inverters. The plant controller is relatively slow compared to the inverter controllers which are very fast. During a dynamic event, the reactive power response of the solar PV plant is a combination of the slow response of the supervisory plant level control and the rapid response of the inverter level control. Under large disturbances, the inverter level control overrides the plant control in providing reactive power response [25].

Commercial and residential (roof-top) level solar PV systems do not have a plant controller. They are directly connected to distribution feeders. The PV inverters themselves provide the required smart inverter functionality at the grid interface, i.e. at the PCC.

The solar PV power plant modeling and validation guideline published by WECC describes the modeling of bulk power system (BPS)-connected solar PV plants for both power flow and dynamic studies [25]. This document describes in detail the modeling techniques for all solar PV resources in transmission and distribution systems. It also discusses the best practices for model validation of utility-scale solar PV systems (≥ 20 MVA) connected to the transmission network (60 kV and above).

The performance of a solar PV plant in providing various smart inverter functions is strongly influenced by the strength of the AC power system, which is characterized by its short circuit ratio (SCR) given as

$$\text{SCR} = \frac{\text{SC}_{\text{sys}}(\text{MVA})}{P_{\text{rated}}(\text{MW})} \tag{3.98}$$

$$\text{SC}_{\text{sys}}(\text{MVA}) = \frac{V_{\text{AC}}^2}{Z_{\text{th}}} \tag{3.99}$$

where, SC_{sys} is the short circuit MVA of the AC system, V_{AC} is the AC voltage at rated power output of the solar PV plant, Z_{th} is the Thevenin's equivalent impedance of the AC system viewed from the POI of the solar PV plant, and P_{rated} is the rated (maximum) power output of the solar PV plant.

Voltages in systems with low SCR tend to be more sensitive to variations in reactive power. It is noted that solar PV plants connected to systems with SCR less than three potentially experience voltage instability issues such as high dynamic overvoltages, voltage flicker, network resonance issues, etc. [25]. Such problems are generally alleviated by the use of synchronous condensers and FACTS controllers, i.e. SVCs, STATCOMs, etc.

The generic models described in the WECC guideline apply to solar PV systems connected to power systems with SCR of 2–3 at the POI [25].

3.7 Modeling Guidelines

A reliability guideline for modeling DERs in transient stability and power flow simulations with detailed load modeling was published by NERC in 2016 [26]. The data collection needs for adequately modeling different types of utility scale DER and retail-scale DER (including residential, commercial, and industrial customers) in stability analyses of the BPS, are described in [27]. A DER_A model has been developed for representing aggregate or stand-alone inverter-based DER resources in stability studies [28]. This DER_A model has capabilities of modeling DERs with both advanced smart inverter features as well as legacy DERs currently installed in North America.

NERC has further published a detailed report on modeling of BPS connected inverter-based resources, validation of models, and system studies [29]. This report describes the modeling needs of inverter-based resources in dynamic stability analyses and studies of their impacts on BPS reliability. One of the studies in this report shows that that reactive current priority with voltage control enabled is the preferred default configuration for BPS-connected solar PV resources in the studied areas. It recommends that detailed studies need to be conducted for determination of the best parameters of control systems for each BPS-connected inverter-based resource. Also, the potential control interactions of inverter controls with control systems of other dynamic devices, i.e. High Voltage DC (HVDC) systems and existing protection settings must be examined.

3.8 Summary

This chapter describes the fundamental concepts of active and reactive power control of smart solar PV inverter systems. The operating principles of different subsystems in the smart inverter system are described. These subsystems include PV panels, power point tracking, VSC, AC filter, isolation transformer, measurement filters, PLL, inner current controllers, and DC-link voltage controller. The implementation methodology of different smart inverter controls is described with smart inverter voltage controller as an example. The components of a utility-scale PV plant and its modeling needs are briefly described. Different smart inverter functions can be implemented in a smart solar PV system based on the controller models described in this chapter. The implementation of these various smart inverter functions will be described in subsequent chapters.

References

1 Rahman, S.A., Varma, R.K., and Vanderheide, T. (2014). Generalised model of a photovoltaic panel. *IET Renewable Power Generation* 8: 217–229.

2 Hohm, D.P. and Ropp, M.E. (2000). Comparative study of maximum power point tracking algorithms using an experimental, programmable, maximum power point tracking test bed. Conference Record

of the Twenty-Eighth IEEE Photovoltaic Specialists Conference – 2000 (Cat. No.00CH37036), 1699–1702.

3 Yazdani, A., Fazio, A.R.D., Ghoddami, H. et al. (2011). Modeling guidelines and a benchmark for power system simulation studies of three-phase single-stage photovoltaic systems. *IEEE Transactions on Power Delivery* 26: 1247–1264.

4 Mohan, N., Undeland, T.M., and Robbins, W.P. (2003). *Power Electronics: Converters, Applications, and Design*. New York: Wiley.

5 Rashid, M.H. (2007). *Power Electronics Handbook: Devices, Circuits, and Applications*. Elsevier.

6 Yazdani, A. and Iravani, R. (2010). *Voltage-Sourced Converters in Power Systems: Modeling, Control, and Applications*. Hoboken: IEEE Press and Wiley.

7 Bolinger, M. and Seel, J. (2018). Utility Scale Solar: Empirical Trends in Project Technology, Cost, Performance, and PPA Pricing in the United States – 2018 Edition. Lawrence Berkeley National Laboratory, Berkeley, CA, USA, *Report*.

8 Fu, R., Feldman, D., and Margolis, R. (2018). U.S. Solar Photovoltaic System Cost Benchmark: Q1 2018. National Renewable Energy Laboratory, Golden, CO, USA. *Technical Report NREL/TP-6A20-72399*.

9 Cherry, L. and Cox, M. (2019). The Global PV Inverter and MLPE Landscape 2019. Wood Mackenzie Power and Renewables, USA, *Report*.

10 Messo, T., Jokipii, J., Puukko, J., and Suntio, T. (2014). Determining the value of DC-link capacitance to ensure stable operation of a three-phase photovoltaic inverter. *IEEE Transactions on Power Electronics* 29: 665–673.

11 Perera, B.K., Ciufo, P., and Perera, S. (2013). Point of common coupling (PCC) voltage control of a grid-connected solar photovoltaic (PV) system. In *Proc. IECON 2013 – 39th Annual Conference of the IEEE Industrial Electronics Society*, 7475–7480.

12 IEEE (2018). *IEEE Standard for Interconnection and Interoperability of Distributed Energy Resources with Associated Electric Power Systems Interfaces*. IEEE Std 1547-2018 (Revision of IEEE Std 1547-2003).

13 Liserre, M., Blaabjerg, F., and Hansen, S. (2005). Design and control of an LCL-filter-based three-phase active rectifier. *IEEE Transactions on Industry Applications* 41: 1281–1291.

14 Gabe, I.J., Montagner, V.F., and Pinheiro, H. (2009). Design and implementation of a robust current controller for VSI connected to the grid through an LCL filter. *IEEE Transactions on Power Electronics* 24: 1444–1452.

15 Cunping,W., Xianggen, Y., Minghao, W. et al. (2010). Structure and parameters design of output LC filter in D-STATCOM. In *Proc. 2010 International Conference on Power System Technology*, 1–6.

16 Parker, S.G., McGrath, B.P., and Holmes, D.G. (2014). Regions of active damping control for LCL filters. *IEEE Transactions on Industry Applications* 50: 424–432.

17 Li, H., Smith, J., and Rylander, M. (2014). Multi-Inverter Interaction with Advanced Grid Support Function. In *Proc. CIGRE US National Committee Grid of the Future Symposium*.

18 Chung, S. (2000). Phase-locked loop for grid-connected three-phase power conversion systems. *IEE Proceedings - Electric Power Applications* 147: 213–219.

19 Hadjidemetriou, L., Kyriakides, E., and Blaabjerg, F. (2013). A new hybrid PLL for interconnecting renewable energy systems to the grid. *IEEE Transactions on Industry Applications* 49: 2709–2719.

20 Timbus, A., Liserre, M., Teodorescu, R. et al. (2005). Synchronization methods for three phase distributed power generation systems - an overview and evaluation. In *Proc. 2005 IEEE 36th Power Electronics Specialists Conference*, 2474–2481.

21 Teodorescu, R., Liserre, M., and Rodriguez, P. (2011). *Grid Converters for Photovoltaic and Wind Power Systems*. Chichester, UK: John Wiley & Sons Limited.

22 Leonhard, W. (2001). *Control of Electrical Drives*. Springer Science & Business Media.

23 Yazdani, A. and Dash, P.P. (2009). A control methodology and characterization of dynamics for a photovoltaic (PV) system interfaced with a distribution network. *IEEE Transactions on Power Delivery* 24: 1538–1551.

24 Mathur, R.M. and Varma, R.K. (2002). *Thyristor-Based FACTS Controllers for Electrical Transmission Systems*. New York: Wiley-IEEE Press.

25 WECC (2019). Solar Photovoltaic Power Plant Modeling and Validation Guideline. Western Electricity Coordinating Council (WECC), Salt Lake City, Utah, USA, *Report*.

26 NERC (2016). Modeling Distributed Energy Resources in Dynamic Load Models. NERC Atlanta, GA, USA, *Reliability Guideline*.

27 NERC (2017). Distributed Energy Resource Modeling. NERC Atlanta, GA, USA, *Reliability Guideline*.

28 NERC (2019). Parameterization of the DER_A Model. NERC Atlanta, GA, USA, *Reliability Guideline*.

29 NERC (2020). BPS-Connected Inverter-Based Resource Modeling and Studies. NERC, Atlanta, GA, USA, *Technical Report*.

4

PV-STATCOM: A NEW SMART PV INVERTER AND A NEW FACTS CONTROLLER

This chapter presents the basic concepts of Flexible AC Transmission System (FACTS) technology and two of its main-shunt-connected member Controllers – the Static Var Compensator (SVC) and STATic Synchronous COMpensator (STATCOM) [1–19]. The focus of this chapter is to present a new technology developed by this book's author for utilizing photovoltaic (PV) solar farms both during nighttime when solar farms are typically idle and during any time of system need during daytime as a STATCOM, termed PV-STATCOM. The different nighttime and daytime operating modes of the PV-STATCOM are presented. The cost of transforming an existing solar PV system into PV-STATCOM as well as the cost of operating the PV-STATCOM are described. Subsequently, the potential of PV-STATCOM technology in providing various benefits in transmission and distribution systems is elucidated.

4.1 Flexible AC Transmission System (FACTS)

When we think of the term "flexible," the first thing that comes to mind is a "rubber band" that can be stretched at will, or a "spring" that can be compressed as required. The term "flexible" in electrical power systems does not imply that transmission lines can be stretched or compressed. Instead, it implies that the electrical behavior of an existing transmission line can be made to resemble that of a longer (figuratively speaking – elongated) or a shorter (figuratively speaking – compressed) line. Changing the electrical behavior of an existing line is done to provide several benefits to power systems, which are not possible with the electrical behavior of the existing line. The Controllers that enable changing the electrical behavior of existing lines to enhance both power transmission capacity and controllability are known as FACTS. Although these Controllers are mainly defined for transmission systems, they have found several applications in distribution systems as well.

The "flexibility of electrical power transmission" is defined by IEEE as "the ability to accommodate changes in the electric transmission system or operating conditions while maintaining sufficient steady state and transient margins" [19]. Correspondingly, FACTS is defined as "alternating current transmission systems incorporating power-electronic based and other static controllers to enhance controllability and increase power transfer capability" [19].

Consider a simple power system depicted in Figure 4.1. The figure illustrates a Single Machine Infinite Bus (SMIB) system in which a synchronous generator is connected to a large power system considered to be an infinite bus through a long transmission line having reactance X.

Smart Solar PV Inverters with Advanced Grid Support Functionalities, First Edition. Rajiv K. Varma.
© 2022 The Institute of Electrical and Electronics Engineers, Inc. Published 2022 by John Wiley & Sons, Inc.

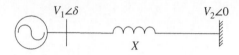

Figure 4.1 Single machine connected to infinite bus.

The steady state active power flow from the synchronous generator (bus 1) to the infinite bus (bus 2) is given by [10],

$$P_{12} = \frac{V_1 V_2}{X} \sin \delta \qquad (4.1)$$

The steady state stability limit of the power flow is obtained for $\delta = 90°$ and is given by,

$$P_{12,\text{max}} = \frac{V_1 V_2}{X} \qquad (4.2)$$

Increase in power flow in a line causes increased I^2R heating in the line, where I is the current flowing in the line which has resistance R. For a given transmission line, the thermal limit is described as the maximum power that can flow in a transmission line such that the sag caused by its I^2R heating is still considered safe according to the prevalent grid codes. While it may appear that this is the maximum power that can be allowed to flow in the line, the power flow is actually constrained by steady state stability limit described in Eq. (4.2). This limit becomes lower when transient stability is considered. In systems where electrical damping is low, the maximum power flow may be restricted by the damping limit, which may be even more constraining than transient stability limit in some power networks. The different limits of power flow in a transmission line including the thermal limit, steady state stability limit, transient stability limit, and damping limit are depicted in Figure 4.2 [10].

FACTS Controllers (also commonly referred as FACTS devices) provide the capability to increase the steady state stability limit, transient stability limit and electrical damping limit up to the thermal limit. Although the lines may not be operated in steady state at the thermal limit, the capability to operate them close to the thermal limit during system disturbances is of great value to power systems in enhancing overall system reliability and in preventing blackouts.

FACTS is a mature technology that continues to be widely used in power systems since more than 50 years. FACTS Controllers provide significant benefits to power systems through their following applications [1–15, 17]:

- Dynamic reactive power compensation
- Dynamic voltage control
- Increasing power transfer capacity of existing transmission lines

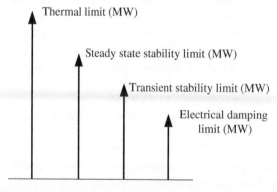

Figure 4.2 Comparison of different power flow limits in a transmission line. *Source:* Mathur and Varma [10].

- Improvement of system transient stability limit
- Enhancement of system damping
- Mitigation of subsynchronous resonance
- Alleviation of voltage instability
- Limiting short circuit currents
- Improvement of High Voltage Direct Current (HVDC) converter terminal performance
- Load compensation
- Grid integration of wind power generation systems, and in some cases solar PV power systems

FACTS Controllers utilize high-power electronic switches, e.g. thyristors and IGBTs. Thyristors can be turned-on but cannot be turned-off by gate pulses, whereas IGBTs can be both turned-on and turned-off by gate pulses. FACTS Controllers utilizing IGBTs are typically configured as Voltage Source Converters (VSCs). Major FACTS Controllers are classified as below, based on the types of semiconductor switches used:

- Thyristor-based FACTS Controllers
    - SVC
    - Thyristor Controlled Series Compensator (TCSC)

- VSC-based FACTS Controllers
    - STATCOM
    - Static Synchronous Series Compensator (SSSC)
    - Unified Power Flow Controller (UPFC)

SVC and STATCOM are connected in shunt with the transmission networks, while TCSC and SSSC are connected in series with transmission networks. UPFC is a combination of both series and shunt connected elements.

It is emphasized that the above listing is only of the most commonly used FACTS Controllers although several other FACTS Controllers have been developed using other semiconductor switches and in other different configurations, whose description is outside the scope of this book.

A brief description of the functionality of two major shunt connected FACTS Controllers – SVC and STATCOM, which are relevant from the perspective of smart inverters, is provided in the next section. Annotated bibliographies of major publications on SVCs and STATCOMs over the period 1991–1997 are compiled in [20–22]. Listings of various publications on SVCs and STATCOMs over the period 1998–2015 are compiled in annual Bibliographies on FACTS prepared by the HVDC and FACTS Bibliography Working Group (under the IEEE Power and Energy Society [PES] Transmission and Distribution Subcommittee, HVDC and FACTS Subcommittee) [23–59]. A list of industrial and utility SVC and STATCOM installations around the world is maintained by the Working Group I4 – SVC (under the IEEE PES Substations Committee, FACTS and HVDC stations Subcommittee) [60].

4.2 Static Var Compensator (SVC)

SVC provides dynamic reactive power compensation by thyristor-based control of physical reactors and capacitors connected in parallel. In the most general form, an SVC comprises a parallel combination of Thyristor-Controlled Reactors (TCRs) and Thyristor-Switched Capacitors (TSCs) [8, 10, 14]. A TCR consists of antiparallel connected thyristors in series with a fixed reactor (split into two halves on either side of the TCR valves), in each of the three phases. The fundamental component of the current through each leg of TCR is continuously controlled by varying the firing angle of the thyristors between 90° (full conduction) and 180° (no conduction). The TSCs, on the other hand, are discretely controlled. They can either be switched on or switched off by selecting the firing angle of the thyristors as above. The parallel combination of the TCRs and TSCs provides continuously controllable reactive current or reactive power over the inductive to capacitive range.

The individual phase TCRs are connected in delta to allow triplen harmonics to circulate within the delta and not to percolate into the power system. A coupling transformer is utilized to connect the medium voltage TSC–TCR combination to the high voltage power transmission system. Inductors of suitable rating are incorporated in the TSC branches to act as fifth and seventh harmonic filters. A high pass filter is separately installed to mitigate high-frequency harmonics generated by the TCRs. A typical SVC installation is depicted in Figure 4.3.

Control System of SVC

A typical control system of SVC is shown in Figure 4.4 [10]. A measurement system measures several power system quantities such as point of interconnection bus voltage (high-side voltage of the coupling transformer) and SVC current which are utilized for dynamic voltage control. The measurement system also measures several auxiliary control signals such as line current, line active power flow, bus frequency, signals from Wide Area Measurement Systems (WAMSs), which are employed for various functions, i.e. power oscillation damping, mitigation of subsynchronous resonance, etc. The measurement systems are equipped with adequate filters both on the AC side and DC side of the transducer to filter out network resonance modes, harmonics, and other high frequencies. These filters are necessary to avoid several types of controller instabilities [10].

The measured voltage is compared with a reference voltage V_{ref} and the error signal is provided to the voltage regulator which is typically a proportional integral controller or an equivalent gain-time constant transfer function. The measured SVC current is multiplied by a constant K_{SL} to implement

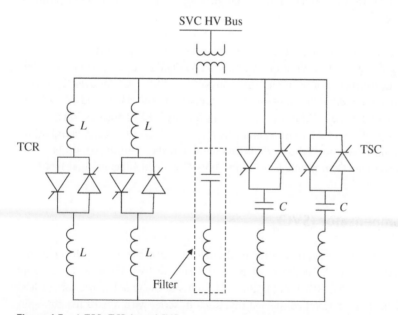

Figure 4.3 A TSC–TCR based SVC.

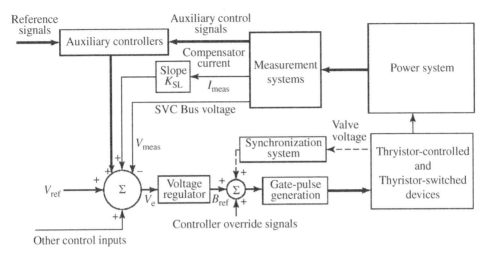

Figure 4.4 Typical control system of SVC. *Source:* Mathur and Varma [10].

the slope or droop in the *V–I* control characteristic of the SVC. The slope is selected between 1% and 5% (typically 3%). The slope provides the following advantages [10]:

- substantially reduces the reactive-power rating of the SVC for achieving nearly the same control objectives;
- prevents the SVC from reaching its reactive-power limits too frequently; and
- facilitates the sharing of reactive power among multiple SVCs operating in parallel.

Auxiliary controllers such as Power Swing Damping Controller (PSDC), Sub Synchronous Damping Controller (SSDC) utilize the measured auxiliary signals and provide a modulating voltage component to be added to the reference voltage V_{ref}. The auxiliary controllers have their respective reference signals based on the electrical quantity to be damped. The magnitude of this modulating voltage component is typically 5–10% of V_{ref}. The damping controllers typically have a gain and lead-lag compensator [10, 61]. The voltage regulator outputs a susceptance reference signal B_{ref}. The gate pulse generation unit utilizes this signal to determine the number of TSC branches to be switched on or off, and the firing angle of the TCR to implement this B_{ref} and the corresponding reactive power output at the point of interconnection of the SVC. These firing angles are applied to the TCR and the TSCs based on the reference provided by the Phase-Locked loop (PLL)-based synchronization system. A detailed description of the SVC control system is provided in [1, 2, 8, 10]. The controller override signals are utilized during severe system disturbances to override the normal SVC control, such as providing either fully inductive, or fully capacitive responses.

4.3 Synchronous Condenser

A synchronous condenser is basically a synchronous motor operating at no load. Figure 4.5 depicts a synchronous motor connected to the power system through a transformer. The motor internal voltage is represented by *E* which is controlled by the DC excitation system of the motor. The bus voltage is indicated by *V*. The combination of motor synchronous reactance and the transformer leakage reactance is represented by *X*.

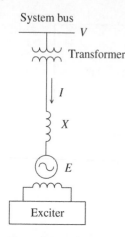

System bus

V

Transformer

I

X

E

Exciter

Figure 4.5 Synchronous condenser.

During no-load condition, there is no flow of active power (assuming machine losses are neglected). In such a situation, the reactive current I flowing into the synchronous condenser is given by

$$I = \frac{V - E}{X} \tag{4.3}$$

The variation of field current results in the variation of armature current which can be represented by a V-curve as shown in Figure 4.6 [10].

Lower values of field current results in lagging power factor operation whereas larger field currents result in leading power factor performance. The reactive power flow in the synchronous motor is given by

$$Q = \frac{1 - \frac{E}{V}}{X} V^2 \tag{4.4}$$

By controlling the excitation of the synchronous motor, and hence the amplitude E of its internal voltage relative to the amplitude V of the system voltage, the reactive power flow can be controlled as follows:

- $E > V$ (over-excitation) results in a leading current, i.e. the machine is "seen" as a capacitor by the AC system.
- $E < V$ (under-excitation) produces a lagging current, i.e. the machine is "seen" as an inductor by the AC system.

Due to this operating mode of a synchronous motor whereby it can inject reactive power into the grid like a condenser (capacitor), the synchronous motor is called a synchronous condenser. However, the synchronous condenser can provide continuously controllable reactive power over lagging to leading range.

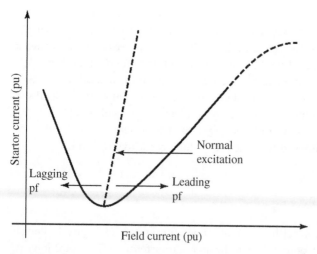

Figure 4.6 Variation of synchronous condenser armature current with change in field current. *Source:* Mathur and Varma [10].

4.4   Static Synchronous Compensator (STATCOM)

The operating principle of STATCOM is similar to that of a synchronous condenser. The variation of machine internal voltage E (as in the rotating synchronous condenser) is accomplished electronically by appropriate switching of semiconductor switches in a VSC.

Figure 4.7a depicts a STATCOM connected to a utility power system through a transformer having leakage reactance X_S. E_S and E_t denote the terminal voltage of the VSC and voltage at the utility grid bus, respectively. I_s represents the current flowing from the VSC to the utility grid. V_{DC} and I_{DC} represent the voltage across the DC capacitor and DC current, respectively.

Figure 4.7 shows the functional representation of the STATCOM system. The STATCOM can exchange reactive power with the utility grid by varying its terminal voltage E_S in a similar manner as a synchronous condenser. The difference in voltage between STATCOM terminals and utility grid bus causes the reactive power to flow in a bidirectional manner as depicted in Figure 4.7c.

- Making $E_s > E_t$ results in a leading current, i.e. the VSC is "seen" as a capacitor by the utility grid system.
- Making $E_s < E_t$ produces a lagging current, i.e. the VSC is "seen" as an inductor by the utility grid system.

Continuous current flow in the VSC causes power losses in the IGBT semiconductor switches. These losses are provided by the DC-link capacitor. Consequently, the capacitor gets discharged. Hence, in order to maintain the capacitor voltage to the desired voltage level V_{DC}, the STATCOM draws a very small amount of active power (typically 1–2% of STATCOM rating) from the utility grid to charge the DC capacitor. This is done by controlling the voltage angle of E_s to be slightly lower than the voltage angle of E_t, through a DC-link voltage controller.

Addition of an energy storage system at the DC input of the VSC allows active power to be exchanged between the VSC and the utility grid. Exchange of active power is achieved by varying the angle of the VSC terminal voltage with respect to the angle of the utility grid bus.

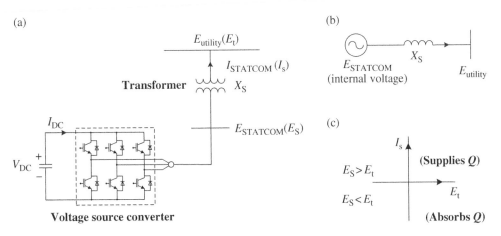

Figure 4.7 STATCOM with DC capacitor (no energy storage). (a) power circuit, (b) equivalent circuit, (c) conditions for reactive power exchange, *Source:* Based on Mathur and Varma [10].

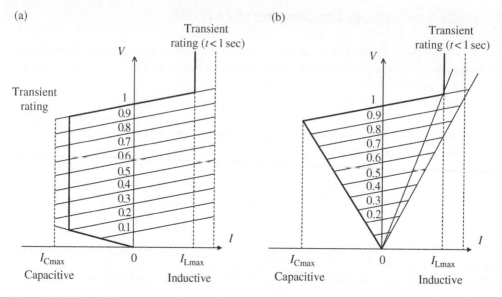

Figure 4.8 Typical Voltage versus Current characteristics of a) STATCOM and b) SVC, on a per unit basis. *Source:* Used with permission from CIGRE [7] ©1999.

- Making the angle of E_s > angle of E_t causes active power to flow from the VSC to the utility grid system.
- Making angle of E_s < angle of E_t causes active power to flow from the utility grid system in to the VSC.

This is also achieved by firing angle control of the VSC semiconductor switches through appropriate controllers.

The Voltage (V) versus Current (I) characteristics of a STATCOM and SVC are illustrated in Figure 4.8 [7].

A slope of 1–10% (typically 3%) is usually incorporated in the V–I characteristic of STATCOM similar to SVC, which provides various advantages as described earlier [10].

However, there are important distinguishing features between the V–I characteristics of STATCOM and SVC. STATCOM can provide rated capacitive current even at very low voltages, because this current is provided by VSC control. Moreover, SVC capacitive current decreases linearly with voltage, since this current is generated by a physical capacitor.

It is noted that STATCOM is typically not operated (firing pulses are blocked) below voltages of 0.2–0.3 pu. This is done to prevent loss of synchronism with the grid by the PLL. Moreover, auxiliary subsystems in STATCOM plant such as cooling systems cannot be operated at such low voltages. On the other hand, SVC is not operated below a voltage (typically 0.6 pu). This voltage level indicates a severe fault condition for which it is not desirable to switch on capacitors (TSCs). Otherwise, transient overvoltages resulting after fault clearance may get worsened if the capacitors are still connected.

For these reasons, the STATCOM is more effective than SVC for voltage control even at low voltages. STATCOM provides a faster response than SVC. This is because STATCOM is based on IGBTs (which can both be turned on and turned off by firing pulses), whereas SVC is based on thyristors (which can only be turned on by firing pulses) but turned off by its current becoming negative. The typical response time of STATCOM is one to two cycles as compared to the SVC response time of two to three cycles. STATCOMS also have a smaller footprint than SVCs since they do not utilize physical inductors and capacitors for reactive power exchange.

Control System of STATCOM

Different types of STATCOM controllers have been described in [7, 8, 14, 62, 63]. The synchronously rotating d-q axis-based control system of a STATCOM as adapted from [63] is illustrated in Figure 4.9. The STATCOM is based on a VSC which has a capacitor C_{DC} connected on its DC side. The voltage across the capacitor and the VSC input DC current are denoted by V_{DC} and I_{DC}, respectively. The VSC is connected through a coupling transformed to the AC power system at the Point of Common Coupling (PCC). The terminal voltage of VSC and PCC voltage is represented by $V_t \angle \rho + \delta$ and $V_s \angle \rho$, respectively. The inductance and resistance of the coupling transformer are denoted by L and R, respectively. R includes the on-state resistance of the VSC switches.

A PLL is utilized to provide synchronism of the VSC terminal voltage with the PCC voltage. The PLL generates the PCC bus angle ρ and the grid frequency ω to the VSC current controller [63]. The current controller produces appropriate modulation indices which are used in Pulse Width Modulation (PWM) block to generate firing pulses for the VSC switches. A harmonic filter is sometimes provided at the VSC terminals to minimize the high-frequency PWM sideband harmonics.

The PCC voltage V_S is passed through a Potential Transformer (PT) and fed to the *abc-dq* transformation block. This block generates the *dq* axis components of the PCC voltage in addition to the VSC voltage V_t and VSC current i_t. A filter is utilized to eliminate any high-frequency components in the measured PCC voltage V_S. The filtered components V_{Sdqf} are subsequently fed to the inner current controller.

The d-q axis control provides independent control of reactive and active power of the STATCOM [63]. The switching losses in the VSC switches causes the DC capacitor voltage V_{DC} to decrease gradually. The DC voltage controller regulates this DC capacitor voltage V_{DC} at its reference value V_{DCref}. The DC voltage controller outputs the active power reference P_{ref}, which is converted to the active current reference signal i_{DCref} for the current controller [63].

STATCOM is mainly used in power systems for reactive power control, voltage control, and damping control. All these functions are achieved through control of reactive current of VSC. The voltage controller, power oscillation damping controller, and subsynchronous oscillation damping controller are depicted in Figure 4.9. The magnitude of PCC voltage $|V_S|$ is obtained from the dq components V_{Sdq} and compared with the reference voltage V_{ref} at the summing junction. A slope/droop in the V–I characteristics of the STATCOM is implemented by multiplying the STATCOM current i_S by a constant K_D and added to the summing junction, similar to SVCs [10, 18]. The slope may also be implemented using STATCOM reactive power Q_S. Different power system signals such as line active power flow P_{line}, line current I_{line}, bus frequency, etc., which contain the system oscillatory modes are provided as input to the power oscillation damping controller. This controller outputs a small voltage modulation component that is added to the voltage summing junction [8, 10, 61]. Similar power system signals including generator rotor frequency ω_r are fed as input to the subsynchronous oscillation damping controller. This controller also generates a small voltage modulation component for mitigating the subsynchronous oscillations, which is added to the voltage summing junction. The error voltage V_e is fed to the voltage regulator which is typically a

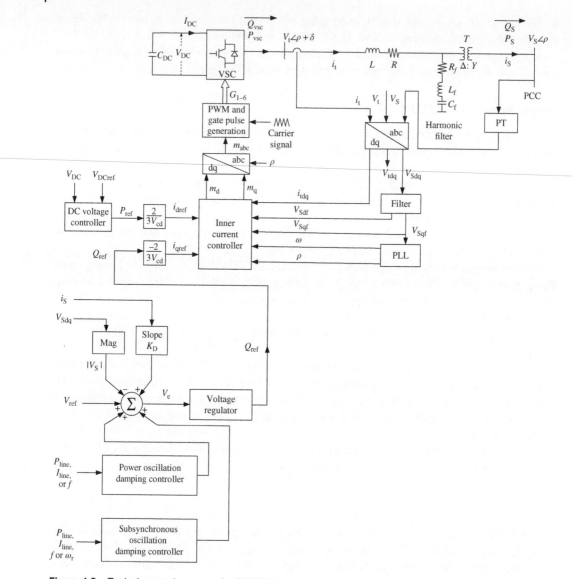

Figure 4.9 Typical control system of a STATCOM. *Source:* Based in part on Yazdani and Iravani [63].

proportional integral controller. The voltage regulator outputs the reactive power reference signal Q_{ref} which is converted to the reactive current reference i_{qref} signal and fed to the current controller of STATCOM.

The typical Low Voltage Ride-Through (LVRT) capability and Over Voltage Ride-Through (OVRT) capability curves of STATCOM are depicted in Figure 4.10a,b, respectively [18]. VSCs are inherently less resilient than TCRs to withstand system overvoltages [18]. In one installation, an SVC has been designed to withstand overvoltages up to 1.5 pu for less than 0.2 seconds [64].

(a)

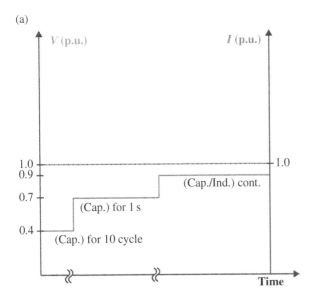

(b)

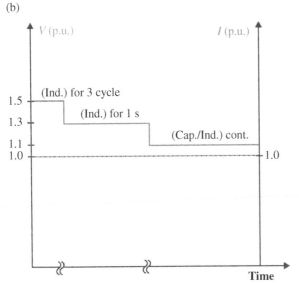

Figure 4.10 (a) Typical low voltage ride-through characteristic; (b) typical over voltage ride-through characteristic of a STATCOM. *Source:* IEEE [18].

Since VSCs are configured having IGBTs with antiparallel diodes, they tend to become rectifiers during high level of overvoltages [65]. This can potentially overcharge the DC-link to prevent which the VSC must be tripped. This problem cannot be resolved by blocking the firing pulses to the IGBTs as diodes provide the conducting path. Hence appropriate overvoltage handling capability needs to be built in the VSCs used in STATCOMs. However, VSCs can be designed to provide higher short time over currents.

4.5 Control Modes of SVC and STATCOM

SVC and STATCOM both provide the following major modes of control:

i) Dynamic voltage regulation
ii) Modulation of bus voltage in response to system oscillations
iii) Load compensation which involves regulating the power factor of a load to a desired value

These control modes allow SVCs and STATCOMs to provide several grid support functions [10]. These modes are described below.

4.5.1 Dynamic Voltage Regulation

In this mode, the SVC dynamically regulates the voltage at a bus to a reference value. This mode allows SVCs and STATCOMs to be utilized for several applications such as (i) increasing power transfer capacity in transmission lines, (ii) improvement of system transient stability, (iii) alleviation of voltage instability, (iv) improvement of HVDC converter terminal performance, and (v) grid-integration of wind/solar power generation systems. In the example below, the dynamic voltage control/regulation function is utilized for increasing the steady state stability limit, and thereby enhancing the power transfer capacity in transmission lines.

In the SMIB system of Figure 4.1, the transmission line is considered to be modeled by lumped pi-sections. The series resistance and shunt conductance of the line are ignored for simplicity of analysis. It is understood that the voltage is regulated at both ends, i.e. at the sending end by the synchronous generator and by the infinite bus at the receiving end. The voltage profiles along the line length for varying levels of power flow as compared to Surge Impedance Loading (SIL), are demonstrated in Figure 4.11. These voltage profiles also apply for any transmission line which is voltage-regulated at both ends.

For power flow less than SIL of the line [61], the reactive power generated by the line charging capacitance is more than the reactive power absorbed by the line series inductance. Therefore, voltage rises in the line with the maximum voltage rise being at the middle of the line. Net reactive power is generated which flows outwards from the line in both directions as shown in Figure 4.11. However, when power flows are more than SIL of the line, reactive power absorbed by the line series inductance becomes more than the reactive power generated by the line charging capacitance. Hence, voltage decreases in the line with the maximum decrease occurring at the line midpoint. Net reactive power is absorbed in the line which flows in from both directions as shown in Figure 4.11.

Assume that a dynamic reactive power compensator is connected at the line midpoint. For power flows less than SIL, the dynamic reactive power compensator functions like a rapidly variable inductor and absorbs reactive power to decrease the midpoint voltage to a constant reference value (1 pu in this case). When power flows are more than SIL, the dynamic reactive power compensator operates like a rapidly variable capacitor and injects reactive power to increase the midpoint voltage to the reference value. The voltage profile gets modified as illustrated in Figure 4.12. It is thus evident that dynamic reactive compensators such as SVC and STATCOM function like a variable inductor in parallel with a variable capacitor and provide voltage regulation at their interconnecting bus. The important thing to note is that the voltage is controlled to a constant reference value. This enables a substantial increase in the power transfer capability of the line as explained below.

Figure 4.11 Voltage profile across a long transmission line.

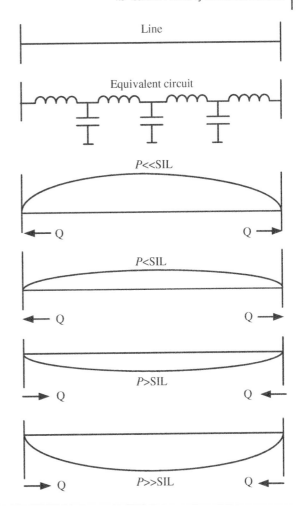

4.5.1.1 Power Transfer Without Midpoint Voltage Regulation

Consider the SMIB system depicted in Figure 4.13, although without the midpoint dynamic reactive power compensator. For ease of analysis, the line charging capacitances are ignored and the line is represented by the series line reactance X.

The power transfer P_{12} in the line is given by

$$P_{12} = \frac{V_1 V_2}{X} \sin \delta \tag{4.5}$$

If both V_1 and V_2 are assumed to be 1 pu, and the angular difference across the line is considered to be 90°, the maximum power flow $P_{12,\text{max}}$ in the line is obtained as,

$$P_{12,\text{max}} = \frac{1}{X} \tag{4.6}$$

4.5.1.2 Power Transfer with Midpoint Voltage Regulation

In Figure 4.13, consider the dynamic reactive power compensator SVC (or STATCOM) is connected at line midpoint providing voltage control at a reference value V_m. The power transfer P_{12} in the line is given by

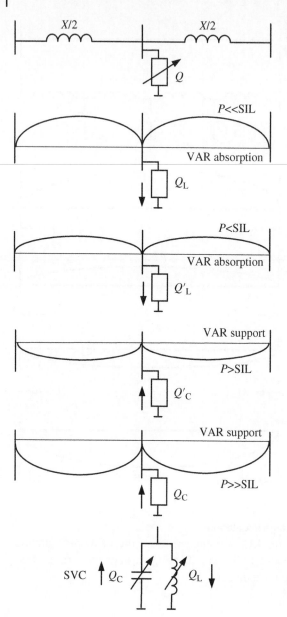

Figure 4.12 Voltage profile across a long transmission line with dynamic reactive power compensation at line midpoint.

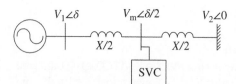

Figure 4.13 Single machine infinite bus (SMIB) system with a mid-line connected dynamic reactive power compensator SVC (or STATCOM).

$$P_{12} = \frac{V_1 V_m}{X/2} \sin \delta/2 \qquad (4.7)$$

If both V_1 and V_m are assumed to be 1 pu, and the angular difference across the line is considered to be 180°, the maximum power flow P_{12} in the line is obtained as,

$$P_{12,\,max} = \frac{2}{X} \qquad (4.8)$$

This demonstrates that the dynamic reactive power compensator SVC (or STATCOM) doubles the maximum power flow or the steady state stability limit in the line. This, however, requires that the reactive power rating of the compensator is $4P_{12,max}$ [3]. In realistic cases, the rating of the dynamic reactive power compensator is much less than $4P_{12,max}$ but still a substantial increase in the steady state stability limit or the power transfer capacity of the line, is achieved.

4.5.2 Modulation of Bus Voltage in Response to System Oscillations

In this mode of control, the dynamic reactive power compensator SVC or STATCOM modulates the bus voltage typically in the range of ± 10–15% around the reference voltage. This voltage modulation is done in response to power oscillations which include inter-area oscillations (typically, 0.1–1.0 Hz), generator inertial mode oscillations (typically, 1–3 Hz), subsynchronous torsional mode oscillations (typically, 10–30 Hz), etc. [10, 61]. The voltage modulation is performed to create an opposing effect on the oscillations leading to their rapid stabilization. SVC provides such bus voltage modulation purely by dynamic reactive power control. On the other hand, a STATCOM can provide this voltage modulation by reactive power control and also by active power control if equipped with an energy storage system. Both the modulation concepts are described with respect to the simple SMIB system shown in Figure 4.13.

Damping of Power Oscillations with Reactive Power Control

In this case, power oscillations are considered to be the synchronous generator inertial mode oscillations caused by a five-cycle three-phase-to-ground fault at generator terminals. The power oscillations comprise alternating periods of acceleration and deceleration of the generator rotor.

The acceleration period implies that the mechanical input power P_m is more than the electrical output power P_e ($= P_{12}$) of the generator. In other words, $d(\Delta\delta)/dt > 0$. This acceleration can be counteracted by performing an appropriate control by the dynamic reactive power compensator which tends to increase P_{12}. Now P_{12} is given by (4.7). The dynamic reactive power compensator appropriately increases V_m by injecting reactive power.

During the deceleration period, the electrical output power P_e ($= P_{12}$) of the generator becomes more than the mechanical input power P_m, or stated alternatively, $d(\Delta\delta)/dt < 0$. To oppose this deceleration, the dynamic reactive power compensator performs a control action to decrease the electrical power P_{12}. It accomplishes this by appropriately reducing V_m by absorbing reactive power.

It is important to note that for power oscillation damping the bus voltage is not regulated to a constant reference value but is modulated to create an opposing effect on the system oscillations. This voltage modulation is performed in response to power system signals which reflect the rotor oscillations. These signals are known as auxiliary control signals and can be both local and/or remote. The locally available signals include line current, line active power flow, bus frequency, etc. The remote signals may be signals such as speed/frequency of a remote generator [66] that are synthesized using locally measurable quantities (bus voltages and line currents). The remote signals may also include signals such as speed of a remote generator, frequency difference between two ends of a transmission line, etc. These remote signals are typically communicated distantly through fiber-optics, PMUs, WAMSs, etc. Auxiliary controllers, also known as PSDC are designed with these control signals for effective damping of the oscillatory modes [14, 61].

The signals utilized for damping of power oscillations must exhibit the following features [10]:

- High sensitivity (observability) of the control signal to power swing modes
- Substantial ability (controllability) of the control signal to control modal accelerating torque, i.e. to damp the power swing modes
- Minimum sensitivity of the signal to the output of the PSDC. In other words, the output of the dynamic reactive power compensator should not contaminate/influence the signal containing information about the oscillatory modes. For instance, line reactive power may not be an effective damping control signal since the reactive power output of SVC or STATCOM will likely influence this signal.
- The damping control signal should be effective not just for one direction of power flow but for different directions of power flow in the lines connected at the dynamic reactive power compensator bus.

The most effective damping signals may be determined from participation factor analysis [61].

4.5.3 Load Compensation

In this mode of control, the reactive power of an SVC or STATCOM is dynamically varied for power factor correction, network balancing, and load balancing [1, 2, 7, 8, 10]. The response time for load compensation is much less than a cycle. This control is generally utilized for controlling power factor of arc furnaces, steel rolling mills, wood-chippers, etc. Power factor control also helps in improving the voltage at the load bus, although in an indirect manner.

4.6 Photovoltaic-Static Synchronous Compensator (PV-STATCOM)

Solar farms are conventionally idle with their expensive assets unutilized in the nighttime. A set of innovative technologies were developed by this book's author over 2009–2020 for utilizing a PV solar farm as a dynamic reactive power compensator – STATCOM, both during nighttime and during critical system needs in the daytime [67–80]. This patented technology was named as PV-STATCOM in 2011 [69].

The nighttime and daytime "PV-STATCOM" technology was demonstrated for the first time in Canada (and perhaps in the world), on 13 December 2016, in the utility network of Bluewater Power, Sarnia, Ontario, Canada [78]. It was demonstrated that a 10 kW PV solar farm autonomously transformed into a PV-STATCOM to stabilize a critical induction motor during a severe system disturbance both during night and day.

A simple understanding of the concept of PV-STATCOM is provided in this section. Figure 4.14a depicts a basic solar PV system with PV solar panels, a VSC with an input DC capacitor, and associated PV control system. Figure 4.14b shows a basic STATCOM with a VSC having an input DC capacitor and associated STATCOM control system. It is seen that both the solar PV system and STATCOM have one critical component in common, which is the VSC. The PV-STATCOM

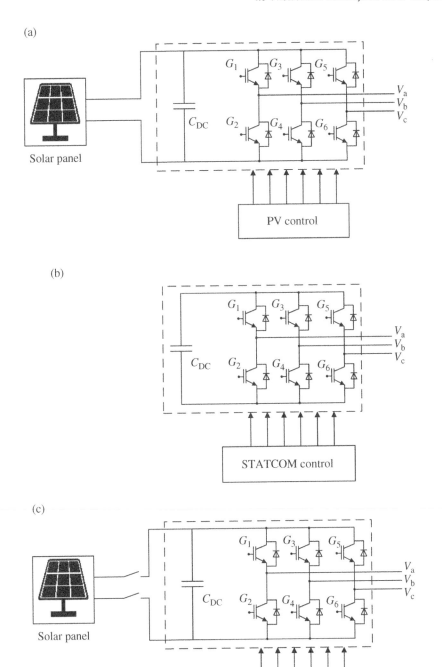

Figure 4.14 (a) A basic PV solar system, (b) A basic STATCOM, and (c) A basic PV-STATCOM.

technology is, therefore, developed as a new set of controls in a solar PV system that not only allow the full functionality of a solar PV system but also the full functionality of a STATCOM, as illustrated in Figure 4.14c.

4.7 Operating Modes of PV-STATCOM

This section describes the different operating modes of a solar PV system as a STATCOM both during night and day. The PV-STATCOM autonomously determines its mode of operation based on the system need.

Figure 4.15 depicts the active power generation P kW (or MW) on a sunny day by a solar PV inverter (or VSC) having a rating S kVA (or MVA). Figure 4.15 also illustrates the reactive power exchange capability $Q\,(=\sqrt{(S^2-P^2)})$ kvar (Mvar) of the PV inverter over a 24 hour period with the unused inverter capacity remaining after active power generation.

It is seen that the entire PV inverter capacity and the solar plant infrastructure are unutilized during nighttime (Region A). During the daytime also, there is a reasonable period of time when the entire PV inverter capacity remains unutilized (Region B). This provides the motivation for utilizing the PV inverter as a STATCOM in different operating modes, which are described below.

4.7.1 Nighttime

The entire inverter capacity (Region A) is utilized for operation as STATCOM for providing various grid support benefits to both transmission and distribution systems where the solar PV plant is connected. This mode is available during nighttime, i.e. when the active power output of solar PV system is zero. This mode is termed "Full-STATCOM" as the entire inverter capacity is used for STATCOM operation.

4.7.2 Daytime with Active Power Priority

The solar PV system is operated as a STATCOM with two different types of priorities – active power priority and reactive power priority, as described below.

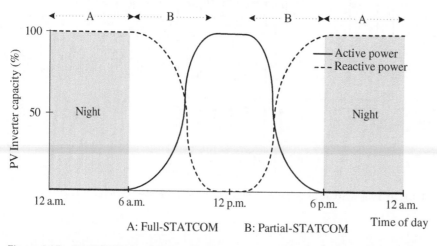

Figure 4.15 PV-STATCOM capability of PV inverter in active power priority mode.

The inverter capacity remaining after active power generation (Region B) is utilized for operation as STATCOM. There is no curtailment of active power in this mode. Since a partial inverter capacity is utilized for STATCOM operation, this mode is termed "Partial STATCOM with active power priority."

If the inverter is not oversized, the remaining inverter capacity when the solar irradiance is maximum will become zero. However, IEEE Standard 1547–2018 [81] requires that Distributed Energy Resource (DER) inverters be able to supply at least 44% reactive power even at rated active power output. Hence, the PV inverter can operate in Partial-STATCOM mode with this additional reactive power capacity, even while producing its rated active power output.

4.7.3 Daytime with Reactive Power Priority

In this operating priority, the PV-STATCOM can operate in two modes – Full STATCOM (utilizing entire inverter capacity) and Partial STATCOM (utilizing partial inverter capacity). However, active power is curtailed to make inverter capacity available for STATCOM operation. The different PV-STATCOM operating modes are described below.

4.7.3.1 Reactive Power Modulation After Full Active Power Curtailment

During any system disturbance, if the need for reactive power support is significant, the PV inverter rapidly curtails its active power to zero and makes the entire inverter capacity available for dynamic reactive power modulation as STATCOM. The PV inverter draws enough active power from the solar panels to maintain required DC-link voltage for exchanging reactive power with full inverter capacity. This is the Full STATCOM mode of operation as the entire inverter capacity is used for STATCOM operation and is illustrated in an expanded manner (Region A1) in Figure 4.16. Once the system need is fulfilled, the PV inverter rapidly restores its active power to the pre-disturbance level. This entire operation of active power reduction and restoration is typically performed in less than a minute. This mode is termed "Daytime Full STATCOM mode."

NOTE: In this case of active power curtailment, it is ensured that while the benefits of enhanced reactive power modulation capability are being realized, no adverse impact is created on the overall system frequency.

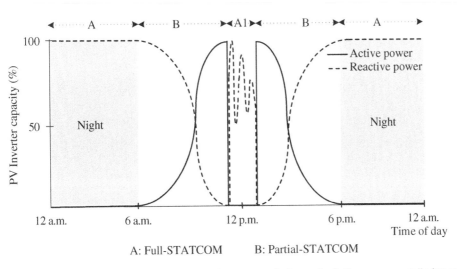

Figure 4.16 PV-STATCOM operation in reactive power priority mode. Active power curtailed to zero. Reactive power modulated with rated inverter capacity during daytime.

4.7.3.2 Reactive Power Modulation After Partial Active Power Curtailment

During a system disturbance, if the need for reactive power is moderate, the active power output of the PV inverter is rapidly curtailed from its pre-disturbance level to an intermediate level, making additional (although less than the rated) inverter capacity available for dynamic reactive power modulation as STATCOM. This intermediate level of curtailment is determined from prior off-line studies for the potential contingencies in the system. This is the Partial STATCOM mode of operation as partial inverter capacity is used for STATCOM operation and is depicted in an expanded manner (Region A2) in Figure 4.17. Once the system need is fulfilled, the PV inverter rapidly restores its active power to the pre-disturbance level. This entire operation of partial active power curtailment and restoration is typically performed in less than a minute. This mode is termed "Partial STATCOM mode with reactive power priority."

NOTE: In this case of partial active power curtailment as well, it is ensured that while the benefits of enhanced reactive power modulation capability are being realized, no adverse impact is created on the overall system frequency.

4.7.3.3 Simultaneous Active and Reactive Power Modulation After Partial Active Power Curtailment

This operation involves dynamic modulation of both active and reactive power (as STATCOM) after curtailing active power from its pre-disturbance level to an intermediate level. This operation is performed in daytime during large disturbances, when there is a critical need for grid support.

In this operating mode, the maximum available active power P_{max} based on the available solar irradiance is curtailed to an intermediate value (e.g. 0.6 P_{max}). Active power is then dynamically modulated between P_{max} and 0.6 P_{max}. Simultaneously, the reactive power is also dynamically modulated between zero and the inverter capacity remaining after active power modulation. This mode is shown in an expanded manner (Region A3) in Figure 4.18. Once the system need is fulfilled, the PV inverter rapidly restores its active power to the pre-disturbance level. This entire operation of partial active power curtailment and restoration is typically performed in less than a minute. This mode is termed "Partial STATCOM mode with simultaneous active and reactive power modulation."

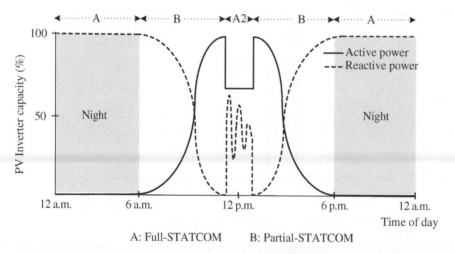

Figure 4.17 PV-STATCOM operation in reactive power priority mode: Active power curtailed to an intermediate level.

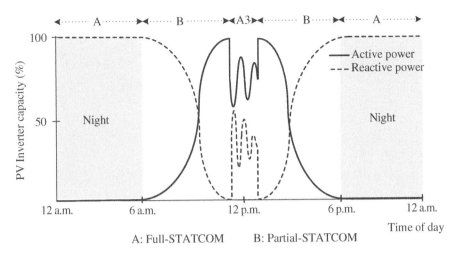

Figure 4.18 Combined modulation of active and reactive power after partial curtailment of active power.

NOTE: In the above case of active power reduction it is ensured that while the benefits of enhanced reactive power modulation capability are realized, no adverse impact is created on overall system frequency.

4.7.3.4 Simultaneous Active and Reactive Power Modulation with Pre-existing Active Power Curtailment

In this mode, the active power is kept curtailed for a substantial duration with the objective of providing frequency regulation services and enhanced reactive power support. It is known that large system disturbances that excite under/overfrequency events also simultaneously stimulate the oscillatory modes of poorly damped systems and tend to diminish system stability [82, 83]. This mode of operation can be utilized for simultaneously providing fast frequency response both during underfrequency and overfrequency events, and dynamic reactive power control, e.g. for power oscillation damping. This operating mode is portrayed in an expanded manner in Figure 4.19.

During underfrequency events, when the system loads exceed the generation, the smart PV inverter system dynamically increases its active power output and modulates it to alleviate system frequency variations. Simultaneously, reactive power is modulated with the inverter capacity remaining after active power modulation. This operation is depicted in Region C1 in Figure 4.19.

During overfrequency events, when the generation exceeds the system loads, the smart PV inverter decreases its active power output and modulates it to mitigate system frequency variations. Simultaneously, reactive power is modulated with the inverter capacity remaining after real power modulation. This operation is depicted in Region C2 in Figure 4.19. This mode is termed "Partial STATCOM mode with simultaneous active and reactive power modulation, with pre-curtailment."

4.7.4 Methodology of Modulation of Active Power

The typical relationship between the DC power generated by a PV panel and the DC voltage across the panel is illustrated in Figure 3.7. The PV panel is generally operated at its Maximum Operating Point (MPP) to produce the maximum amount of power P_{MPP} at a given solar irradiance and temperature. Maximum power is produced when the voltage across the PV panel is V_{MPP}. Increasing

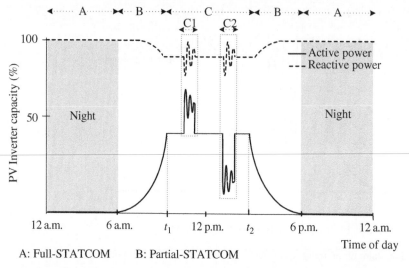

A: Full-STATCOM   B: Partial-STATCOM

C: Curtailed mode (C1: Under-frequency support; C2: Over-frequency support)

Figure 4.19 Simultaneous modulation of active power and reactive power with pre-existing power curtailment.

the voltage beyond V_{MPP} gradually decreases the power output until it becomes zero at the open-circuit voltage V_{OC}.

Active power output from the solar PV system is modulated by the DC-link voltage controllers of the PV inverters by varying the DC voltage across PV panels. Active power variation can be both slow or extremely fast (response time one to two cycles). Fast frequency response can be provided by solar PV systems by modulating the active power output very rapidly based on the needs of the power system. Active power can only be modulated downwards (decreased) from the maximum power output level. If the solar PV system is required to modulate power upwards, it is operated at an Off-MPP (non-maximum power point) leaving a headroom to increase power up to the maximum power level. Variation of PV panel voltage between V_{MPP} and V_{OC} allows the power output to be modulated continuously between maximum and zero, depending on the needs of power system.

Modulation of active power may also be done together with selective switching of some solar panels. In a large solar PV plant, the active power can also be varied by switching some inverters. Once the need for grid support is fulfilled, the PV solar system rapidly returns to its normal active power generation mode.

4.8 Functions of PV-STATCOM

4.8.1 A New Smart Inverter

Around 2011–2014, EPRI and Smart Inverter Working Group (SIWG) published several papers and reports defining smart inverters as inverters that can provide control of both active and reactive power output to provide grid support functions [84–86]. In addition, they can respond to variations in system voltage and frequency either autonomously or in accordance with utility communicated signals.

The PV-STATCOM, although developed in 2009–2013 [67–69], comes under the family of smart inverters as it also provides modulation of both active and reactive power. It responds autonomously and can also respond to utility signals.

However, the novel features of PV-STATCOM technology, which distinguish it from the currently available (in 2020) smart inverter functions described above, are as follows:

i) Nighttime grid support functions are provided, which are not defined for present smart inverters.
ii) Several STATCOM functions are provided during daytime, which are not envisaged presently for smart inverters.
iii) The dynamic reactive power exchange is very rapid (~1–2 cycles) on a continuous basis, as compared to typically 5–10 s for volt-var, volt-watt functions of smart inverters.
iv) Rated reactive current and its dynamic modulation are provided even at highly reduced voltages similar to STATCOM. This feature is presently (2020) optional in IEEE Standard 1547–2018 [81] but is available in some Grid Codes [87, 88] under LVRT requirements.

4.8.2 A New FACTS Controller

The PV-STATCOM is also a new FACTS controller [78] as it can provide:

i) dynamic voltage control with a similar response time as a STATCOM, both during night and day [73]
ii) night and day (24/7) functionality of a STATCOM for different applications as listed in Section 4.1 and described in subsequent chapters [67–72, 74–77, 79, 80, 89]

4.9 Cost of Transforming an Existing Solar PV System into PV-STATCOM

The motivation for transforming a solar PV system into a STATCOM comes from mainly three considerations:

i) Both PV system and STATCOM are based on VSCs, i.e. inverters.
ii) Both have similar electrical and civil infrastructure (substation, transformer, switchgear, buswork, protection systems, etc.).
iii) The entire PV system infrastructure is typically idle at night (as there is no active power production at night).

Therefore, if the inverter controls of an existing PV system can be modified to provide STATCOM functionalities both during night and day, the modified inverter controls together with the existing electrical/civil infrastructure can be significantly economical compared to a completely new STATCOM installation. In other words, if a utility has identified the need for installing a STATCOM to provide any of its functionalities, and if an equivalent size PV solar farm is available nearby, only an additional PV-STATCOM controller with its measurement and protection circuitry needs to be augmented to the existing solar system inverters and their controls to transform it into a STATCOM, termed PV-STATCOM.

While the above makes intuitive sense, the potential savings and the delivery of complete STATCOM functionalities need to be determined carefully. The following section provides some insights into this determination.

4.9.1 Constituents of a PV System

The different categories of items which constitute the total cost of an installed utility-scale PV system (5–100 MW) are described in [90–93]. The categories considered in the Q1 2018 U.S. benchmark costing of 5–100 MW utility-scale fixed-tilt or single-axis tracker PV system total cost (Engineering, Procurement and Construction [EPC] + developer) are listed below [91]:

1) PV module
2) Inverter (DC : AC ratio = 1.3)
3) Structural Balance of Systems (BOS), including racking, etc.
4) Electrical BOS corresponding to 1500 Vdc, including conductors, conduit and fittings, transition boxes, switchgear, panel boards, onsite transmission, etc.
5) Installation, labor, and equipment
6) EPC Overhead – Costs associated with EPC Selling, General and Administrative expense (SG&A), warehousing, shipping, and logistics (calculated as % of equipment costs, material costs except the transmission line, and labor costs)
7) Sales tax (if any)
8) Land acquisition
9) PII: Permitting + inspection + interconnection (testing, commissioning, etc.)
10) Transmission line (if any)
11) Developer overhead (includes overhead expenses such as payroll, facilities, travel, legal fees, administrative, business development, finance, and other corporate functions)
12) Contingency (3%) – estimated as markup on EPC cost
13) EPC/developer net profit (applies a percentage margin to all costs including hardware, installation labor, EPC overhead, developer overhead, etc.)

4.9.2 Costing of PV-STATCOM

This section presents the additional costs likely to be incurred on existing PV systems to transform them into PV-STATCOMs. The actual cost of different components of a PV system is taken from the Q1 2018 US benchmark for utility-scale PV systems [91]. Reasonable percentage additional costs are applied to the costs of relevant PV system constituents to calculate the additional cost of transforming the PV system into a PV-STATCOM.

The control system of PV inverters constitutes typically about 5% of the inverter cost. Furthermore, the cost of PV solar plant level controls is typically about 1% of total PV plant cost (information obtained from personal communication with PV plant developers). Since major modification is needed in the control system of the inverters and the PV plant control, a 10% additional cost is considered for incorporating the STATCOM functionality. It is important to conduct system studies a priori to demonstrate that the incorporated STATCOM control is adequate to meet the desired system performance requirements. Therefore a 10% additional cost is considered under the permitting, interconnection, and interconnection category. A 5% additional cost is considered for electrical BOS for installation of the STATCOM controls, and associated communication and protection systems. Also, a 5% additional cost is included under installation and labor charges. No enhancement of structural BOS is required for incorporating the STATCOM functionality. Five percent additional costs are included under the overheads, contingency, profits, and sales tax. It is emphasized that these percentages are intended to be conservative estimates, from author's perspective based on personal communications with some vendors and PV plant developers/operators, as there is no published source of information available. Admittedly, inverter manufacturers, EPC and System developers will be able to provide more exact percentages.

4.9.2.1 Cost of 5 Mvar PV-STATCOM

The cost of a 5 MW utility-scale single-axis PV system is presented in Table 4.1 based on Q1 2018 data provided in [91]. Additional percentage costs as described above, are incorporated in the Table to compute the projected cost of transforming the existing 5 MW PV system into an equivalent size PV-STATCOM.

It is seen that the calculated cost of 5 Mvar PV-STATCOM based on the PV system cost data in Q1-2018, utilizing the selected percentages and assumptions is US$235 000.

There is no information about the cost of 5 Mvar STATCOM available in public domain. However, based on personal communication with some STATCOM vendors, the cost of a 5 Mvar STATCOM (excluding switchgear, cables, components, and installation) is estimated to be about US$2.4 million. It is evident from the above analysis that transforming an existing 5 MW PV system into a PV-STATCOM is about 10 times more economical compared to a new installation of equivalent sized STATCOM. The factor of 10 can vary if more exact percent additional costs are applied. Still,

Table 4.1 Cost of a 5 MW utility-scale single-axis tracker PV system with estimates for transforming the PV system into PV-STATCOM.

No.	Cost categories	Cost (US $/Wdc)	Percentage cost of total project cost (%)	Cost for 5 MW PV system (US$)	Percentage additional cost of transforming into PV-STATCOM (%)	Additional cost of transforming into PV-STATCOM (US$)
1.	PV module	0.47	32.19	2 350 000	0	0
2.	Inverter	0.05	3.43	250 000	10	25 000
3.	Structural balance of systems (BOS)	0.17	11.64	850 000	0	0
4.	Electrical balance of systems (BOS)	0.17	11.64	850 000	5	42 500
5.	Installation labor and equipment	0.13	8.90	650 000	5	32 500
6.	EPC overhead	0.1	6.85	500 000	5	25 000
7.	Sales tax	0.05	3.43	250 000	5	12 500
8.	(a) Land acquisition + (b) permitting+ inspection + interconnection + (c) transmission line (if any)	0.07	4.80	350 000	10	35 000
9	Developer overhead	0.14	9.58	700 000	5	35 000
10.	Contingency (3%)	0.04	2.74	200 000	5	10 000
11.	EPC/developer net profit	0.07	4.80	350 000	5	17 500
	Total Cost	**1.46**	**100%**	**7 300 000**		**235 000**

Source: Based on Fu et al. [91].

it can be reasonably concluded that a PV-STATCOM (realized from transforming an existing PV system) is significantly cheaper than a fresh STATCOM installation.

4.9.2.2 Cost of 100 Mvar PV-STATCOM

A similar computation of a 100 Mvar PV-STATCOM based on 100 MW utility-scale single-axis tracker PV system cost data for Q1 2018 provided in [91], is demonstrated in Table 4.2. The same conservative percentage additional costs are applied as depicted in Table 4.2. It is assumed that these conservative percentages incorporate the economies of scale.

The computations in Table 4.2 are performed based on a turnkey installed cost of US$1.13/Wdc for utility-scale PV systems with single-axis tracking in Q1 2018 in USA [91]. Based on this data and the conservative percentage additional costs assumptions, the cost of transforming an existing 100 MW PV system into an equivalent size PV-STATCOM is US$3.35 Million.

Table 4.2 Cost of a 100 MW utility-scale single-axis tracker PV system with estimates for transforming the PV system into PV-STATCOM.

No.	Cost categories	Cost (US $/Wdc)	Percentage cost of total project cost (%)	Cost for 100 MW PV system (US$)	Percentage additional cost of transforming into PV-STATCOM (%)	Additional cost of transforming into PV-STATCOM (US$)
1.	PV module	0.47	41.59	47 Million	0	0
2.	Inverter	0.05	4.43	5 Million	10	500 000
3.	Structural balance of systems (BOS)	0.13	11.50	13 Million	0	130 000
4.	Electrical balance of systems (BOS)	0.08	7.08	8 Million	5	400 000
5.	Installation labor and equipment	0.10	8.85	10 Million	5	500 000
6.	EPC overhead	0.06	5.31	6 Million	5	300 000
7.	Sales tax	0.05	4.43	5 Million	5	250 000
8.	(a) Land acquisition + (b) permitting + inspection + interconnection + (c) transmission line (if any)	0.09	7.96	9 Million	10	900 000
9	Developer overhead	0.02	1.77	2 Million	5	100 000
10.	Contingency (3%)	0.03	2.65	3 Million	5	150 000
11.	EPC/developer net profit	0.05	4.43	5 Million	5	250 000
	Total cost	**1.13**	**100%**	**113 Million**		**3.35 Million**

Source: Based on Fu et al. [91].

It is noted that the turnkey cost of utility-scale single-axis tracking PV systems (US$1.13/Wdc in Q1 2018) has dropped to 0.95 US$/Wdc in Q1 2020 [94]. Similarly, the turnkey installed cost of fixed-tilt utility-scale PV systems in USA has dropped from 1.06 US$/Wdc in Q1 2018 [91] to 0.82 US$/Wdc in Q1 2020 [94].

Extrapolating the data from Table 4.2 with the US$0.95/Wdc cost of utility-scale single-axis PV systems in 2020, the additional cost for transforming a 100 MW single-axis tracking PV system to a corresponding size PV-STATCOM is estimated to be US$2.82 million. Again no information on the cost of a 100 Mvar STATCOM is available in published literature. However, in a personal communication of the author with a US utility (name withheld due to confidentiality) in June 2020, the quoted cost of ±100 Mvar STATCOM to the utility was mentioned to be US$35 million. This STATCOM also had the capability of acting as a filter for ambient harmonics.

It is thus seen from the above analysis, based on the assumed percentage additional costs, that the cost of transforming a 100 MW PV system into an equivalent size PV-STATCOM is about 12 times more economical than installation of a new STATCOM of same size. Once again it is emphasized that more exact percentages can be provided by vendors of PV inverters and PV system developers, which will yield a more accurate price of PV-STATCOM. Still, the PV-STATCOM is expected to be significantly economical compared to a fresh STATCOM installation.

4.9.3 Cost of a STATCOM

In general, the investments costs for FACTS controllers consist of the following two components [95]:

i) Equipment cost
ii) Infrastructure cost

4.9.3.1 Equipment Cost

This cost is directly related to the size of the FACTS Controller equipment being considered. This includes the cost of converters, control systems, protection systems, cooling systems, transformers, reactors, capacitors, communication systems, etc. This cost additionally depends on special requirements including:

- Redundancy needs with respect to control and protection systems
- Redundancy requirements of main components, i.e. transformers, reactors, and capacitors
- Seismic conditions
- Ambient temperature and pollution levels
- Communication needs with the substation control center or utility/regional control center.

4.9.3.2 Infrastructure Costs

These costs are dependent upon the location of substation where the FACTS Controller needs to be installed. These costs include the following:

- Acquisition of additional land if the space in substation is not adequate
- Modifications are needed in the existing substation to accommodate the FACTS Controller, i.e. new switchgear, etc.

- Building construction for indoor equipment including control systems, protection systems, thyristor, IGBT valves, auxiliaries, etc.
- Air-conditioning for indoor equipment
- Fire protection systems
- Civil works in the substation yard including foundations, grading, drainage, etc.
- Integration of the communication systems for FACTS with the existing substation communication systems

The costs of STATCOM and SVC are proprietary information of the manufacturers and are not available in public domain. However, valuable information about typical trends and costs of STATCOM and SVC around 2008 are provided in [95] based on Siemens AG database. These data are summarized in Table 4.3.

Table 4.3 shows that costs are higher for lower rating FACTS Controllers but are lower for higher rating FACTS Controllers. The costs indicated can vary from −10% to +30% due to the factors described in Sections 4.9.3.1 and 4.9.3.2, as well as due to different taxes and duties in different countries and jurisdictions.

As mentioned above, the costs of FACTS Controllers listed in Table 4.3 are obtained from a 2008 publication [95]. In this Table, the cost of 100 Mvar STATCOM is depicted as US$13 Million, whereas the quoted cost of a 100 Mvar STATCOM in 2020 (as obtained from a personal communication with a US utility) is US$35 Million. This large difference is attributed to the inflation and rise in cost of different components of STATCOM over the period 2008–2020.

An important information obtained from Table 4.3, even though it corresponds to 2008, is that the infrastructure cost of a STATCOM and of SVC is between 30% and 40% of the total project cost. The infrastructure costs in 2020 for STATCOM are about 25% (as obtained through personal communication with a utility and some consultants). This cost becomes a saving when an existing PV solar system is transformed into a STATCOM, since the infrastructure is already pre-existing.

Table 4.3 Costs of STATCOMs and SVCs.

No.	FACTS Controller	Rating (Mvar)	Equipment cost (US $/kvar)	Infrastructure cost (US $/kvar)	Ratio of infrastructure cost to total cost (US $/kvar) (%)	Total cost (US $/kvar)
1.	STATCOM	100	90	40	31	130
		200	75	40	35	115
		300	70	40	36	110
		400	60	40	40	100
2.	SVC	100	60	40	40	100
		200	50	30	38	80
		300	42	28	40	70
		400	40	20	33	60

All indicated costs are exclusive of taxes and duties.
Source: Based on Habur and O'Leary [95].

4.10 Cost of Operating a PV-STATCOM

4.10.1 Nighttime Operating Costs

The technical costs for operating a solar PV system as PV-STATCOM at night are evaluated in [96]. These costs relate to the amount of active power that needs to be purchased from grid to supply the losses in different components of the PV solar system including VSC, tie-reactor, AC filter, and the coupling transformer. The converter losses include both switching and conduction losses. The transformer losses include core losses and copper losses in the windings. Losses in the filter and air-core tie-reactor relate to the I^2R heating in their resistive components.

Simulation studies are conducted using data from a 1.5 MW solar PV farm. It is seen that the reactive power support capability of PV-STATCOM improves with the increase in DC-link voltage V_{DC}. The total losses are evaluated as a function of DC-link capacitor voltage which is varied between V_{DCmin} and V_{DCmax} to provide the required reactive power output from the solar PV system. It is ensured that the constraints on maximum VSC current, maximum VSC terminal voltage, and maximum DC current are not violated in providing the desired reactive output. In the worst-case scenario of constant $V_{DC} = 1400$ V, the total losses can become 7.7% of the rated power. It is shown that for nighttime reactive power dispatch, operation with variable V_{DC} can reduce the total losses by 8.9% as compared to constant V_{DC}. It is recommended that optimal values of DC-link voltages which minimize total losses corresponding to different reactive power outputs be stored in the PV-STATCOM control. These DC voltages may be then be used to provide the reactive power dispatch signals received from the utility.

4.10.2 Daytime Operating Costs

Technical costs have also been evaluated for operating a solar PV plant as PV-STATCOM during daytime [97]. These are the cost of additional power required to supply the difference in losses with and without reactive power support from the PV plant. The PV-STATCOM is operated with active power priority, using the inverter capacity remaining after active power generation for reactive power support. Similar component losses are included in this daytime study as in nighttime [96]. The VSC losses are the major constituent of the total losses.

The technical losses are seen to decrease with increasing solar irradiance while providing a constant level of reactive power capability. This is because the DC-link voltage increases with increasing solar irradiance to provide the maximum power output. In the worst case, the technical costs can become 5% of solar plant rating for minimum irradiance of 15 W/m^2, but decrease to less than 1% for an irradiance of 800 W/m^2 (maximum irradiance is assumed to 1000 W/m^2).

4.10.3 Additional Costs

Costs related to the operation and maintenance of the solar PV plant at night will also be incurred. In addition, consideration needs to be given to DC-link capacitor lifespan, inverter component lifespans, etc. due to extended operation at nighttime [65].

Such estimation of the technical and operating costs will be helpful for the owners of solar farms to determine the financial benefits of providing reactive power based ancillary services.

4.10.4 Technical Considerations of PV-STATCOM and STATCOM

This section describes some technical considerations in understanding the performance of PV-STATCOM and a STATCOM. It is assumed that an existing PV system is transformed into a PV-STATCOM as described in previous sections.

4.10.4.1 Number of Inverters

The inverters in a PV system transformed into a utility-scale PV-STATCOM are large in number and are located over a large geographic area. Meanwhile, the inverter modules in a STATCOM are less in number and are located in a centralized manner. In a PV system, the control signals from a centralized PV plant controller need to be distributed to a large number of inverters, thereby involving communication delays. This delay can be reduced if the PV-STATCOM control is executed at inverter level instead of plant level.

4.10.4.2 Ability to Provide Full Reactive Power at Nighttime

Some design limitations may constrain the delivery of full reactive capability at low or zero active power output. These may be due to heating losses as described above, hardware and software limitations and any other limitations imposed during the design phase of the plant. For these reasons, the amount of reactive power capability at nighttime or at zero active power output may be slightly less than 100% of inverter or PV plant rating [65].

4.10.4.3 Transient Overvoltage and Overcurrent Rating

The Category II PV system inverters compliant with IEEE 1547-2018 are required to have permissive operation capability for a 1.2 pu voltage for little more than 0.16 seconds (approx. 10 cycles), and continuous operation for voltage up to 1.1 pu [81]. The PV inverters can also potentially ride through a voltage of 1.3 pu for up to 0.16 seconds (approx. 10 cycles) [81]. On the other hand, STATCOM can withstand overvoltage of 1.5 pu for three cycles, i.e. 0.5 seconds [18]. Thus STATCOM has a somewhat higher overvoltage withstand capability than PV inverters (transformed to PV-STATCOM).

STATCOMs have a transient overcurrent rating which may be two to three times rated current for up to two seconds [98, 99]. This is made possible by the elaborate cooling systems of IGBT valves. Hence, the overcurrent capability of STATCOM is much higher than the overcurrent current capability of PV inverters, and consequently of PV-STATCOM.

The purpose of PV-STATCOM is to provide dynamic reactive power compensation. While PV-STATCOMs may not match the overcurrent capability of STATCOMs with the current state of the art of PV inverters, they can provide several FACTS functionalities where such overcurrent capability may not be required. Several such applications of FACTS functionalities provided by PV-STATCOM are described in Chapters 5 and 6. If there is a need of high overcurrent capability, then a derating of PV-STATCOM may be considered, i.e. a 10 MW solar PV system transformed to PV-STATCOM can be considered to be equivalent to 5 Mvar STATCOM with 200% overcurrent rating.

4.10.4.4 Low Voltage Ride-through

STATCOMs continue to provide rated current even at voltages as low as 0.25 pu, i.e. when enough AC power is not available to keep the DC-link capacitor charged to sufficient voltage level to meet the switching losses, and the voltage is not enough to operate auxiliary systems in the STATCOM plant. Such capability requires more robust inverter designs, control power supplies, and redundant control and protection systems. In some cases, redundant inverters are also provided.

SVCs typically do not operate if the voltage goes below typically 0.5 pu but restart their operation as the voltage goes above this limit.

Category III PV inverters are required to have mandatory operation capability for voltages as low as 0.5 pu although dynamic reactive current injection is optional [81]. The German BDEW Standard requires PV inverters to inject a minimum 2% dynamic reactive current per 1% voltage drop for voltages below 0.9 pu to support the grid during transients [88]. It can be seen that with appropriate modifications in PV inverter controls, the PV-STATCOM can provide LVRT functionalities similar to those of SVC and STATCOM.

4.10.5   Potential of PV-STATCOM

The PV-STATCOM technology developed in 2009–2013 holds promise for the following reasons:

- The PV-STATCOM can provide FACTS functionalities through dynamic reactive power compensation similar to a STATCOM or SVC.
- The advantage of PV-STATCOM is that it capitalizes the hitherto unutilized PV system infrastructure at night.
- It is significantly cheaper (about 10 times) than an equivalent size STATCOM.
- Considering that SVCs are 1.3–1.5 times cheaper than equivalent sized STATCOMs based on Table 4.3, the PV-STATCOM is also substantially more economical than an SVC.
- The time required to transform an existing solar PV system into a PV-STATCOM is significantly shorter than that needed for constructing afresh an equivalent size STATCOM, including the time required to obtain all the needed regulatory and building permissions.
- Obviating the need for fresh construction of STATCOM also prevents deforestation (hurting the environment) and eliminates associated greenhouse gas emissions (no vehicle usage for moving people, equipment, etc.)
- A STATCOM is always installed at the desired location based on system needs. However, a solar PV system that can potentially function as PV-STATCOM may not be located at the desired site. Hence it may appear that the PV-STATCOM may not be able to provide the desired functionality. With the rapid growth of solar PV systems, it is quite likely that a solar PV system may be located in the same network in the vicinity of the desired location. Studies in Chapters 5 and 6 have shown that the PV-STATCOM can still be effective in achieving the desired STATCOM performance even if it is located at a substantial distance from the required STATCOM site.
- If a utility has identified the need for FACTS functionalities, and if an appropriate size PV system is nearby, the PV system can provide the required FACTS functionalities at substantially lower cost compared to a new STATCOM or SVC installation. Of course, extensive studies need to be conducted to confirm that the PV-STATCOM of the available size can perform the required FACTS functionalities in a satisfactory manner over diverse set of operating conditions and contingencies.
- Even if the PV-STATCOM with the current technology of PV inverters and their cooling systems is not able to match the overcurrent criteria of same-size STATCOMs, a derating factor may be considered for PV-STATCOMs. For instance, a 100 Mvar PV-STATCOM can provide similar magnitude of overcurrent as a 40 Mvar STATCOM, for a relatively shorter period of time, only in cases where overcurrent requirement is critical. With the widespread growth in PV systems worldwide, a derated PV-STATCOM can still be highly advantageous in the above scenario.

- The highcost benefit of PV-STATCOM over an equivalent size STATCOM or SVC allows for investing in more advanced controls for PV-STATCOM to perform functions as close as possible to SVC and STATCOM.
- Transforming a PV system into PV-STATCOM requires some additional finances. Moreover, operating PV-STATCOM incurs more heating losses compared to simple PV system operation. However, markets are evolving around the world to provide financial compensation for dynamic reactive power support. Such support is expected to more than compensate (even over-compensate) PV-STATCOM for the additional costs incurred in transforming the solar PV system into PV-STATCOM and for its various potential STATCOM services.
- The FACTS functionalities provided by PV-STATCOM, and in dedicated manner in some cases, are expected to reduce the need for expensive FACTS Controllers such as STATCOMs and SVCs required to deliver the same functionalities for grid support. This will expectedly bring significant savings for concerned utilities and consequently a substantial revenue making opportunity for solar farm owners providing dedicated STATCOM service as PV-STATCOM.

In 2017, a report was published by National Renewable Energy Laboratory (NREL) on field demonstration of reliability services by a 300 MW solar PV plant in California [100]. Among its different conclusions, the report states that fast response by PV inverters coupled with plant-level controls make it possible to develop other advanced controls, such as STATCOM functionality, power oscillation damping controls, subsynchronous oscillations damping and mitigation, etc. One of the future plans of the project team includes demonstrating true PV STATCOM functionality during nighttime hours.

A 2018 NERC Report [65] has stated that appropriately configured PV inverters can function like a STATCOM at night. The report also describes different benefits of dynamic reactive power support from PV inverters during periods of no active power output, i.e. during nighttime for PV inverters and during periods of zero wind for wind farm inverters. The report states that reactive power support during such times may be able to decrease the transmission investments that would otherwise be required in absence of such dynamic reactive power support capabilities [65]. According to inverter manufacturers, the incremental cost to enable this capability at the solar PV plant is significantly lower than a transmission-connected dynamic reactive power resource. However, there will be some minor costs associated with supplying the additional heating losses in the inverter during zero power output conditions, added operations and maintenance costs, in addition to some degradation in lifespan of the inverter components [65].

4.11 Summary

This chapter introduces the concept of a new patented control of PV solar farms both during night and day as a dynamic reactive power compensator (STATCOM) named PV-STATCOM, developed by this book's author. At first, the controls and operating principles of two major shunt FACTS controllers – SVC and STATCOM are described. It is then shown how an existing solar PV system can be transformed into PV-STATCOM. The different operating modes of PV-STATCOM for providing STATCOM/SVC functionalities both during night and day are described. The PV-STATCOM can also provide active power modulation to augment reactive power based STATCOM functions. A detailed analysis for determining the cost of transforming existing solar PV systems of different sizes into STATCOM is presented. The additional costs of operating a solar PV system as a

STATCOM both during night and day are described. The technical considerations in the use of PV-STATCOM compared to a STATCOM are discussed.

It is shown that the PV-STATCOM is more than 10 times economical than constructing an equivalent size STATCOM afresh. Even if a derating has to be applied for meeting the overload characteristics typical of STATCOM/SVC, the PV-STATCOM is significantly cost-effective than an equivalent sized STATCOM/SVC.

The PV-STATCOM is thus shown to be a new smart inverter and also a new member of the FACTS family, which can provide the functionalities of shunt connected FACTS Controllers. The FACTS functionalities provided by PV-STATCOM, and in a dedicated manner in some cases, is expected to reduce the need for expensive FACTS Controllers such as STATCOMs and SVCs required to deliver the same functionalities for grid support. This will bring substantial financial savings for utilities and open potentially new revenue streams for solar systems by providing FACTS functionalities.

References

1 Miller, T.J.E. (1982). *Reactive Power Control in Electric Systems*. New York, NY: John Wiley & Sons Inc.

2 CIGRE (1986). Static Var Compensators. Paris, France: CIGRE *Technical Brochure 025*.

3 IEEE (1987). *Application of SVC for System Dynamic Performance*. New York, NY: *IEEE Special Publication 87TH0187-5-PWR*.

4 Andersen, B.R, and Nilsson, S., eds. (2020). *Flexible AC Transmission Systems*. CIGRE Green Book, Heidelberg, Germany: Springer International Publishing

5 IEEE PES/CIGRE Publication 95-TP108 (1995). *FACTS Overview*. New York, NY: IEEE Press.

6 IEEE PES Publication 96-TP-116-0 (1996). *FACTS Applications*. New York, NY: IEEE Press.

7 CIGRE (1999). Static Synchronous Compensator (STATCOM). CIGRE, Paris, France, *Technical Brochure No. 144*.

8 Hingorani, N.G. and Gyugyi, L. (1999). *Understanding FACTS*. Piscataway, NJ: IEEE Press.

9 Song, Y.H. and Johns, A.T. (1999). *Flexible AC Transmission Systems (FACTS)*. London: IEE Press.

10 Mathur, R.M. and Varma, R.K. (2002). *Thyristor-Based FACTS Controllers for Electrical Transmission Systems*. New York, NY: Wiley-IEEE Press.

11 Acha, E., Fuerte-Esquivel, C.R., Ambriz-Pérez, H., and Angeles-Camacho, C. (2004). *FACTS: Modelling and Simulation in Power Networks*. London: Wiley.

12 Sood, V.K. (2004). *HVDC and FACTS Controllers: Applications of Static Converters in Power Systems*. London: Springer.

13 Zhang, X.-P., Rehtanz, C., and Pal, B. (2006). *Flexible AC Transmission Systems: Modelling and Control*. Berlin: Springer.

14 Padiyar, K.R. (2007). *FACTS Controllers in Power Transmission and Distribution*. New Delhi: New Age International Publishers.

15 Sen, K.K. and Sen, M.L. (2009). *Introduction to FACTS Controllers: Theory, Modeling, and Applications*. Wiley-IEEE Press.

16 IEEE (2011). *IEEE Guide for the Functional Specification of Transmission Static Var Compensators*, 1–89. IEEE Std 1031-2011 (Revision of IEEE Std 1031-2000).

17 Varma, R.K. and Paserba, J. (2012). Flexible AC transmission systems (FACTS). In: *Electric Power Engineering Handbook on Power System Stability and Control*, vol. 3 (ed. P. Kundur). USA: CRC Press/Taylor & Francis.

18 IEEE (2019). *IEEE Guide for Specification of Transmission Static Synchronous Compensator (STATCOM) Systems*, 1–115. IEEE Std 1052-2018.

19 Edris, A., Adapa, R., Baker, M.H. et al. (1997). Proposed terms and definitions for Flexible AC Transmission System (FACTS). *IEEE Transactions on Power Delivery* 12: 1848–1853.

20 Litzenberger, W.H. (1994). *An Annotated Bibliography of High-Voltage Direct-Current Transmission and Flexible AC Transmission (FACTS) Devices, 1991–1993*. Portland, OR: Bonneville Power Administration and Western Area Power Administration.

21 Litzenberger, W.H. and Varma, R.K. (1996). *An Annotated Bibliography of High-Voltage Direct-Current Transmission and FACTS Devices, 1994–1995*. Portland, OR: Bonneville Power Administration and U.S. Department of Energy.

22 Litzenberger, W.H., Varma, R.K., and Flanagan, J.D. (1998). *An Annotated Bibliography of High-Voltage Direct-Current Transmission and FACTS Devices, 1996–1997*. Portland, OR: Electric Power Research Institute and Bonneville Power Administration.

23 Axente, I., Varma, R.K., and Litzenberger, W.H. (2011). Bibliography of FACTS: 1999 IEEE working group report. In *Proc. 2011 IEEE Power & Energy Society General Meeting*, 1–8.

24 Axente, I., Varma, R.K., and Litzenberger, W.H. (2011). Bibliography of FACTS: 2000 – Part I IEEE working group report. In *Proc. 2011 IEEE Power & Energy Society General Meeting*, 1–7.

25 Axente, I., Varma, R.K., and Litzenberger, W.H. (2011). Bibliography of FACTS: 2000 – Part II IEEE working group report. In *Proc. 2011 IEEE Power & Energy Society General Meeting*, 1–6.

26 Varma, R.K., Litzenberger, W., and Berge, J. (2007). Bibliography of FACTS: 2001 – Part I IEEE working group report. In *Proc. 2007 IEEE Power Engineering Society General Meeting*, 1–5.

27 Varma, R.K., Litzenberger, W., and Berge, J. (2007). Bibliography of FACTS: 2001 – Part II IEEE working group report. In *Proc. 2007 IEEE Power Engineering Society General Meeting*, 1–6.

28 Varma, R.K., Litzenberger, W., and Berge, J. (2007). Bibliography of FACTS: 2002 – Part I IEEE working group report. In *Proc. 2007 IEEE Power Engineering Society General Meeting*, 1–7.

29 Varma, R.K., Litzenberger, W., and Berge, J. (2007). Bibliography of FACTS: 2002 – Part II IEEE working group report. In *Proc. 2007 IEEE Power Engineering Society General Meeting*, 1–6.

30 Varma, R.K., Litzenberger, W., and Berge, J. (2007). Bibliography of FACTS: 2003 – Part I IEEE working group report. In *Proc. 2007 IEEE Power Engineering Society General Meeting*, 1–8.

31 Varma, R.K., Litzenberger, W., and Berge, J. (2007). Bibliography of FACTS: 2003 – Part II IEEE working group report. In *Proc. 2007 IEEE Power Engineering Society General Meeting*, 1–7.

32 Varma, R.K., Litzenberger, W., Auddy, S. et al. (2006). Bibliography of FACTS: 2004–2005 – Part I IEEE working group report. In *Proc. 2006 IEEE Power Engineering Society General Meeting*, 1–7.

33 Varma, R.K., Litzenberger, W., Auddy, S. et al. (2006). Bibliography of FACTS: 2004–2005 Part II IEEE working group report. In *Proc. 2006 IEEE Power Engineering Society General Meeting*, 1–8.

34 Varma, R.K., Litzenberger, W., Auddy, S. et al. (2006). Bibliography of FACTS: 2004–2005 Part III IEEE working group report. In *Proc. 2006 IEEE Power Engineering Society General Meeting*, 1–6.

35 Varma, R.K., Litzenberger, W., Ostadi, A., and Auddy, S. (2007). Bibliography of FACTS: 2005–2006 Part I IEEE working group report. In *Proc. 2007 IEEE Power Engineering Society General Meeting*, 1–8.

36 Varma, R.K., Litzenberger, W., Ostadi, A., and Auddy, S. (2007). Bibliography of FACTS: 2005–2006 Part II IEEE working group report. In *Proc. 2007 IEEE Power Engineering Society General Meeting*, 1–7.

37 Varma, R.K., Axente, I., and Litzenberger, W.H. (2010). Bibliography of FACTS: 2006–2007 Part I IEEE working group report. In *Proc. 2010 IEEE PES General Meeting*, 1–6.

38 Axente, I., Varma, R.K., and Litzenberger, W.H. (2010). Bibliography of FACTS: 2006–2007 Part II IEEE working group report. In *Proc. 2010 IEEE PES General Meeting*, 1–6.

39 Axente, I., Varma, R.K., and Litzenberger, W.H. (2010). Bibliography of FACTS: 2006–2007 Part III IEEE working group report. In *Proc. 2010 IEEE PES General Meeting*, 1–5.

40 Varma, R.K., Shah Arifur, R., and Litzenberger, W.H. (2010). Bibliography of FACTS: 2008 Part I IEEE working group report. In *Proc. 2010 IEEE PES General Meeting*, 1–7.

41 Shah Arifur, R., Varma, R.K., and Litzenberger, W.H. (2010). Bibliography of FACTS: 2008 Part II IEEE working group report. In *Proc. 2010 IEEE PES General Meeting*, 1–8.

42 Varma, R.K., Shah Arifur, R., and Litzenberger, W.H. (2010). Bibliography of FACTS: 2008 Part III IEEE working group report. In *Proc. 2010 IEEE PES General Meeting*, 1–7.

43 Varma, R.K., Berge, J., and Litzenberger, W.H. (2010). Bibliography of FACTS 2009 – Part 1 IEEE working group report. In *Proc. 2010 IEEE PES General Meeting*, 1–6.

44 Berge, J., Varma, R.K., and Litzenberger, W.H. (2010). Bibliography of FACTS 2009 – Part 2 IEEE working group report. In *Proc. 2010 IEEE PES General Meeting*, 1–6.

45 Berge, J., Varma, R.K., and Litzenberger, W.H. (2010). Bibliography of FACTS 2009 – Part 3 IEEE working group report. In *Proc. 2010 IEEE PES General Meeting*, 1–4.

46 Berge, J., Varma, R.K., and Litzenberger, W.H. (2011). Bibliography of FACTS 2009–2010: Part I IEEE working group report. In *Proc. 2011 IEEE Power & Energy Society General Meeting*, 1–8.

47 Berge, J., Varma, R.K., and Litzenberger, W.H. (2011). Bibliography of FACTS 2009/2010 – Part II IEEE working group report. In *Proc. 2011 IEEE Power & Energy Society General Meeting*, 1–6.

48 Berge, J., Rangarajan, S.S., Varma, R.K., and Litzenberger, W.H. (2011). Bibliography of FACTS 2009–2010: Part III IEEE working group report. In *Proc. 2011 IEEE Power & Energy Society General Meeting*, 1–10.

49 Berge, J., Rangarajan, S.S., Varma, R.K., and Litzenberger, W.H. (2011). Bibliography of FACTS 2009–2010: Part IV IEEE working group report. In *Proc. 2011 IEEE Power & Energy Society General Meeting*, 1–10.

50 Rahman, S.A., Varma, R.K., and Litzenberger, W.H. (2011). Bibliography of FACTS applications for grid integration of wind and PV solar power systems: 1995–2010 IEEE working group report. In *Proc. 2011 IEEE Power & Energy Society General Meeting*, 1–17.

51 Rahman, S.A., Varma, R.K., Litzenberger, W.H., and Berge, J. (2012). Bibliography of FACTS 2011: Part I IEEE working group report. In *Proc. 2012 IEEE Power & Energy Society General Meeting*, 1–6.

52 Berge, J., Varma, R.K., and Litzenberger, W.H. (2012). Bibliography of FACTS 2011: Part II IEEE working group report. In *Proc. 2012 IEEE Power & Energy Society General Meeting*, 1–6.

53 Berge, J., Varma, R.K., and Litzenberger, W.H. (2012). Bibliography of FACTS 2011: Part III IEEE working group report. In *Proc. 2012 IEEE Power & Energy Society General Meeting*, 1–5.

54 Rahman, S.A., Moharana, A., Varma, R.K., and Litzenberger, W.H. (2013). Bibliography of FACTS 2012: IEEE working group report. In *Proc. 2013 IEEE Power & Energy Society General Meeting*, 1–21.

55 Rahman, S.A., Mahendra, A.C., Varma, R.K., and Litzenberger, W.H. (2014). Bibliography of FACTS 2012–2013: IEEE working group report. In *Proc. 2014 IEEE PES General Meeting*, 1–35.

56 Rahman, S.A., Maleki, H., Mohan, S., Varma, R.K., and Litzenberger, W.H. (2015). Bibliography of FACTS 2013–2014: IEEE working group report. In *Proc. 2015 IEEE Power & Energy Society General Meeting*, 1–5.

57 Rahman, S.A., Varma, R.K., and Litzenberger, W.H. (2014). Bibliography of FACTS applications for grid integration of wind and PV solar power systems: 2010–2013 IEEE working group report. In *Proc. 2014 IEEE PES General Meeting*, 1–20.

58 Subramanian, S.B., Mohan, S., Akbari, M. et al. (2018). Control of STATCOMs – a review. In *Proc. 2018 IEEE Power & Energy Society General Meeting*, 1–5.

59 Axente, I., Varma, R.K., and Litzenberger, W.H. (2011). Bibliography of FACTS: 1998 IEEE working group report. In *Proc. 2011 IEEE Power & Energy Society General Meeting*, 1–7.

60 (2020). SVC & VSC User List 2015 REV24.xlsx. https://ewh.ieee.org/cmte/substations/sci0/wgi4/basefile.htm (accessed 6 April 2021).

61 Kundur, P. (1994). *Power System Stability and Control*. New York, NY: McGraw-Hill.

62 Schauder, C. and Mehta, H. (1993). Vector analysis and control of advanced static VAr compensators. *IEE Proceedings C – Generation, Transmission and Distribution* 140: 299–306.

63 Yazdani, A. and Iravani, R. (2010). *Voltage-Sourced Converters in Power Systems: Modeling, Control, and Applications*. Hoboken, NJ, USA: IEEE Press/Wiley.

64 (2011). SVC for increased power interchange capability between Canada and USA. ABB *Technical Report*.

65 NERC (2018). BPS-Connected Inverter-Based Resource Performance. NERC, Atlanta, GA, USA, *Reliability Guideline*.

66 Padiyar, K.R. and Varma, R.K. (1991). Damping torque analysis of static VAR system controllers. *IEEE Transactions on Power Systems* 6: 458–465.

67 Varma, R.K., Khadkikar, V., and Seethapathy, R. (2009). Nighttime application of PV solar farm as STATCOM to regulate grid voltage. *IEEE Transactions on Energy Conversion (Letters)* 24: 983–985.

68 Varma, R.K., Rahman, S.A., and Seethapathy, R. (2010). Novel control of grid connected photovoltaic (PV) solar farm for improving transient stability and transmission limits both during night and day. In *Proc. 2010 World Energy Conference*, Montreal, Canada.

69 Varma, R.K., Das, B., Axente, I., and Vanderheide, T. (2011). Optimal 24-hr utilization of a PV solar system as STATCOM (PV-STATCOM) in a distribution network. In *Proc. 2011 IEEE Power & Energy Society General Meeting*, 1–8.

70 Varma, R.K., Rahman, S.A., Mahendra, A.C., Seethapathy, R., and Vanderheide, T. (2012). Novel nighttime application of PV solar farms as STATCOM (PV-STATCOM). In *Proc. 2012 IEEE Power & Energy Society General Meeting*, 1–8.

71 Varma, R.K., Rahman, S.A., and Vanderheide, T. (2015). New control of PV solar farm as STATCOM (PV-STATCOM) for increasing grid power transmission limits during night and day. *IEEE Transactions on Power Delivery* 30: 755–763.

72 Varma, R.K., Rahman, S.A., Atodaria, V. et al. (2016). Technique for fast detection of short circuit current in PV distributed generator. *IEEE Power and Energy Technology Systems Journal* 3: 155–165.

73 Varma, R.K. and Siavashi, E.M. (2018). PV-STATCOM: A new smart inverter for voltage control in distribution systems. *IEEE Transactions on Sustainable Energy* 9: 1681–1691.

74 Varma, R.K. and Siavashi, E.M. (2019). Enhancement of solar farm connectivity with smart PV inverter PV-STATCOM. *IEEE Transactions on Sustainable Energy* 10: 1161–1171.

75 Varma, R.K. and Salehi, R. (2017). SSR Mitigation with a new control of PV solar farm as STATCOM (PV-STATCOM). *IEEE Transactions on Sustainable Energy* 8: 1473–1483.

76 Varma, R.K. and Maleki, H. (2019). PV Solar system control as STATCOM (PV-STATCOM) for power oscillation damping. *IEEE Transactions on Sustainable Energy* 10-4: 1793–1803.

77 Varma, R.K. and Akbari, M. (2020). Simultaneous fast frequency control and power oscillation damping by utilizing PV solar system as PV-STATCOM. *IEEE Transactions on Sustainable Energy* 11-1: 415–425.

78 Varma, R.K., Siavashi, E.M., Mohan, S., and Vanderheide, T. (2019). First in Canada, night and day field demonstration of a new photovoltaic solar based Flexible AC Transmission System (FACTS) device PV-STATCOM for stabilizing critical induction motor. *IEEE Access* 7: 149479–149492.

79 Varma, R.K. and Mohan, S. (2020). Mitigation of fault induced delayed voltage recovery (FIDVR) by PV-STATCOM. *IEEE Transactions on Power Systems* 35-6: 4251–4262.

80 Varma, R.K., Mohan, S., and McMichael-Dennis, J. (2020). Multi-mode control of PV-STATCOM for stabilization of remote critical induction motor. *IEEE Journal of Photovoltaics,* 10-6: 1872–1881.

81 IEEE (2018). *IEEE Standard for interconnection and interoperability of distributed energy resources with associated electric power systems interfaces.* IEEE Std 1547-2018 (Revision of IEEE Std 1547-2003).

82 Liu, Y., Gracia, J.R., Hadley, S.W., and Liu, Y. (2013). Wind/PV Generation for Frequency Regulation and Oscillation Damping in the Eastern Interconnection (EI). Oak Ridge National Laboratory, Oak Ridge, TN, USA, *Techn. Rep. ORNL/TM-2013/587.*

83 NERC (2017). Forced Oscillation Monitoring & Mitigation. NERC, Atlanta, GA, USA, *Reliability Guideline.*

84 Smith, J.W., Sunderman, W., Dugan, R., and Seal, B. (2011). Smart inverter volt/var control functions for high penetration of PV on distribution systems. In *Proc. 2011 IEEE/PES Power Systems Conference and Exposition*, 1–6.

85 EPRI (2014). Common functions for smart inverters, version 3. EPRI, Palo Alto, CA, USA, *Techn. Rep. 3002002233.*

86 SIWG (2014). *Recommendations for updating the technical requirements for inverters in distributed energy resources.* Smart Inverter Working Group, California, USA

87 (2013). *Technical guideline: generating plants connected to the medium-voltage network. (Guideline for generating plants' connection to and parallel operation with the medium-voltage network)*, BDEW (Bundesverband der Energie- und Wasserwirtschaft e.V.), Berlin, Germany (June 2008), revised 2013.

88 (2006). *Grid code for high and extra high voltage.* E.ON Netz GmbH, Bayreuth, Germany.

89 Varma, R.K. and Akbari, M. (2018). A novel reactive power based frequency control by PV-STATCOMs during day and night. In *Proc. 2018 IEEE Power & Energy Society General Meeting*, 1–5.

90 Moskowitz, S. (2018). Trends in solar technology and system prices. Wood Mackenzie (GTMresearch) *Report.*

91 Fu, R., Feldman, D., and Margolis, R. (2018). U.S. Solar Photovoltaic System Cost Benchmark: Q1 2018. National Renewable Energy Laboratory, Golden, CO, USA, *Techn. Rep. NREL/TP-6A20-72399.*

92 Bolinger, M. and Seel, J. (2018). Utility Scale Solar: Empirical Trends in Project Technology, Cost, Performance, and PPA Pricing in the United States – 2018 Edition. Lawrence Berkeley National Laboratory, Berkeley, CA, USA, *Report.*

93 Cherry, L. and Cox, M. (2019). The Global PV Inverter and MLPE Landscape 2019. Wood Mackenzie Power and Renewables, USA, *Report.*

94 Davis, M., Smith, C., White, B. et al. (2020). US solar market insight – Executive Summary Q2 2020. Wood Mackenzie and Solar Energy Industries Association, USA, *Report.*

95 Habur, K. and O'Leary, D. (2008). FACTS For Cost Effective and Reliable Transmission of Electrical Energy. Germany: Power Transmission and Distribution Group, Siemens *Report.*

96 Lourenço, L.F.N., Salles, M.B.d.C., Monaro, R.M., and Quéval, L. (2019). Technical cost of operating a photovoltaic installation as a STATCOM at nighttime. *IEEE Transactions on Sustainable Energy* 10: 75–81.

97 Lourenço, L.F.N., Salles, M.B.d.C., Monaro, R.M., and Quéval, L. (2017). Technical cost of PV-STATCOM applications. In *Proc. 2017 IEEE 6th International Conference on Renewable Energy Research and Applications (ICRERA)*, 534–538.

98 (2014). PCS100 STATCOM Dynamic Reactive Power Compensation. *ABB Technical Catalogue 2UCD180000E002 rev. C*.

99 (2014). Dynamic volt-amp reactive (D-VAR) compensation solution. *AMSC Catalog No. DVAR_DS_A4_0214*.

100 Loutan, C., Morjaria, M., Gevorgian, V. et al. (2017). Demonstration of Essential Reliability Services by a 300-MW Solar Photovoltaic Power Plant. National Renewable Energy Laboratory, Golden, CO, USA, *Techn. Rep. NREL/TP-5D00-67799*.

5

PV-STATCOM APPLICATIONS IN DISTRIBUTION SYSTEMS

This chapter presents applications of PV-STATCOM technology for providing different grid support functions related to distribution systems. These include dynamic voltage control, enhancing connectivity of photovoltaic (PV) solar farms (SFs), increasing connectivity of neighboring wind farms (WFs), and stabilization of critical motors. These are the functions for which typically static Var compensators (SVCs) or static synchronous compensators (STATCOMs) are employed, which are quite expensive. A PV-STATCOM application for reduction of line losses is also presented.

Case studies of different applications, which are extracted from the relevant publications, are described. The PV-STATCOM control system for each application is elucidated and the unique controller components relevant to the specific PV-STATCOM application are explained. The standard controller components such as the maximum power point tracking (MPPT) control, DC-link voltage controller, inner loop current controllers, phase-locked loop (PLL), *abc-dq* transformation, etc. are already explained in Chapter 3 and hence, not described here.

5.1 Nighttime Application of PV Solar Farm as STATCOM to Regulate Grid Voltage

In 2009, a novel patented control of utilizing solar PV farm as a FACTS Controller – STATCOM – was proposed to regulate the point of common coupling (PCC) voltage during night-time when the SF is not producing any active power [1]. This concept, although general, is presented for the scenario of a distribution feeder which has both PV SF and WF connected to it. During nighttime, the feeder loads are usually much lower compared to daytime, while the WFs produce more power due to increased wind speeds. This potentially causes reverse power flow from the PCC toward the main grid resulting in feeder voltages to rise above allowable limits, typically ±5% [2]. To allow further distributed generators (DG) connections, utilities need to install expensive voltage regulating devices (e.g. SVC, STATCOM, voltage regulators, etc.) [2, 3]. Voltage source converters (VSCs) (inverters) are essential components of SFs, which provide solar power conversion during daytime (normal operation). However, SFs are practically inactive during nighttime and do not produce any active power output. This study was the pioneering application of a solar PV system as STATCOM at night to regulate voltage variations at the PCC due to increased and intermittent WF power and/ or by load variations [1]. The proposed control is expected to enable increased connections of renewable energy sources, i.e. WFs in the grid at nighttime.

Smart Solar PV Inverters with Advanced Grid Support Functionalities, First Edition. Rajiv K. Varma.
© 2022 The Institute of Electrical and Electronics Engineers, Inc. Published 2022 by John Wiley & Sons, Inc.

5.1.1 Modeling of Solar PV System

Figure 5.1 shows the single-line diagram of the study system. A WF and a SF are integrated in a 12.7 kV distribution network. The WF is modeled as a fully controlled converter-inverter-based permanent magnet synchronous generator. The SF is represented by a VSC with a DC-link capacitor. The load is considered to be of passive R–L type.

5.1.2 Solar Farm Inverter Control

Figure 5.2 shows the block diagram of the control scheme used to achieve the proposed PV-STATCOM control. The controller is composed of two proportional–integral (PI) based voltage-regulation loops. One loop regulates the PCC voltage while the other maintains the DC bus voltage

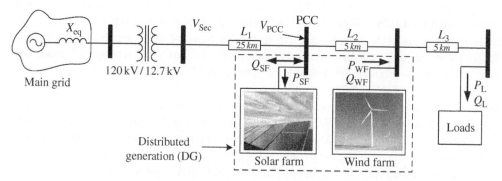

Figure 5.1 Single-line diagram of the distributed generation study system. *Source:* Varma et.al. [1].

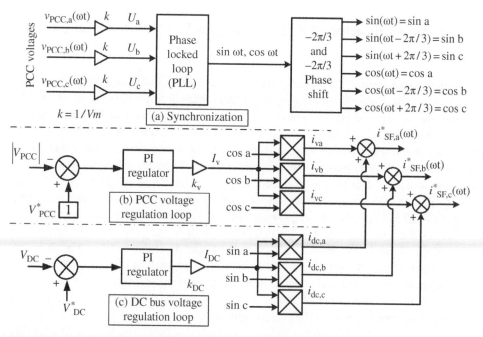

Figure 5.2 Solar farm as STATCOM – controller diagram. (a) synchronization, (b) PCC voltage regulation loop, (c) DC bus voltage regulation loop. *Source:* Varma et.al. [1].

across SF DC-link capacitor at a constant level. The PCC voltage is regulated by providing leading or lagging reactive power during bus voltage drop and rise, respectively. A PLL based control approach is used to maintain synchronization with PCC voltage [4]. A hysteresis current controller is utilized to perform switching of VSC switches. To facilitate reactive power exchange, the voltage across the DC-link capacitor of SF is regulated by drawing a small of active power from the grid to compensate for the losses in the VSC switches.

5.1.3 Simulation Study

MATLAB/Simulink simulation results for the study system during steady-state and transient conditions are depicted in Figure 5.3. All parameters are in per unit (pu) with base values of 10 MVA and 12.7 kV. Three scenarios of WF generation are considered: (i) generation less than

(a)

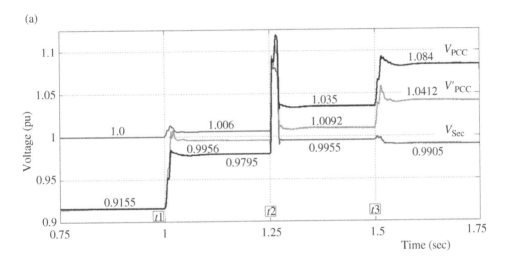

(b)

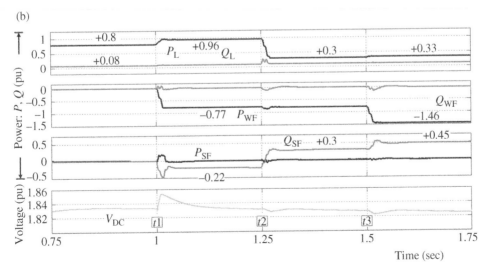

Figure 5.3 Solar farm as STATCOM – PCC voltage regulation. (a) PCC voltage profile without and with STATCOM operation. (b) Active power and reactive power of load, wind farm, solar farm; and DC bus voltage. *Source:* Varma et.al. [1].

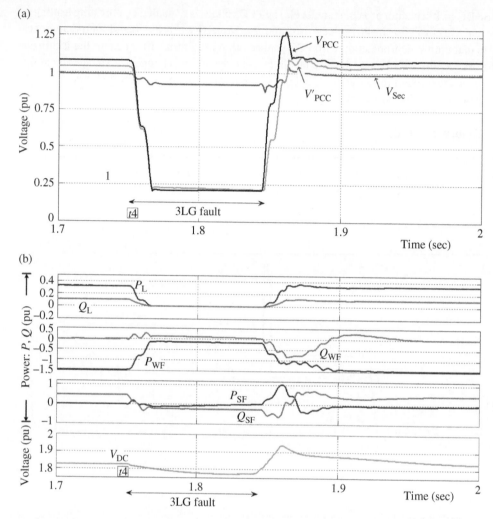

Figure 5.4 Solar farm as STATCOM – transient performance during 3LG fault. (a) PCC voltage profile without and with STATCOM operation. (b) Active power and reactive power of load, wind farm, solar farm; and DC bus voltage. *Source:* Varma et.al. [1].

load, (ii) generation approximately 2.5 times load, and (iii) generation much higher than the load ($\approx$ 5 times the load). The different time instants of interest are t_1 at which WF ($P_{WF} = 0.77$ pu) is connected to network, t_2 when load is changed from $0.8 + j0.08$ pu to $0.3 + j0.08$ pu, and t_3 when WF power generation increases to 1.46 pu. Initially, SF is not controlled as STATCOM. Figure 5.3a depicts the voltage profile of PCC voltage V_{PCC} without any dynamic reactive power compensation. A voltage drop during 1–1.25 seconds due to heavy loading and voltage rise during 1.25–1.75 seconds due to reverse power flow are observed. With increased reverse power flow, PCC voltage rises up to 8.4%, exceeding the utility specification of ±5%.

The SF inverter is now operated as STATCOM. The regulated PCC voltage V'_{PCC} is shown in Figure 5.3a. The voltage variation is only ±3% (between t_1 and t_3) and is within the utility limits. Even during significant amount of reverse power flow (after t_3), SF maintains the PCC voltage at

1.0412 pu. The distribution transformer secondary side voltage V_{Sec} also does not vary more than $\pm 1\%$. The per unit active and reactive powers of the load (P_L, Q_L), the WF (P_{WF}, Q_{WF}), and the SF (P_{SF}, Q_{SF}) are illustrated in Figure 5.3b. The reactive power exchanged between SF and the distribution network (leading/lagging), and the self-supporting DC bus voltage V_{DC} profile (Figure 5.3b) demonstrate the proposed concept of utilization of SF as STATCOM to regulate the PCC voltage.

The system performance during a three-line to ground (3LG) fault condition (for 6 cycles) is illustrated in Figure 5.4a, b. It is evident that the SF as STATCOM helps achieve improved voltage recovery subsequent to fault clearance.

5.1.4 Summary

A novel concept of optimal utilization of an idle PV SF during night-time as a STATCOM is proposed and validated through MATLAB/Simulink simulations. The SF VSC is used to regulate the distribution voltage at PCC within utility specified limits even during wide variations in WF output and loads. The proposed strategy of SF control will expectedly facilitate integration of more wind plants in the system without needing additional voltage regulating devices. This novel strategy implies operating PV solar plant as a generator during the day (providing megawatt [MW]) and ancillary services provider at night (providing Mvars). This work was published in 2009 when the grid codes did not allow SFs to control PCC voltages during daytime, and SFs typically remained idle during nighttime [1].

5.2   Increasing Wind Farm Connectivity with PV-STATCOM

High penetration of wind power and solar power-based DGs presents several challenges especially related to steady-state voltage rise and transient overvoltages (TOVs) [5, 6]. Flexible AC transmission systems (FACTS) Controllers such as SVC and STATCOM have been predominantly used to alleviate voltage issues in grid integration of Type 1 and Type 2 WFs. Even though Type 3 and Type 4 WFs have inherent capability of reactive power control, this may not be adequate in all applications and require external reactive power support from FACTS Controllers. SVCs and STATCOMs have been actually installed in utility systems for providing dynamic reactive power support for voltage control for grid integration of Type 1 [6–9], Type 3 [10–12], and Type 4 [13–15] WFs. Several other installations of FACTS Controllers for WF integration are reported in [16, 17]. It is, however, noted that FACTS Controllers are expensive.

PV solar systems are growing at an unprecedented rate and are expected to provide 35% of world's electricity by 2050 [18]. Due to such rapid growth of PV solar systems, it is quite likely that they may get installed on same or neighboring transmission/distribution lines as WFs for instance in [19–20].

This study presents a novel PV-STATCOM application for increasing the penetration level of a neighboring WF in a realistic distribution system (in Ontario). The PV-STATCOM provides control of both steady-state voltage rise and TOV due to reverse power flow caused by increasing amount of wind power generation. PSCAD software studies are performed for different PV system operating conditions, feeder loading levels, and distances between solar and WF.

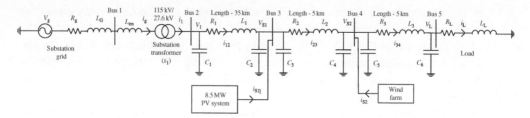

Figure 5.5 Single-line diagram of a realistic feeder in Ontario. *Source:* Based on Subramanian, Varma et al. [21].

5.2.1 Study System

The single-line diagram of the study system adapted from an actual distribution feeder of Ontario [22] is shown in Figure 5.5. It consists of a 27.6 kV, 45 km radial distribution feeder. The load on the feeder is represented by a lumped equivalent load at bus 5. The daytime peak load is 4.82 MW and 2.2 Mvar. The nighttime peak load is 1.6 MW and 0.73 Mvar. An 8.5 MW PV SF and a WF are connected at buses 3 and 4, respectively. The PV SF is controlled as a PV-STATCOM for which the operating modes are described in Section 4.7.2.

5.2.2 Control System

The PV-STATCOM consists of two subsystems, i.e. the PV power circuit and the control circuit, both of which are depicted in Figure 5.6. The PCC voltage control strategies for steady-state voltage rise mitigation and TOV control are similar to those described in [23].

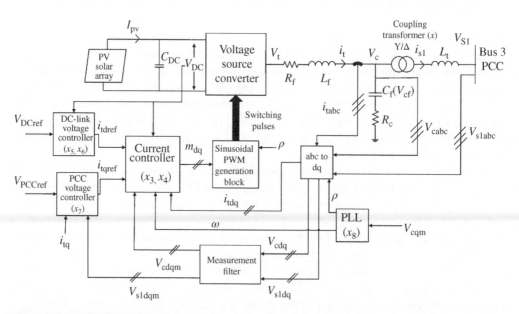

Figure 5.6 PV-STATCOM control system. *Source:* Based on Subramanian, Varma et al. [21].

5.2.3 Model of Wind Farm

The WF is modeled as a controlled current source for injecting required active power. The mechanical dynamics and the electrical machine dynamics are not considered. The controller of WF consists of PLL for grid synchronization and current controller for power regulation. The overall wind system model is based on modeling concepts provided in [24].

5.2.4 Simulation Studies

In this section, the mitigation of steady-state voltage rise and TOV by PV-STATCOM for increasing wind power penetration is studied. The simulation studies are carried out using PSCAD during night and day.

During night, the PV-STATCOM operates in Full STATCOM mode. During daytime, the PV-STATCOM operates in Partial STATCOM mode in which reactive power is provided utilizing inverter capacity remaining after active power generation. The reactive power availability, therefore, varies depending on its operating irradiance level. Studies are carried out at different irradiance levels of 0.75, 0.5, 0.25, and 0 kW/m^2. This study considers a conservative approach in which no reactive power capability is available from the solar system at its rated power output during daytime. It is, however, noted that the IEEE Standard 1547-2018 mandates a minimum reactive power availability of 44% at the rated power output of DERs [25].

5.2.4.1 Mitigation of Steady-state Voltage Rise

a) *Nighttime Studies.* Figure 5.7 shows the variation in PCC voltage of wind farm (i.e. at bus 4) with increasing wind farm capacity (i.e. power generation) without and with PV-STATCOM. Without PV-STATCOM control, the maximum capacity of wind farm that can be grid-integrated without

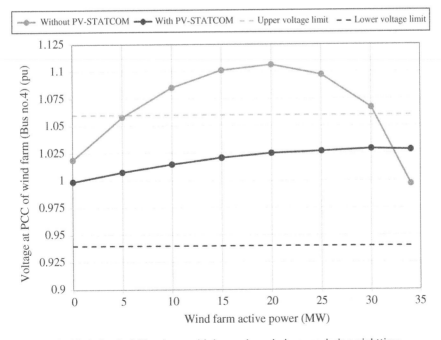

Figure 5.7 Variation in PCC voltage with increasing wind power during nighttime.

violating utility's steady-state voltage limits (0.94–1.06 pu, [26]) is 5.25 MW. With PV-STATCOM control, wind farm capacity is increased to 34 MW without violating the utility voltage limits and feeder thermal limit of 665 A.

b) *Daytime Studies.* Figure 5.8a, b illustrates the variation in PCC voltage of wind farm with increasing wind farm capacity without and with PV-STATCOM, respectively, for various irradiance levels. It is seen that the PCC voltage exceeds the steady-state voltage limit of feeder as wind farm capacity is increased without PV-STATCOM control. With PV-STATCOM operation, the PCC voltage is regulated well within the steady-state voltage limits. The wind farm capacity can be increased until the thermal limit of feeder. The maximum wind farm capacities achievable from steady-state voltage consideration without and with PV-STATCOM are compiled in Table 5.1.

(a)

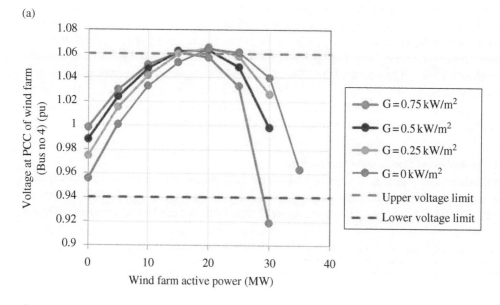

(b)

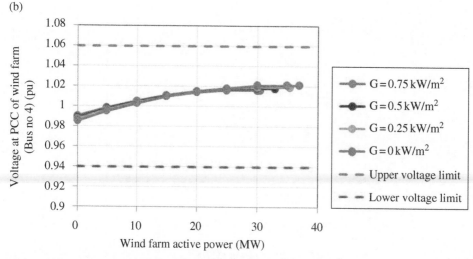

Figure 5.8 Variation in PCC voltage with increasing wind power during daytime – (a) without PV-STATCOM control and (b) with PV-STATCOM control.

Table 5.1 Increase in capacity of wind farm.

Time			Daytime					Nighttime
PV-STATCOM irradiance level (G in kW/m²)		0 kW/m²	0.25 kW/m²	0.5 kW/m²	0.75 kW/m²	1 kW/m²	0 kW/m²	
Wind farm output (MW)	Steady-state voltage control by PV-STATCOM	Without control	17.5	15.5	14	14.5	11	5.25
		With control	37	35.5	33	30.5	—	34
	Transient Overvoltage control by PV-STATCOM	Without control	11	9.5	8	9	8.5	3.25
		With control	19.5	18	16	16.5	—	12.5

5.2.4.2 Control of Transient Overvoltage

a) *Nighttime Studies.* A single line-to-ground (SLG) fault in phase C is applied at $t = 0.95$ s for 100 ms at the load end of the feeder. Without any voltage control by PV-STATCOM, the maximum capacity of WF that can be grid-integrated without violating the utility's TOV limit of 1.3 pu [26] is 3.25 MW. With TOV control by PV-STATCOM, the WF capacity is increased to 12.5 MW. Figure 5.9a, b shows the three-phase voltages at load end without and with PV-STATCOM operation, respectively for a WF capacity of 12.5 MW. As shown in Figure 5.9b, the TOV is within 1.3 pu during fault due to PV-STATCOM operation.

b) *Daytime Studies.* Similar studies are carried out and the maximum WF capacities without and with TOV control by PV-STATCOM at different irradiance levels are listed in Table 5.1.

5.2.4.3 PV-STATCOM Reactive Power Requirement

In the realistic feeder under consideration, the capacity of PV-STATCOM is 8.5 MVA. From the studies conducted, it is found that the maximum reactive power of PV-STATCOM is 6.2 Mvar and 3.51 Mvar for TOV and steady-state voltage rise mitigation, respectively, during nighttime. During daytime, the maximum reactive power of PV-STATCOM is 5.29 Mvar and 2.315 Mvar for TOV and steady-state voltage rise mitigation, respectively, for PV system irradiance level of 0.5 kW/m². Hence, no over sizing of PV-STATCOM is required for all the considered cases.

5.2.4.4 Effect of Distance of PV-STATCOM from Wind Farm

The objective of this study is to show that the available capacity of PV-STATCOM (8.5 MVA) is sufficient to mitigate steady-state voltage rise and increase WF penetration even if it is located at a larger distance from PV-STATCOM. For simplicity, studies are conducted using a simple load flow model of the system. Distance between the SF and WF is varied from 5 to 30 km and the maximum WF penetration that can be obtained with available reactive power capability of PV-STATCOM is studied. Figure 5.10 shows the relation between distance of WF from PV-STATCOM and the maximum reactive power requirement of PV-STATCOM. It is observed that the maximum reactive

(a)

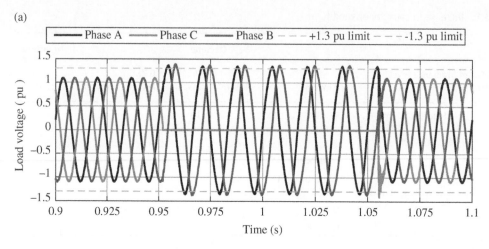

(b)

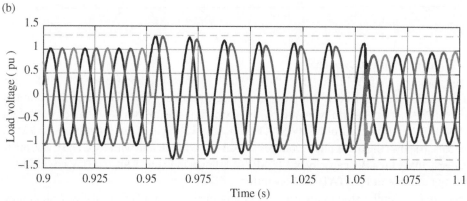

Figure 5.9 Transient overvoltage at load end during nighttime condition (a) without PV-STATCOM operation and (b) with PV-STATCOM operation.

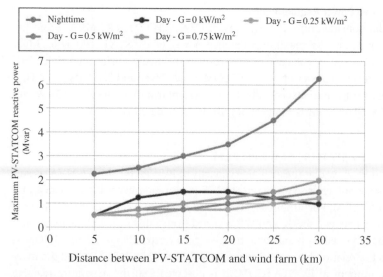

Figure 5.10 Relation between distance of wind farm from PV-STATCOM and reactive power requirement.

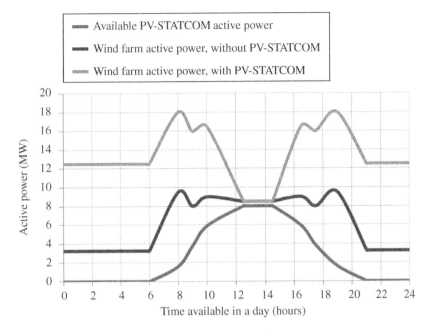

Figure 5.11 Overall increase in wind farm penetration.

power requirement is within 8.5 Mvar even as the distance increases to 30 km. This capacity is sufficient to increase WF penetration close to the thermal limit of the line.

Although this study is limited to steady-state voltage rise mitigation by PV-STATCOM, the effectiveness of PV-STATCOM control even if it is located far from wind farms, is demonstrated.

5.2.4.5 Increase in Wind Farm Connectivity

Figure 5.11 shows the overall increase in WF penetration without and with PV-STATCOM support along with the available MW capacity of PV-STATCOM for a typical solar power profile of Ontario [27]. It is seen that during nighttime, 12.5 MW WF capacity can be integrated with the support of PV-STATCOM as compared to 3.25 MW without PV-STATCOM support. This results in a net increase in WF capacity of 9.25 MW at nighttime when wind blows the most. Substantial increase in WF connectivity is also achieved during daytime. This clearly demonstrates the benefits of employing PV-STATCOM for increasing WF connectivity.

5.2.5 Summary

The application of PV-STATCOM to increase WF penetration level on a realistic Ontario distribution feeder by solving issues of steady-state voltage rise and TOV during both daytime and nighttime is demonstrated. The studies demonstrate that even if the SF is far from the WF (30 km in this case), the capability of the PV-STATCOM technology for increasing WF connectivity does not diminish. Implementation of PV-STATCOM technology on SFs in the vicinity of WFs can reduce the need for expensive SVCs or STATCOMs which are typically used for this purpose. This new PV-STATCOM control can be expected to bring significant cost savings for utilities and WF developers.

5.3 Dynamic Voltage Control by PV-STATCOM

This section describes a PV-STATCOM application in which a PV inverter is controlled as a STAT-COM for voltage control and symmetrical load compensation [28]. The PV-STATCOM is utilized to provide voltage control during critical system needs on a 24/7 basis. In the nighttime, the entire inverter capacity is utilized for STATCOM operation. During a critical system disturbance in the daytime, the smart inverter curtails its active power completely (for about a few seconds) and makes available its entire inverter capacity for STATCOM operation. Once the disturbance is cleared and the need for grid voltage control is fulfilled, the SF returns to its pre-disturbance active power production. The low-voltage ride-through (LVRT) performance of the PV-STATCOM is further demonstrated through PSCAD simulations.

5.3.1 Study System

The single-line diagram of the study system is depicted in Figure 5.12. The study system comprises a 10 kW PV solar system operating as PV-STATCOM connected through a Δ-Y isolation transformer to a 208 V_{L-L} distribution system equivalent model having impedance parameters (R_g, L_g).

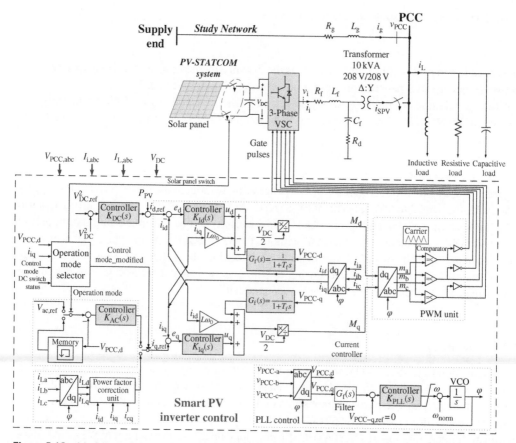

Figure 5.12 Modeling of the study system and PV-STATCOM controller components. *Source:* Varma and Siavashi [28].

A constant-impedance RLC load of total 10 kVA rating (at a nominal voltage of 208 V) is connected at the PCC. The PV system utilizes a 10 kVA two-level six-pulse IGBT-based VSC operating with a switching frequency of 10 kHz. An LCL filter is used to mitigate the harmonics generated by the inverter. In Figure 5.12, R_f represents the sum of IGBT ON-state resistance and internal resistance of filter inductor, while L_f models the filter inductance. C_f represents the filter capacitor in Delta configuration with a damping resistor R_d. C_f is chosen to limit the reactive power exchange below 0.05 pu of the inverter rating [29].

5.3.2 Control System

Figure 5.12 illustrates the control system of PV-STATCOM [28]. This is also interchangeably referred as smart PV inverter system control. The PV-STATCOM operation is based on the control modes described in Section 4.7.3.1. The controller components relevant to this application are described below:

5.3.2.1 DC-link Voltage Control

The DC-link capacitor provides active power to compensate the power loss of the inverter IGBT switches. Consequently, the DC-link capacitor voltage gets reduced gradually. The inverter needs to absorb small amount of active power to keep the DC-link capacitor charged. When sun is available, the smart inverter control utilizes a small amount of DC power from the solar panels to keep the capacitor charged, while the rest of the solar power is injected into the grid. During night, the inverter control absorbs a small amount of active power from the grid and charges the capacitor through inverter diodes.

5.3.2.2 AC Voltage Control

The AC voltage control utilizes an integral controller of the form:

$$K_{AC}(s) = \frac{K_{gain,AC}}{s} \tag{5.1}$$

5.3.2.3 Power Factor Control (PFC)

The power factor is dynamically controlled through the reactive current reference i_{qref}. The reference value is chosen to provide appropriate reactive power at the terminals to the PV inverter to achieve the desired power factor. During conventional active power generation termed as "Full PV mode", the reactive current reference i_{qref} is selected to provide unity power factor (UPF) operation.

5.3.3 Operation Mode Selector

Figure 5.13 depicts the flowchart of the PV-STATCOM control during nighttime and daytime, which is explained below. Index "M" denotes the specific operating mode.

  i) Full PV Mode: $M = 0$
 ii) Partial STATCOM Mode: $M = 1$
iii) Full STATCOM Mode: $M = 2$

Both during night and day, the PCC voltage control (VC) smart inverter function has the higher priority. Power factor correction (PFC) function is performed only if the PCC voltage is within the utility acceptable range.

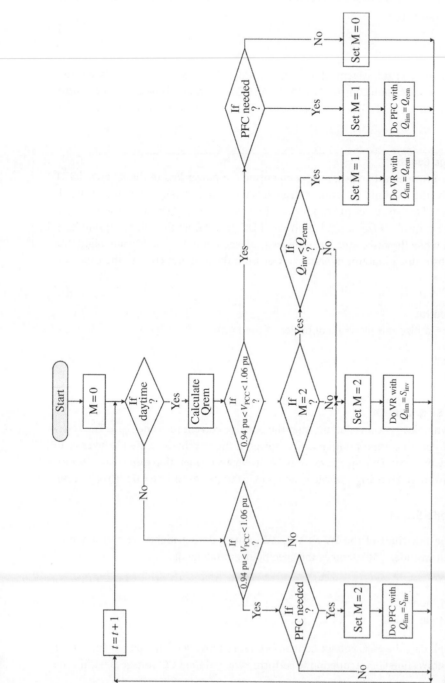

Figure 5.13 Flowchart of the PV-STATCOM operating modes. *Source:* Varma and Siavashi [28].

a) *Daytime Operation.* During daytime, Q_{rem} – the inverter capacity remaining after active power generation based on available solar insolation is computed at every time step. If at any time due to any system disturbance (e.g. fault), the bus voltage violates the utility specified limit, the operating mode is switched to Full STATCOM mode ($M = 2$). Voltage control is then performed utilizing reactive power exchange up to the full inverter capacity S_{inv}. Hence, $Q_{lim} = S_{inv}$, where Q_{lim} is the reactive power limit. If during such voltage control, the amount of needed reactive power becomes less than Q_{rem}, the operating mode is switched to Partial STATCOM mode ($M = 1$) with $Q_{lim} = Q_{rem}$. This implies that the available active power from solar insolation can still be made available to the grid while voltage regulation is being performed.

   If the voltage is successfully regulated to within the utility specified range, and if PFC is needed, it is performed in Partial STATCOM mode ($M = 1$) with reactive power up to the remaining inverter capacity ($Q_{lim} = Q_{rem}$). If PFC is not needed, the solar system reverts to Full PV mode of operation ($M = 0$).

b) *Nighttime Operation.* If any system disturbance causes the PCC voltage to violate the utility specified range, the operating mode is switched to Full-STATCOM mode ($M = 2$). Voltage control is then performed with reactive power exchange up to the entire inverter capacity ($Q_{lim} = S_{inv}$). If the voltage is successfully controlled to within the utility specified range, power factor control if needed, may be performed utilizing the entire inverter capacity for reactive power exchange ($Q_{lim} = S_{inv}$).

5.3.4 PSCAD Simulation Studies

This section presents the PSCAD software-based simulation studies of the PV system with the proposed smart inverter controls in a distribution system as depicted in Figure 5.12. System studies for the following smart PV inverter operation modes are described for different system operating conditions:

i) "Full STATCOM" mode for voltage control during both forward and reverse power flow conditions during day.

ii) "Full STATCOM" mode for voltage control during night.

In all the simulation results, the PCC voltage is denoted by V_{PCC}. Grid current and load current are represented by i_{grid} and i_{Load}, respectively. The PV system currents before and after harmonics filter are represented by $i_{inverter}$ and i_{PV}, respectively.

It is noted that a small amount of reactive power is exchanged with the grid by the harmonics filter and partly by the interface transformer of the PV system. The harmonics filter generates reactive power while the interface transformer absorbs reactive power. These reactive power components are included in the smart inverter reactive power expressed as Q_{spv}. Hence, "reactive power of the grid Q_{grid}" together with the "reactive power of the smart PV system Q_{spv}" balance the "reactive power of the load Q_{load}" at all times.

5.3.4.1 Full STATCOM Mode – Daytime

Figure 5.14a–h demonstrates the per-unit value of the PCC voltage ($V_{PCC,pu}$), the instantaneous PCC voltage (V_{PCC}), grid current (i_{grid}), smart PV system current (i_{spv}), inverter current ($i_{inverter}$), load current (i_{load}), power, respectively.

$t < 1\,s$: The smart PV system (PV-STATCOM) is not connected, and hence the active and reactive power of load and grid are respectively, equal.

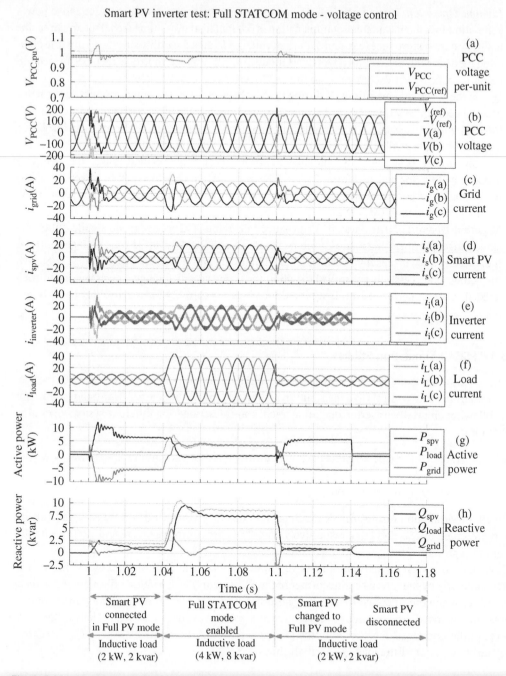

Smart PV inverter test: Full STATCOM mode - voltage control

Figure 5.14 Simulation results for Full STATCOM mode with voltage control during daytime. (a) PCC voltage (pu), (b) PCC voltage, (c) Grid current, (d) smart PV current, (e) inverter current, (f) load current, (g) active power, and (h) reactive power. *Source:* Varma and Siavashi [28].

t = 1 s: Full PV Mode enabled: The PV system is connected to the grid when it generates 6 kW active power. Initially, 2 kW active and 2 kvar reactive loads are connected to the grid. Active power injection (P_{spv}) by the PV system increases the PCC voltage ($V_{PCC,pu}$) to 0.97 pu. The PV system supplies active power of the load (P_{load}) and the surplus power flows into the grid in reverse direction. Therefore, active power of the grid (P_{grid}) becomes negative. The reactive power output of the inverter is kept zero by the controller in this Full PV mode. However, some reactive power is generated by the filter capacitor. In this mode, the 2 kvar reactive load (Q_{load}) is supplied by the grid (Q_{grid}) and the small amount of reactive power generated by the harmonics filter of the PV-STATCOM (Q_{spv}).

t = 1.04 s: Full STATCOM Mode enabled: A large reactive load of 2 kW, 6 kvar is connected to the grid. This large load reduces the voltage to 0.91 pu, which is below the acceptable voltage range of the utility. The total load becomes 4 kW active and 8 kvar reactive. The PV-STATCOM controller curtails in entire active power output in this situation and controls the PCC voltage to its pre-fault value with reactive power generation in Full STATCOM mode. The PCC voltage is successfully regulated to its pre-fault value of 0.97 pu within one cycle. Since the active power of PV-STATCOM has become zero, the entire active component of load is supplied by the grid. A large part of load reactive power is supplied by PV-STATCOM while it operates in Full STATCOM operation mode. The reactive power of the smart PV system Q_{spv} together with the reactive power of the grid Q_{grid} equal the reactive power of the load Q_{load}.

t = 1.10 s: Full PV Mode enabled: The large load is removed and the smart inverter controller returns to Full PV operation mode. In other words, the controller restores active power output of the inverter to its predisturbance value. The voltage control mode is deactivated as the PCC voltage is within acceptable range. Since active power generation of the PV-STATCOM system is more than the active power consumption of the load, the surplus power flows in reverse direction in the grid. The reactive power of inverter is zero during Full PV mode of operation, although some reactive power is generated by the filter capacitor. The reactive power of the grid Q_{grid} together with the reactive power of the smart PV system (primarily by the filter capacitor) Q_{spv} balance the reactive power of the load Q_{load}.

t = 1.14 s: PV system disconnected: The disconnection of PV-STATCOM causes the PCC voltage to drop slightly. The PV-STATCOM current and inverter current fall to zero. The grid current increases to supply active and reactive loads. Hence, the active power and reactive power of the load become equal to the active power and reactive power supplied by the grid, respectively.

5.3.4.2 Full STATCOM Mode – Nighttime

During nighttime, the inverter is operated in Full-STATCOM mode with the entire inverter capacity. The control objective is selected as PCC voltage control. The load power is kept 3 kW for this study. Figure 5.15a–f demonstrates the per-unit PCC voltage ($V_{PCC,\,pu}$); instantaneous PCC voltage (V_{PCC}); grid current (i_{grid}); PV-STATCOM current (i_s); active power of load (P_{load}), grid power (P_{grid}), PV system active power output ($P_{statcom}$); and reactive power of load (Q_{load}), grid reactive power (Q_{grid}), PV system reactive power output ($Q_{statcom}$), respectively.

t = 1 s: Full STATCOM Mode enabled: The PV system is connected to the PCC in Full STATCOM mode. The initial reference voltage is kept at 1.06 pu.

The PV-STATCOM control follows the reference value and increases PCC voltage from 1 to 1.06 pu in less than a cycle. Due to additional STATCOM current, the grid current is increased. The PV-STATCOM absorbs some active power to keep the DC-link capacitor charged. At the

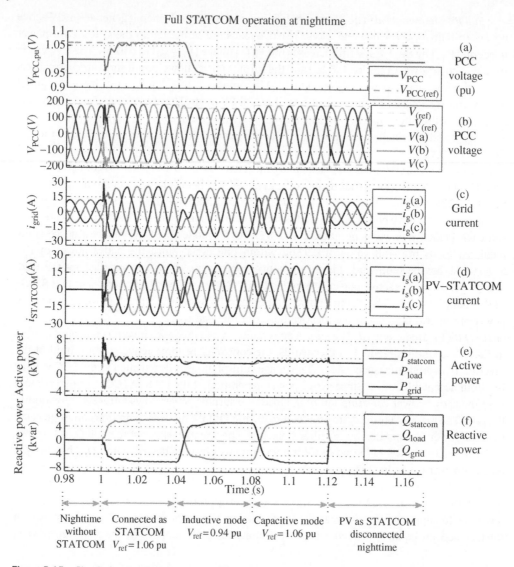

Figure 5.15 Simulation results for Full STATCOM mode with voltage control during nighttime. (a) PCC voltage (pu), (b) PCC voltage, (c) grid current, (d) PV-STATCOM current, (e) active power, and (f) reactive power. *Source:* Varma and Siavashi [28].

instant of connection of the PV-STATCOM system to the grid, the inrush current for charging capacitor creates a transient in inverter active power.

$t = 1.04\,s$: *Operation mode changed from capacitive to inductive:* The reference value of the voltage controller is changed from 1.06 to 0.94 pu. The PV-STATCOM control reduces the bus voltage and regulates it to 0.94 pu in less than one cycle. The STATCOM changes its operation mode from capacitive to inductive, with the phase of STATCOM current changing from 90° lead to 90° lag, to reduce PCC voltage.

$t = 1.08\,s$: *Operation mode changed from inductive to capacitive:* The voltage reference is changed in reverse from 0.94 to 1.06 pu. The STATCOM control changes the PCC voltage (V_{PCC}) to 1.06 pu within one cycle. This test verifies the rapid performance of PV-STATCOM.

t = 1.12 s: PV-STATCOM system disconnected: The PV-STATCOM current (i_s) goes to zero instantaneously. Since the previous operation mode of the PV-STATCOM was capacitive, the disconnection of the PV-STATCOM reduces the PCC voltage ($V_{PCC,\ pu}$), correspondingly.

5.3.4.3 Low-voltage Ride-through (LVRT)

The LVRT performance of the smart inverter PV-STATCOM while simultaneously providing dynamic reactive power compensation as STATCOM is now demonstrated. Figure 5.16 displays the LVRT performance of PV-STATCOM controller during daytime as simulated using PSCAD software. For this LVRT test, the active power output of PV system is about 8 kW and duration of large load connection is 1.5 seconds. Figure 5.16a–e demonstrates the per-unit value of PCC voltage ($V_{AC,pu}$), the instantaneous PCC voltage (V_{PVS}), smart PV system current (i_{PVS}), PV system active power P_{PVS}, PV system reactive power Q_{PVS}, and the status of the large load switch, respectively. The overall performance is described below:

t = 1 s: Full PV Mode enabled. The PV system is connected to the PCC in Full PV mode. The smart PV system generates 8 kW and a small amount of reactive output (0.5 kvar) due to filter capacitor. The initial PCC voltage increases from 0.97 to 0.99 pu because of active power generation by the PV system.

t = 1.10 s: Partial STATCOM Mode enabled. The control objective is set to voltage control with a reference voltage of 1 pu. The controller uses the remaining capacity of the inverter to regulate the PCC voltage ($V_{AC,pu}$) to the reference value 1 pu in about one cycle. While the active power output (P_{PVS}) continues to be 8 kW the reactive power (Q_{PVS}) changes to 2 kvar.

t = 1.30 s: Full STATCOM Mode enabled. The large load is suddenly connected at the PCC voltage for 1.5 seconds. Since this causes the PCC voltage to dip to 0.85 pu, the Full STATCOM mode is enabled according to the flow chart of Figure 5.13. The controller changes the reference value of DC-link voltage to open-circuit voltage of solar panel. This causes the controller to reduce the active power output to zero. In this test, however, the DC-link reference voltage is chosen to be slightly lower (430 V) than the open-circuit voltage (440 V) to ensure controller stability. Consequently, the active power reduces from 8 to 0.8 kW.
The available capacity of the inverter (which is almost equal to the full inverter capacity) is then used for reactive power generation to control the PCC voltage. The PCC voltage ($V_{AC,pu}$) is successfully regulated to 0.98 pu with the smart PV system generating about 10 kvar reactive power.

t = 2.80 s: Full PV Mode enabled: The large load is disconnected and the system returns to its predisturbance state. The controller restores the power output of the inverter (by changing the DC-link reference voltage to the MPP voltage) and the smart PV inverter starts generating 8 kW active power. Although, the inverter reactive power is zero, the PV system reactive power has a very small value due to the reactive power generated by the filter capacitor.

t = 2.90 s: PV system disconnected: The smart PV current (i_{PVS}) goes to zero, instantaneously. The disconnection of the PV system reduces the PCC voltage ($V_{AC,pu}$), correspondingly.

5.3.5 Summary

The above studies clearly demonstrate that the performance of the proposed smart inverter PV-STATCOM is similar to a STATCOM. The response time of 1–2 cycles of the PV-STATCOM matches that of an actual STATCOM. The PV-STATCOM not only meets the LVRT requirement of IEEE Standard 1547-2018 [25] but surpasses it by providing dynamic reactive power compensation as STATCOM and successfully regulating the PCC voltage to within the utility acceptable

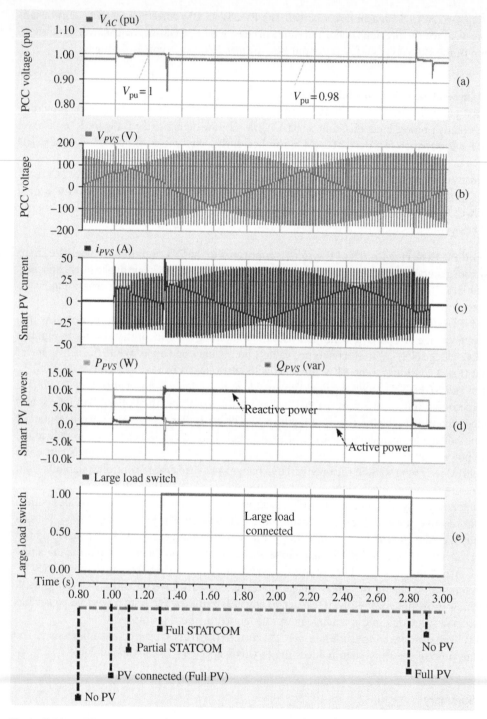

Figure 5.16 Simulation results for LVRT test with smart PV system during daytime. (a) PCC voltage (pu), (b) PCC voltage, (c) PV current, (d) PV active and reactive powers, (e) Large load switch status. *Source:* Varma and Siavashi [28].

range. The LVRT tests further demonstrate that the PV-STATCOM control system continues to remain stable despite transitioning between widely different operating modes.

5.4 Enhancement of Solar Farm Connectivity by PV-STATCOM

High penetration of solar PV systems is known to cause reverse power flows resulting in overvoltages and TOV at PCC which potentially limit any future DG installations, as described in Chapter 1. Conventional voltage control devices such as shunt capacitor banks (SCs), on-load tap changers (OLTC), and step type voltage regulators (SVRs) are slow acting with a response time ranging from seconds to few minutes. Moreover, these devices (OLTC and SVR) operate based on unidirectional flow of power and cannot operate reliably during bidirectional power flows caused by DG e.g. SFs. FACTS Controllers such as SVC and STATCOM having response times of 1–3 cycles have been utilized in several such cases for voltage regulation.

The Ontario Independent Electricity System Operator (IESO) proposed the installation of a −33/ +48 Mvar SVC to provide dynamic reactive power requirement of the 100 MW PV SF at the Grand Renewable Energy Park (GREP) in Haldimand County, Nanticoke, Ontario [30]. Four STATCOMs in the range 1–2 Mvar have been installed in Massachusetts, the United States, at various locations to mitigate feeder voltage rise and voltage fluctuations in a distribution feeder caused by variability of power output from a cluster of 13 MW PV solar systems [31]. An SVC is proposed to prevent rise in customer voltage caused by interconnection of PV solar systems in a Japanese distribution feeder [32]. Small size STATCOMs are also proposed in Japan to mitigate voltage rise issues due to surplus power injected by PV SFs in distribution feeders [33]. Studies for determining optimal size and location of SVC for improving voltage regulation at different operating conditions of residential PV solar systems connected in an Egyptian distribution feeder are described in [34]. A 6 Mvar STATCOM is installed at the substation of 52.5-MW Shams Ma'an PV solar project in Jordan to maintain a smooth voltage profile under different network conditions and variable PV output [35]. A D-STATCOM is proposed to improve voltage regulation due to variability of power output from PV solar systems in an Australian distribution network [36].

This section presents an innovative smart PV inverter control as PV-STATCOM for obviating the need for a physically connected STATCOM in a distribution network for controlling steady-state voltage and TOVs resulting from unsymmetrical faults. Two 10 MW PV solar systems are already connected in the distribution feeder of a utility in Ontario, Canada. A STATCOM is installed to prevent the steady-state voltage issues arising from the connection of a third 10 MW PV SF at the same bus. It is demonstrated below from PSCAD electromagnetic transient studies that if the proposed PV-STATCOM control is implemented on the incoming third 10 MW PV SF, both the steady-state voltage rise and TOV issues are mitigated satisfactorily as required by the utility grid code [26]. This proposed smart inverter PV-STATCOM control, therefore, eliminates the need for the physical STATCOM, saving an enormous cost for utilities dealing with voltage rise and TOV issues with grid-connected PV systems.

5.4.1 Study System

Figure 5.17 shows a 44 kV feeder in a utility distribution network in Ontario, Canada (name and location withheld for confidentiality reasons) [37]. The study feeder system includes three 10 MW PV systems with a total capacity of 30 MW connected about 35 km away from the utility transformer station (TS). The 30 MW PV plants are connected to the distribution system through a

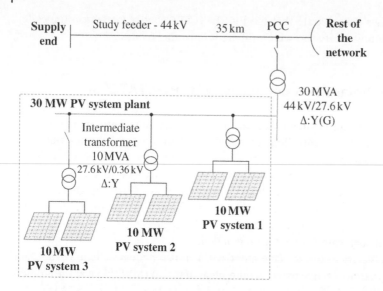

Figure 5.17 Single-line diagram of the study system. *Source:* Varma and Siavashi [23].

30 MVA interface transformer, although each 10 MW PV system uses an intermediate transformer prior the interface transformer. Two solar systems with 10 MW generation are already connected to the PCC. Installation of the additional 10 MW PV system causes increased reverse power flow during light-load conditions resulting in steady-state overvoltages. A 3.5 Mvar STATCOM is installed at PCC to mitigate these steady-state overvoltages. The TOV is also observed to exceed the permissible limits during single line-to-ground faults (SLGF) or line-to-line ground (L-L-G) fault scenarios. According to an Ontario utility requirement [26], the TOV caused by a DG facility should be less than 1.25 pu and in no circumstance exceed 1.30 pu.

5.4.2 System Modeling

The model of the study system is depicted in Figure 5.18. The system behind substation is represented as an equivalent voltage source rated 1.05 pu supplying the 44 kV feeder. The 35 km line from substation to PCC is represented by a π model in which the shunt admittance (e.g. line charging) is neglected. In Figure 5.18, R_g and L_g represent the line resistance and inductance, respectively. The electrical load is considered to be a constant-power static RL load. At nominal voltage of $44 \text{ kV}_{\text{L-L}}$, the total load is considered to be 30 MVA. The peak-time active and reactive loads are considered to be 27 MW and 6 Mvar, respectively, whereas during off-peak hours, these loads are 6 MW and 1.5 Mvar, respectively. All three PV systems are utilized with 10 MVA two-level six-pulse IGBT-based voltage source inverters (VSI). This study proposes that the additional (third) 10 MW PV system be equipped with the PV-STATCOM control. The PV-STATCOM operation is based on the operating modes described in Section 4.7.3.1. The other two PV systems use only conventional controllers to generate active power at UPF.

5.4.3 Control System

During daytime, the smart PV control operates as a conventional PV system, i.e. in Full PV mode. If steady-state voltage control is required in all three phases, together with active power generation,

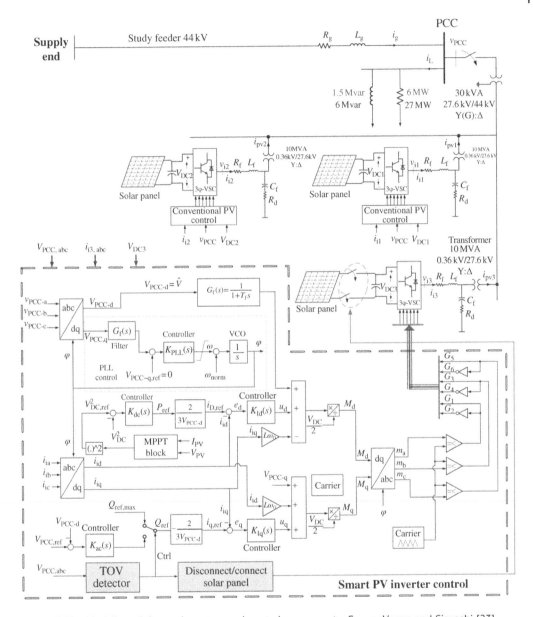

Figure 5.18 Modeling of the study system and control components. *Source:* Varma and Siavashi [23].

Partial STATCOM mode is activated. The Full STATCOM mode is activated when a TOV occurs due to unsymmetrical faults. MPPT based on incremental conductance method [38] is utilized during Full PV mode and Partial STATCOM mode. In Full-STATCOM mode, the MPPT mode is disabled and the active power generation is made zero by making the voltage across PV panels equal to their open-circuit voltage. The entire inverter capacity is then utilized to absorb reactive power in order to reduce the phase voltage. After the TOV is mitigated, power production from the solar panels is restored to the predisturbance level and the control mode is switched to Partial STATCOM mode. The controller components relevant to the PV-STATCOM control are described below [23]. The other standard controller components are described in Chapter 3.

The PCC voltage is controlled by the AC voltage controller. The reference value of reactive current control loop is either dictated by the output of the AC voltage controller or selected as the maximum reactive current (i.e. in Full-STATCOM mode). The current controller in q-axis regulates the reactive current to its reference value. It is noted that the TOV detector unit switches between voltage control mode and TOV mitigation mode. Also, this unit generates the command to enable or disable the power production from PV solar panels.

5.4.3.1 Operation Mode Selector

Figure 5.19 shows the flowchart of the smart PV inverter PV-STATCOM control for selecting the operation mode. During daytime, the voltages in three phases are measured. If any phase voltage exceeds the TOV limit while the voltages in other phase/phases decrease substantially, the output of TOV detector unit is trigged "ON", and Full STATCOM mode is activated. The controller keeps absorbing reactive power to reduce TOV until the phase voltages reach an acceptable value.

After the fault is cleared all the phase voltages will return to their normal values. The controller thus recognizes that TOV is mitigated. It subsequently restores power generation from the solar panels and switches to Partial STATCOM mode for steady-state voltage control. In Partial

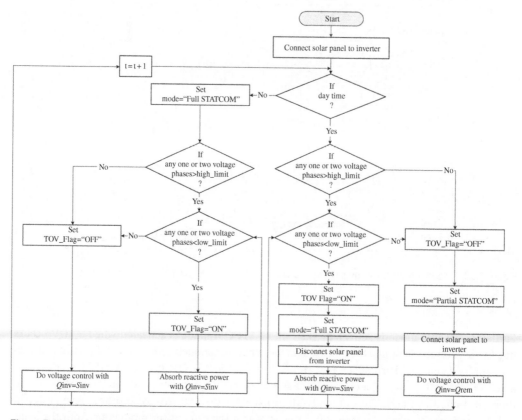

Figure 5.19 Flowchart of the smart PV inverter PV-STATCOM operating modes. *Source:* Varma and Siavashi [23].

STATCOM mode, the controller regulates the PCC voltage with Q_{rem}, which is the inverter capacity remaining after active power generation.

During nighttime, the PV solar system operates in Full STATCOM mode to control either the steady-state voltage or TOV. The smart PV inverter control thus autonomously determines its operation mode and prioritizes between active power generation and reactive power exchange based on the system requirements, nature of transient/disturbance, time of the day, and remaining inverter capacity.

5.4.3.2 PCC Voltage Control

In voltage control mode, the AC voltage controller defines the reactive current reference for current controller. However, in TOV reduction mode the reference value for reactive current controller is set to the maximum value as determined from the inverter kVA/MVA rating. This is done to reduce the TOV in the most effective manner with largest possible inductive current.

5.4.3.3 TOV Detection Block

The TOV Detection block triggers the TOV flag when a TOV occurs. When TOV flag is "ON", the controller transforms the PV system into Full STATCOM mode by changing the DC voltage to the open-circuit voltage level of the solar panels thereby reducing their power output to zero. Reactive power is then absorbed using the full inverter capacity. When the fault is cleared and PCC voltage has reached an acceptable value, the TOV flag is triggered "OFF" and the smart PV controller transforms into the Partial STATCOM mode for voltage control.

Figure 5.20 depicts the structure of the TOV Detection block. This block includes three different sections, voltage rise detection (VRD), voltage fall detection (VFD), and fault detection (FD) units. Each unit uses RMS blocks to obtain the RMS value of each phase and then converts it to its per-unit value. The VRD unit compares the per-unit voltage of each phase with its hysteresis band limits. If one or two of the phase voltages exceed the high band limit, the output of VRD unit is set to "1" otherwise the output is "0." Based on the Ontario grid connection requirements [26], the low and high band limits of the hysteresis block for VRD unit are chosen as 1.20 and 1.25 pu, respectively. The VFD unit detects the fall in voltage. The output of VFD is triggered to "1" if one or two-phase voltages pass the lower limit of the hysteresis band. The low and high bands of VFD unit are selected to be 0.8 and 0.85 pu, respectively. When the output of both VRD and VFD blocks is "1" the TOV detection block recognizes the TOV event and triggers the output to "1." In other words, the TOV event is detected when one or two-phase voltages are larger than 1.25 pu and voltages of other phase/phases are below 0.8 pu.

After the fault is cleared all phase voltages will be above a certain value which is chosen to be 0.85 pu. The FD unit triggers the TOV flag to become "0" when fault is cleared. The look-up table in each case operates such that it provides a True "1" signal if the desired condition is met only in either one or two phases, but will give a False "0" output if the desired condition is met on all or none of the three phases. The hysteresis limits for different units are as below:

$$
\begin{cases}
\text{VRD unit} \ \rightarrow \ H_1 = 1.20 \, \text{pu}, \ H_2 = 1.25 \, \text{pu} \\
\text{VFD unit} \ \rightarrow \ L_1 = 0.80 \, \text{pu}, \ L_2 = 0.85 \, \text{pu} \\
\text{FD unit} \ \rightarrow \ L_3 = 0.50 \, \text{pu}, \ L_4 = 0.55 \, \text{pu}
\end{cases}
\tag{5.2}
$$

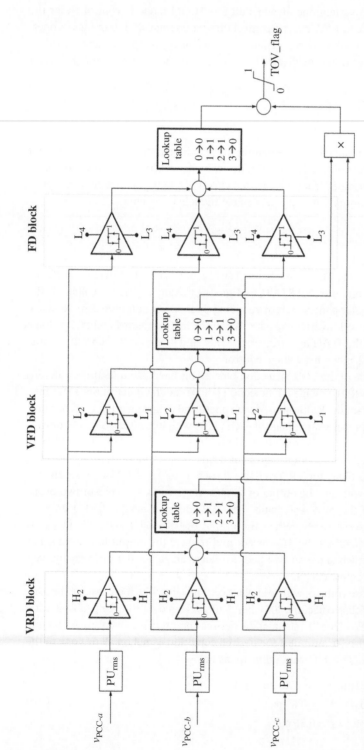

Figure 5.20 Structure of TOV detection block. *Source:* Varma and Siavashi [23].

5.4.4 Simulation Studies

The performance of the smart PV inverter PV-STATCOM while fulfilling two control objectives, voltage control and TOV reduction, are presented in this section. The PSCAD software is used for the simulation of the study system shown in Figure 5.17. In all these studies, light (small) load is defined as 6 MW and 2 Mvar, whereas a heavy (large) load is considered to be 27 MW and 9 Mvar. All the three SFs are considered to be producing 7 MW each during light-load conditions, i.e. a total of 21 MW power.

5.4.4.1 Conventional PV System (Without PV-STATCOM Control)

In this study, the incoming third 10 MW PV solar system does not have a smart PV inverter control. It operates as a conventional PV solar system with active power generation at unity power. Figure 5.21 illustrates the PCC voltages in the three phases and their per-unit RMS values when PV systems generate their rated power during small load conditions at the initiation of a SLGF.

Before the connection of PV systems, the PCC voltage is about 1.04 pu which is acceptable according to [26]. However, after the connection of the PV systems, the voltage rises to 1.10 pu which is unacceptable [26].

At $t = 0.56\,s$ a SLG fault is initiated on phase "A." This causes the voltage of phase "A" to fall to zero whereas the voltages in the other two phases reach 1.35 pu during the fault. This TOV during SLG fault is beyond the utility specified limit of 1.25 pu [26]. Therefore, there is a need to control both the steady-state overvoltage and the TOV. Additional studies reveal that during large load condition the steady-state voltage is 1.01 pu whereas the TOV is 1.23 pu during SLG fault, both of which are within utility specified limits. These studies demonstrate that there is no need for either voltage regulation or TOV reduction during heavy loading condition.

5.4.4.2 PV-STATCOM and Two Conventional Solar PV Systems

In this study, instead of using an external STATCOM, the incoming third PV system is equipped with the proposed smart PV inverter PV-STATCOM controller, while the other two PV systems operate as conventional PV systems. The proposed PV-STATCOM controller regulates the PCC

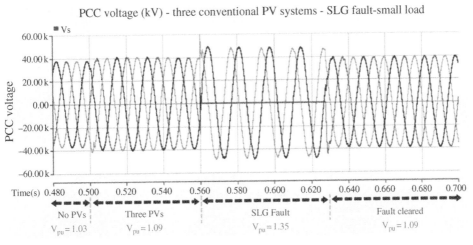

Figure 5.21 Performance of three conventional PV systems during small load and single-line-to-ground (SLG) fault. *Source:* Varma and Siavashi [23].

voltage in steady state with the remaining capacity of the inverter and also transforms the PV system operation to the Full STATCOM mode during a TOV event. Two different faults, SLG fault and line-to-line-ground (LLG) fault, are considered to demonstrate the performance of the proposed controller.

(a) Single Line-to-Ground (SLG) Fault

Figure 5.22a–h demonstrates the per-unit value of the PCC voltage ($V_{\mathrm{PCC,pu}}$), the three-phase instantaneous PCC voltage (V_{PCC}), smart PV system current (i_s), output powers (P_{PV}, Q_{PV}), reactive current (I_q), active current (I_d), DC-link voltage (V_{DC}), angular frequency, PLL angle output, and TOV flag status, respectively.

$t < 0.5\,s$: The smart PV system is not connected, and hence the active and reactive power of PV system are both zero.

$t = 0.5\,s$: *Three conventional PV system connected.* Due to 21 MW active power generation of PV systems, the PCC voltage increases from 1.04 to 1.10 pu which is unacceptable. The DC-link voltage is controlled at its reference value by controlling the active current output. The reactive power is controlled at zero.

$t = 0.54\,s$: *Partial STATCOM mode enabled.* This operating mode for voltage control reduces the voltage to an acceptable range in less than one cycle utilizing the remaining capacity of the inverter. The reactive power output of the inverter reaches 7 Mvar capacitive from zero to maintain the voltage at acceptable range. However, the active power remains unchanged since the reactive power control utilizes the remaining capacity of the inverter. The active and reactive currents follow their reference values to satisfy the voltage control objective.

$t = 0.58\,s$: *SLG fault initiated.* The SLG fault causes the voltage of phase "A" to fall to zero, whereas the other phase voltages experience TOV. The TOV detection unit detects this TOV event and triggers the TOV flag. Hence, the smart inverter autonomously switches from Partial STATCOM mode for voltage control to Full STATCOM mode for TOV reduction. In this situation, the controller changes the DC-link voltage of the inverter to a value equal to the open-circuit voltage of the solar panels (Figure 5.22g). It subsequently absorbs reactive power with full capacity of the inverter to reduce the voltages of phase "B" and phase "C."

On comparing with Figure 5.22a, it is revealed that the proposed smart PV inverter control reduces the PCC voltage from 1.35 to 1.23 pu which is below the TOV limit specified in [29]. Figure 5.22d illustrates that active power output reaches zero and the entire capacity of the inverter is used for reactive power absorption.

$t = 0.63\,s$: *SLG fault cleared.* The fault is cleared and the controller returns to Partial STATCOM mode for voltage control while generating active power. The PCC voltage is controlled to an acceptable level (1.03 pu) by using the remaining capacity of the inverter.

(b) Line-to-Line-to-Ground (LLG) Fault

The performance of the proposed smart inverter controller during an LLG fault is demonstrated in Figure 5.23. As in the previous case of SLG fault, the smart PV inverter controls the PCC voltage to its reference value during steady state.

$t = 0.58\,s$: *LLG fault initiated.* The voltages of two phases phase "A" and phase "B" fall to zero and a TOV is caused in phase "C" due to LLG fault. The TOV detection unit triggers the TOV flag and the controller changes its mode from Partial STATCOM mode for voltage control to Full

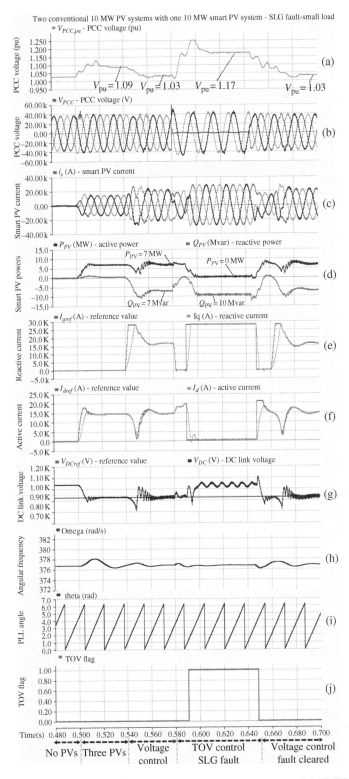

Figure 5.22 Performance of the third 10 MW PV system as PV-STATCOM, together with two conventional PV systems, during small load and SLG fault, (a) PCC voltage (pu), (b) PCC voltage, (c) smart PV current, (d) active and reactive powers, (e) reactive current, (f) active current, (g) DC-link voltage, (h) angular frequency of PCC voltage, (i) angle output of PLL, and (j) TOV flag status. *Source:* Varma and Siavashi [23].

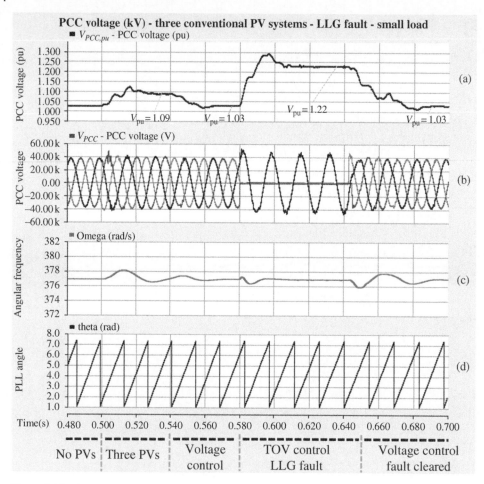

Figure 5.23 Performance of one PV system with proposed smart inverter control, together with two conventional PV systems, during small load and LLG fault (a) PCC voltage (pu), (b) instantaneous PCC voltage, (c) angular frequency of PCC voltage, and (d) angle output of PLL. *Source:* Varma and Siavashi [23].

STATCOM mode for TOV reduction. This smart inverter control effectively reduces TOV in the healthy phase to an acceptable value of 1.22 pu in about a cycle. It is noted that the designed PLL performs in a stable manner both during steady state and during SLG and LLG faults.

The proposed smart inverter PV-STATCOM provides TOV mitigation using the full inverter capacity both during night and anytime during the day as needed by the grid. It also provides steady-state voltage control throughout the night. It is assumed in this study that light-load conditions do not arise during full noon hours, i.e. when the solar system is producing its rated power. Hence there is some remaining inverter capacity available for steady-state voltage regulation in Partial STATCOM mode. It is noted that according to IEEE Standard 1547-2018, the PV inverters are also expected to provide reactive power injection and absorption capability of 44% of nameplate apparent power rating at their rated power output [25]. This capability will provide additional room for steady-state voltage control in the Partial PV-STATCOM mode even when the PV inverters are generating their rated power output. In the worst case, if light-load conditions do arise due to a disturbance during noon hours and reactive power requirement exceeds 44% of the nameplate rating, the SF can switch to Full PV-STATCOM mode for voltage control as long as the disturbance lasts. In other words, the proposed control offers the full functionality of a STATCOM for voltage control both during day or night.

5.4.5 Summary

This study presents an innovative PV-STATCOM control, for reduction of steady-state overvoltage and more importantly, mitigation of TOVs. The PV-STATCOM in Partial STATCOM mode regulates the steady-state overvoltage to the desired reference value within one and half cycle. Further, in Full STATCOM mode, it successfully reduces the TOV caused during both single-line-to-ground fault and line-to-line-to-ground fault to within utility acceptable values, within one cycle. The PV-STATCOM thus provides the full function of a STATCOM for voltage control on a 24/7 basis.

For the actual distribution system studied, the proposed PV-STATCOM control can help integrate the third 10 MW PV SF thereby eliminating the need for the actually installed STATCOM for the same purpose. The proposed PV-STATCOM control is expected to be at least 10 times cheaper than a conventional STATCOM. This control can, therefore, bring a significant saving for the concerned utility. Such a control can also help in increasing the hosting capacity of PV SFs on distribution systems which may be restricted due to voltage issues.

5.5 Reduction of Line Losses by PV-STATCOM

Transmission and distribution networks experience I^2R heating losses due to the flow of current I through the resistance R of the lines. In North American systems, the line losses are typically 3–4% of the total power delivered, while there are several countries where the line losses can be even more than 20% [39, 40]. This results in a significant loss of revenue for the system operator. It is, therefore, important to consider strategies to reduce these losses.

In 2012, a simple real-time, night, and day control of PV SFs as STATCOM (PV-STATCOM) for reduction of line losses in distribution feeders was patented by the author of this book. Studies based on this patent were conducted in [41] and are described below. The required reactive power for optimal bus voltage control for loss reduction is derived utilizing the entire inverter capacity in the nighttime and from the capacity remaining after active power generation during the day. The different solar PV farms connected in the network are operated as PV-STATCOM at optimal voltage setpoints computed locally or provided to all participating SFs by the network operator performing a central optimal power flow (OPF) [42] for the entire system. The inverter losses are accounted for in the control strategy. It is assumed that the actual real-time generation from the solar PV systems is available through communication channels to the system operator, as well as the load data is obtainable through smart meters.

5.5.1 Concept of PV-STATCOM Voltage Control for Line Loss Reduction

A PV solar system connected to a radial distribution feeder is depicted in Figure 5.24. The PV-STATCOM controller is also illustrated in Figure 5.24. It follows the PV-STATCOM operation mode with active power priority as described in Section 4.7.2. Only the relevant controller components are described here. To operate the PV inverter as PV-STATCOM for controlling the PCC bus voltage a "PCC voltage controller block" is added to the outer control loop to compare the measured voltage

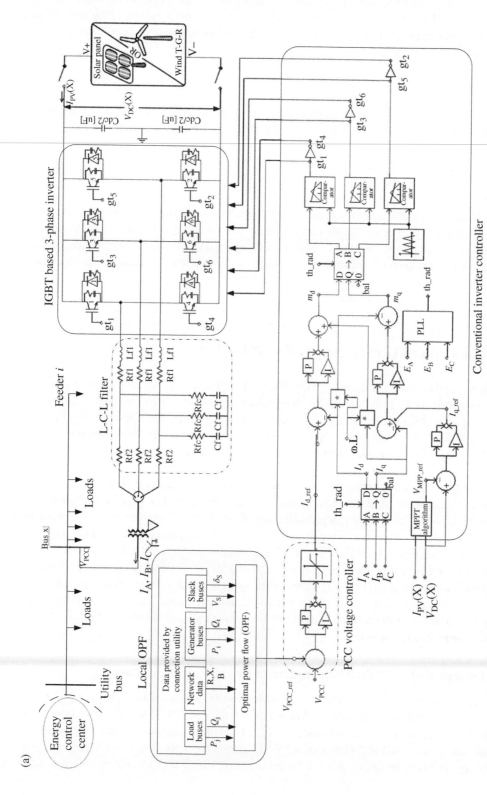

Figure 5.24 PV-STATCOM control for line loss minimization (a) optimal power flow performed locally; (b) optimal power flow performed centrally at the energy control center.

(b)

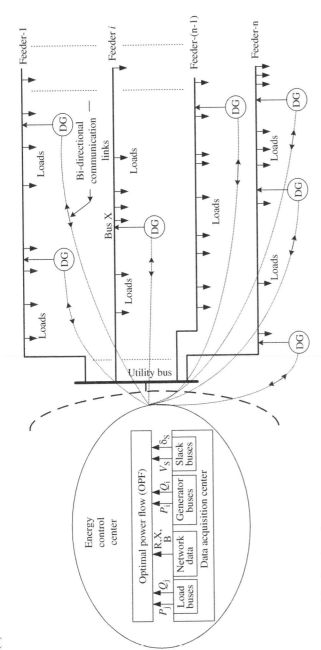

Figure 5.24 (Continued)

signal at PCC, V_{PCC}, with the reference voltage value of PCC, V_{PCC_ref}, as demonstrated in Figure 5.24a. The output of this block sets the reference value of "I_d" which subsequently controls the reactive power flow from the inverter. A limiter is used to constrain the amount of reactive power flow from the inverter within the remnant capacity of the inverter after active power production. A DC-link voltage controller is included in the conventional PV inverter control.

5.5.1.1 Determination of Optimal Voltage Setpoints

Line losses can be minimized by any of two strategies: (i) minimization of line losses at a feeder level with just one PV SF, and (ii) minimization of line losses at the system level with multiple PV SFs. In case (i), the PV system operator conducts OPF [42] locally as depicted in Figure 5.24a, whereas, in case (ii), the network operator at the energy control centre (ECC) performs the OPF centrally for the entire network for all the PV systems participating in line loss minimization as illustrated in Figure 5.24b. The OPF provides the optimal voltage setpoint for the PCC, V_{PCC_ref} for the individual PV system, or several PV systems as the case may be. In case (ii), the optimal voltage setpoints are transmitted to different PV systems through respective communication channels. The optimal voltage setpoints can be provided to participating SFs as frequently as the communication systems allow the network operator to receive updated load and generation information, and as frequently as an OPF can be conducted with state-estimated data and its results be transmitted to the SFs.

The PV SF is typically modeled as a PQ bus with zero Q output for UPF operation. However, for PV-STATCOM operation the PV SF is considered to provide voltage regulation as a *PV* bus with reactive power limits defined by the remaining inverter capacity for the specific active power "P" output. Here, $Q_{limit} = \sqrt{(S^2 - P^2)}$ where "Q_{limit}" is the upper (+ve) and lower (−ve) limits of reactive power capacity available after transferring all the available active power "P" to the grid and "S" is the total MVA capacity or apparent power rating of the PV inverter. During nighttime, the entire inverter capacity, "S" is available for reactive power exchange.

The PSS/E software is utilized to conduct the OPF solution. In OPF formulation, the active power losses P_{loss} is a function of (P_R, Q_R, V_S, R, X), where, "P_R" and "Q_R" represent the net power balance at a particular bus, V_S is the slack or sending end bus voltage, R and X are the line resistance and reactance parameters, respectively [42]. All the quantities are constant for a given node or bus with known active power generation and loads except the required net "Q_R." The corresponding bus voltage set points are found by conducting OPF solution with an objective function of reducing line active power losses while satisfying the available reactive power capacity constraints of the PV inverters and other network constraints.

5.5.1.2 Inverter Operating Losses

As PV SFs transfer power through inverters, there are active power losses associated with the inverter operation both in conventional SF inverter operation as well as in the PV-STATCOM inverter operation. These losses are associated with IGBT conduction, switching and snubber losses, and expressed as a polynomial [43]:

$$P_{loss\_inv} = C_S + C_V.S + C_R.S^2 \text{ (pu)} \tag{5.3}$$

where, S is the apparent power rating of the solar PV system expressed in pu, coefficient C_S is associated with the standby fixed loss, coefficient C_V is associated with voltage-dependent losses and coefficient C_R is associated with resistive losses for IGBT conduction. C_S, C_V, and C_R are determined through a curve fitting technique for a typical inverter efficiency curve with the help of the efficiency expression for both daytime and nighttime operation.

The inverter efficiency η is given by

$$\eta = S/(S + P_{\text{loss\_inv}}) \tag{5.4}$$

where $P_{\text{loss\_inv}}$ is the magnitude of inverter active power loss. For high efficiency inverters typically around 98% efficiency, the values of C_S, C_V, and C_R are 0.2414%, 0.00364%, and 0.0002%, respectively [44].

5.5.2 Simulation Studies

Two case studies for reduction of line losses with PV-STATCOM control are presented below:
Case Study I: A simple two bus radial system with flat load,
Case Study II: IEEE 33 Bus system [45] with variable load profile as extracted from IESO of Ontario, Canada [46].

5.5.2.1 Case Study 1: Two Bus Radial System with Constant Load

Figure 5.25 illustrates a 48 km long feeder comprised of overhead conductors and connected to the "Utility Bus" for two different scenarios based on the location of DGs as follows [47]:

a) Scenario 1: 12 MW PV SF connected at the end of feeder on load bus,
b) Scenario 2: 12 MW PV SF connected at the middle of feeder.

In each scenario, a constant load of 4 MW, 2 Mvar is connected at the end of this feeder.

The line losses are investigated for these two different scenarios by conducting OPF in Power-World Simulator [48] and PSS/E. In this study, the PV SF is controlled as PV-STATCOM to regulate the PCC bus voltage within the limits of 0.94–1.06 pu as specified under the technical interconnection requirements of the Ontario utility Hydro One [26], while minimizing line losses. This study is performed for a typical case of PV power generation data obtained from IESO, Ontario, [46] for a specific day and illustrated in Figure 5.26. This figure depicts both the hourly active power generation and the corresponding available inverter reactive power capacity for a sunny day.

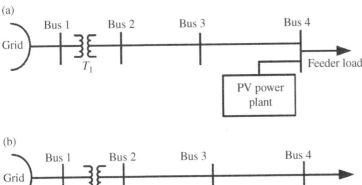

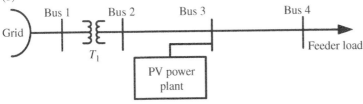

Figure 5.25 One line diagram of (a) Scenario 1 and (b) Scenario 2. *Source:* Based on Varma et. al. [47].

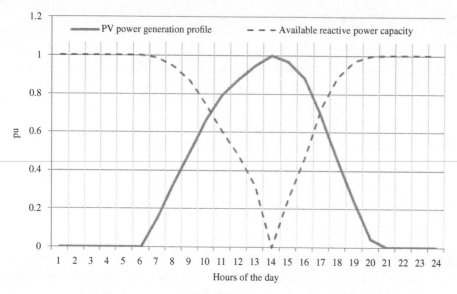

Figure 5.26 PV power generation profile and available reactive power capacity of inverter. *Source:* Independent Electricity System Operator (IESO).

Table 5.2 System losses and additional loss reduction for Scenario 1.

PV active power generation (MW)		Loss with conventional PV operation (kW)		Proposed PV-STATCOM operation				
		Line loss [X]	PV inverter loss	PCC voltage set point (pu)	Reactive power (Q_PV) (Mvar)	Line loss (kW) [Y]	STATCOM loss (kW) [Z]	Net loss reduction (kW) LR = [X–Y–Z]
Night	0	250.7	29	0.994	2.415	176.1	29	45.60
Day	2	92.1	29.04	1.010	2.021	42.1	0.030 0	49.97
	4	42.1	29.11	1.030	1.918	0.0	0.016 0	42.08
	6	83.7	29.19	1.052	2.026	38.8	0.012 0	44.89
	8	207.0	29.27	1.060	1.738	152.8	0.006 8	54.19
	10	406.5	29.35	1.060	1.272	347.3	0.003 0	59.20
	12	680.4	29.43	—	—	—	—	—

Source: Based on Rahman [41].

The study results for Scenario-1 are summarized in Table 5.2 for different PV power generation levels – 0 MW (min) to 12 MW (max), during daytime and nighttime. In this table, $[X]$ denotes the line loss in the study system with conventional UPF operation of PV SF. $[Y]$ represents the line losses when the PV solar system operates as PV-STATCOM. $[Z]$ indicates the additional inverter losses associated with the PV-STATCOM operation based on Eq. (5.3). PV inverter losses during conventional operation are also indicated in the table for reference. The net line loss reduction is expressed as LR $= [X] - [Y] - [Z]$. This table clearly demonstrates the net loss reduction with PV-STATCOM control as compared with conventional PV operation at all levels of power generation except at rated power generation at 12 noon, when the inverter does not have any remaining capacity to provide reactive power at this level. It is noted that this was a conservative study performed in 2012 whereas presently the IEEE Standard 1547-2018 mandates at least 44% reactive power capability at rated power output [25].

Since this is a conceptual study, it is assumed for ease of analysis that the PV system is changing its power output not continuously but in a stepwise manner on an hourly basis. Hence the average value of power output at a particular hour is utilized to compute the different quantities [X], [Y], and LR. (It is noted that presently, detailed quasi-static time-series (QSTS) studies are performed for optimizing different quantities including daytime line losses, at very frequent time intervals e.g. 1 minute, as described in Chapter 7.)

Similar results are also obtained for Scenario 2, although not shown. Finally, a comparison of net line loss reduction over 24 hours for Scenario 1 and Scenario 2 is presented in Table 5.3. It is observed that the PV-STATCOM located adjacent to the load is more effective than the PV-STATCOM located at the midpoint of the line. Placing a distributed generator (PV SF in this case) next to the load reduces the import of power over the distribution line and hence reduces line losses. However, control of PV SF as STATCOM (PV-STATCOM) reduces the line loss even further, both during night and day.

5.5.2.2 Case Study II: IEEE 33 Bus System with Variable Load

In this case, three PV SFs are considered to be connected in the IEEE 33 Bus system at optimal locations with optimal sizes as described in [45]. The single-line diagram of the modified IEEE 33 Bus system with the three PV SFs is depicted in Figure 5.27. The SFs are operated in conventional mode at UPF with the optimal sizes of 0.9, 0.9, and 0.72 MW and are optimally located at Buses 6, 12, and 31, respectively [45].

Figure 5.28 presents a typical load profile for a 24 hour period in Ontario as obtained for a specific day from IESO, Ontario [46]. The maximum load (considered 1 pu) for the IEEE 33 Bus system as given in [45] is now mapped on the Ontario's load profile depicted in Figure 5.28. The objective is to examine the effectiveness of PV-STATCOM control in IEEE 33 bus system with a loading pattern similar to an actual loading profile in the province [46]. It is assumed that when SFs produce their maximum active power around 2 p.m. as shown in Figure 5.26, the system load is 1.0 pu. This assumption results in peak system load of 1.08 pu at 7:30 p.m.

5.5.2.3 Improvement in Loss Reduction with PV-STATCOM

To investigate the improvement in line loss reduction with the proposed PV-STATCOM control on the three PV SFs, OPF study is performed in PSS/E software to determine the optimal voltage set points for the three PV SFs with the objective of minimizing line losses as described previously. The apparent power rating "S" of the PV SFs are taken to be the optimal ratings of the PV SFs reported in [45] and are considered to be 1.0 pu on their own power bases of 0.9, 0.9, and 0.72 MW, respectively.

Table 5.3 Summary of energy savings for different cases.

Cases and scenarios	Energy savings (kWh/day)	Energy demand (homes)	Cost savings ($/year)
Case I, Scenario-1	1006	33	27 528
Case I, Scenario-2	558	18	15 272
Case II, IEEE 33 Bus	493	16	13 496

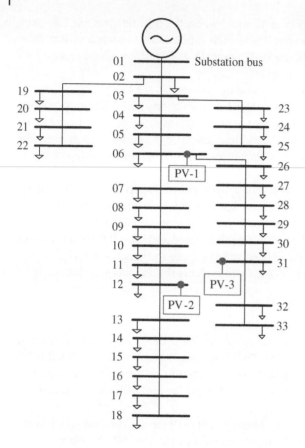

Figure 5.27 Modified one line diagram of IEEE 33 Bus system with PV solar farms.

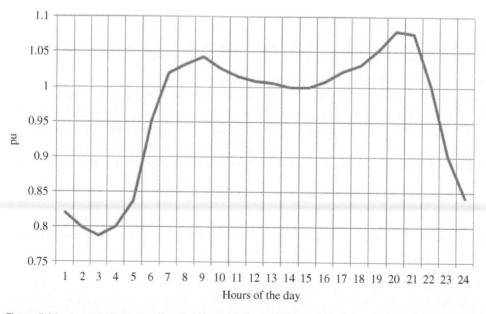

Figure 5.28 Load profile for typical day as created from IESO data. *Source:* Independent Electricity System Operator (IESO).

The available reactive power capacity of each of the three PV solar inverters operating as PV-STATCOM is given by $Q_{\text{limit}} = \sqrt{(S^2 - P^2)}$ pu and is assumed to be similar to that plotted in Figure 5.26. For OPF studies, the above available reactive power capacities are used as constraints on reactive power for the three PV SFs.

Figure 5.29a depicts the line losses (active power losses) over the 24 hour period for the cases (i) no PV solar farms, (ii) three PV solar farms operating in conventional UPF mode, and

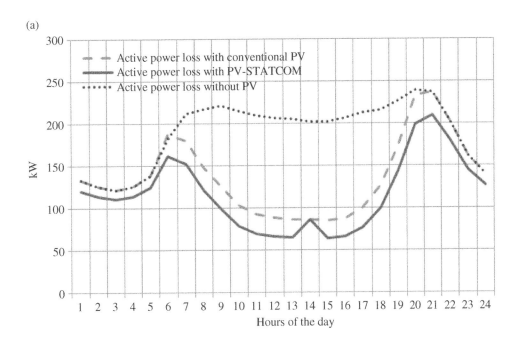

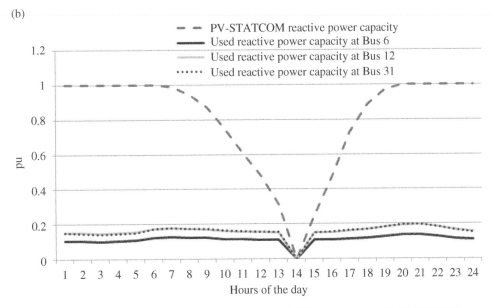

Figure 5.29 (a) Active power loss without PV systems, with PV systems and with PV-STATCOMs and (b) reactive power contribution from each PV-STATCOM.

(iii) all three PV solar farms operating as PV-STATCOM and regulating their PCC voltages at optimal voltage set points. The active power losses with PV-STATCOM operation in case (iii) also includes the inverter operating losses as per Eq. (5.3). It is evident from Figure 5.29a that while line losses are reduced significantly with the optimal placement of PV SFs, the use of PV-STATCOM reduces the line losses even further, despite accounting for inverter losses. Figure 5.29b illustrates the per unit reactive power used by the three PV SF inverters as PV-STATCOM at buses 6, 12, and 31, respectively. In each case, the required reactive power is much less than the available reactive power capacity of inverter.

Figure 5.30a–c illustrates the voltages at buses 6, 12, and 31 both without any SFs connected, with conventional UPF operation of PV SFs, and with SFs operating as PV-STATCOM. The voltages with the use of PV-STATCOM are obtained through OPF for minimizing losses over a period of 24 hours for the actual system load profile. It is evident from this study that the optimal voltage set points are higher than the network voltages without any PV SFs, thereby improving the voltage profile. The PV-STATCOM operation improves the voltages even during nighttime and consequently reduces the line losses.

In Ontario, the electricity rate is dependent upon the time of the day (off-peak, mid-peak, and on-peak loading conditions), and on the type of customers [46]. Even though an exact calculation of the financial saving based on the time of use rates is not done, but a conservative average energy rate of 7.5 cents/kWh is used in estimating the cost of energy savings. The energy savings is also expressed in terms of the number of equivalent homes that can be supplied annually with the saved energy. For this calculation, the average household energy consumption is assumed to be 867 kWh/month [49].

It is seen from Table 5.3 that in case of a two bus radial system with constant load, a 12 MVA PV SF controlled as PV-STATCOM can save enough energy to power 33 homes annually [47]. This energy saving is worth $27 528/year. On the other hand, three PV SFs with a total rating of 2.52 MW operating with the proposed PV-STATCOM control in the IEEE 33 Bus system with a realistic load profile, can save adequate energy to serve 16 homes annually in North America. This energy savings is worth $13 496/year.

5.5.3 Summary

This study presents a novel PV-STATCOM control to reduce line losses. The proposed control can potentially reduce line losses in a significant manner, even after accounting for inverter losses. The saved energy can be utilized to power a large number of homes annually (33 homes in a single feeder and 16 homes in the IEEE 33 bus system). This is a conservative study in which no reactive power availability during peak power production hours is assumed. However, with the implementation of IEEE Standard 1547-2018, even further line loss reduction can be achieved. Such a saving has been demonstrated on very small systems. In utility systems with multiple feeders having PV systems and especially long feeders with relatively inadequate voltage profiles, the proposed line loss reduction with PV-STATCOM operation can bring substantial savings.

In countries, which have high line losses even up to 15–20% [39], the implementation of proposed PV-STATCOM control on the ever-increasing number of PV installations in such networks can be especially advantageous.

It is understood that FACTS Controllers such as STATCOM are recommended and installed only for dynamic voltage regulation and not for steady-state voltage control. However, in this case, the unused remaining inverter capacity on existing PV SFs is utilized for PV-STATCOM control for optimal voltage regulation.

(a)

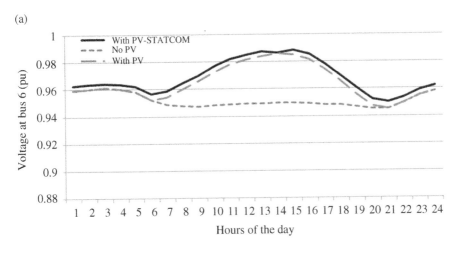

(b)

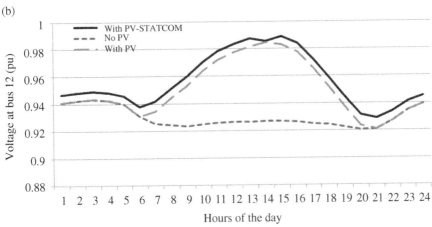

(c)

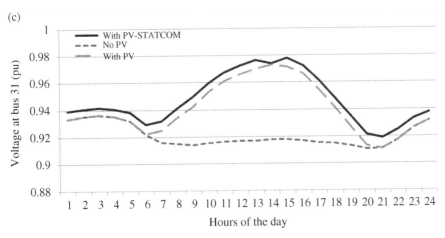

Figure 5.30 Voltage profile without PV system, with conventional PV system operation, and with PV-STATCOM at buses (a) 6, (b) 12, and (c) 31 to minimize line losses over 24 hours.

The proposed novel control of PV solar farms for line loss reduction can greatly benefit utilities in terms of asset lifespan and energy savings. Reduction in line heating losses decreases the aging process in transmission and distribution lines thus extending their lifespan. In several jurisdictions, line losses are passed on the customers. With the night and day operation of solar PV systems as PV-STATCOM, a decrease in line losses can also result in reducing customer bills.

5.6 Stabilization of a Remotely Located Critical Motor by PV-STATCOM

Induction motors (IMs) are widely used in industries such as petrochemicals, process control, rolling mills, mining, automotive, etc. [50, 51]. These industries are typically fed through long distribution feeders, thus reducing system strength at motor terminals. This makes the motors susceptible to stalling even for minor system disturbances due to inadequate reactive power support [52, 53]. Shutdown of these critical IMs, even for a short duration of few minutes, can result in substantial economic loss to the industrial facilities as the entire batch of materials (e.g. petrochemicals, automotive parts, chemicals, medicinal products) being transported/served by these motors may get damaged [54, 55]. In an Ontario petrochemical industry (name withheld for confidentiality), the financial loss estimated for just one hour of motors' shutdown is $1 Million.

Shunt-connected FACTS Controllers, such as SVC and STATCOM, have been conventionally employed in power systems for providing dynamic reactive power support to such critical motors during system contingencies [56, 57]. It is shown that SVC and STATCOM are superior than IM drives in ensuring faster starting with reduced transients and improved dynamic behavior [58]. The capability of SVC to prevent stalling of IM is described in [59]. The superiority of SVC controls over a breaker switched capacitor and a thyristor-controlled tap changer in preventing IM instability are demonstrated in [60, 61], respectively.

Compared to SVC, a STATCOM can ensure quicker recovery of IM speed after a disturbance [58]. This is due to faster response of STATCOM and its ability to provide rated capacitive current even at very low bus voltages [62]. It is shown that voltage overshoots present with SVC control during asymmetrical faults are alleviated by STATCOM [58].

Due to the above benefits, STATCOMs are installed at several industrial premises to support critical motors. Six 4.5 Mvar STATCOMs are installed in a gold mine in Matachewan, Ontario to provide dynamic reactive power support for motors fed by long cables [63]. A 16.8 Mvar STATCOM is connected in a petrochemical plant in Gulf Coast, Texas, to prevent voltage instability of large motors [63]. A 5 Mvar STATCOM is connected at Seattle Iron and Metal Corp. to provide dynamic support to critical motors [64]. The application of STATCOM to prevent stalling of critical motors in a refinery in Southern California Edison system is presented in [65]. It is estimated that the return of investment in just first year with STATCOM installation is 24% [65]. Thus, it is evident that SVC and STATCOM are widely employed in industries for preventing instability of critical motors. However, these Controllers are quite expensive.

Since PV solar plants are growing at an enormous rate worldwide, it is quite likely that they may be located on same networks as critical IMs, although not necessarily close to them. This section presents a novel cost-effective smart PV inverter control for providing motor stabilization similar to STATCOMs and SVCs, even though the motor may be remote from the solar plant.

The concept of smart inverters for distributed energy resources (DERs) was introduced in [66, 67], whereby smart PV solar inverters can provide multiple functions of active and reactive power control either autonomously or in response to utility issued commands for grid support. Smart inverter functions, such as volt–var, volt–watt, LVRT, etc., have been incorporated in the recent IEEE Standard 1547-2018 [25]. The German grid code (BDEW) [69] requires PV inverters to inject 2% dynamic reactive current per 1% voltage drop for voltages below 0.9 pu to support the grid during transients. However, this dynamic reactive current support is optional in [25]. A PV plant control for providing reactive power support per German grid code is proposed in [70], whereas a strategy for ensuring maximum reactive power support during LVRT is presented in [71].

In 2009, an autonomous smart PV inverter technology named PV-STATCOM for utilizing PV solar plants as a STATCOM both in the night and day, for providing various grid support functions with a response time of 1–2 cycles, was patented by the author of this book. The PV-STATCOM control is distinct from conventional smart inverter functions such as volt–var, volt–watt, etc. [25, 72, 73], which respond in 1–2 seconds and are defined only for daytime operation and not during nighttime. The PV-STATCOM technology utilizes the full inverter capacity both during nighttime and also during periods of system disturbances in the daytime for providing various STATCOM (or SVC) functions as described earlier in this chapter as well as in Chapter 6.

This section presents a new multi-mode control of PV-STATCOM for stabilizing a remote critical motor [74], which:

i) provides enhanced voltage support utilizing communication of critical motor terminal voltage to the PV system location, and high voltage ride-through (HVRT) capability of PV system inverter.
ii) is field validated on a 10 kW PV solar system located in the utility network of Bluewater Power Distribution Corporation, Sarnia, Ontario, Canada.
iii) performs as effectively as an actual STATCOM connected locally at motor terminal, although at much lower cost than the local STATCOM.

The other novel features of the proposed PV-STATCOM control compared to conventional smart inverters are that it:

i) provides motor stabilization both during night and day whereas conventional smart inverters provide LVRT only during day [25, 68, 72, 73, 75].
ii) ensures continuous stable operation of a remote motor during a large disturbance whereas the same solar plant providing reactive current support per German grid code [69] is unable to do so.

In addition, the proposed PV-STATCOM control provides voltage support with much lower active power curtailment, than in previously reported PV-STATCOM controls [28].

5.6.1 Study Systems

5.6.1.1 Study System 1

The MATLAB Simulink-based simulation studies and field demonstration are performed on the 10 kW solar PV plant located in the utility network of Bluewater Power Distribution Corp., Sarnia, Ontario, Canada. This field demonstration system, termed Study System 1, is portrayed in Figure 5.31. The system consists of a PV array rated at 10 kW, 280 V, and 49 A. PV power is fed into the distribution system using an existing three-phase inverter rated at 10 kVA, 600 V, and 9.6 A,

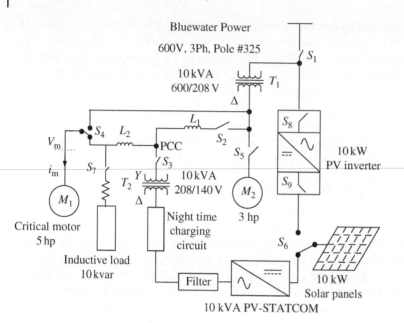

Figure 5.31 Single-line diagram of the Study System 1. *Source:* Varma et.al. [74].

which is connected to the Bluewater distribution network. A 5 hp IM M_1 is considered as the critical motor for this study.

Modifications are made in the existing PV system site for testing the operation of PV-STATCOM under various grid conditions. The PV-STATCOM is constructed using a three-phase two-level inverter of 10 kVA rating. The PV-STATCOM controller is developed in dSPACE Control Suite and is used to generate switching pulses for the 10 kVA PV-STATCOM inverter. Switch S_6 is used to disconnect the existing commercial inverter and connect the PV-STATCOM. As the MPPT voltage of the solar panel is 280 V, a 140/208 V transformer is additionally installed to connect the PV-STATCOM inverter output to the 208/600 V utility transformer located at the site. An LCL filter is used to limit the THD of the inverter output within limits specified by [25].

Although the critical motor is physically located at the site of the solar plant, its effective distance from the solar plant is varied by using a variable inductor L_2 which is inserted through switch S_4. The delay in communication of motor terminal voltage signal to the solar plant is implemented in software. A large system disturbance is created by turning on a large inductive load of 10 kvar at the motor terminal using switch S_7. The inductor L_1 is used to vary system short circuit ratio. A 3 hp motor M_2 is used for tracking the PV panels.

5.6.1.2 Study System 2

PSCAD simulation studies are conducted on a realistic 45 km long 27.6 kV feeder in Ontario [22, 76] to study the impact of key system parameters on PV-STATCOM performance. The single-line diagram of study system 2 is depicted in Figure 5.32a. Transformer T_1 is rated 115 kV/27.6 kV, 32 MVA with 5% impedance. The distribution lines are represented by equivalent π models. A 4-MW PV solar plant is connected to the feeder at bus 3 that is 20 km away from the IMs. A 5.3 MVA load at 0.9 power factor is connected at feeder end through 27.6 kV/4 kV transformer T_2. The motor to be stabilized is rated 2.65 MVA at 0.9 pf. The static load is also 2.65 MVA at 0.9 pf.

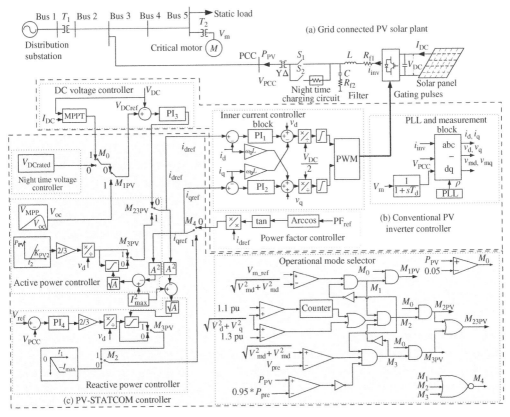

Figure 5.32 A PV system connected to Study System 2 with the proposed PV-STATCOM controller. (a) Grid connected PV solar plant, (b) Conventional PV inverter controller, (c) PV-STATCOM controller. *Source:* Varma et.al. [74].

5.6.2 Study System with PV-STATCOM Control

5.6.2.1 Grid-Connected Solar PV Plant

The PV inverter (as PV-STATCOM) is connected to the PCC bus in both Study Systems 1 and 2. Figure 5.32a illustrates the PV inverter connected in study system 2 at PCC bus. The solar panel output is fed to the inverter. The inductor L, capacitor C, and inductance of $\Delta - Y$ transformer constitute the LCL filter. R_{f1} and R_{f2} represent the resistance of the inductor and capacitor, respectively. To provide both night and day PV-STATCOM operation, modifications are made in the PV inverter system. In the day, the DC-link capacitor is charged from the PV solar panels during start-up. However, at night, the PV panels are idle, and therefore a charging circuit is added for nighttime DC-link capacitor charging. This is accomplished by keeping switch S_2 on and S_1 off. Once the capacitor voltage reaches the peak of AC phase voltage, switch S_1 is turned on, and S_2 is turned off. The PV inverter then functions as STATCOM (PV-STATCOM) at night.

The voltage at motor terminal is denoted as V_m, which is measured and communicated (through a fiber optic cable) to the PV-STATCOM controller, incurring a delay in the process. The overall delay represents: (i) the communication delay between motor terminal and PV-STATCOM, and (ii) the communication delay between master control and individual inverter controllers in an actual solar PV plant. The latency for fiber optic cable is 3.34 μs/km [77]. The communication delay incurred in a PV solar plant control is approximately 100 ms [78]. Thus, in both study systems where the distances between the IM and PV solar plant are 15.6 and 20 km, respectively, the delays are 100.05 and 100.067 ms, respectively, both of which are approximated as 100 ms.

5.6.2.2 Conventional PV Inverter Control

Figure 5.32b depicts the structure of a conventional PV inverter control used in studies for both Study Systems 1 and 2. It consists of DC-link voltage controller and power factor controller, inner current controller, and PLL [79]. Sinusoidal Pulse Width Modulation (SPWM) is used with a carrier frequency of 10 kHz for generating the gating pulses for the inverter. These gating signals are generated through two current control loops in d-q coordinates [79]. The inverter current i_{inv}, PCC voltage V_{PCC}, and IM terminal voltage V_m are transformed to d-q reference frame using the PLL, and are fed to the other controller blocks through a measurement filter [79]. The d-q frame components of i_{inv}, V_{PCC}, and V_m are i_d and i_q, v_d and v_q, and v_{md} and v_{mq}, respectively. The DC voltage controller and power factor controller generate the active and reactive current references for the inverter inner current controller [79]. The Power factor controller generates i_{qref} as below:

$$i_{qref} = i_{dref} * \tan\left(\cos^{-1} PF_{ref}\right) \tag{5.5}$$

where PF_{ref} is the power factor set point; and i_{qref} and i_{dref} are the reactive current reference and active current reference, respectively.

5.6.2.3 PV-STATCOM Controller

The proposed PV-STATCOM controller used in both Study Systems 1 and 2 is illustrated in Figure 5.32c. It consists of: (i) operation mode selector, (ii) active power controller, (iii) reactive power controller, and (iv) nighttime voltage controller. During a system disturbance, the PV-STATCOM controller generates the reference signals for active current (i_{dref}) and reactive current (i_{qref}) for the inner current controller based on V_{PCC} and V_m, to ensure fast voltage recovery of the critical IM. The functioning of different blocks of the proposed PV-STATCOM controller is described below.

Operation Mode Selector

The active power output and reactive power capability of a PV solar plant on a sunny day are illustrated in Figure 5.33a. A system disturbance is considered to occur at $t0$. The controller immediately initiates PV-STATCOM operation to prevent stalling of the critical motor. The hatched region $t0$–$t3$ indicates PV-STATCOM operation, which ranges typically between 1 to 3 seconds. At $t3$, the critical motor is stabilized. Also, by that time, the active power output of PV plant reaches its pre-disturbance level. At this time, the PV-STATCOM operation is stopped. The enlarged version of the hatched region ($t0$–$t3$) is depicted in Figure 5.33b. The proposed PV-STATCOM control functions in three modes: Modes 1, 2, and 3. Mode 4 is the conventional UPF operating mode of a PV plant. The conceptual active and reactive power during the different operating modes are illustrated in Figure 5.33b.

The modes of operation are decided by the Operation Mode Selector block based on V_{PCC}, V_m, and active power output of PV solar plant (P_{PV}). The operation mode selector compares P_{PV} with a small value (selected as 0.05 pu) to determine if PV-STATCOM operation is being performed during day ($P_{PV} > 0.05$ pu) or night ($P_{PV} < 0.05$ pu), setting flag M_0 to 1 or 0, respectively.

Mode 1 (represented by flags M1 and M1PV). This mode provides enhanced voltage control at the PV plant. In this reactive power priority mode, active power is curtailed from the pre-disturbance active power level P_{pre} (1 pu in Figure 5.33) to zero, and rated maximum reactive current is injected

(a)

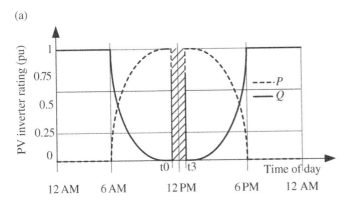

(b)

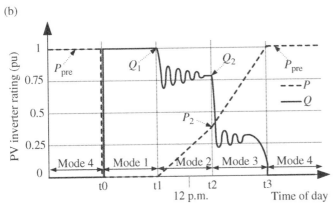

Figure 5.33 (a) Active power output (*P*) and reactive power capability (*Q*) of a PV solar plant for a sunny day. Hatched region indicates proposed PV-STATCOM operation for typically less than 3 seconds around noontime; (b) Enlarged version of the hatched region in (a) illustrating active (*P*) and reactive (*Q*) power output of PV-STATCOM in different operating modes, around noon time. *Source:* Varma et.al. [74].

utilizing the entire inverter capacity. This mode is initiated when motor voltage V_m goes below a predetermined threshold value V_{m_ref} (considered 0.8 pu for this study) due to a system disturbance.

As the IM is located remotely from the PV plant, regulation of PV plant PCC voltage simply to its upper steady-state limit may not lead to the recovery of remote IM terminal voltage to a sufficient level to stabilize the IM. Hence, an enhanced voltage support is provided as follows:

Ideal Mode 1 operation (in accordance with HVRT [25]):

i) The PV plant voltage is first made to increase up to a voltage level V_A (say, 1.2 pu [25]).
ii) A counter (representing enhanced voltage support) is then activated for a duration of T_C and the voltage is allowed to increase beyond V_A, but not to exceed V_B which is selected to be the HVRT voltage limit of 1.3 pu [25].

The HVRT characteristics specify that the PV plant may ride through for a voltage between 1.2 and 1.3 pu for no more than 160 ms [25]. Hence the counter (T_C) is set to 150 ms (slightly less than 160 ms, for safety). Mode 1 operation with maximum reactive current injection is ceased after 150 ms. However, if V_{PCC} exceeds 1.3 pu at any time during this interval, Mode 1 operation is stopped instantaneously by providing an override signal to counter.

Conservative Mode 1 Operation:

Due to specific network conditions and PV plant size, it is quite likely that voltage may not go up to 1.2 pu even with the injection of rated reactive current. Thus, it is up to the PV plant operator (together with the concerned utility) to select voltage setpoints V_A and V_B and T_C, in accordance with the corresponding HVRT requirements for those voltage levels.

In Study system 1, the PCC voltage of PV system does not exceed 1.08 pu. Hence, V_A and V_B are selected to be 1.07 and 1.08 pu, respectively. In Study system 2, V_{PCC} does not exceed 1.13 pu. Hence, V_A and V_B are selected to be 1.07 and 1.13 pu, respectively. T_C is set as 250 ms for both study systems. The HVRT criteria specify that PV system can ride through a voltage less than 1.1 pu indefinitely, and a voltage between 1.1 and 1.2 pu for 13 seconds [25]. Thus, the chosen value of T_C is well within the HVRT limit.

It is noted that the above selection of voltage setpoints and counter timing is quite conservative with respect to those permissible under HVRT requirements. However, it is shown from studies in this section that this conservative Mode 1 operation is still very effective in stabilizing the critical IM.

Mode 1 operation is enabled when V_m goes below V_{m_ref} due to a disturbance, by setting $M_1 = 1$, and correspondingly $M_{1PV} = 1$ during daytime. Rated reactive power is injected into the system using full inverter capacity by setting $i_{qref} = -1$. During nighttime, flag M_{1PV} is disabled to 0. V_{DC} is set to $V_{DCrated}$ which is chosen to be twice the magnitude of V_{PCC} to ensure proper operation of SPWM modulated inverter [79].

Mode 2 (represented by flags M_2 and M_{2PV}). This is also a reactive power priority mode. However, in this mode, controlled (but not maximum) reactive power is injected into the grid with the objective of maintaining V_{PCC} at its steady-state voltage limit of V_{ref} (1.05 pu in this study). In Figure 5.33 (b), the reactive power required to maintain V_{PCC} at V_{ref} is Q_2. Simultaneously, active power (P_{PV}) is ramped as close as possible to the pre-disturbance level (P_{pre}) by using inverter capacity remaining after reactive power injection. In Figure 5.33 (b), active power is ramped till P_2. This mode is initiated once Mode 1 operation ceases and continued till V_m reaches an acceptable limit (i.e. pre-disturbance voltage V_{pre}).

Mode 2 operation is initiated by setting flag $M_2 = 1$, and correspondingly $M_{2PV} = 1$ during daytime. The DC voltage controller is disabled and i_{dref} is calculated by the power ramp function in active power controller as below:

$$i_{dref} = \frac{2P_{ramp}}{3v_d} \tag{5.6}$$

P_{ramp} is the instantaneous value of active power ramp. i_{dref} calculated using Eq. (5.6) is passed through the saturation block to limit it to within the available inverter capacity. The saturation block limits (i_{dlimit}) are calculated as below.

$$i_{dlimit} = \pm \sqrt{I_{max}^2 - i_{qref}^2} \tag{5.7}$$

where I_{max} is the rated inverter current, and i_{qref} is the reactive current reference generated by PCC voltage controller to maintain V_{PCC} at V_{ref}.

During nighttime $M_0 = 0$. Hence, M_{2PV} (and M_{3PV}) get disabled and $M_{23PV} = 0$. The DC voltage controller regulates V_{DC} to $V_{DCrated}$.

Mode 3 (represented by flag M_{3PV}). This is an active power priority mode in which active power (P_{PV}) is ramped to its pre-disturbance level (P_{pre}). Simultaneously, remaining inverter capacity is utilized for dynamic reactive power support to maintain V_{PCC} at V_{ref}. The simultaneous reactive power control reduces voltage oscillations during active power ramp up and thus helps in restoring active power at a fast rate. Mode 3 operation is initiated at the end of Mode 2 (i.e. when V_m reaches V_{pre}) and continued till active power P_{PV} reaches P_{pre}. In Figure 5.33b, the active power is ramped from P_2 to P_{pre} (1 pu). One practical way for checking is that P_{PV} becomes at least 0.95 times P_{pre}.

Mode 3 operation is enabled by setting $M_{3PV} = 1$, and consequently $M_{23PV} = 1$. Active power P_{PV} is ramped to P_{pre} using active power controller. The DC voltage controller is disabled and i_{dref} is calculated by the power ramp function in active power controller using Eq. (5.6). The remaining inverter capacity is simultaneously used for dynamic reactive power support to regulate V_{PCC} at V_{ref}. i_{qref} generated by PCC voltage controller is limited by the saturation block using Eq. (5.8), to ensure that maximum inverter current rating is not violated.

$$i_{qlimit} = \pm \sqrt{I_{max}^2 - i_{dref}^2} \tag{5.8}$$

Mode 4 (represented by flag M_4). This is the conventional operating mode of PV system, in which the PV system generates maximum active power at utility specified power factor, which is typically UPF. This mode is initiated once P_{PV} reaches P_{pre}. The PV system operates in this mode and remains on standby to operate as STATCOM, when needed.

Mode 4 operation is initiated by setting $M_4 = 1$. During daytime, i_{dref} and i_{qref} are generated by the DC voltage controller and power factor controller respectively. During nighttime, this mode regulates V_{DC} at $V_{DCrated}$.

Active Power Controller

The active power controller block generates i_{dref} for the proposed control strategy. This block consists of a DC voltage reference generator and active power ramp up functions. During Mode 1 operation, the DC voltage reference generator sets V_{DCref} to the open-circuit voltage of PV panels (V_{oc}) to curtail active power generation. Such a curtailment of PV plant output power to zero by increasing the voltage across the solar panels to their open-circuit voltage has no adverse effect on the PV panels [80, 81]. The active power ramp-up function also generates i_{dref} during Modes 2 and 3.

Reactive Power Controller

The reactive power controller block determines i_{qref} for the different operating modes of proposed PV-STATCOM control. It consists of PCC voltage controller and a reactive current reference generator. During Mode 1, the reactive current reference generator sets $i_{qref} = -1$ to utilize the full inverter capacity to provide reactive power support. In Modes 2 and 3, the PCC voltage controller regulates V_{PCC}.

Nighttime Voltage Controller

During nighttime, this block masks the operation of active power controller block. It maintains V_{DC} at $V_{DCrated}$, and operates PV inverter as a STATCOM, as needed.

The design of various PV-STATCOM controllers is described in [74].

5.6.3 Simulation Studies on Study System 1

The PV-STATCOM controller performance in stabilizing the motor is first studied using MATLAB Simulink software.

5.6.3.1 Performance of the Proposed PV-STATCOM Controller

A large disturbance is created at the terminal of critical motor by connecting a 10 kvar inductive load for 750 ms (45 cycles), and the results are depicted in Figure 5.34. Two cases are illustrated in this figure. The solid lines show the performance of IM and the PV solar plant with PV-STATCOM controller (stable motor operation), and dashed lines show the operation without PV-STATCOM control (unstable motor operation). Figure 5.34 (a) – (h) depicts PCC RMS voltage (V_{PCC}) and motor terminal RMS voltage (V_m), reactive and active power output of inverter, IM speed, IM torque, i_q, i_d, V_{DC}, and converter terminal voltage, respectively.

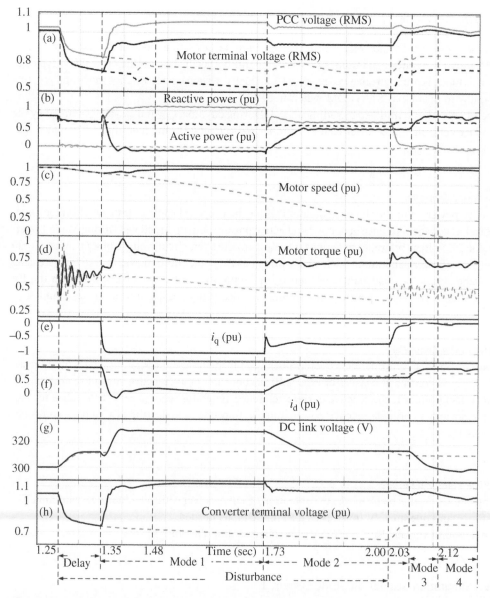

Figure 5.34 Response of induction motor with and without PV-STATCOM control. (a) PCC rms voltage and motor terminal voltage, (b) reactive power and active power output of inverter, (c) IM speed, (d) IM torque, (e) i_q, (f) i_d, (g) DC link voltage, (h) Converter terminal voltage. *Source:* Varma et.al. [74].

$t < 1.25\,s$. V_{PCC} and V_m are 1.04 and 1.02 pu, respectively. The PV inverter is operating in Mode 4 by injecting 0.8 pu active power ($P_{pre} = 0.8$ pu) at UPF. The IM is providing 0.75 pu load torque at 2% slip.

$t = 1.25\,s$. The large disturbance is applied as described above causing V_m to drop below 0.8 pu. The active power injected by PV solar plant reduces slightly due to the drop in V_{PCC}. Motor speed starts to reduce, and the motor torque becomes oscillatory due to voltage reduction. Without PV-STATCOM control, the PV system continues operating at UPF. The voltages at PCC and motor terminal decrease. The IM torque declines, and the IM eventually stalls.

$t = 1.35\,s$. A 100 ms communication delay is considered between the voltage measurement at motor terminal and its receipt at the PV solar plant. Due to this delay, the controller detects the drop in V_m at $t = t_0 = 1.35\,s$, and initiates Mode 1 operation by injecting rated reactive power ($Q_1 = 1$ pu). Due to injection of constant rated reactive power, V_{PCC} and V_m increase and the motor speed starts to recover.

$t = 1.48\,s$. V_{PCC} reaches V_A (1.07 pu), and the counter is triggered with T_C set to 250 ms. During this Mode 1 operation, the transients in IM torque stabilize and the pre-disturbance load torque of almost 0.75 pu is restored.

$t = 1.73\,s$. V_{PCC} reaches level V_B (1.08 pu) while the counter reaches its setpoint T_C. Mode 1 operation is stopped and Mode 2 is initiated at $t = t_1 = 1.73\,s$. Thereafter, the PV-STATCOM regulates V_{PCC} to V_{ref} (1.05 pu) by injecting 0.8 pu reactive power ($Q_2 = 0.8$ pu). Active power is ramped up from zero to 0.6 pu ($P_2 = 0.6$ pu) utilizing the remaining inverter capacity. The IM is able to provide steady-state torque at an acceptable slip of 6%, even with $V_m \approx 0.9$ pu. The DC-link voltage and converter terminal voltage stay within their limits.

$t = 2\,s$. The large load is disconnected causing V_m to increase.

$t = 2.03\,s$. V_m reaches its pre-disturbance voltage level V_{pre} and PV-STATCOM switches to Mode 3 operation at $t = t_2 = 2.03$ s. Active power is ramped up to the pre-disturbance level P_{pre} ($= 0.8$ pu), while simultaneously regulating V_{PCC} to V_{ref} (1.05 pu) through dynamic reactive power control by PV-STATCOM. This voltage control ensures fast power ramp-up without causing any undesirable oscillations in V_{PCC}.

$t = 2.12\,s$. The controller switches to Mode 4 at $t = t_3 = 2.12$ s and the inverter starts operating at UPF in Mode 4.

This study shows that the proposed PV-STATCOM control stabilizes the remote IM which would have stalled otherwise. The key features of the proposed control are that it:

i) provides enhanced reactive power support to control PV plant voltage up to the HVRT limit. Studies have shown (although not reported here) that the motor could not be stabilized if this enhanced control were not used.

ii) continues dynamic reactive power support at PV plant till the motor voltage recovers to its pre-disturbance value. The PV-STATCOM control is thus based on both motor voltage and solar plant PCC voltage. It is seen from studies (though not included) that if the PV-STATCOM control was based only on regulating PCC voltage (without consideration of motor voltage) the critical motor could not be stabilized.

iii) curtails active power to zero for a shorter duration and then ramps it up over the duration of PV-STATCOM control. The previously described PV-STATCOM control [28] curtailed active power to zero throughout the control period. Hence, energy curtailment with the proposed controller is 35% lower than in [28].

5.6.3.2 Comparison of PV-STATCOM and STATCOM Operation

The effectiveness of the proposed PV-STATCOM controller for stabilizing a remotely located IM is now compared with a STATCOM of same capacity located at the motor terminal itself. The responses of the IM during a large disturbance with: (a) a 10 kvar STATCOM located at motor terminal and (b) a 10 kVA PV-STATCOM located 15.6 km away from the motor terminal with 100 ms delay are depicted in Figure 5.35. The STATCOM for this study is modeled according to [62]. A large disturbance is created by connecting a large load of 10 kvar at IM terminal at $t=1.25$ s for 750 ms (45 cycles). Figure 5.35a–d represents the motor terminal voltage V_m, STATCOM or PV-STATCOM reactive power output, IM speed, and IM torque, respectively. The STATCOM located at IM terminal starts injecting reactive power immediately to support V_m, and continues this support till V_m recovers. The response of the remote PV-STATCOM controller is similar to that described in previous Section.

This study shows even if the PV-STATCOM is located 15.6 km away from the critical motor, it can stabilize the motor during a large disturbance in a similar manner as a STATCOM located at the motor terminal.

5.6.4 Field Validation Tests on Utility Solar PV System

The effectiveness of the proposed PV-STATCOM control in stabilizing a remotely located 5 hp critical IM is then field-demonstrated on a 10 kW PV solar system in the utility network of Bluewater Power Distribution Corporation (Study System 1). Three studies are presented (i) PV plant operating at UPF, (ii) PV plant operation with reactive power support according to German grid code [69], (iii) PV plant control as PV-STATCOM at night. In all these studies, the distance between the PV

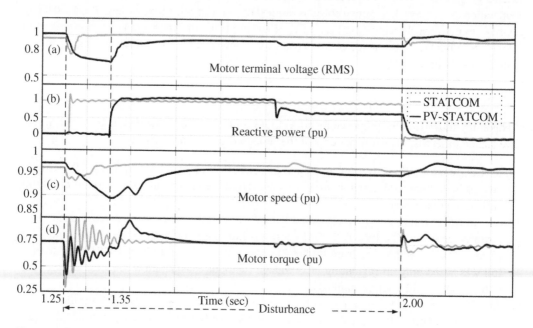

Figure 5.35 Performance comparison of remotely located PV-STATCOM and locally sited STATCOM in stabilizing IM during a large disturbance. (a) Motor terminal voltage, (b) reactive power, (c) motor speed, (d) motor torque. *Source:* Varma et.al. [74].

solar plant and the motor is kept at 15.6 km. The large disturbance is created by connecting an inductive load of 10 kvar for 750 ms. The communication delay between voltage measurement at motor terminal and its receipt at the PV system is taken as 100 ms for all cases.

5.6.4.1 PV Solar Plant Without PV-STATCOM Control

Figure 5.36a–c shows the motor terminal phase voltage, V_m, active (P_{PV}) and reactive power (Q_{PV}) output of the PV solar system, and motor current, respectively. The studies show that postdisturbance, the critical IM draws a large current and finally stalls. The motor electrical torque reduces due to the drop in motor terminal voltage, as torque is proportional to the square of voltage. The mismatch between electrical and mechanical torques leads to a decline in rotor speed. As the electrical torque reduces, the rotor consumes a higher active current to increase the electrical torque toward the mechanical torque. Also, as the speed drops, the motor consumes a higher reactive current. The increase in both active current and reactive current consumption leads to an increase in the total motor current. Thus, the IM current during a disturbance (voltage sag) is much higher as compared to steady-state operation. To prevent the motor from getting damaged due to this large sustained current, an overcurrent protection system typically disconnects the motor (not shown here).

5.6.4.2 PV Solar System Operation According to German Grid Code

The German grid code [69] requires PV inverters to inject 2% reactive current per 1% voltage drop for voltages below 0.9 pu to support the grid during transients. The performance of PV solar system with dynamic reactive current injection according to [68, 69] is depicted in Figure 5.37.

Figure 5.37a–c shows V_m, P_{PV}, Q_{PV} and motor current respectively. The PV solar system is initially operating at UPF by generating 4.8 kW. The motor is operating in steady state and consuming 8 A at $V_m = 1.02$ pu.

$t = 0$ s. The large disturbance is applied which causes V_m to decrease from 1 to 0.58 pu as shown in Figure 5.37(a). The PV system rides through the disturbance and injects maximum available active power as illustrated in Figure 5.37(b).

$t = 0.1$ s. As V_m reduces to 0.58 pu, the inverter injects 0.64 pu reactive current as per [69]. This causes voltage to increase rapidly up to 0.69 pu. At this time, the inverter modifies the reactive

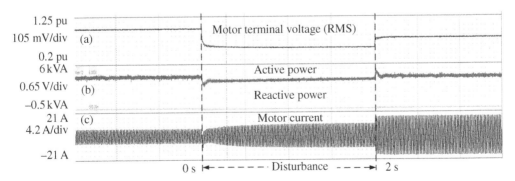

Figure 5.36 PV solar system operating at unity power factor (without PV-STATCOM control). (a) Motor terminal voltage, (b) active power and reactive power, (c) motor current. *Source:* Varma et.al. [74].

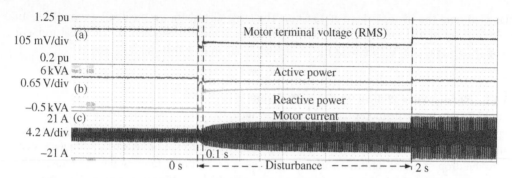

Figure 5.37 Response of PV operating according to German grid code. (a) Motor terminal voltage, (b) active power and reactive power, (c) motor current. *Source:* Varma et.al. [74].

current injection to 0.62 pu. The inverter active power remains almost constant, but the motor consumes a large current of 16 A during this period.

$t = 2\,s$. The large load is disconnected. The motor terminal voltage increases but only up to 0.82 pu (due to increased motor current). PV inverter injects 0.16 pu reactive current as V_m is below 0.9 pu.

This field test demonstrates that reactive current injection by PV inverter according to [69] is not adequate to stabilize a remotely located IM during a large disturbance. In contrast, the PV-STATCOM successfully stabilizes the remotely located motor under same conditions.

5.6.4.3 PV Solar System Operating as PV-STATCOM at Night

Figure 5.38a–c shows the motor terminal phase voltage V_m, active (P_{PV}) and reactive power (Q_{PV}) output of the PV solar system, and motor current, respectively, for PV-STATCOM operation during nighttime.

Initially, the PV system is idle. As soon as it detects a drop in V_m due to the large disturbance described earlier, the PV-STATCOM controller initiates dynamic reactive power support and ensures continuous stable operation of IM. When the large load is switched off, the PV system stops reactive power support. The PV-STATCOM control thus successfully stabilizes the remote IM at night during a large disturbance which would have otherwise destabilized the motor.

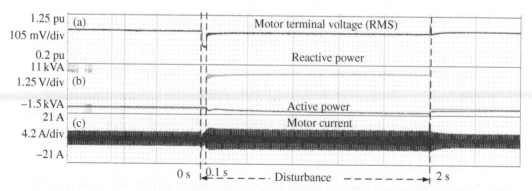

Figure 5.38 Motor stabilization by PV-STATCOM operation at night. (a) Motor terminal voltage, (b) active power and reactive power, (c) motor current. *Source:* Varma et.al. [74].

5.6.5 Simulation Studies on Study System 2

The PV-STATCOM performance is now investigated on Study System 2 through PSCAD simulation studies. The PV-STATCOM performance is evaluated over a wide range of key system parameters/conditions in terms of critical clearing time (CCT). The CCT is defined as the maximum fault duration for a three-phase fault at Bus 1 for which the IM will not stall.

The CCT is seen to decrease with:

i) increasing distance resulting in enhanced inductive impedance between PV plant and IM, thereby reducing effectiveness of voltage control at PV system bus;
ii) decreasing X/R ratio, because the impact of reactive power on voltage control reduces as resistance of line increases;
iii) increasing system strength (SCR), because reactive power control is more effective in weak system; and
iv) increasing voltage sag depth, because the motor decelerates faster when the voltage sag is larger.

Despite these wide variations in key system parameters, the proposed PV-STATCOM successfully stabilizes the motor.

In this 27.6 kV study system, a 4-MW PV solar plant is able to successfully stabilize a 2.65 MVA IM, even at a distance of 40 km. The stabilization of motor is dependent on adequate reactive power support from the PV solar plant for voltage regulation. Therefore, in any given system, the maximum distance from the PV plant for safely operating the motor will depend on the rating of PV plant, size of motor, the system impedance, and X/R ratio of the distribution line between the PV plant and the motor.

5.6.6 Summary

With the extensive growth of PV solar plants, it is quite likely that solar plants may get located in same distribution feeders as critical motor loads. This section presented a multi-mode control strategy for PV-STATCOM operation of PV solar plant for stabilizing remotely located critical IMs during system disturbances. The proposed PV-STATCOM control strategy provides a novel enhanced voltage support utilizing HVRT limits. The effectiveness of PV-STATCOM control strategy is demonstrated through (i) MATLAB Simulink software simulations, and subsequent field validation on a 10 kW PV solar system in the utility network of Bluewater Power Distribution Corp., Sarnia, in Canada, and (ii) PSCAD simulations of a realistic utility distribution feeder in Canada. The PV-STATCOM control:

1) stabilizes a critical IM even up to 40 km away, and also over a wide range of system conditions.
2) is effective even if the delay in communication of motor voltage control signal to PV plant site is 750 ms, which is seven times more than actual delay in real conditions.
3) ensures stable operation of critical IM during both day and night, even for the case in which reactive current support from PV systems required under German grid code fails to do so.
4) is as effective as a local STATCOM at motor terminal.

5.7 Conclusions

This chapter presented different night and day applications of PV-STATCOM technology in power distribution systems, each providing significant benefits. The PV-STATCOM can provide dynamic voltage control both during night and day as a STATCOM with a response time of 1–2 cycles that is typical of a STATCOM. The PV-STATCOM not only meets the LVRT requirement of IEEE Standard 1547-2018 but additionally provides rated dynamic current injection as STATCOM in the LVRT region.

Implementation of PV-STATCOM control on a SF is shown to significantly increase the connectivity of a WF connected on the same distribution feeder during nighttime when wind blows the most (in Ontario, Canada). The PV-STATCOM control is effective even if it is located 30 km away from the WF. Typically FACTS Controllers such as SVCs or STATCOMs are employed to provide dynamic voltage support for WFs and help their integration.

PV-STATCOM control on SFs can also help in reducing line losses both during night and day, resulting in substantial energy savings and potential reduction in customers' electricity bills.

Furthermore, in an actual distribution system with two 10 MW SFs, the proposed PV-STATCOM control is shown to help integrate a third 10 MW PV SF thereby eliminating the need for the actually installed STATCOM for reducing steady-state overvoltage and mitigating TOVs. The PV-STATCOM control on a solar PV plant is also shown to stabilize critical IMs located more than 15 km away. Typically SVCs or STATCOMs are utilized for stabilizing such critical motors. The PV-STATCOM control is thus shown to obviate or reduce the need for expensive FACTS Controllers such as SVC or STATCOM.

References

1 Varma, R.K., Khadkikar, V., and Seethapathy, R. (2009). Nighttime application of PV solar farm as STATCOM to regulate grid voltage. *IEEE Transactions on Energy Conversion (Letters)* 24: 983–985.

2 Zavadil, R., Miller, N., Ellis, A. et al. (2007). Queuing up. *IEEE Power and Energy Magazine* 5: 47–58.

3 Mathur, R.M. and Varma, R.K. (2002). *Thyristor-Based FACTS Controllers for Electrical Transmission Systems*. New York, NY: Wiley-IEEE Press.

4 Khadkikar, V., Chandra, A., Barry, A.O. et al. (2006). Application of UPQC to protect a sensitive load on a polluted distribution network. In *Proc. 2006 IEEE Power Engineering Society General Meeting*, 1–6.

5 Masters, C.L. (2002). Voltage rise: the big issue when connecting embedded generation to long 11 kV overhead lines. *Power Engineering Journal* 16: 5–12.

6 Ackermann, T. (2005). *Wind Power in Power Systems*. Chichester: Wiley.

7 Camm, E. and Edwards, C. (2008). Reactive compensation systems for large wind farms. In *Proc. 2008 IEEE/PES Transmission and Distribution Conference and Exposition*, 1–5.

8 Grünbaum, R. and Willemsen, N. (2012). FACTS for distribution grid integration of large scale wind generation. *CIRED 2012 Workshop: Integration of Renewables into the Distribution Grid*, 1–4.

9 Bostrom, A., Hassink, P., Thesing, M. et al. (2009). Voltage stabilization for wind generation integration in Western Texas grid. In *Proc. 2009 CIGRE/IEEE PES Joint Symposium Integration of Wide-Scale Renewable Resources Into the Power Delivery System*.

10 Piątek, K., Firlit, A., and Wojciechowski, D. (2015). Field experience with STATCOM in application to wind farms. In *Proc. 2015 International School on Nonsinusoidal Currents and Compensation (ISNCC)*, 1–5.

11 Ronner, B., Maibach, P., and Thurnherr, T. (2009). Operational experiences of STATCOMs for wind parks. *IET Renewable Power Generation* 3: 349–357.

12 Lake Turkana Wind Power (2016). https://www.rxhk.co.uk/solutions/facts/maxivar-statcom/lake-turkana-wind-power (accessed 9 April 2021).

13 Burbo Bank Extension DRC (2016). https://www.rxhk.co.uk/solutions/facts/maxivar-statcom/burbo-bank-extension-drc.

14 All systems go for Pen y Cymoedd Wind Energy Project Grid Connection (2016). Pen y Cymoedd Wind Farm, South Wales. https://www.power-technology.com/projects/pen-y-cymoedd-wind-farm-south-wales (accessed 9 April 2021).

15 Race Bank Offshore Wind Farm DRCs (2017). https://www.rxhk.co.uk/solutions/facts/maxivar-statcom/race-bank-offshore-wind-farm-drc.

16 Rahman, S.A., Varma, R.K., and Litzenberger, W.H. (2011). Bibliography of FACTS applications for grid integration of wind and PV solar power systems: 1995–2010 IEEE working group report. In *Proc. 2011 IEEE Power & Energy Society General Meeting*, 1–17.

17 Rahman, S.A., Varma, R.K., and Litzenberger, W.H. (2014). Bibliography of FACTS applications for grid integration of wind and PV solar power systems: 2010–2013 IEEE Working Group report. In *Proc. 2014 IEEE PES General Meeting*, 1–20.

18 Global Energy Transformation (2019). A roadmap to 2050. *International Renewable Energy Agency (IRENA) Report.*

19 Gullen Solar Farm (2017). https://www.gullensolarfarm.com

20 Grand Renewable Energy Park (2015). Haldimand County, Ontario. https://www.power-technology.com/projects/grand-renewable-energy-park-haldimand-county-ontario

21 Subramanian, S.B., Varma, R.K., and Vanderheide, T. (2020). Impact of grid voltage feed-forward filters on coupling between DC-link voltage and AC voltage controllers in smart PV solar systems. *IEEE Transactions on Sustainable Energy* 11: 415–425.

22 Varma, R.K., Rahman, S.A., Mahendra, A.C. et al. (2012). Novel nighttime application of PV solar farms as STATCOM (PV-STATCOM). In *Proc. 2012 IEEE Power & Energy Society General Meeting*, 1–8.

23 Varma, R.K. and Siavashi, E.M. (2019). Enhancement of solar farm connectivity with smart PV inverter PV-STATCOM. *IEEE Transactions on Sustainable Energy* 10: 1161–1171.

24 Muljadi, E., Singh, M., and Gevorgian, V. (2014). User Guide for PV Dynamic Model Simulation Written on PSCAD Platform. *NREL, Golden, CO, USA, Technical Rep. NREL/TP-5D00-62053.*

25 IEEE (2018). *IEEE Standard for Interconnection and Interoperability of Distributed Energy Resources with Associated Electric Power Systems Interfaces.* IEEE Std 1547-2018 (Revision of IEEE Std 1547-2003).

26 Hydro One Networks Inc. (2013). Distributed Generation Technical Interconnection Requirements. *Report DT-10-015 R3.*

27 Varma, R.K., Rahman, S.A., Vanderheide, T., and Dang, M.D.N. (2016). Harmonic impact of a 20-MW PV solar farm on a utility distribution network. *IEEE Power and Energy Technology Systems Journal* 3: 89–98.

28 Varma, R.K. and Siavashi, E.M. (2018). PV-STATCOM: a new smart inverter for voltage control in distribution systems. *IEEE Transactions on Sustainable Energy* 9: 1681–1691.

29 Liserre, M., Blaabjerg, F., and Hansen, S. (2005). Design and control of an LCL-filter-based three-phase active rectifier. *IEEE Transactions on Industry Applications* 41: 1281–1291.

30 Independent Electricity System Operator (IESO) (2011). System Impact Assessment Report – Grand Renewable Energy Park Project. *Report No. CAA ID 2010-399.*

31 ABB Report (2015). VArProTM STATCOM case study solar PV cluster in Massachusetts voltage regulation through reactive power support. https://library.e.abb.com/public/9ae411b2ed304cbba37115159e81fb5d/9AAK10103A2230_PV-Cluster_STATCOM-CS-LR.pdf

32 Iioka, D., Sakakibara, K., Yokomizu, Y. et al. (2009). Distribution voltage rise at dense photovoltaic generation area and its suppression by SVC. *Electrical Engineering in Japan* 166-2: 47–53.

33 Kabasawa, Y., Noda, T., Fukushima, K. et al. (2012). Consumer voltage regulation using coordinated control of distributed static synchronous compensators – μSTATCOMs. In *Proc. 2012 3rd IEEE PES Innovative Smart Grid Technologies Europe (ISGT Europe)*, 1–7.

34 Aly, G.E.M., El-Zeftawy, A., El-Hefanwy, A., and Eraky, S.A. (1999). Reactive power control on residential utility-interactive PV power systems. *Electric Power Systems Research* 51-3: 187–199.

35 ABB (2019). ABB technology stabilises grid and improves power quality at the largest solar power plant in Jordan. http://www.abb.pl/cawp/seitp202/c3a70ee0924b6062c12580ec00529aa0.aspx.

36 Roy, N.K., Pota, H.R., Mahmud, M.A. et al. (2013). D-STATCOM control in distribution networks with composite loads to ensure grid code compatible performance of photovoltaic generators. In *Proc. 2013 IEEE 8th Conference on Industrial Electronics and Applications (ICIEA)*, 55–60.

37 Tang, L., Yan, A., Narang, A. (2014). Mitigation for connecting distributed generators beyond power distance limitation. In *Proc. 2014 CIGRE Canada Conference*, Toronto, ON, Canada.

38 Yazdani, A., Fazio, A.R.D., Ghoddami, H. et al. (2011). Modeling guidelines and a benchmark for power system simulation studies of three-phase single-stage photovoltaic systems. *IEEE Transactions on Power Delivery* 26: 1247–1264.

39 Pedro Antmann (2009). Reducing Technical and Non-Technical Losses in the Power Sector. *Background Paper* for the World Bank Group Energy Sector Strategy.

40 World Bank (2017). Reduction of Technical and Non-Technical Losses in Distribution Networks – CIRED Overview, Working Group on Losses Reduction CIRED WG CC-2015-2 *Final Report*.

41 Rahman, S.A. (2012). *Novel Controls of Photovoltaic (PV) Solar Farms*. PhD Thesis, Electrical and Computer Engineering, The University of Western Ontario, London, ON, Canada.

42 Carpentier, J. (1979). Optimal power flows. *International Journal of Electrical Power and Energy Systems* 1-1: 3–15.

43 Luque, A. and Hegedus, S. (2011). *Handbook of Photovoltaic Science and Engineering*, 2e. Wiley.

44 Emerson Solar Utility Scale PV Inverter Systems (2012). *SPV Product Guide: 145kVA-1590kVA*.

45 Hung, D.Q. and Mithulananthan, N. (2013). Multiple distributed generator placement in primary distribution networks for loss reduction. *IEEE Transactions on Industrial Electronics* 60: 1700–1708.

46 Independent Electricity System Operator (IESO). Ontario. http://ieso.ca (accessed 9 April 2021).

47 Varma, R.K., Siavashi, E., Maleki, H. et al.(2016). PV-STATCOM: A novel smart inverter for transmission and distribution system applications. *Poster Paper, 7th International Conference on Integration of Renewable and Distributed Energy Resources*, Niagara Falls, Canada.

48 Powerworld Software. http://www.powerworld.com (accessed 9 April 2021).

49 US Energy Information Administration (2017). How much electricity does an American home use?. https://www.eia.gov/tools/faqs/faq.php?id=97&t=3

50 Jarc, D.A. and Robechek, J.D. (1982). Static induction motor drive capabilities for the petroleum industry. *IEEE Transactions on Industry Applications* IA-18: 41–45.

51 Bristow, R. (2003). Induction motors and their controllers as part of energy reduction strategies within the pulp and paper industry. *Paper Technology* 44: 25–35.

52 Gomez, J.C., Morcos, M.M., Reineri, C.A., and Campetelli, G.N. (2002). Behavior of induction motor due to voltage sags and short interruptions. *IEEE Transactions on Power Delivery* 17: 434–440.

53 Potamianakis, E.G. and Vournas, C.D. (2006). Short-term voltage instability: effects on synchronous and induction machines. *IEEE Transactions on Power Systems* 21: 791–798.

54 Thorsen, O.V. and Dalva, M. (1995). A survey of faults on induction motors in offshore oil industry, petrochemical industry, gas terminals, and oil refineries. *IEEE Transactions on Industry Applications* 31: 1186–1196.

55 EPRI (2013). The Cost of Power Disturbances to Industrial and Digital Economy Companies. EPRI, Palo Alto, CA, USA, *Rep. No. 3002000476*.

56 Jose Luis Olabarrieta Rubio, Pablo Eguia (2018). FACTS brings power stability to mining equipment. https://new.abb.com/news/detail/7821/facts-brings-power-stability-to-mining-equipment (accessed 9 April 2021).

57 ABB Report. Static Var Compensators for Mining. https://library.e.abb.com/public/96b201ac4a74016583257d5b003f5cd7/FACTS%20-%20SVC%20for%20Mining-20140811-LowRes.pdf (accessed 9 April 2021).

58 Hedayati, M., Mariun, N., Hizam, H. et al. (2010). Performance study of drive systems and shunt FACTS for the operation of induction motors. In *Proc. 2010 IEEE Symposium on Industrial Electronics and Applications (ISIEA)*, 463–468.

59 Ebadian, M. and Alizadeh, M. (2010). Effect of static VAR compensator to improve an induction motor's performance. In *Proc. 2010 IPEC*, 209–214.

60 Hammad, A.E. and El-Sadek, M.Z. (1989). Prevention of transient voltage instabilities due to induction motor loads by static VAr compensators. *IEEE Transactions on Power Systems* 4: 1182–1190.

61 Tan, O.T. and Thottappillil, R. (1994). Static VAr compensators for critical synchronous motor loads during voltage dips. *IEEE Transactions on Power Systems* 9: 1517–1523.

62 Hingorani, N.G. and Gyugyi, L. (1999). *Understanding FACTS*. New York: IEEE Press.

63 ABB (2015). *VarPro STATCOM Dynamic Reactive Power Compensation Power Quality Solutions for Heavy Industry*. New Berlin, WI: ABB.

64 Seattle Iron & Metals Corp. (2000). D-STATCOM Project. Mitsubishi Electric Power Products, Tokyo, Japan, *Report*.

65 Teleke, S., Yazdani, A., Gudimetla, B. et al. (2011). Application of STATCOM for power quality improvement. In *Proc. 2011 IEEE/PES Power Systems Conference and Exposition*, 1–6.

66 Rule 21 Smart Inverter Working Group. https://www.cpuc.ca.gov/general.aspx?id=4154#:~:text=The%20Smart%20Inverter%20Working%20Group,distributed%20energy%20resources%20(DERs) (accessed 9 April 2021).

67 EPRI (2014). Common Functions for Smart Inverters, 3e. EPRI, Palo Alto, CA, USA, *Techn. Rep. 3002002233*.

68 BDEW (2008). *Technical guideline: generating plants connected to the medium-voltage network*. (Guideline for Generating plants' Connection to and Parallel Operation with the Medium-Voltage Network). BDEW (Bundesverband der Energie- und Wasserwirtschaft e.V.), Berlin, Germany (June 2008, revised January 2013).

69 E.ON Netz GmbH (2006). *Grid Code for High and Extra High Voltage*. E.ON Netz GmbH, Bayreuth, Germany.

70 Bae, Y., Vu, T., and Kim, R. (2013). Implemental control strategy for grid stabilization of grid-connected PV system based on German grid code in symmetrical low-to-medium voltage network. *IEEE Transactions on Energy Conversion* 28: 619–631.

71 Tang, C., Chen, Y., and Chen, Y. (2015). PV power system with multi-mode operation and low-voltage ride-through capability. *IEEE Transactions on Industrial Electronics* 62: 7524–7533.

72 CAISO (2014). *Rule 21: Generating Facility Interconnections*. California, USA.

73 EPRI (2016). Common Functions for Smart Inverters, 4e. EPRI, Palo Alto, CA, USA, *Techn. Rep. 3002008217*.

74 Varma, R.K., Mohan, S., and McMichael-Dennis, J. (2020). Multi-mode control of PV-STATCOM for stabilization of remote critical induction motor. *IEEE Journal of Photovoltaics* 10-6: 1872–1881.

75 NERC (2020). NERC standard PRC-024-3 *Frequency and voltage protection settings for generating resources*.

76 AC, Mahendra (2013). *Novel Control of PV Solar and Wind Farm Inverters as STATCOM for Increasing Connectivity of Distributed Generators*. MESc. Thesis, Electrical and Computer Engineering, The University of Western Ontario, London, ON, Canada, 2013.

77 Coffey, J. (2017). *Latency in Optical Fiber Systems*. Hickory, NC: CommScope *NREL/TP-5D00-67799*.

78 NREL (2017). Demonstration of Essential Reliability Services by a 300 MW Solar Photovoltaic Power Plant. NREL, Golden, CO, USA, *Report NREL/TP-5D00-67799*.

79 Yazdani, A. and Iravani, R. (2010). *Voltage Sourced Converters in Power Systems Modeling, Control and Applications*. New York: IEEE Press/Wiley.

80 NREL (2014). Wind and Solar Energy Curtailment: Experiences and Practices in the United States. NREL, Golden, CO, USA, *Rep. No. NREL/TP-6A20-60983*.

81 Solar Power (2015). *100% Self – Consumption Solution with the SMA Power Control Module*. SMA, USA.

6

PV-STATCOM APPLICATIONS IN TRANSMISSION SYSTEMS

This chapter describes different night and day grid support functions provided by PV-STATCOM in transmission systems. These include improving power transfer capacity in transmission lines, damping of power oscillations, and alleviation of Fault Induced Delayed Voltage Recovery (FIDVR). These functionalities are provided by reactive power modulation at night and by a combination of active and reactive power modulation during daytime. PV-STATCOM applications are presented for mitigation of subsynchronous oscillations (SSOs) in synchronous generators and induction generator (IG)-based wind farms (WFs) connected to series compensated transmission lines. The PV-STATCOM is also shown to provide fast frequency response (FFR) and power oscillation damping (POD), simultaneously.

The unique PV-STATCOM control system for each functionality together with its case study is presented. The standard controller components such as the maximum power point tracking (MPPT) control, DC-link voltage controller, inner loop current controllers, PLL, *abc-dq* transformation, etc. are already explained in Chapter 3 and hence, not described here.

6.1 Increasing Power Transmission Capacity by PV-STATCOM

Transient stability and POD are recognized as two major factors that limit power transfer over long transmission lines [1]. Conventionally, low-frequency electromechanical power oscillations (typically 0.1–2 Hz) are damped by Power System Stabilizers (PSSs) integrated with synchronous generators [1]. However, Flexible AC Transmission System (FACTS) Controllers have been extensively utilized in power systems for improving both transient stability and POD, thereby enhancing power transfer capability in transmission lines [2–12].

Large-scale solar photovoltaic (PV) plants in excess of 100 MW are being increasingly connected worldwide. Some of these large solar farms in 2020 are Qinghai (2220 MW) in China [13], Pavagada (2000 MW) and Kamuthi (648 MW) in India [14]; Longyangxia Solar-Hydro (697 MW) in China; Rancho Cielo Solar Farm (600 MW), Solar Star I and II (579 MW), Topaz Solar Farm (550 MW), Agua Caliente Solar Project (295 MW), and California Valley Solar Ranch Farm (250 MW) in USA [15]. However, there is a growing concern that significant penetration of such inertia-less PV solar plants will have an adverse impact on power system stability [16–18].

Solar PV systems are typically idle during nighttime. In 2010, a novel night and day control of solar PV systems as STATCOM, termed PV-STATCOM, was proposed by the author of this book for increasing power transfer capacity in transmission systems [19], and further elaborated in

Smart Solar PV Inverters with Advanced Grid Support Functionalities, First Edition. Rajiv K. Varma.
© 2022 The Institute of Electrical and Electronics Engineers, Inc. Published 2022 by John Wiley & Sons, Inc.

[20, 21]. As described above, the sizes of solar PV plants are increasingly becoming comparable (even exceeding) those of commercially used Static Var Compensators (SVCs) and STATCOMs. Hence, potential implementation of PV-STATCOM technology on such solar PV plants can be cost-effectively utilized for increasing power transfer capacity of existing transmission networks and obviate the concerns about adverse impacts of solar PV plants on system stability.

This section describes a novel voltage control together with auxiliary damping control for a grid-connected PV solar farm inverter to act as PV-STATCOM both during night and day for increasing transient stability and consequently the power transmission limit [21]. The PV-STATCOM provides POD utilizing the entire solar farm inverter capacity at night (Full STATCOM mode) and the inverter capacity remaining after active power generation during the day (Partial STATCOM mode with active power priority) as described in Sec. 4.7.2. Similar STATCOM control functionality can also be implemented in inverter-based wind turbine generators during no-wind or partial wind scenarios for improving the transient stability of the system. Studies are performed for two variants of a Single Machine Infinite Bus (SMIB) system. One SMIB system uses only a single solar PV farm as PV-STATCOM connected at the midpoint whereas the other system uses a combination of a PV-STATCOM and another PV-STATCOM or an inverter-based wind Distributed Generator (DG) with similar STATCOM functionality. Three-phase fault studies are conducted using the electromagnetic transient software PSCAD, and the improvement in power transmission limit is investigated for different combinations of STATCOM controllers on the solar and WF inverters, both during night and day.

6.1.1 Study Systems

The single line diagrams of two study systems – study system I and study system II are depicted in Figure 6.1a,b, respectively. Both systems are SMIB systems in which a large equivalent synchronous generator (1110 MVA) supplies power to the infinite bus over a 200 km, 400 kV transmission line. This line length is typical of a long line carrying bulk power in Ontario. In study system I, a 100 MW PV solar farm (DG) as Static Synchronous Compensator (STATCOM) (PV-STATCOM) is connected at the midpoint of the transmission line. In study system II, two 100 MVA inverter-based DGs are connected at 1/3rd (bus 5) and 2/3rd (bus 6) of line length from the synchronous generator. The DG connected at bus 6 is a PV-STATCOM and the other DG at bus 5 is either a PV-STATCOM or a WF with STATCOM functionality. In this case, the WF employs Permanent Magnet Synchronous Generator (PMSG)-based wind turbine generators with full AC–DC–AC converter. It is understood that both the solar DG and wind DG employ several inverters. However, for this analysis, each DG is considered to have a single equivalent inverter with the rating equal to the total rating of solar DG or wind DG, respectively. The wind DG and solar DG are considered to be of the same rating, hence can be interchanged in terms of location depending upon the studies being performed. All the system parameters are given in [4, 21].

6.1.2 System Model

Figure 6.2 presents the block diagrams of various subsystems of the two equivalent DGs. The synchronous generator is represented by a detailed sixth-order model and a DC1A type exciter [4]. The transmission line segments TL1, TL2, TL11, TL12, and TL22 shown in Figure 6.1 are represented by lumped pi-circuits. The PV solar DG, as shown in Figure 6.2, is modeled as an equivalent voltage source inverter along with controlled current source as the DC source which follows the $I–V$ characteristics of PV panels [22]. The wind DG is likewise modeled as an equivalent voltage sourced inverter. In the solar DG, the DC power is provided by the solar panels, whereas in the full

(a)

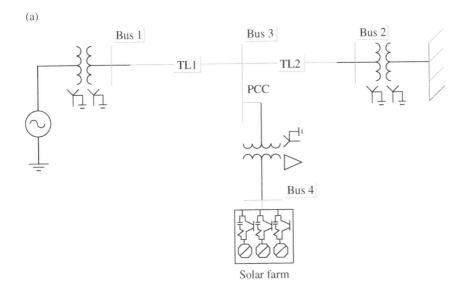

(b)

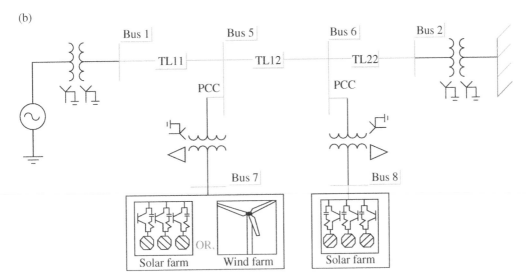

Figure 6.1 Single line diagram of (a) study system I with single solar farm (DG) and (b) study system II with a solar farm (DG) and a solar/wind farm (DG). *Source:* Varma et al. [21].

converter-based wind DG, the DC power comes out of a controlled AC–DC rectifier connected to the PMSG wind turbines, depicted as "Wind Turbine-Generator-Rectifier (T-G-R)." The DC power produced by each DG is fed into the DC bus of the corresponding inverter, as illustrated in Figure 6.2. An MPPT algorithm based on incremental conductance algorithm [23] is used to operate the solar DGs at their maximum power point (MPP) all the time and is integrated with the inverter controller. The wind DG is also assumed to operate at its MPP, as the proposed control utilizes only the inverter capacity left after the MPP operation of both the solar DG and wind DG.

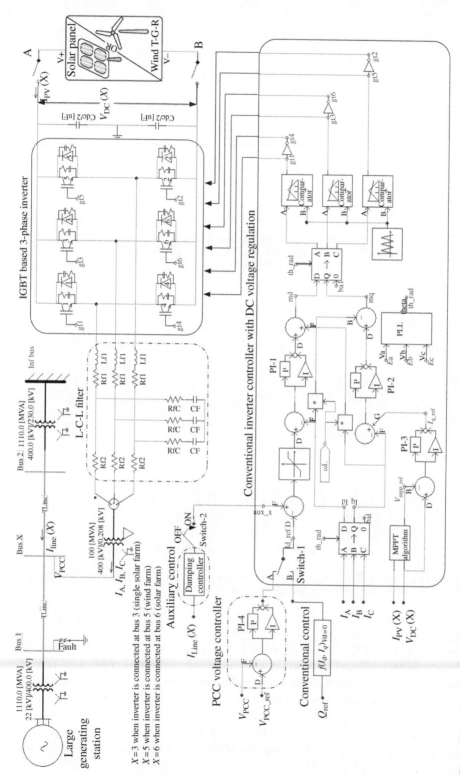

Figure 6.2 Overall DG (solar/wind) system model with damping controller and PCC voltage control system. *Source:* Varma et al. [21].

For PV-STATCOM operation during nighttime, the solar panels are disconnected from the inverter and a small amount of active power is drawn from the grid to charge the DC capacitor. The voltage source inverter in each DG is composed of six IGBTs and associated snubber circuits as shown in Figure 6.2. An appropriately large DC capacitor of size 200 F is selected to reduce the DC side ripple [24]. Each phase has a pair of IGBT devices which converts the DC voltage into a series of variable width pulsating voltages, using the sinusoidal pulse width modulation (SPWM) technique [25]. An L–C–L filter is also connected at the inverter AC side [24].

6.1.3 Control System

6.1.3.1 Conventional Reactive Power Control
The conventional reactive power control only regulates the reactive power output of the inverter such that it can perform unity power factor (UPF) operation along with the DC-link voltage control [26]. The switching signals for the inverter switching are generated through two current control loops in d-q coordinate system [25, 26]. The inverter operates in conventional controller mode only provided that "Switch-2" is in "OFF" position. In this simulation, the voltage vector is aligned with the quadrature axis, i.e. $V_d = 0$ [25, 26], hence, Q_{ref} is only proportional to I_d which sets the reference I_{d_ref} for the upper control loop involving PI1. (Please note that this alignment of axes is different from that described in Chapter 3, although it does not affect the overall results.) Meanwhile, the quadrature axis component I_q is used for DC-link voltage control through two PI controllers (PI-2 and PI-3) [25, 27] shown in Figure 6.2b according to the set point voltage provided by the MPPT and injects all the available active power "P" to the network [26]. To generate the proper IGBT switching signals (gt1, gt2, gt3, gt4, gt5, and gt6), the d-q components (md and mq) of the modulating signal are converted into three-phase sinusoidal modulating signals and compared with a high frequency (5 kHz) fixed magnitude triangular wave (carrier signal).

6.1.3.2 PCC Voltage Control
In the *PCC voltage control* mode of operation, the PCC voltage is controlled through reactive power exchange between the DG inverter and the grid. The conventional "Q" control channel is replaced by the PCC voltage controller in Figure 6.2, simply by switching the "Switch-1" to the position "A." Hence, the measured signal, V_{PCC}, at the PCC is compared with the preset reference value V_{PCC_ref} and is passed through the PI regulator, PI-4, to generate I_{d_ref}. The rest of the controller remains unchanged. The upper current control loop is used to regulate the PCC voltage whereas the lower current control loop is used for DC voltage control and as well as for supply of DG power to the grid. The amount of reactive power flow from the inverter to the grid depends on set point voltage at PCC. The parameters of the PCC voltage controller are tuned by systematic trial and error method to achieve the fastest step response, least settling time, and a maximum overshoot of 10–15%.

6.1.3.3 Damping Control
An auxiliary *damping controller* is added to the PV control system and shown in Figure 6.2. This controller utilizes line current magnitude as the control signal. The output of this controller is added with the signal I_{d_ref}. The transfer function of this damping controller is expressed as in [1]:

$$F_D(s) = G \left[\frac{sT_w}{1 + sT_w} \right] \left[\frac{1 + sT_1}{1 + sT_2} \right] \tag{6.1}$$

The transfer function comprises a gain, a washout stage, and a first-order lead–lag compensator block. This controller is utilized to damp the rotor mode oscillations of the synchronous generator and thereby improve system stability. The damping controller is activated by toggling "Switch-2" to the "ON" position. This damping controller can operate in conjunction with either the conventional reactive power control mode or with the PCC voltage control mode by toggling the "Switch-1" to position "B" or "A," respectively.

At first, the base case generator operating power level is selected for performing the damping control design studies. This power level is considered equal to the transient stability limit of the system with the solar farm being disconnected at night. At this operating power level, if a three-phase fault occurs at bus 1, the generator power oscillations decay with a damping ratio of 5%. The solar farm is now connected and operated in the PV-STATCOM mode. The parameters of the damping controller F_D are selected as follows. The washout time constant T_w is chosen to allow the generator electromechanical oscillations in the frequency range up to 2 Hz to pass through [1]. The gain G, time constants T1 and T2 are sequentially tuned to get the fastest settling time of the electromechanical oscillations at the base case generator power level through repetitive PSCAD simulations. Thus the best combination of the controller parameters is obtained with a systematic hit and trial technique [21]. It is emphasized that these controller parameters are not optimal and better parameters could be obtained by following more rigorous control-design techniques [1]. However, the objective of this study is only to demonstrate a new concept of using a PV solar farm inverter as a STATCOM using these reasonably good controller parameters. In this controller, although line current magnitude signal is used, other local or remote signals which reflect the generator rotor mode oscillations [4] may also be utilized.

6.1.4 Power Transfer Studies for Study System I

The effectiveness of different PV-STATCOM controls is evaluated in terms of the power transfer limit achievable with each type of control. Stability studies are carried out using PSCAD simulation software for both the study systems during night and day, by applying a 3-Line to Ground (3LG) fault at bus 1 for five cycles. The power transfer limit is considered to be that level of generator power output for which: (i) the system remains stable after the fault, (ii) the postfault power oscillations decay with a damping ratio of 5% [4, 28], and (iii) the peak overshoot of PCC voltage is limited within 1.1 pu of nominal voltage. The damping ratio is used to express the rate of decay of the oscillation [28]. For an oscillatory mode having an eigenvalue of $\sigma + j\omega$, the damping ratio ξ is defined as

$$\xi = -\frac{\sigma}{\sqrt{\sigma^2 + \omega^2}}, \text{and } \sigma = \frac{1}{\tau} \tag{6.2}$$

where, τ is the time constant.

Therefore, for a 5% damping ratio of the rotor mode having oscillation frequency of 0.95 Hz, as considered in this study, the postfault settling time of the oscillations to come within 5% (typically within three times the time constant) of its steady state value is almost 10 seconds. Simulation studies for determining the maximum power transfer limits for different PV-STATCOM controls of the solar DG in study system I, are described below.

6.1.4.1 Nighttime Operation of Solar PV System as PV-STATCOM

a) *Solar system idle:* The maximum stable power output from the generator P_g is 731 MW when the solar DG is simply sitting idle during night and is disconnected from the network. This power flow level is chosen to be the base value against which the improvements in power flow with different proposed controllers are compared in Table 6.1.

Table 6.1 Increase in stable power transfer limit (MW) for study system I with different PV-STATCOM controls.

PV STATCOM Control	Night	Day	
		Solar power output 19 MW	Solar power output 91 MW
Voltage control	102	85	7
Damping control	119	121	142
Voltage control with damping control	168	93	36

Source: Varma et al. [21].

b) *PV-STATCOM operation with voltage control:* This study is performed to evaluate the transient stability limit when the PV-STATCOM provides voltage control at PCC. No damping control is considered. It is noted that the increase in power transfer limit is dependent upon the choice of reference values for PCC voltage V_{PCC}. In the best-case scenario when V_{PCC} is regulated to 1.01 pu, the maximum power output from the generator increases to 833 MW, i.e. 102 MW above the base case.

c) *PV-STATCOM operation with damping control:* For this control, the quantities P_g, P_{inf}, P_{solar}, and Q_{solar} are illustrated in Figure 6.3a. The power transfer limit is seen to increase to 850 MW, i.e. 119 MW more than the base case. The damping controller utilizes the full rating of the DG inverter at night to provide controlled reactive power Q_{solar} and effectively damps the generator rotor mode oscillations. The voltages at generator bus V_g and at PCC bus $V_{rms(PCC)}$ are depicted in Figure 6.3b. A very small amount of negative power flow from the solar farm P_{solar} is observed during nighttime. This reflects the losses in the inverter IGBT switches, transformer, and filter resistances caused by the flow of active current from the grid into the solar farm inverter to charge the DC-link capacitor and maintain its voltage constant while operating the PV inverter as STATCOM with the damping controller (or even with voltage controller). During nighttime, the reference DC-link voltage V_{mpp_ref} is chosen around the typical daytime rated MPP voltage. The oscillations in the solar PV power output during nighttime, as seen in Figure 6.3, are due to the active power exchanged by the solar inverter both during the charge and discharge cycles in trying to maintain a constant voltage across the DC-link capacitor, thereby enabling the inverter to operate as a STATCOM.

d) *PV-STATCOM operation with combination of voltage control and damping control:* The generator and infinite bus power are depicted in Figure 6.4a, and corresponding voltages are shown in Figure 6.4b. Although rotor mode oscillations settle faster, the power transfer cannot be improved beyond 899 MW (i.e. 168 MW more than base case) due to high overshoot in voltages.

6.1.4.2 Daytime Operation of Solar PV System as PV-STATCOM

a) *Solar system operation at unity power factor:* The conventional control of a PV solar DG does not seem to alter the power transmission limit of 731 MW in any appreciable manner. The base case power transfer limit during daytime is, therefore, taken as 731 MW.

b) *PV-STATCOM operation with voltage control:* The power transfer increases for both low (19 MW) and high (91 MW) power output from the solar farm are seen to be highly sensitive to the PCC bus voltage setpoint. It is also noted that there is a decrease in the availability of reactive power capacity $Q = \sqrt{(S^2 - P^2)}$, after active power production P, where S is the inverter rating of the solar DG. Hence, the ability to change the bus voltage is limited, which leads to a lower increase in power transmission capacity. The increases in power transfer limit with this PV-STATCOM control at both low (19 MW) and high (91 MW) PV power outputs, above the base case, are compiled in Table 6.1.

(a)

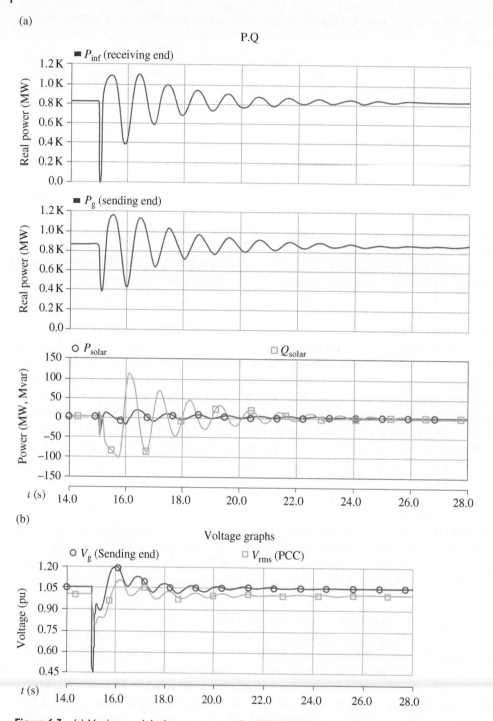

Figure 6.3 (a) Maximum nighttime power transfer (850 MW) from generator with PV-STATCOM damping controller, (b) voltages at generator terminal and PCC of PV-STATCOM. *Source:* Varma et al. [21].

(a)

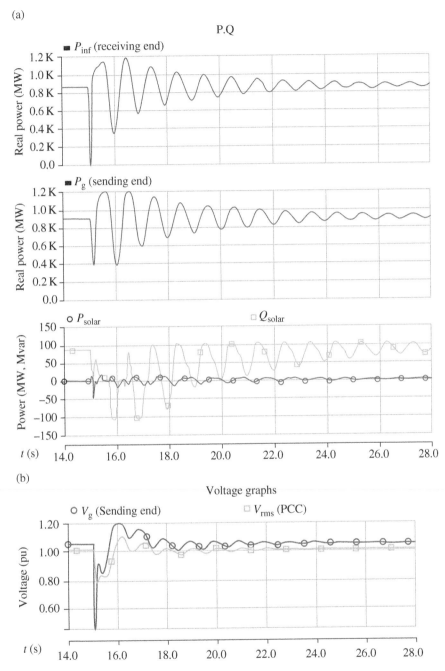

Figure 6.4 (a) Maximum nighttime power transfer (899 MW) from generator with combined damping control and voltage control provided by PV-STATCOM, (b) voltages at generator terminal and PV-STATCOM PCC. *Source:* Varma et al. [21].

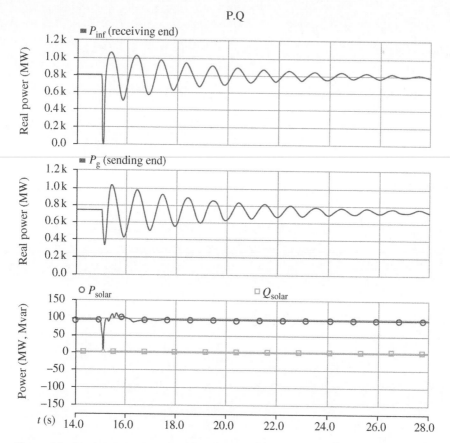

Figure 6.5 Maximum daytime power transfer (719 MW) from generator with solar DG generating 91 MW active power at unity power factor. *Source:* Varma et al. [21].

c) *PV-STATCOM operation with damping control:* The quantities P_g, P_{inf}, P_{solar}, and Q_{solar} are shown for the cases without damping controller and with PV-STATCOM damping controller in Figures 6.5 and 6.6, respectively. The available inverter capacity after real power generation of 91 MW is, $Q = \sqrt{(S^2 - P^2)} = 41.5$ Mvar, which is used for damping oscillations during daytime.

The power transfer capacity increase in the daytime is expected to be lower than the nighttime, as only a part of the total inverter capacity is available for damping control during the day. However, it is actually observed to be higher, i.e. 850 MW. This is 142 MW more than the base case. This is because of the additional constraint that while increasing the power transfer, the overshoot in PCC voltage should not exceed 1.1 pu. If the power transfer is allowed until its damping ratio limit of 5% regardless of voltage overshoot, the maximum nighttime power transfer is found to be 964 MW (although not plotted here). The increases in power transfer limit with this PV-STATCOM control at both low (19 MW) and high (91 MW) PV power outputs, above the base case, are compiled in Table 6.1.

d) *PV-STATCOM operation with combination of voltage control and damping control:* A further increase in power transfer is observed when both voltage control and damping control are employed, as compared to voltage control alone. The increases in power transfer limit with this combined PV-STATCOM control at both low (19 MW) and high (91 MW) PV power outputs, above the base case, are listed in Table 6.1. However, these increases in power transfer limits are lower than those achieved by PV-STATCOM damping control acting alone.

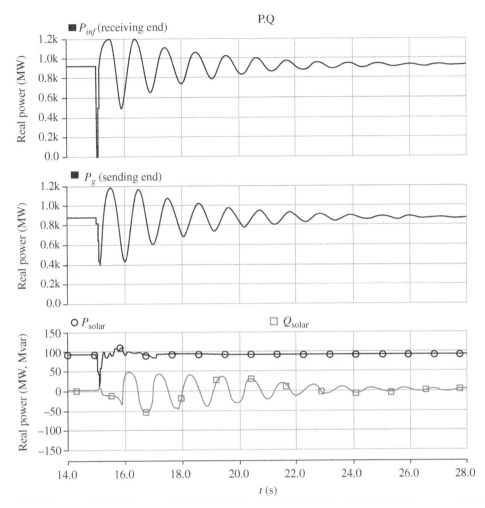

Figure 6.6 Maximum daytime power transfer (861 MW) from generator with solar DG generating 91 MW active power and using PV-STATCOM damping controller. *Source:* Varma et al. [21].

The maximum increase in power transfer limit during nighttime is achieved with a combination of voltage control and damping control, whereas the same during daytime is accomplished with damping control alone. This is because, in the night the entire MVA rating of the solar DG inverter is available for reactive power exchange, which can be utilized for achieving the appropriate voltage profile at PCC conducive for increasing the power transfer, as well as for increasing the damping of oscillations.

During daytime, firstly, the generation of active power from the solar DG tends to increase the voltage at PCC, and secondly, the net reactive power availability also gets reduced especially at large solar active power outputs. Therefore, it becomes difficult with limited reactive power to accomplish the appropriate voltage profile at PCC for maximum power transfer and also to impart adequate damping to the oscillations. However, if only damping control is exercised during daytime, power transfer limits appear to improve with higher active power outputs from the solar DG. This is because active power generation increases the PCC voltage which can be potentially helpful in increasing the power transfer capacity.

6.1.5 Power Transfer Studies for Study System II

In this study, the effectiveness of the proposed PV-STATCOM damping control strategy is shown for study system II depicted in Figure 6.1b. A three phase to ground fault of five cycles is applied to generator bus at $t = 8$ s and power transfer limits are obtained through PSCAD studies for the following eight cases. The net increases in power transfer limits accomplished with the proposed damping control for different active power outputs from both DGs as compared to those attained with the conventional UPF operation of both DGs are depicted in Table 6.2.

6.1.5.1 Nighttime Operation of Solar DG and Wind DG as STATCOM

i) *Case 1 – None of the DGs generate active power:* The maximum power transfer limit is 731 MW as obtained in the previous study

ii) *Case 2 – Only wind DG generates active power. Both DGs operate at unity power factor:* The power transfer limit decreases slightly with increasing wind power output.

iii) *Case 3 – None of the DGs generate active power but both operate with proposed STATCOM damping control:* The different variables, generator power P_g, infinite bus power P_{inf}, real power of wind DG P_{wind}, reactive power of the wind DG Q_{wind}, real power of the solar DG P_{solar}, and the reactive power of the solar DG Q_{solar} are illustrated in Figure 6.7. Even though the entire ratings (100 Mvar) of the wind DG and solar DG inverters are not completely utilized for damping control, the power transfer limit increases significantly to 960 MW, i.e. 229 MW more than the base case.

iv) *Case 4 – Only wind DG generates active power but both DGs operate on damping control:* There is only a marginal improvement in power limit with decreasing power output from the wind DG.

6.1.5.2 Daytime Operation of Solar DG and Wind DG as STATCOM

i) *Case 5 – Both DGs generate real power:* The power transfer limit from the generator decreases as the power output from both DGs increase.

ii) *Case 6 – Only solar DG generates power:* The power transfer limit from the generator decreases as the power output from the solar DG increases. However, no substantial changes in power limits are observed as compared to the case when both DGs generate power (Case 5).

Table 6.2 Increase in power transfer limits for study system II with different DG power outputs.

DG real power outputs (MW)	Power limit increase (MW)
Night	
$P_{solar} = 0$; $P_{wind} = 0$	229
$P_{solar} = 0$; $P_{wind} = 20$	219
$P_{solar} = 0$; $P_{wind} = 95$	220
Day	
$P_{solar} = 20$; $P_{wind} = 20$	197
$P_{solar} = 95$; $P_{wind} = 95$	230
$P_{solar} = 20$; $P_{wind} = 0$	214
$P_{solar} = 95$; $P_{wind} = 0$	219

Source: Varma et al. [21].

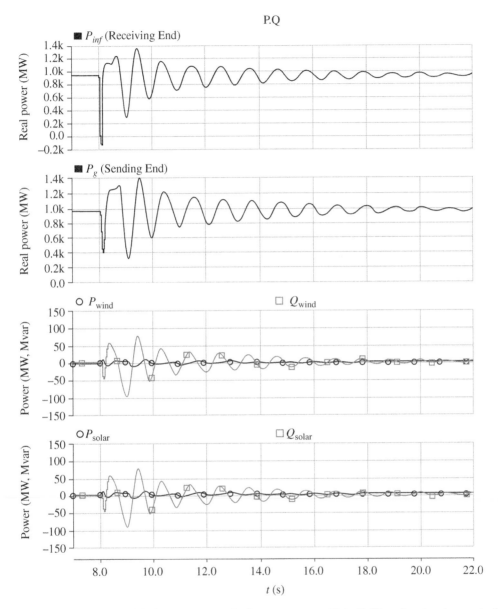

Figure 6.7 Maximum nighttime power transfer from generator with both DGs using damping controller but with no real power generation. *Source:* Varma et al. [21].

iii) *Case 7 – Both DGs generate real power and operate on damping control:* This case is illustrated by different variables P_g, P_{inf}, P_{wind}, Q_{wind}, P_{solar}, and Q_{solar} in Figure 6.8. The power limit does not change much with increasing power output from both DGs.

iv) *Case 8 – Only solar DG generates active power but both the DGs operate with STATCOM damping control:* The power limit does not appear to change much with increasing power output from the solar DG.

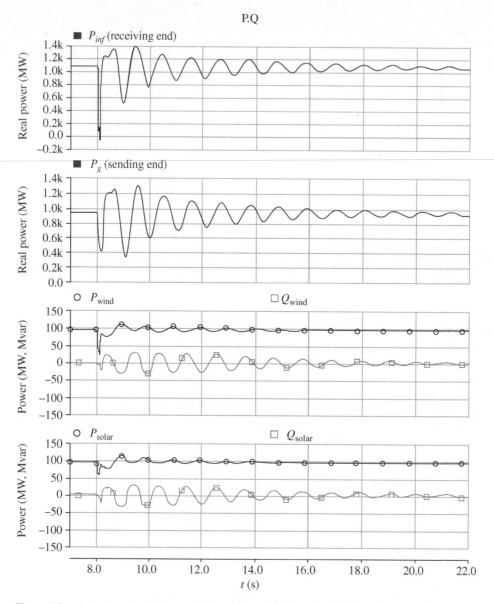

Figure 6.8 Maximum daytime power transfer from generator while both DGs generate 95 MW each using STATCOM damping controller. *Source:* Varma et al. [21].

The proposed damping control on the two DGs producing zero active power output in the night increases the power transfer limits substantially by about 220 MW. This is expected as in the night the entire inverter MVA rating (100 MVA) of both the DGs is available for damping control. The improvement is slightly less when wind DG produces high amount of active power. This is expected as the reactive power availability decreases with the wind DG power output. During daytime, the proposed damping control on both the DGs also increases the power transfer limits substantially. A greater increase is seen during high power generation by any DG, as high power output improves the PCC voltage profile which assists in increasing the power transfer capacity.

6.1.6 Summary

Solar farms are idle during nights. A novel patented control paradigm of PV solar farm as PV-STATCOM is presented whereby they can be operated as a STATCOM in nighttime with full inverter capacity and during daytime with inverter capacity remaining after real power generation, for providing significant improvements in the power transfer limits of transmission systems. The effectiveness of the proposed PV-STATCOM controls is demonstrated on two study SMIB systems: system I having one 100 MW PV-STATCOM and system II having one 100 MW PV-STATCOM and another 100 MW PV-STATCOM or 100 MW WF controlled as STATCOM. Three different types of STATCOM controls are proposed for both the PV solar DG and inverter-based wind DG. These are voltage control, damping control, and a combination of voltage control and damping control. The following inferences are made:

1) In study system I, the power transfer can be increased by 168 MW during nighttime and by 142 MW in daytime even when the solar DG is generating a high amount (~90%) of active power.
2) In study system II, the transmission capacity in the night can be increased substantially by 229 MW if no DG is producing active power. During both night and day, the power transfer can be increased substantially by 200 MW, even when the DGs are generating high real power.

It is noted that significant increase in power transfer is attained during daytime in Partial STATCOM operating mode with inverter capacity remaining after active power generation. If reactive power is made available at rated power generation for transmission-connected solar PV systems (as provided by IEEE Standard 1547-2018 for distribution connected DERs) even higher improvements in power transfer can be achieved.

6.2 Power Oscillation Damping by PV-STATCOM

The PV-STATCOM control described in the previous section had a limitation that during daytime the dynamic reactive power control capability declined with increasing active power output from solar farm [19–21]. An eighth-order POD controller for large PV solar farm was proposed in [29], whereas an energy function-based design of POD controller was presented in [30]. Both these controllers are relatively complex in design. All the POD controls in the above papers [21, 29–31] are based on remaining inverter capacity during daytime. Hence, the proposed POD capability of solar farm is limited during day, indeed becoming zero during hours of full sun.

This section describes an enhanced patented PV-STATCOM control termed "Daytime Full STATCOM mode" (Section 4.7.3) for POD [32]. In this proposed control, if any disturbance occurs in the power system causing undesirable power oscillations, the PV solar farm autonomously curtails its active power to zero for a short period (typically less than a minute) and makes its entire inverter capacity available for operating as STATCOM to damp power oscillations through reactive power modulation. As soon as the power oscillations are reduced to an acceptable level, the solar farm restores its power output to its predisturbance level in a ramped manner. Another novel feature of this PV-STATCOM control is that the POD function is kept activated during the ramp up of power to its predisturbance value utilizing the inverter capacity remaining after active power generation. This prevents any recurrence of power oscillations and also allows a much faster ramp-up than typically specified by present Standards where such a damping function during ramp-up is not envisaged.

The proposed PV-STATCOM control thus allows POD with full inverter capacity both during night and day. The effectiveness of the proposed PV-STATCOM for POD is demonstrated on an SMIB system [33], Two-area power system [34], and the 12 bus FACTS power system [35] through detailed electromagnetic transients studies using PSCAD software.

This section presents a case study of POD by PV-STATCOM in a two-area system [32].

6.2.1 Study System

The two-area system having four generators connected with the 220 km tie-line [1] is depicted in Figure 6.9. A 100 MW PV system is connected at the midpoint of the tie-line between buses 7 and 9. In both study systems, the synchronous generators are represented by their detailed sixth-order model and DC1-A-type exciter [1]. No PSS is installed on generators. The parameters for the two-area system are provided in [1]. The two-area system exhibits both local inertial mode and inter-area mode of oscillations in the power flow.

6.2.2 PV-STATCOM Control System

Figure 6.10 portrays the different components of the PV system and the PV-STATCOM controller [32]. Only the controller components relevant for PV-STATCOM control are described below. The PV-STATCOM operating modes are described in Section 4.7.

6.2.2.1 DC Voltage Controller

The DC voltage controller has two components: (i) the MPPT block with a PI controller, and (ii) a DC voltage controller [36]. During conventional PV operating mode, based on the *VI* characteristic of the PV panels, the MPPT block utilizes V_{DC} and I_{DC} to generate the reference voltage $V_{DC\text{-ref}}$, which eventually produces I_{d_ref} for the inner-loop controller. In STATCOM control mode, S_1 changes to position 2 and the DC voltage V_{DC} is regulated to PV panel open circuit voltage to disable active power injection from the PV solar panels [22]. The open-circuit voltage is not a constant and depends on the solar irradiance and temperature. For the specifically utilized PV panel in the solar farm, the largest open-circuit voltage obtained from various (manufacturer supplied) power-voltage characteristics for different realistically prevalent temperatures and solar irradiance [22] is chosen as V_{ref} for the DC-link voltage controller module in Figure 6.10.

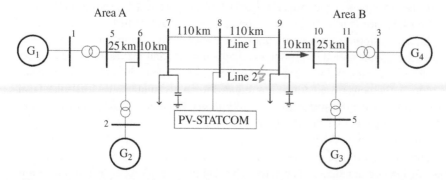

Figure 6.9 Single-line diagram of two-area system with 100 MW PV plant connected midline. *Source:* Varma and Maleki [32].

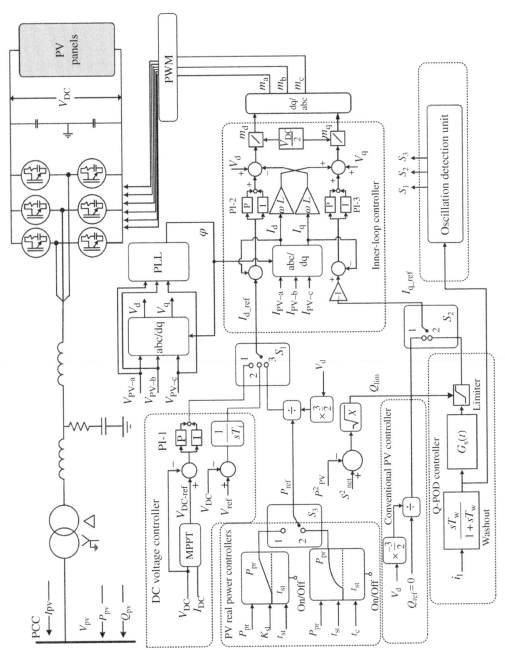

Figure 6.10 PV-STATCOM controller. *Source:* Varma and Maleki [32].

6.2.2.2 Conventional PV Controller

The conventional PV controller regulates the inverter reactive power such that PV power output is at UPF [25]. This controller has been adapted from [36, 37] and is utilized only during normal operation of the power system in which UPF is required for PV systems. In this control, Q is set to zero during steady state operation resulting in $I_{q\_ref} = 0$. It is clarified that this controller is deactivated during disturbances, i.e. during power oscillations. In this situation, I_{q_ref} is generated by the Q-POD controller in a closed-loop manner.

6.2.2.3 Q-POD Controller

The Q-POD controller controls the reactive power output of PV-STATCOM to damp the low-frequency electromechanical oscillations. In this study, the magnitude of the line current at the PCC of solar farm is selected as the control signal for POD [1, 21]. In the study system, i_l represents the line current between buses 9 and 10. The i_l signal is fed to the washout filter to remove its steady state component. The POD controller transfer function is selected as:

$$G_s(t) = G \times \frac{1 + sT_{lead}}{1 + sT_{lag}} \tag{6.3}$$

where G represents the controller gain; and T_{lead} and T_{lag} denote the *lead* and *lag* time constants, respectively.

The operating principle of POD controller is described in [1]. This controller effectively adds adequate phase lead or phase lag to a selected system oscillatory mode to enhance its damping. In this study, the oscillatory modes for the study system are selected and the corresponding optimized lead–lag controllers are designed based on an optimization process described later. This POD controller generates the reference signal I_{q_ref} for PV inverter inner loop controller to control PV reactive power.

6.2.2.4 Oscillation Detection Unit (ODU)

The ODU autonomously detects the occurrence of unacceptable low-frequency electromechanical power oscillations caused by any grid disturbance such as faults. The ODU operates based on the flow chart depicted in Figure 6.11 and generates the On/Off status signals for switches S1, S2, and S3. The oscillatory component of line current Δi_l is compared with a predefined value ε which is chosen as 5% in this study. If the variation is more than ε the Daytime Full STATCOM mode is activated for POD control and the PV active power is reduced to zero.

6.2.2.5 PV Active Power Controllers

These controllers are responsible for the restoration of active power output of the PV solar farm to its predisturbance level after power oscillations are damped with the Full STATCOM mode of operation. Grid codes such as [38, 39] require the solar farms to restore their power with a prespecified ramp rate so that any voltage and power oscillations can be prevented. No damping function is envisaged in these grid codes during the process of power ramp-up.

A novel power restoration technique is proposed [32], according to which the solar farm continues to perform POD during the entire power restoration process in the Partial STATCOM mode. This prevents the recurrence of power oscillations while power is being restored to its predisturbance level. The proposed technique allows a much faster ramp rate to be achieved since power oscillations continue to be damped during the entire restoration process. Two power restoration techniques are implemented in the PV active power controllers shown in Figure 6.10 and described below.

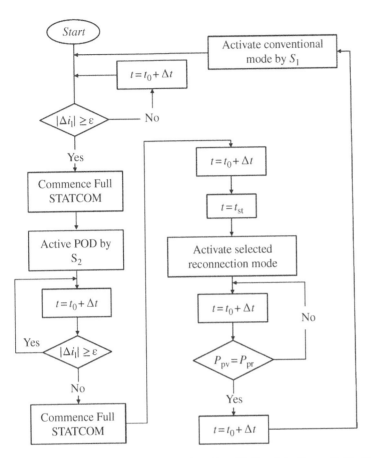

Figure 6.11 Flowchart of the operation of oscillation detection unit. *Source:* Varma and Maleki [32].

Power Restoration in a Ramped Manner

In this mode, the controller changes the PV active power output from zero to the predisturbance PV power level in a ramped manner with a ramp rate of K_{sl} starting at time $t = t_{st}$. This is the normal recommended mode for restoration of solar farms by grid codes [38]. No damping function is envisaged during ramp-up [38, 40].

Power Restoration in the Partial STATCOM Mode with POD Control Active

In this mode, the controller changes the PV active power output from zero to the predisturbance PV power level in a ramped manner with a ramp rate of K'_{sl} starting at time $t = t_{st}$. The solar farm is operated in the Partial STATCOM mode with POD control active.

A variant of this technique is also studied, according to which the power is restored from zero to the predisturbance level in a nonlinear mode starting at $t = t_{st}$ with an exponential time constant t_c. This time constant can be determined based on the decay time constants of the ambient power oscillatory modes.

During active power restoration process, the solar farm performs POD in Partial PV-STATCOM mode with reactive power capacity available after active power generation at that time instant. The reactive power limit Q_{lim} which continuously keeps declining as the active power gets restored to its original predisturbance level is given by $Q_{\text{lim}} = \sqrt{S^2 - P^2}$, where, S represents the total inverter capacity, P is the inverter active power output and Q_{lim} is the maximum available inverter capacity during power restoration.

6.2.2.6 Design of POD Controller

The POD controller parameters – *Gain, Lead*, and *Lag* time constant are determined by the Simplex optimization technique [41] embedded in the PSCAD software [42], to minimize the low-frequency power oscillations in-line current. The corresponding Objective Function (OF) is defined as:

$$OF = \int_{T_1}^{T_2} (i_1 - i_{1_{\text{ref}}})^2 dt \tag{6.4}$$

where i_{1_ref} is the reference value of the midline current i_1. T_1 and T_2 are the start and end time instants of the current oscillations after the fault, respectively. The main concept of the embedded optimization is to run a *Slave* simulation to determine the value of the (OF) in the i_{th} run. The results are then sent to a *Master* project to check if the results are converging and a new set of POD controller parameters are generated for the *i + 1th* iteration. The OF converges in about 40 and 59 runs for study systems I and II, respectively.

6.2.2.7 Small Signal Studies of the POD Control

The efficacy of the PV-STATCOM for POD at different locations is demonstrated through small-signal residue analysis [43] in MATLAB. The magnitude of residue is an indicator of the effectiveness of POD controller [44]. The higher the magnitude of the residue the better the location of PV-STATCOM for POD. In the two-area power system, five different locations considered for PV system placement are at buses 6, 7, 8, 9, and 10. Figure 6.12 shows the residues for different

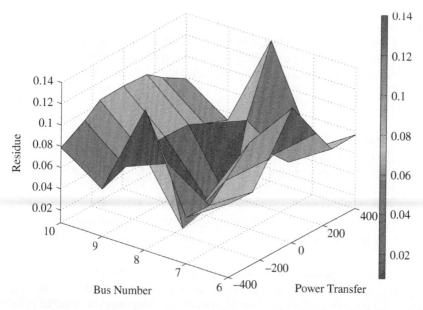

Figure 6.12 Residue analysis for PV-STATCOM POD controller. *Source:* Varma and Maleki [32].

PV-STATCOM locations with varying levels of power transfer from area *A* to *B* and vice versa. The highest residue for the maximum midline power transfer of 400 MW is observed at bus 8 and hence the PV-STATCOM is connected to bus 8.

6.2.3 Simulations Studies

6.2.3.1 Power Transfer without PV-STATCOM Control

In this case study for the two-area power system, the tie-line power is transferred from area *A* to area *B* equally through lines 1 and 2 under normal operation. A three phase to ground fault is initiated at $t = 2$ s for five cycles in line 2 close to bus 9. The circuit breakers disconnect the faulted line 2 and the entire tie-line power is subsequently transferred through line 1. The midline connected PV solar farm is considered to produce its rated 100 MW power at noon under maximum solar irradiance.

Figure 6.13 shows the midline active power and the PV solar power for this study. As soon as the fault occurs the PV solar farm may go into momentary cessation due to voltage limit violations as required by various grid codes. Severe oscillations in power flow are observed if the prefault tie-line power is considered to be 430 MW. The maximum tie-line power that can be stabilized post fault with a damping ratio of 5% [1, 28] is 230 MW. In this study, the objective is to increase the line power transfer limit from 230 to 430 MW.

6.2.3.2 Power Transfer with Full STATCOM Damping Control and Power Restoration in Normal Ramped Manner

As soon as power oscillations are initiated, the solar power output is curtailed to zero and the solar farm is transformed to Full STATCOM with POD control. The proposed PV-STATCOM control then utilizes its entire inverter capacity for POD to stabilize the power system and increase the power transfer. Figure 6.14a illustrates the midline power and the PV active power. Figure 6.14b,c shows the PV-STATCOM reactive power and PCC bus voltage, respectively.

The PV-STATCOM POD function successfully stabilizes the power oscillations to within acceptable limits in about 10 seconds (just before $t = 12$ s). The PCC voltage oscillations are also mitigated rapidly. The power restoration is then commenced at $t = 15$ s, after a 2 s delay for safety purpose. It is recommended [45] that power restoration from a PV solar farm from zero to its rated level may be done with a typical ramp rate of 10% of rated capacity in one minute to avoid any power oscillations. In this case study, the fastest ramp rate which will expectedly not cause any resurrection of power oscillations is determined from simulations to be 5.5 MW/s. The solar power is, therefore, ramped up to its predisturbance level of 100 MW at a rate of 5.5 MW/s in about 18 seconds.

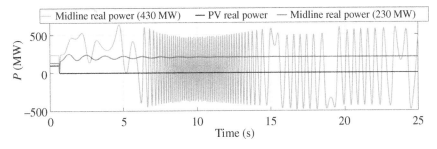

Figure 6.13 Midline and PV active power in two-area system (230 and 430 MW power transfer). *Source:* Varma and Maleki [32].

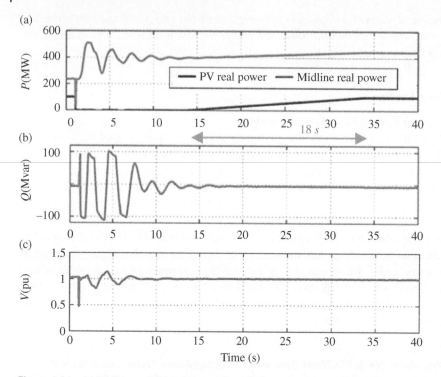

Figure 6.14 (a) Midline and PV active power, (b) PV reactive power, (c) midline voltage during POD, and power restoration in a normal ramped manner. *Source:* Varma and Maleki [32].

6.2.3.3 Power Transfer with Full STATCOM Damping Control and Ramped Power Restoration with POD Control Active in Partial STATCOM Mode

This study is performed to illustrate the efficacy of the proposed restoration technique for the same system operating conditions as in the previous case.

Figure 6.15a depicts the midline power flow and the PV solar power output whereas Figure 6.15b, c demonstrates the reactive power of the PV-STATCOM and PCC voltage, respectively. It is evident that POD with Partial PV-STATCOM mode of operation reduces the time of PV power restoration to its predisturbance level of 100 MW to just 5 seconds. This is 3.5 times faster than without the proposed restoration technique.

6.2.3.4 Nighttime Power Transfer Enhancement with Full STATCOM POD Control

The effectiveness of the proposed Full STATCOM-based POD control subsequent to the same fault as in previous case, during nighttime is presented in this study.

Figure 6.16a portrays the behavior of power flow at the levels of 230 MW and 430 MW in the tie-line without the PV-STATCOM control. Figure 6.16b,c demonstrates the tie-line power and PV system reactive power. The maximum power transfer in the tie line was initially only 230 MW. It is seen that the proposed control increases the power transfer capability of the tie-line from 230 to 430 MW, i.e., by 200 MW.

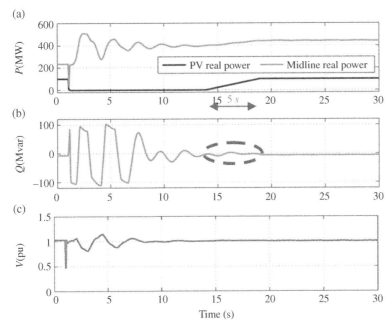

Figure 6.15 (a) Midline and PV active power, (b) PV reactive power, (c) Midline voltage during POD, and power restoration in a fast ramped manner. *Source:* Varma and Maleki [32].

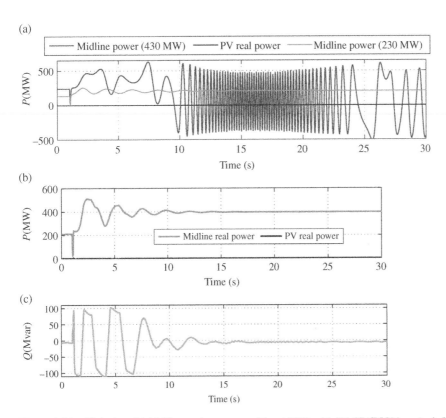

Figure 6.16 Nighttime (a) Midline active power without POD with PV-STATCOM control, (b) Midline active power with Full STATCOM POD Control, (c) PV-STATCOM reactive power. *Source:* Varma and Maleki [32].

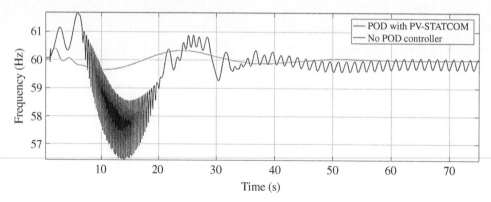

Figure 6.17 Effect of PV-STATCOM control on system frequency in two-area power system. *Source:* Varma and Maleki [32].

6.2.3.5 Effect of PV-STATCOM Control on System Frequency

The proposed POD utilizing PV-STATCOM reactive power control is not expected to create any adverse impact on system frequency. The proposed PV-STATCOM control provides only damping of power oscillations. This may indirectly alleviate the frequency deviations as well. Figure 6.17 depicts the power system frequency without and with POD control with PV-STATCOM in the two-area power system considering a tie-line power flow of 430 MW. It is evident that the proposed POD control of PV-STATCOM not only damps power oscillations but also reduces the frequency oscillations that would be caused in the absence of such a control.

6.2.4 Summary

This study presents a novel PV-STATCOM control for damping power oscillations and thereby increasing the power transfer capacity of the two-area power system, substantially. The proposed control also provides POD through reactive power modulation utilizing the entire inverter capacity during nighttime. During daytime, the solar farm curtails its active power to zero very briefly (about 15 seconds) and utilizes its entire inverter capacity for POD. It subsequently restores power generation to its predisturbance level in a gradual manner while keeping the POD function activated utilizing the remaining inverter capacity.

PSCAD simulation studies demonstrate that in the two-area system a 100 MW PV solar system increases the power transmission limit by 200 MW. Moreover, the proposed power restoration technique keeping POD activated is more than three times faster than that specified by grid codes (without POD function). The temporary (about 18 seconds) curtailment of active power for POD is not seen to cause any adverse impact on system frequency.

6.3 Power Oscillation Damping with Combined Active and Reactive Power Modulation Control of PV-STATCOM

FACTS Controllers such as SVCs [1, 4, 46], static synchronous compensators (STATCOM) [47, 48] are widely utilized in power systems for POD. These controls are all based on reactive power. POD with active power control has been reported by STATCOM coupled with energy storage systems [49, 50]. Superconducting magnetic energy storage system (SMES) for damping inter-area oscillations has been presented in [51]. The concept of controlling distributed generator inverters as virtual synchronous generator (VSG) for POD is described in [52]. The VSG is shown to produce virtual inertia from energy storage over a short duration and damp power oscillations. However, none of the above devices provide combined control of active and reactive power for POD.

This study presents a novel patented smart inverter control of PV solar farms as a STATCOM (PV-STATCOM) for performing POD at any time of the day with full inverter capacity using either individual modulation of active and reactive power or more importantly, their combined modulation. During any system contingency, the active power of the PV solar farm is curtailed for a short duration (typically, less than 1 minute) and the above modulations are performed till power oscillations are damped out. Subsequently, the active power output of solar farm is restored rapidly to its pre-disturbance level.

Three different POD control strategies for PV-STATCOM are presented based on active power, reactive power, and combined active and reactive power modulation. The POD controllers are designed through small-signal residue analysis and further tuned by detailed PSCAD simulations. The effectiveness of different POD controls for varying power outputs and locations of a 100 MW solar farm in the two-area power system is demonstrated.

6.3.1 Modes of PV-STATCOM Control

The different operating modes of PV-STATCOM for POD as described in Section 4.7.3.3 are utilized in this proposed control.

6.3.1.1 Partial STATCOM

In Partial STATCOM mode, the PV inverter capacity remaining after active power generation or modulation during daytime is utilized for POD.

6.3.1.2 Full STATCOM

This mode is utilized during a critical system need during the day. In this mode, the PV active power is temporarily curtailed to zero or any desired level thereby making the remaining inverter capacity available for reactive power modulation. Three variants of this mode are utilized for POD.

i) *Reactive Power Modulation-Based POD Control (Q-POD):* The active power is curtailed to zero or an intermediate value. The reactive power is then modulated between zero and the remaining inverter capacity or between zero and the rated inverter capacity to accomplish POD. No active power modulation is involved. This mode is available both during day and night.

ii) *Active Power Modulation-Based POD Control (P-POD):* The maximum active power production based on available solar irradiance is reduced to half. The active power is then modulated between zero and the maximum available value for POD. No reactive power modulation is involved. This mode is available during daytime.

iii) *Combined Active and Reactive Power Modulation-Based POD Control (PQ-POD):* The maximum active power production based on available solar irradiance is reduced to half. The active power is then modulated between zero and the maximum value for POD. Simultaneously reactive power is also modulated between zero and the inverter capacity remaining after active power modulation. This mode is available during daytime.

6.3.2 Study System

Figure 6.18 shows the single line diagram of two-area power system [1] with a large utility-scale 100 MW PV solar farm connected at bus 10. Four synchronous generators in two distinct areas are connected via 220 km tie-line. Synchronous generators are represented by the detailed sixth-order model and DC1-A-type exciter [1] in PSCAD software. PSSs are not considered on generators. Lines are represented by their pi-circuit model.

6.3.3 PV-STATCOM Control System

Figure 6.19 portrays the detailed model of various PV-STATCOM components in PSCAD software.

6.3.4 PV Reactive Power Controllers

6.3.4.1 Conventional Reactive Power Control

In conventional PV system operation, the aim is to convert DC power to AC with UPF. Hence, Q_{ref} is set to zero, and reactive power control is not performed [25].

6.3.4.2 Q-POD Controller

The magnitude of the line current signal i_l from bus 9 to bus 10, which has the highest participation factor in selected interarea mode of oscillation [1] is used as a control signal. i_l is subsequently passed through a washout filter. The oscillatory component of i_l is fed as the input signal for compensator with a transfer function $G_{\text{Q-POD}}$:

$$G_{\text{Q}-\text{POD}}(s) = G \times \frac{1 + sT_{\text{lead}}}{1 + sT_{\text{lag}}} \tag{6.5}$$

where G is the controller gain; and T_{lead} and T_{lag} are the *lead* and *lag* time constants, respectively. A hard limiter is used on the compensator output to limit the $i_{\text{q-ref}}$ based on:

$$Q_{\text{lim}} = \sqrt{S_{\text{pv}}{}^2 - P_{\text{pv}}{}^2} \tag{6.6}$$

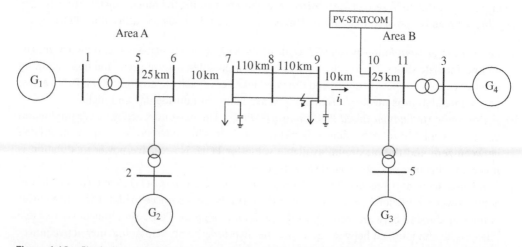

Figure 6.18 Single-line diagram of two-area power system with PV-STATCOM connected at bus 10. *Source:* Modified from Kundur [1].

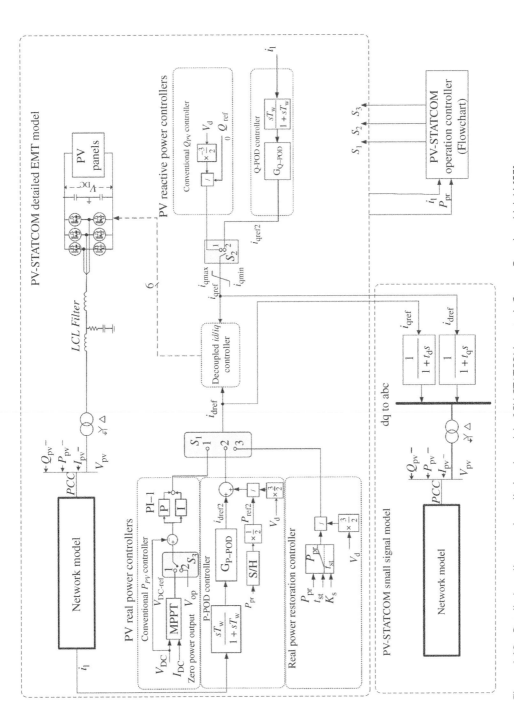

Figure 6.19 Detailed nonlinear and small-signal model of PV-STATCOM control. *Source:* Based on Maleki [53].

$$i_{q\max} = \pm \frac{2}{3 \times V_d} Q_{\lim} \tag{6.7}$$

The conventional PV operating mode is changed to Q-POD control mode as follows. $V_{DC\text{-ref}}$ is changed to V_{op} by changing S_3 to position 2. Further, i_{qref} is changed to i_{qref2} by switching S_2 to position 2.

6.3.5 PV-Active Power Controllers

6.3.5.1 Conventional Active Power Control

In conventional PV operation mode, the MPPT [25] unit generates DC voltage reference signal V_{dref}. The PV-STATCOM DC side voltage V_{DC} is controlled at $V_{DC\text{-ref}}$ to extract the maximum solar power.

6.3.5.2 P-POD Controller

The i_{dref} is controlled to i_{dref2}. To activate P-POD controller, the active power set point is reduced to half of its prefault value (P_{pr}) using sample and hold (S/H) and divider block. The active power is controlled around P_{ref2} through P-POD controller. I_{dref2} is calculated as:

$$i_{dref2} = \frac{2}{3 \times V_d} P_{ref2} + i_{P-POD} \tag{6.8}$$

where i_{P-POD} represents the current output from P-POD controller with G_{P-POD} transfer function:

$$G_{P-POD}(s) = G \times \frac{1 + sT_{lead}}{1 + sT_{lag}} \tag{6.9}$$

6.3.5.3 PQ-POD Controller

In this mode of operation, both P-POD and Q-POD are activated to enhance the PV-STATCOM POD performance. The i_{dref} is controlled at i_{dref2} and i_{qref} is controlled to i_{qref2}. The generated i_{qref2} from Q-POD controller is limited through the hard limit block based on the available PV system inverter capacity. During daytime, the PQ-POD controller can be activated by switching S_1 to position 2 and S_2 to position 2.

6.3.5.4 Active Power Restoration Controller

Once the power oscillations are damped, at $t = t_{st}$ the PV active power is restored to its prefault value P_{pr} with a ramp function having K_{st} ramp rate. After power restoration is completed, PV controllers switch back to conventional PV operation mode. A novel feature of this control is that the POD function is kept activated in the Partial STATCOM mode while active power is being restored to its predisturbance value P_{pr}. This technique can significantly reduce the restoration interval than that specified by different standards [38, 54]. It is noted that these grid codes do not envisage damping control during the ramp-up period.

6.3.6 Small Signal PV-STATCOM Modeling

A simplified small-signal PV model is developed in MATLAB software. In this model, the detailed model of solar panels is replaced by a controlled current source. The inverter is modeled as a first-order transfer function having steady-state gain of unity with t_d, $t_q = 15$ ms [25] as depicted in Figure 6.19. This is done because the model of an inverter as first-order function with time constant of the dominant pole of closed-loop transfer function is considered to provide an acceptable response [55]. The *LCL* filter is ignored. The small-signal model is validated by detailed electromagnetic transients simulation studies with PSCAD. There are only minor differences between the detailed and small-signal model results during transients that are not relevant in the time-scale of study of low-frequency oscillations.

6.3.7 Selection of PV-STATCOM Controller Mode

The PV-STATCOM operation controller block shown in Figure 6.19 is explained here. The flow-chart for PV-STATCOM operation mode selection is illustrated in Figure 6.20. The specific mode of operation is selected based on the desired mode of operation. The magnitude of i_l deviation is

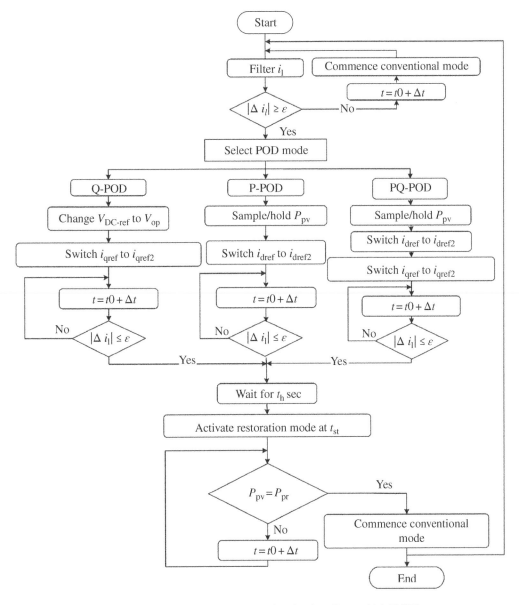

Figure 6.20 Flowchart of PV-STATCOM operation mode selection. *Source:* Maleki [53].

compared with a predefined limit ε. If oscillations in i_1 signal are less than ε (chosen to be 5%), the conventional PV mode of operation is activated. If not, POD mode is activated as follows:

a) Q-POD is activated by changing $V_{DC\text{-ref}}$ to V_{op}, and i_{qref} to i_{qref2}.
b) P-POD is activated by changing i_{dref} to i_{dref2}.
c) PQ-POD is activated by changing $V_{DC\text{-ref}}$ to V_{op}, i_{qref} to i_{qref2}, and i_{dref} to i_{dref2}.

In these POD techniques, once the magnitude of oscillations in i_1 stabilizes within ε, the active power restoration mode is activated after a two seconds safety delay. When P_{pv} reaches P_{pr}, the conventional PV mode of operation is reactivated.

6.3.8 Design of POD Controllers

Controllers are designed in two steps. First, POD controllers are designed to satisfy the small-signal stability of the power system based on *Residue* technique [43, 44]. After the lead–lag controller parameters are designed for both P-POD and Q-POD controllers through small-signal studies, these values are used as initial conditions for optimal POD controller design using the Nonlinear Simplex optimization technique implemented in PSCAD software [42]. In this optimization technique, the optimized controller parameters, i.e. Gains, *Lead,* and *Lag* time constant are obtained through geometrical minimization of OF [42]. Here, the aim is to minimize the low-frequency power oscillations in-line current i_1. The corresponding OF is defined as:

$$ \text{OF} = \int_{T_1}^{T_2} (i_1 - i_{1ref})^2 dt \tag{6.10} $$

where i_{1ref} is the reference value of the current. T_1 and T_2 represent the start and end time instants, respectively, of the current oscillations after the contingency.

6.3.9 Effect of PV-STATCOM Placement on Effectiveness of POD Techniques

Residue analysis is used to determine the effectiveness of the PV-STATCOM control at different locations [44]. The magnitude of residue has a direct relation with $\Delta\lambda$ (change in eigenvalue of the selected inter-area mode to be damped) and thereby with the effectiveness of specific POD controller at particular location. Hence, the magnitudes of residues associated with $\frac{\Delta i_1}{\Delta P_{PV}}$ and $\frac{\Delta i_1}{\Delta Q_{PV}}$ are obtained by calculating $\Delta\lambda$. In the two-area power system, five different locations (buses 6, 7, 8, 9, and 10) are considered for PV system placement.

6.3.9.1 Residue Analysis for PV-STATCOM with Q-POD
Figure 6.21 depicts the residue analysis results for PV-STATCOM interconnection at different buses for different levels of power transfer from Area A to B vice versa. The highest residue associated with PV-STATCOM in Q-POD control mode is obtained considering the midline power transfer is at its maximum (430 MW) and PV-STATCOM is connected to bus 10.

6.3.9.2 Residue Analysis for PV-STATCOM with P-POD
The results of the residue analysis for PV-STATCOM with P-POD controller are presented in Figure 6.22. If the PV-STATCOM is connected at the midline of the two-area power system (bus 8), PV active power does not have a significant effect on the interarea mode of oscillations. The highest residue for this study is achieved when the PV-STATCOM is connected at bus 6 or 10.

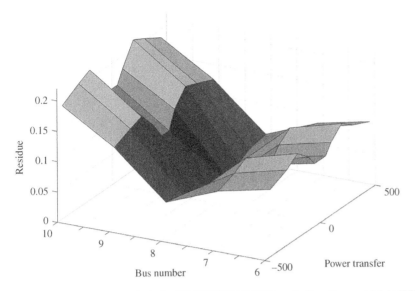

Figure 6.21 Residue analysis for PV-STATCOM Q-POD controller. *Source:* Maleki [53].

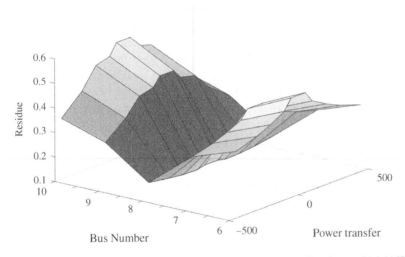

Figure 6.22 Residue analysis for PV-STATCOM P-POD controller. *Source:* Maleki [53].

Based on the results obtained in Figures 6.21 and 6.22, the best location for PV-STATCOM PQ-POD controller is bus 10 in which both Q-POD and P-POD have the highest residue magnitude for the relevant interarea mode of oscillation.

6.3.10 Simulation Studies

Three case studies based on the location of PV-STATCOM are presented in this chapter. In case study 1, the PV-STATCOM is connected at bus 10 which is the best bus candidate for PQ-POD control. The performances of the proposed POD control techniques are examined through PSCAD simulations. Also, the effect of various PV active power generation levels on each POD technique is evaluated. In case studies 2 and 3, the PV-STATCOM location is changed from bus 10 to bus 8 and 6, respectively. The effect of the location of PV-STATCOM on the performance of the proposed POD techniques is examined and results are compared with those obtained by residue analysis.

6.3.10.1 POD by PV-STATCOM

In case study 1, it is assumed that the PV-STATCOM is connected at the best location, i.e. bus 10. A power flow of 430 MW is considered from Area A to B. A three-phase to ground fault is initiated at $t = 1$ s for 5 cycles near bus 9. Due to the fault, growing low-frequency power oscillations occur in-line current. Figure 6.23 shows the power transfer capability of the two-area power system after three-phase fault initiation. It is seen that the maximum stable power transfer of the line is 250 MW. The aim now is to increase the power transfer capability of the same line to 430 MW with the proposed POD controllers.

Figure 6.24 depicts the results of POD achieved by different POD techniques.

In Figure 6.24a, "No POD" refers to the case in which PV plant goes into momentary cessation after fault initiation. It is observed that Q-POD, P-POD, and PQ-POD controllers damp the power oscillations to within acceptable limits in 10, 10, and 6 seconds, respectively. This demonstrates that the best POD is achieved by PQ-POD control.

Figure 6.24b illustrates the PV-STATCOM active and reactive power after the fault for Q-POD control. The PV active power is reduced to zero within 0.3 seconds after the fault occurs. The entire PV-STATCOM inverter capacity is then made fully available for Q-POD control. The oscillations in-line current i_l stabilize within ε by $t = 10$ s. Due to decoupled P-Q control, the active power of the PV system continues to be zero during this period. Subsequently after the safety time delay of two seconds, at $t = 12$ s, the PV active power is restored to $P_{pr} = 100$ MW with a ramp rate of 20 MW/s.

Figure 6.24c shows the PV-STATCOM active and reactive power after the fault for P-POD control. The PV system active power is reduced to half of its prefault value (50 MW) and P-POD control is performed by controlling the PV-STATCOM active power around 0–100 MW. Due to the decoupled control, the reactive power of the PV system continues to be almost zero. Subsequently after the safety time of two seconds, at $t = 12$ s, the PV active power is restored to $P_{pr} = 100$ MW with a ramp rate of 20 MW/s. The effectiveness of P-POD control is observed to be similar to the Q-POD control.

Figure 6.24d portrays the PV-STATCOM active and reactive power after the fault for PQ-POD control. Following the fault, the PV-STATCOM active power is reduced to half of its prefault value.

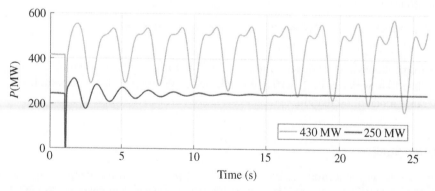

Figure 6.23 Maximum power transfer capability of the two-area power system. *Source:* Maleki [53].

Both P-POD and Q-POD controllers are activated. In this mode, P-POD is the primary control function and Q-POD is done with inverter remnant capacity. The oscillations in-line current i_l stabilize within ε by $t = 6$ s. Hence after a safety delay time of two seconds, at $t = 8$ s, the PV active power is restored to $P_{pr} = 100$ MW within three seconds.

As stated earlier, the POD control, i.e. Q-POD, P-POD, or PQ-POD remains activated during the restoration interval. This novel technique allows a much faster ramp rate to be achieved than the rates specified in [56]. This ramp rate is 2 MW/min in Hawaiian standard [57] and maximum of 10%/min for Germany [38].

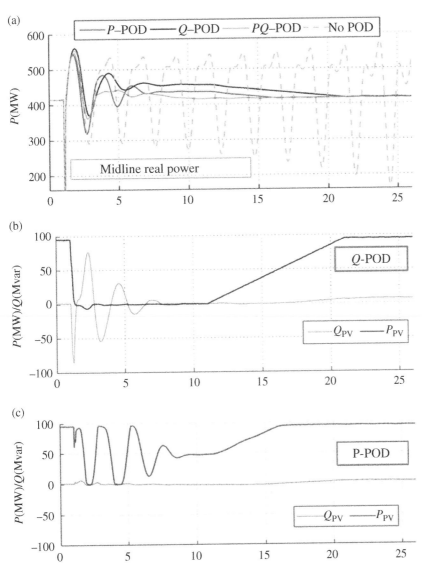

Figure 6.24 Midline active power; and PV-STATCOM active power, reactive power, and DC voltage for case study 1. *Source:* Maleki [53].

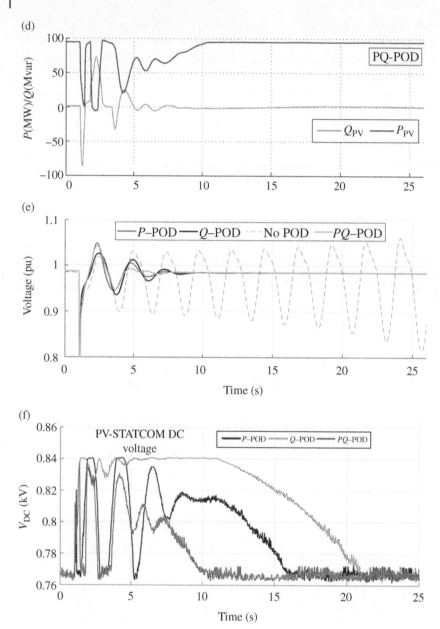

Figure 6.24 (Continued)

Figure 6.24e depicts the voltage at the PCC for different POD controls. It is clear that none of the POD controls cause the PCC voltage to violate typical utility specified limits.

Figure 6.24f illustrates the PV-STATCOM DC voltage modulation based on the selected POD mode of operation. In Q-POD mode of operation, the DC voltage is controlled at 840 V to reduce the PV active power to zero based on VI characteristic of PV modules. In P-POD mode, the DC voltage is controlled around 820 V (50 MW). This DC voltage variation results in 0–100 MW PV active power variation. In PQ-POD mode of operation, the DC voltage of the PV system is controlled around 830 V to reduce the PV active power set point to 50 MW. It is shown that the DC voltage variation in PQ-POD control is smaller than both the Q-POD and P-POD control techniques. This reduces the stress on the DC-link capacitor.

6.3.10.2 Effect of POD Controllers on System Frequency
Figure 6.25 illustrates the impact of the proposed POD controls on power system frequency after the fault. If no POD control is activated the power system frequency variation violates the limits set by Standards [58]; in fact, the system becomes unstable. It is evident from Figure 6.25 that none of the POD controls utilized in this study destabilize the frequency. It is seen that the largest frequency variation is experienced with the Q-POD control since the active power is suddenly reduced to zero to release the entire inverter capacity for reactive power modulation. P-POD causes lower frequency oscillations since power is reduced by only 50% while PQ-POD results in the lowest frequency excursion.

6.3.10.3 Effect of PV Active Power Output on POD Controls
Studies reveal that the effectiveness of P-POD controller reduces due to lower availability of PV active power (although not shown here). The performance of Q-POD control technique is, however, not affected by variation in PV system active power. It is further noted that although the effectiveness of P-POD control technique is influenced by available PV system active power, the PQ-POD control provides the most effective damping in comparison to both Q-POD and P-POD controls.

6.3.10.4 POD by PV-STATCOM Connected at Other Buses
Case studies 2 and 3 (similar to case study 1) considering PV-STATCOM connected to bus 8 and 6, respectively, are performed using PSCAD. Table 6.3 illustrates the power oscillation settling times with different POD control techniques implemented on solar PV farms at different locations. The fastest settling time is achieved with PQ-POD control when the PV-STATCOM is connected at bus 10. These results validate the conclusions from residue analysis.

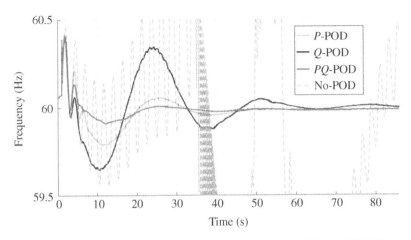

Figure 6.25 Power system frequency for No POD, Q-POD, P-POD, and PQ-POD control techniques. *Source:* Maleki [53].

Table 6.3 Settling time of power oscillations with different PV-STATCOM locations and POD control techniques.

PV-STATCOM location	Settling time (s)		
	Q-POD	P-POD	PQ-POD
Bus 6	13	15	9
Bus 8	15	15	12
Bus 10	10	10	6

Source: Maleki [53].

6.3.11 Summary

This study presents a novel PV-STATCOM control for damping power oscillations in the two-area system through modulation of either: (i) reactive power (Q-POD), (ii) active power (P-POD), or (iii) combination of active and reactive power (PQ-POD). A new ramp-up technique is further presented, wherein the above damping controls are kept activated while ramping up power from solar PV farms.

The following inferences are made:

1) The best POD is achieved if PV-STATCOM active and reactive power are used together for damping power oscillations. Although P-POD controller is affected by the amount of PV active power availability, PQ-POD control results in best POD among all the three POD controls for all levels of power transfer. The PQ-POD control also causes the least stress on the DC-link capacitor.
2) The proposed power ramp-up technique with POD controls kept active, results in much faster power restoration times than those stipulated by grid codes. This ramp-up time may be longer when plant-level controls of the PV solar farm are modeled but is still expected to be lower than the grid-code values.
3) None of the three POD controls have any adverse impact on the system frequency in the study system. Among all, PQ-POD control has the least impact on power system frequency.
4) The 100 MW PV solar farm with PV-STATCOM controls can increase the tie-line power transfer substantially by 180 MW in the two-area power system in which the power transfer is limited by inadequate electrical damping.

6.4 Mitigation of Subsynchronous Resonance (SSR) in Synchronous Generator by PV-STATCOM

Series capacitive compensation is an effective means for increasing the power transfer capacity of transmission lines. However, the potential of subsynchronous resonance (SSR) in such lines must be examined apriori and adequately addressed [59, 60, 61]. SSR is an unstable electrical power system condition that can manifest as steady state induction generator (IG) effect, torsional interaction (TI), or as transient torque amplification [60, 61, 62]. SSR resulted in damage to generator shaft in the steam turbine-driven synchronous generator in Mohave plant, Nevada, USA, in 1970 and 1971 [59]. SSR was also reported due to adverse interaction between High Voltage DC (HVDC) converter controls and generator torsional system at Square Butte, North Dakota, USA, in 1977 [63, 64].

Several countermeasures have been investigated and utilized to alleviate SSR since its first occurrence, including filters, excitation system control, and protection systems [65]. Shunt-connected FACTS Controllers, such as SVC and STATCOM are known to be effective solutions for preventing SSR [2–4, 66–70]. Reactive power from the dynamic reactive power compensator is modulated in response to the turbine generator rotor oscillations causing a corresponding voltage modulation at the interconnecting bus. This modulation is performed in a manner that subsynchronous currents are produced in phase opposition, which tend to cancel the original SSR causing subsynchronous currents caused by the interaction between series capacitors and transmission network inductance.

This study presents a novel PV-STATCOM control for alleviation of SSR in a steam turbine-driven synchronous generator connected to a series compensated transmission line [71]. The PV-STATCOM operating modes are depicted in Section 4.7.3.1. During nighttime, the PV solar farm can operate as a STATCOM with its entire inverter capacity for SSR mitigation (Full-STATCOM mode). During daytime, if a system fault triggers SSR, the solar farm autonomously curtails its entire active power generation and releases its rated inverter capacity to operate as PV-STATCOM for SSR prevention (Daytime Full-STATCOM mode). Once the SSOs are damped, the solar farm restores its normal active power production. Electromagnetic transients studies using PSCAD are performed to demonstrate that a solar farm connected at the terminals of synchronous generator in the IEEE First SSR Benchmark system [72] can damp all the four torsional modes at all the four critical levels of series compensation, and return to normal PV power production in less than half a minute.

6.4.1 Study System

Figure 6.26 illustrates the study system in which the IEEE First SSR benchmark system [72] is augmented with a PV solar farm at the generator bus. The mechanical system is modeled fully by its six mass-spring system: the high pressure (HP) turbine, the intermediate pressure (IP) turbine, the low-pressure turbines (LPA and LPB), the generator (GEN), and the exciter (EXC). The mechanical damping is considered zero in all modes to represent the worst damping condition [68]. This radial system produces SSR with four torsional modes: 15.71 Hz (Mode 1), 20.21 Hz (Mode 2), 25.55 Hz (Mode 3), and 32.28 Hz (Mode 4).

The synchronous generator and the entire network are modeled in PSCAD software according to parameters provided in [72]. No AVR is considered on the synchronous generator [25]. The synchronous generator although rated at 892.4 MVA [25] is operated at 500 MW. However, a 300 MVA PV solar inverter is connected at generator bus to make the total power generation at the generating end match with that in [72]. The series capacitive reactance (X_C) of the system is varied to excite the different torsional modes. The values of X_C (pu) for which different torsional modes have their largest destabilization are as follows: Mode 1 (0.47 pu), Mode 2 (0.38 pu), Mode 3 (0.285 pu), and Mode 4 (0.185 pu) [68].

6.4.2 Control System

The proposed control of the PV solar farm as PV-STATCOM is illustrated in Figure 6.26. It is known that a STATCOM with voltage control alone is unable to damp the torsional SSR oscillations and hence an auxiliary damping controller is required [4, 66, 68, 69]. It is possible to mitigate SSR by damping controllers utilizing electrical signals from the PCC, such as line current [73, 74]. However, in this study rotor speed deviation is selected as the control signal for the auxiliary SSR damping controller of PV-STATCOM. This is because rotor speed contains information about all the torsional modes of oscillation [66, 68]. Also, it has been shown to be an effective signal to mitigate

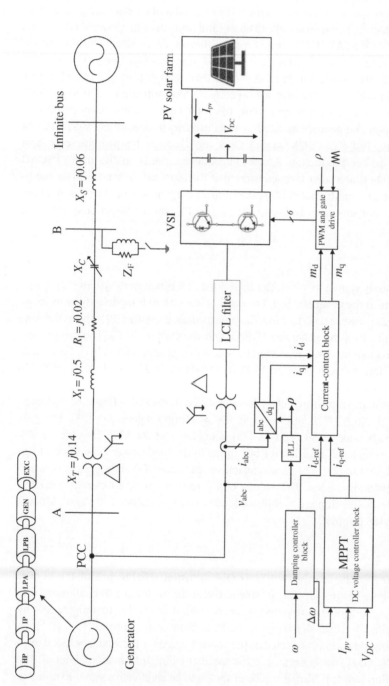

Figure 6.26 Study system involving a PV solar farm connected at the synchronous generator terminal in the IEEE First SSR Benchmark System. *Source*: Varma and Salehi [71].

SSR by SVCs and STATCOMs connected at the terminals of the turbine generator [65, 66, 68, 69]. The controller components relevant to PV-STATCOM control are described below.

6.4.3 SSR Damping Controller

Figure 6.27 illustrates the configuration of the proposed SSR damping controller of the PV-STATCOM. Since it is intended to utilize the entire STATCOM inverter capacity only for damping SSR, the voltage controller typically employed in the STATCOM is not implemented. The damping controller block utilizes the generator speed signal to produce the *d*-axis reference current $i_{d\text{-ref}}$ for current controller, for damping SSR. The PV-STATCOM is connected at the terminals of the turbine-driven synchronous generator. It is, therefore, expected that the generator rotor speed signal will be available to the PV-STATCOM control without any appreciable delay. This is the approach adopted by almost all the papers dealing with SSR mitigation by FACTS Controllers connected at the terminals of the turbine generators [66, 69]. Hence, the same approach has also been adopted in this study.

The generator speed is continuously measured and passed through the washout block to obtain the generator speed deviation which reflects the SSR occurring in the generator. It is enhanced by a gain factor K and phase shifted by 180° to produce the *d*-axis reference $i_{d\text{-ref}}$ for the current loop controller. This controller produces $i_{d\text{-ref}}$ in a manner that the corresponding PV-STATCOM reactive power exchange can damp the SSR (Note that in this study the *dq* axis are chosen such that the *d*-axis current component controls reactive power.)

The best controller parameters are obtained through a systematic hit-and-trial method to result in a minimal settling time and acceptable overshoot (less than 10%) in generator speed. There are analytical approaches for designing the damping controllers by FACTS Controllers such as [69, 70], which are more efficient than gain selection through trial and error. However, as the objective of this study is to demonstrate a new concept of SSR mitigation by PV solar farms control as STATCOM (PV-STATCOM), a simpler controller parameter selection through systematic trial-and-error method was chosen.

6.4.4 DC Voltage Controller

Figure 6.28a illustrates the conceptual DC voltage controller. It is comprised of the MPPT block and a PI controller. The MPPT block is simulated in PSCAD software based on an incremental conductance algorithm [36]. The MPPT block produces $V_{DC\text{-ref}}$ to control the active power generated by PV solar farm. The measured DC voltage is compared with $V_{DC\text{-ref}}$ to create an error signal. The PI controller processes this error signal and generates the *q*-axis reference $i_{q\text{-ref}}$ for the current loop controller. The PI controller parameters are tuned by a hit-and-trial method in the same manner as the PI controller of the current-loop controller.

The flow chart of the proposed DC voltage controller for PV-STATCOM operation is portrayed in Figure 6.28b. The DC voltage controller constantly monitors if the system is operating in a healthy manner and no SSOs are initiated, i.e. the rotor speed deviation $\Delta\omega$ is less than a prespecified

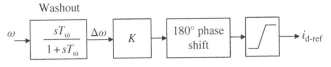

Figure 6.27 Damping controller configuration. *Source:* Varma and Salehi [71].

(a)

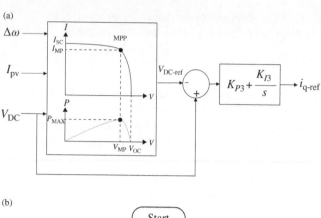

(b)

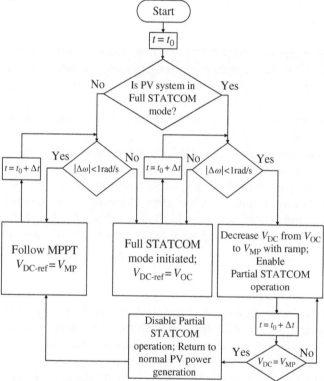

Figure 6.28 (a) DC voltage controller, (b) flowchart of DC voltage controller operation. *Source:* Varma and Salehi [71].

quantity which is chosen to be 1 rad/s. In this case, the Full STATCOM mode is not activated and the system operates in normal PV power generation mode with $V_{\text{DC-ref}}$ set to V_{MP}, the MPP voltage. If SSOs are caused due to any system disturbance or fault, and the rotor speed deviation $\Delta\omega$ exceeds 1 rad/s, the Full STATCOM mode is initiated. This is accomplished by setting $V_{\text{DC-ref}}$ to V_{OC} which is the open-circuit voltage of the solar panels. The active power generated by the solar panels is rapidly curtailed to zero and the entire inverter capacity is released for STATCOM operation to damp SSOs.

Once SSOs are mitigated, i.e. $\Delta\omega < 1$ rad/s, the DC voltage controller gradually resumes normal solar power generation by decreasing the DC voltage to its prefault value V_{MP} in a ramped manner

[40]. Simultaneously, the Partial STATCOM mode with active power priority is enabled. In this mode while the active power generation is being ramped up, damping of SSO is continued with the inverter capacity remaining after active power generation. This ensures that the PV power can be restored to its prefault value without recurrence of SSO. This ramp-up of power with SSR damping control in operation is a novel feature of the proposed PV-STATCOM control which is not specified in any grid code, presently. Once V_{DC} becomes equal to V_{MP}, the Partial STATCOM operation is disabled and Full PV power generation is resumed.

6.4.5 Simulation Studies

Studies for damping SSO using PV-STATCOM control are performed using PSCAD software. These studies are reported for the most stringent case when both the synchronous generator and the PV solar system are producing their rated power representing a similar power flow as [72]. A three-line-to-ground (3LG) fault for five cycles is initiated at bus B at $t = 5$ s. These fault studies are performed for four critical levels of series compensation when the four respective torsional oscillatory modes are most undamped, as described in [68]. The detailed responses are reported only for the damping of Modes 1 and 4, which are more destabilized as compared to Modes 2 and 3.

Figure 6.29 depicts the postfault behavior of the study system when the solar farm functions normally and is *not controlled* as the proposed PV-STATCOM. This figure illustrates the system response for the series compensation level when Mode 1 is critically excited. The responses of the PCC RMS voltage, PV-STATCOM reactive power; generator active and reactive powers, rotor speed, and the torque in the LPB-GEN section are depicted. It is evident from this study that the system becomes highly unstable due to growing SSO.

According to the voltage ride-through (VRT) criteria of existing grid codes and Standards, the PV solar farm may be disconnected due to its large voltage excursions. For instance, NERC standard "PRC-024-3" requires the generating unit to have the capability of VRT [54]. Based on VRT time duration curve and Table in Attachment 2 of this Standard, whenever the PCC voltage goes under 0.65 pu for a duration of more than 0.3 seconds the generating unit may be disconnected from the grid. Furthermore, whenever the voltage goes above 1.2 pu the generating unit may be tripped instantaneously. As evident from Figure 6.29, the PCC voltage after the occurrence of the fault goes under 0.65 pu for a duration of 0.8 seconds. Therefore, the PV solar farm should be disconnected at $t = 5.835$ s. In addition, the voltage goes above 1.2 pu at $t = 5.86$ s, which is another criterion for the solar farm to be disconnected at $t = 5.86$ s, although the previous criterion will take precedence for disconnection.

ERCOT Nodal Operating Guides [75] needs all intermittent renewable resources (IRRs) to conform to VRT requirements. Based on Figure 6.1 of Section 6.2 of this operating guide, the IRR shall be disconnected whenever the PCC voltage goes under 0.6 pu for 1.25 seconds, or immediately whenever it goes above 1.2 pu. Therefore, based on ERCOT Nodal Operating Guides and Figure 6.29, the PV solar farm shall be disconnected at $t = 5.86$ s when the PCC voltage goes above 1.2 pu. It is, therefore, emphasized that the solar farm will *anyway be disconnected* due to the VRT criteria of the grid codes, and not because of the proposed PV-STATCOM control. The issue of frequency control will arise, but that will be need to addressed separately by the system operator.

It is shown in the next section that the novel proposed PV-STATCOM control goes a step beyond, and instead of remaining idle in the disconnected mode, utilizes its inverter capacity to successfully damp SSR. This control further returns the solar farm to its normal prefault power production and restores the system frequency to its nominal value. It performs this function autonomously without

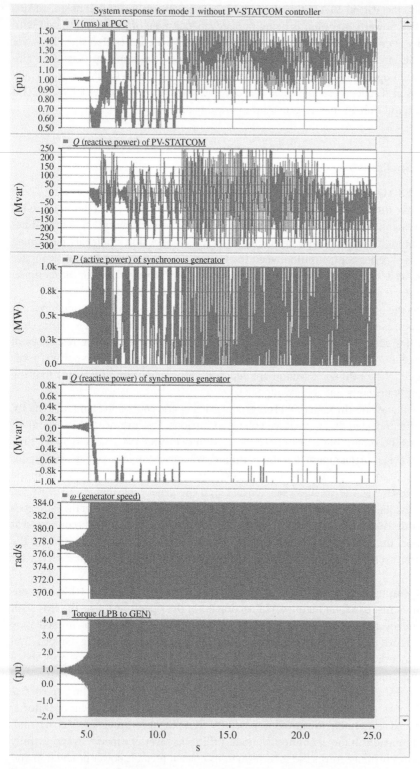

Figure 6.29 System response for Mode 1 SSR without PV-STATCOM controller. *Source:* Varma and Salehi [71].

any support from the system operator. Without this proposed control, the system will become unstable and the system operator will have to take due measures to restores system stability and frequency.

6.4.5.1 Damping of Critical Mode 1 (67% Series Compensation)

The stabilizing impact of SSR damping control of the PV-STATCOM is now investigated. Figure 6.30 depicts the different PV-STATCOM variables which include inverter active power P, reactive power Q, apparent power S, DC side current I_{PV}, and the DC side voltage V_{DC}. Figure 6.31 illustrates the different synchronous generator variables for this case, which include the active power, reactive power, generator speed ω, and torques between HP-IP, IP-LPA, LPA-LPB, LPB-GEN, and GEN-EXC sections, respectively. The corresponding transmission system parameters for this case study are portrayed in Figure 6.32. These consist of RMS voltage at PCC, active power flow, and reactive power flow.

As soon as the onset of SSO in rotor speed is sensed at $t = 5$ s by the damping controller, the DC voltage controller increases the DC bus voltage to its open circuit value V_{OC} to make the inverter active power P and the DC current I_{PV} go to zero, however, with a slight delay based on the time constant of the involved controllers. The entire capacity of the inverter is thereby made available for the Daytime Full STATCOM operation for SSR damping. The rotor oscillations are successfully damped to within 1 rad/second at $t = 9.68$ s. Once this happens, the PV solar farm restores power generation by decreasing the DC voltage in a ramped manner, while continuing to damp SSR in the Partial STATCOM mode of operation. The prefault level of power generation is achieved without any voltage/power oscillations in about 5 seconds at $t = 14.68$ s. The entire process of damping Mode 1 SSR takes only 9.68 seconds utilizing a maximum of 200 Mvar STATCOM reactive power.

It is observed from Figure 6.31 that torsional oscillations occur in all the mechanical sections, however, the highest torque in excess of 4 pu is experienced in the LPA-LPB shaft section. The PV-STATCOM successfully damps SSO in the entire torsional system. It is noted from Figure 6.32 that during the damping operation of Mode 1 by PV-STATCOM operation, the line active power reduces as the power output of the solar farm is made to go to zero but subsequently returns to the prefault value. The PCC voltage rises slightly above 1 pu during the transient.

6.4.5.2 Ramp Up without PV-STATCOM Control

Grid codes such as [38, 40, 58] stipulate that active power sources including PV solar farms should ramp up their power output slowly during the connection process in order to avoid voltage or power oscillations due to the sudden injection of a large amount of active power into the grid. The grid codes further specify a range for the rate at which the power may be ramped up depending upon system characteristics. A typical specification is that the increase in active power supplied to the network by the power source must not exceed a maximum gradient of 10% of the agreed active connection power per minute [38]. The appropriate ramp-rate for a specific system may be determined from off-line system studies.

This study has proposed a novel fast method of reconnection of PV solar farm while keeping the PV-STATCOM SSR damping function activated. It is demonstrated from Figures 6.31, 6.32, 6.33 that with this proposed technique the active power can be ramped up from zero to 300 MW in about five seconds without resumption of SSR.

To demonstrate the effectiveness of this proposed technique, a new study is performed. In this study, after SSR has been mitigated and the generator rotor speed has stabilized to within acceptable limits at $t = 10$ s in Figure 6.31, the power ramping up is performed without the damping control in Partial STATCOM mode. The ramp-up rate gets slowed down three times, i.e. the power is

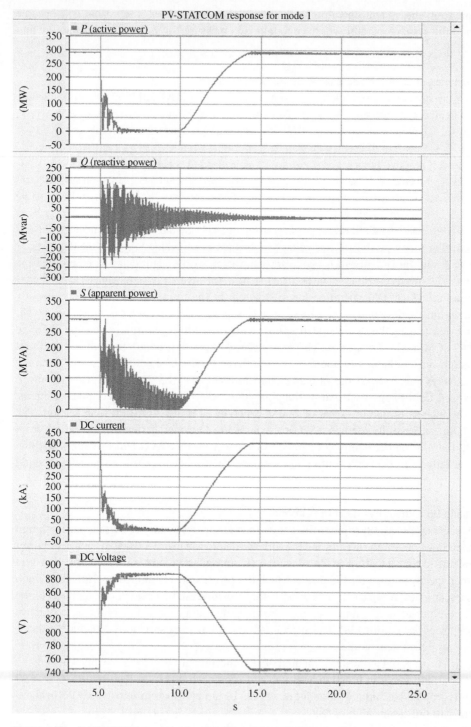

Figure 6.30 PV-STATCOM response for damping of Critical Mode 1 SSR. *Source:* Varma and Salehi [71].

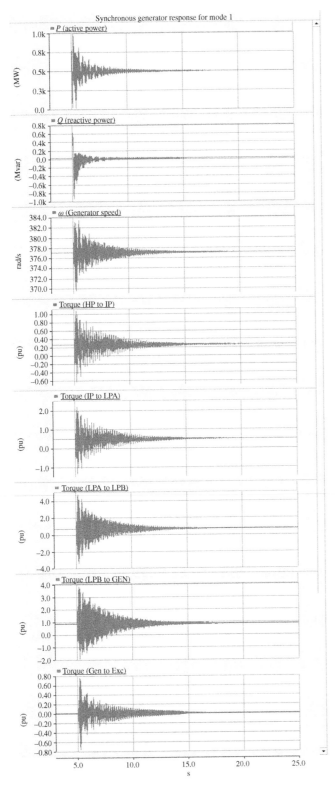

Figure 6.31 Synchronous generator response for damping of Critical Mode 1 SSR. *Source:* Varma and Salehi [71].

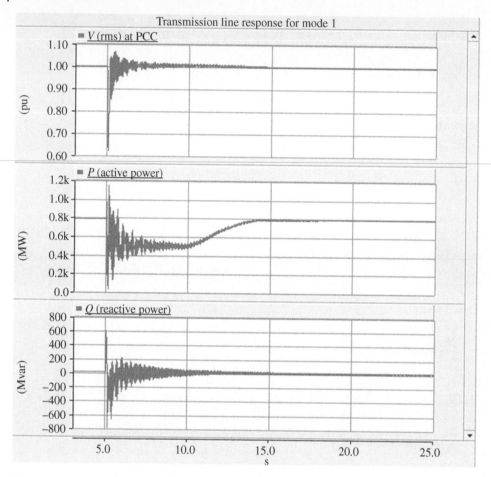

Figure 6.32 Transmission system response for damping of Critical Mode 1 SSR. *Source:* Varma and Salehi [71]

ramped up from zero to 300 MW over 15 seconds instead of 5 seconds as in Figure 6.32 with PV-STATCOM damping control in operation. Figure 6.33 portrays the response of active power of the synchronous generator, the active power of solar farm, and generator rotor speed. The system is seen to become unstable due to recurrence of SSO. This clearly shows the efficacy of the proposed ramp up with PV-STATCOM damping control active.

6.4.5.3 Damping of Critical Mode 2 (54% Series Compensation)
In this case study, the fault is initiated as before at $t = 5$ s. The PV-STATCOM damps the torsional oscillations ($\Delta\omega < 1$ rad/second) at $t = 7.96$ s, and the solar power output is restored to its prefault value without any SSR at $t = 12.96$ s (not shown). Thus the total time required for SSR alleviation is 7.96 seconds.

6.4.5.4 Damping of Critical Mode 3 (41% Series Compensation)
In this case study, the fault is initiated as before at $t = 5$ s. The PV-STATCOM damps the torsional oscillations ($\Delta\omega < 1$ rad/s) at $t = 6.81$ s, and the solar power output is reinstated to its prefault value without any SSR at $t = 11.81$ s (not shown). Thus the total time required for SSR alleviation is 6.81 seconds. The maximum PV-STATCOM reactive power needed for this purpose is also about 200 Mvar.

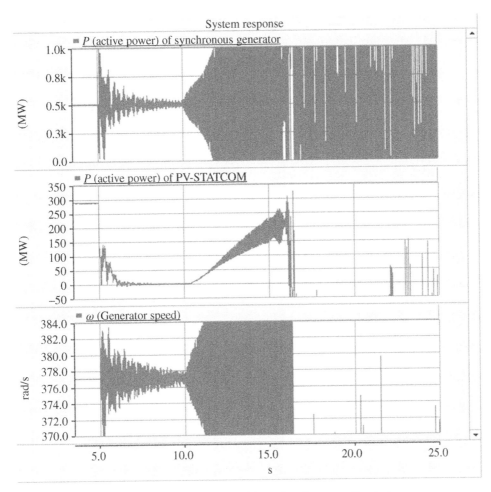

Figure 6.33 System response for Mode 1 SSR without damping controller during ramp-up.
Source: Varma and Salehi [71].

6.4.5.5 Damping of Critical Mode 4 (26% Series Compensation)

For this case, Figure 6.34 illustrates the PCC RMS voltage, PV-STATCOM reactive power; generator active and reactive powers, rotor speed, and the torque in the LPB-GEN section. The PCC bus voltage is about 1.05 pu, which is higher than in the case of Mode 1 damping (Figure 6.31). This is because, during the PV-STATCOM operation, when PV solar system does not generate any active power, the synchronous generator reactive power is largely capacitive, as opposed to being largely inductive as in case of Mode 1 damping (Figure 6.31).

The maximum subsynchronous torque of about 3 pu is experienced in the LPB-GEN section. The PV-STATCOM successfully mitigates all the torsional oscillations ($\Delta\omega < 1$ rad/second) at $t = 6.77$ s, and the solar power output is restored to its prefault value without any SSR at $t = 11.77$ s. The total time required for SSR alleviation is 6.77 seconds and the maximum PV-STATCOM reactive power needed for this purpose is about 200 Mvar.

6.4.6 Potential of Utilizing Large Solar Farms for Damping SSR

With increasing electrical power generation to meet the load demand, series compensation of lines is expected to be increasingly considered as an economical alternative to construction of new

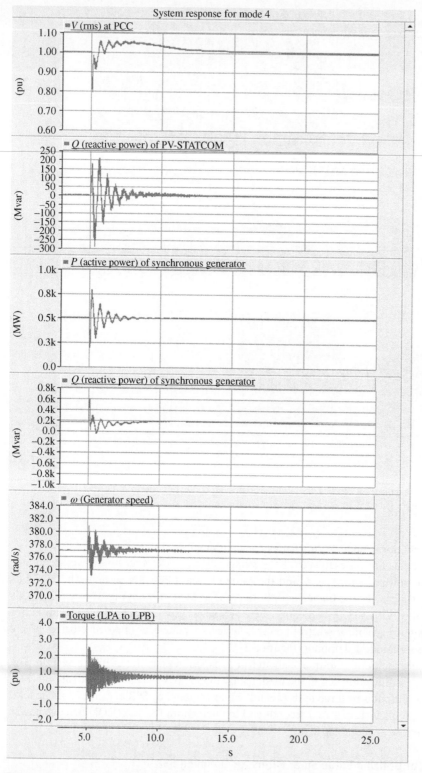

Figure 6.34 System response for damping of Critical Mode 4 SSR. *Source:* Varma and Salehi [71].

transmission lines, worldwide. This is likely to make the associated generation systems more prone to SSR issues. On the other hand, due to the rapid adoption of renewable energy systems coupled with the drop in PV panel prices, large-scale PV solar farms with rating in excess of 100 MW are increasingly being deployed in transmission systems worldwide. Furthermore, these sizes of large utility-scale solar farms are becoming comparable to transmission level shunt-connected FACTS Controllers such as SVCs and STATCOMs. It is understood that large-scale PV solar farms may be installed at locations which will be determined from nontechnical considerations, such as availability of cheap land, etc. However, it is quite a possibility that such transmission-connected solar farms may be connected in networks that are series compensated and also may be in close vicinity of synchronous generators that may be SSR prone. For instance, the 290 MW Agua Caliente Solar Project constructed by First Solar is connected to Hassayampa-North Gila 500 kV transmission line that is series compensated [76].

Considering the above growing developments of large solar farms and their possible connection in series compensated lines, such large solar farms can potentially become candidates for implementation of PV-STATCOM SSR damping function.

While synchronous generator-based devices and other mechanisms for damping SSR do exist, FACTS Controller-based SSR mitigation measures have been widely utilized as described earlier [2–4]. Therefore, in cases where generator-based SSR damping devices are not considered or found to be inadequate, the PV-STATCOM technology can become both economic and feasible solution for SSR mitigation.

6.4.7 Summary

Rapidly emerging, large utility-scale PV solar farms are likely to find themselves being connected in transmission systems that are series compensated. This study presents a PV-STATCOM control for mitigating SSR in steam turbine-driven synchronous generators connected to series compensated transmission lines. The proposed PV-STATCOM control provides solar farm the capability to mitigate SSR both in the night and anytime during the day. In the night, since the solar farm is idle, the entire inverter capacity is utilized for PV-STATCOM operation. During the day, at any time, if SSR is triggered due to any system disturbance, the solar farm autonomously curtails its normal active power generation and transforms into a STATCOM with the full inverter capacity for SSR mitigation. Once the subsynchronous oscillations are reduced below an acceptable level, the solar farm autonomously returns to its prefault active power generation level in a ramped manner.

Studies are conducted on the IEEE First SSR Benchmark system [72] having a large 300 MVA solar farm connected at the generator terminals producing its rated power, to simulate similar study conditions as in [72]. The following conclusions are drawn:

1) The PV-STATCOM successfully mitigates all the four torsional modes at all the four critical levels of series compensation.
2) The total time taken by the PV-STATCOM from the autonomous initiation of damping control to mitigate the different subsynchronous mode oscillations and return to prefault normal solar power production is 9.68 seconds for Mode 1, 7.96 seconds for Mode 2, 6.81 seconds for Mode 3 and 6.77 seconds for Mode 4, oscillations, respectively. This time of interruption of PV solar power generation is comparable to a cloud passing event.
3) The PV-STATCOM reactive power required to damp SSR for all the critical series compensation levels in this study is <200 Mvar.
4) The novel concept of power ramp-up keeping the SSR damping function of PV-STATCOM activated allows a ramp-up rate much faster than that specified in various grid codes.

6.5 Alleviation of Subsynchronous Oscillations (SSOs) in Induction-Generator-Based Wind Farm by PV-STATCOM

SSR can also occur in wind farms connected to series compensated lines. SSR may be caused in IG-based Type-1 WFs due to: (i) interaction between electrical network and generator stator, called induction generator effect (IGE) and (ii) interaction between electrical network and the mechanical system of wind generator termed TI [77]. TI can occur in Type-1 WFs only at very high levels of series compensation due to the low shaft inertia. However, SSR due to IGE can occur even at realistic levels of series compensation in transmission lines [78].

SSO also gets initiated in doubly fed induction generator (DFIG)-based Type-3 WFs connected to series compensated lines. This may lead to an adverse interaction between DFIG controller and series compensated electrical network known as subsynchronous control interaction (SSCI) [79]. This interaction has resulted in severe overvoltages and damage to wind generator control circuits in a Type-3 WFs in Minnesota (USA) in 2007 [79], in the Electric Reliability Council of Texas (ERCOT) in 2009 [80], and in north China in 2012 [81].

The recent trend is to employ Type-3 or full converter-based Type-4 WFs [79, 82]. However, there are still a significant number of large Type-1 WFs around the world, such as Alta Wind Energy Center (1170 MW) in California, USA; Brahmanvel WF (549 MW) in India, Biglow Canyon WF (450 MW) in Oregon, USA; Twin Groves WF (396 MW) in Illinois, USA; Papalote Creek WF (380 MW) in Texas, USA; Clyde WF (350 MW) in the UK, Hallett WF (348 MW) in Australia, Maple Ridge WF (321 MW) in New York, USA; Pioneer Prairie WF (300 MW) in Iowa, USA; Tarfaya WF (300 MW) in Morocco, Stateline WF (300.96 MW) at the border of Washington and Oregon, Fowler Ridge WF (300.3 MW) in Indiana, USA, etc. [83, 84].

This study deals with SSO alleviation in IG-based Type-1 WFs since (i) there are a substantial number of Type-1 WFs around the world, (ii) some of these WFs are already connected to series compensated networks, e.g. Stateline WF (300.96 MW) [78] or may likely be connected to such networks in future, or (iii) these WFs may find themselves radially connected to series compensated lines due to some postfault line switching scenarios, as occurred in Texas [85].

A novel PV-STATCOM control is described below for alleviation of SSO-induced instability in IG-based Type-1 WFs [86]. This concept is demonstrated in a modified IEEE First SSR Benchmark system in which the solar farm may be located either at WF terminal or remotely from the WF. The SSO damping controller is based on WF generator speed and local transmission line current when the PV solar farm is located at the above two locations, respectively. During daytime, at the occurrence of SSO, the PV system curtails entire active power generation and utilizes the rated inverter capacity to operate as STATCOM. The effectiveness of the PV-STATCOM is demonstrated both when it is connected locally at WF terminal or remotely at the line midpoint. Furthermore, the PV-STATCOM control is shown to successfully mitigate SSR for different: (i) WF sizes, (ii) WF output power levels, (iii) series compensation levels, and (iv) fault locations; both during day and night.

6.5.1 Study System

The IEEE First SSR Benchmark System [72] is modified by replacing the 892.4 MW synchronous generators by a 500-MW IG-based Type-1 WF. This modified system is considered to be the study system and is depicted in Figure 6.35a. The WF is connected to the infinite bus through a series compensated transmission line. The series compensation level (k) is defined as

$$k = \frac{X_C}{X_T + X_l + X_S} \tag{6.11}$$

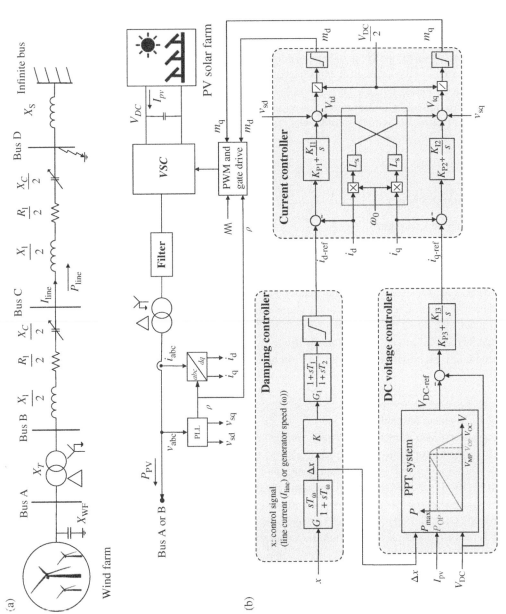

Figure 6.35 Study system: (a) modified IEEE First SSR Benchmark System with integrated wind farm and (b) PV solar farm plant and its control system. *Source:* Salehi [86].

where X_C is the series capacitive reactance, X_T is the transformer reactance, X_l is the transmission line reactance, and X_S is the equivalent system reactance at infinite bus [72]. Figure 6.35b illustrates the 300-MW PV plant and its control system for operation as PV-STATCOM. The PV plant is connected to the series compensated network of Figure 6.35a at either of the two locations: (i) at the WF generator (Bus A) or (ii) at the middle of transmission line (Bus C). The entire system is modeled in PSCAD software according to data provided in [72, 78]. The 500-MW WF is constituted of an aggregation of 217 identical 2.3-MW double-cage IGs. It is assumed that all the generators are of the same type and size and are subjected to the same wind speed [78]. Unlike synchronous generator, the IG does not have any source of excitation and needs to absorb inductive reactive power for its operation. A shunt capacitor (X_{WF}) is, therefore, connected at the terminal of the WF to keep its power factor close to unity. To model the dynamics of mechanical system, an aggregated two-mass wind turbine is considered [78]. The mechanical damping is assumed to be zero to represent the worst-case scenario [78].

The PV solar farm model includes a voltage source converter (VSC) which is connected to a large capacitor and a controlled current source on its DC side. The DC source follows the I-V characteristic of PV panels of the solar farm [21]. The AC side of the VSC is connected to the transmission line through an LCL filter and a coupling transformer. The PV system utilizes a Power Point Tracking (PPT) system which provides the DC-link voltage reference V_{DC-ref} for the current loop controller.

6.5.2 Control System

6.5.2.1 Current Controller
Figure 6.35b illustrates the proposed PV-STATCOM control system for damping SSO. This control system includes a damping controller with DC voltage controller and the conventional current controller [25]. All the controllers are modeled in dq-reference frame. The angle of the grid voltage is estimated and provided by a phase-locked loop (PLL) block [21, 25]. In this study, the voltage vector is aligned with the quadrature axis and V_d equals zero [21]. Thus, reactive power of the VSC (Q_{PV}) is controlled with d-axis control loop whereas active power generation of PV solar farm (P_{PV}) is controlled by the q-axis loop. The current controller regulates the output current of VSC by generating appropriate switching signals for VSC's switches as explained in [25].

6.5.2.2 Damping Controller
The proposed SSO damping controller consists of a washout filter block, controller gain block, phase compensator (lead–lag compensator), and a limiter block. When the PV system is connected at WF terminal, i.e. at Bus A, the generator speed is utilized as the control signal for damping controller of PV-STATCOM. However, when the PV system is connected at line midpoint, i.e. at Bus C, local line current (I_{line}) is employed as a control signal. This signal further helps avoid the impact of communication delay if ω were to be selected as the control signal. The damping control signal, ω or I_{line}, is measured and passed through the washout filter. The output of the washout filter contains the oscillatory component of the control signal which reflects SSO in the system. This signal is amplified with the controller gain, and its phase is compensated by the phase compensator to provide i_{d-ref} for current controller. The damping controller modulates the reactive power Q_{PV} of PV-STATCOM to damp SSO by controlling i_{d-ref}, appropriately.

6.5.2.3 DC Voltage Controller

As Figure 6.35b illustrates, the DC voltage controller includes a PPT block and a PI controller. The PPT block provides the reference voltage ($V_{DC\text{-ref}}$) to control PV-STATCOM active power output P_{PV}. In conventional UPF operation of the PV solar farm, the PPT block adjusts the DC side voltage (V_{OP}) to the MPP voltage (V_{MP}) at which maximum active power is extracted from the PV panels [25]. However, the PPT block operates in the non-MPP mode for PV-STATCOM operation, as described below [71]. The PI controller processes the error between $V_{DC\text{-ref}}$ and measured voltage of DC-link and provides the $i_{q\text{-ref}}$ for current control block to control P_{PV}.

The DC voltage controller continuously monitors the oscillatory component of control signal (x). Whenever undamped SSO occurs due to any system fault or disturbance, the PV system switches to Full-STATCOM mode of operation. In this mode, the DC voltage controller increases V_{OP} to V_{OC} (the open-circuit voltage of PV panels) and makes $P_{PV} = 0$. This makes the entire converter capacity available for reactive power modulation. The damping controller modulates Q_{PV} as required and up to the maximum converter capacity for mitigating SSR. After SSO is damped to an acceptable limit, the PV system restores P_{PV} to its predisturbance level in a ramped manner. For this purpose, the DC controller decreases V_{OP} from V_{OC} to V_{MP} gradually.

A novel aspect is that during the period when P_{PV} is increased in a ramped manner, the VSC is made to operate in the Partial-STATCOM mode. The VSC capacity remaining after active power production is utilized for Q_{PV} modulation to prevent any recurrence of SSO due to ramp-up of P_{PV} [71]. Such damping functionality during power ramp-up is not a requirement of grid-codes [38, 87].

An optimization block [42] resident in PSCAD software is used to determine the parameters of the damping controller, DC voltage controller, and current controllers. An OF is defined to provide least settling time and fast step response, as below.

$$OF = \int_0^{t_s} |\Delta x| \, t \, dt \tag{6.12}$$

where x is the control signal and t_s is the simulation time.

6.5.3 Simulation Studies

6.5.3.1 Daytime Case Study: PV System Connected at Wind Farm Terminal

This section presents the daytime performance of PV-STATCOM in alleviation of SSO, as obtained from PSCAD simulation studies. A three-line-to-ground (3LG) fault is initiated at Bus D (remote from the WF terminals) for five cycles at $t = 5$ s. For all the studies, the size of the PV plant is 300 MW, i.e. $P_{PV} = 300$ MW.

Figure 6.36a–c depicts the electrical torque (T_e) of WF, shaft torque (T_{sh}), and rotor speed (ω), respectively, when WF power output P_{WF} is 500 MW, without PV-STATCOM damping control. k is selected to be 55%, as this is a realistic level of series compensation in transmission system. It excites SSO with frequency of 24 Hz in this radial system for different levels of P_{WF} [78]. The system is seen to be unstable due to growing SSO in T_e and ω. Although SSO is reflected in T_{sh}, but it does not make T_{sh} unstable in a similar time frame as T_e.

Figure 6.36d–f illustrates the WF variables while PV-STATCOM damping controller is implemented in PV system. The variables shown are correspondingly similar to Figure 6.36a–c. Comparison of these figures clearly demonstrates the effectiveness of PV-STATCOM in damping SSO.

Figure 6.3 presents different variables of the system when $P_{WF} = 500$ MW, $k = 55\%$, and PV-STATCOM control is implemented on the PV plant. Figure 6.3a c depicts active power generation

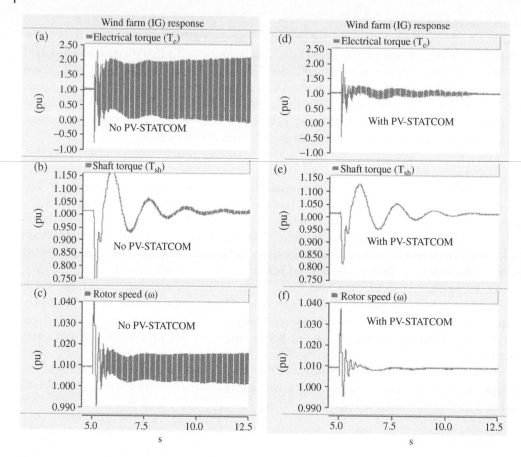

Figure 6.36 Windfarm response, without and with PV-STATCOM controller, P_{WF} = 500 MW, PV system at bus A, and k = 55%. *Source:* Salehi [86].

of PV system (P_{PV}), reactive power generation of PV system (Q_{PV}), and DC side voltage (V_{DC}), respectively. Before fault occurrence, the PV system operates in its conventional MPP mode and $P_{PV} = 300$ MW (Figure 6.3a). Once fault takes place in the system at $t = 5$ s, the DC voltage controller increases the DC voltage to V_{OC} making $P_{PV} = 0$ although with a slight delay based on the controller time constant. Consequently, the entire VSC capacity is made available for Full-STATCOM operation. Once SSOs are damped and $\Delta\omega$ becomes less than 1 rad/s at $t = 7.53$ s, the PV system starts increasing P_{PV} gradually to its prefault value in four seconds. Based on grid codes [29, 30], a gradual ramping of P_{PV} is required to prevent initiation of low frequency voltage/power oscillations. The DC voltage controller commences power restoration process by decreasing the DC voltage in a ramped manner. During power ramp-up, the PV plant continues to operate in Partial STATCOM mode utilizing the remaining capacity of the VSC to avoid recurrence of SSO. The PV plant resumes its prefault power generation at $t = 11.53$ s, i.e. just 6.53 seconds after fault occurrence.

Figure 6.37d–f depicts transmission line variables including PCC voltage (V_A), active power flow (P_{line}), and reactive power flow (Q_{line}), respectively, with PV-STATCOM controller implemented. After fault is initiated, the PCC voltage and reactive power undergo a short transient but stabilize quickly. P_{line} decreases due to cessation of P_{PV}. After alleviation of SSO, P_{line} returns to its prefault condition due to restoration of P_{PV} to its prefault level.

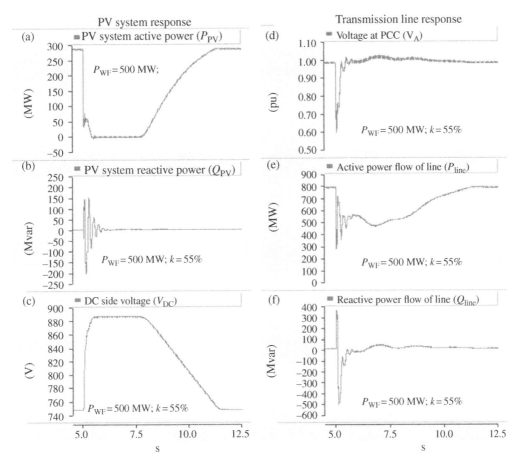

Figure 6.37 System response with PV-STATCOM controller, P_{WF} = 500 MW, PV system at bus A, and k = 55%. *Source:* Salehi [86].

6.5.3.2 Daytime Case Study: PV System Connected at Line Midpoint

The SSO damping performance of PV-STATCOM when PV system is connected to the middle of the line is studied here. To avoid communication delays, line current (as a local signal) is chosen as the damping control signal. PSCAD studies are conducted for 500 MW rated WF producing P_{WF} = 500 MW and P_{WF} = 300 MW when k = 55%.

Figure 6.38a,b depicts the system variables for the case P_{WF} = 500 MW, whereas Figure 6.38e–h depicts the system variables when P_{WF} = 300 MW. It is seen from Figure 6.38a,e that without PV-STATCOM controller, the system is unstable due to both relatively large and undamped SSO in T_e. However, Figure 6.38b,f demonstrates that the PV-STATCOM with local line current-based damping control can successfully mitigate these large magnitude SSO. At the occurrence of fault, due to PV active power curtailment, the active power flow in the line P_{line} reduces (Figure 6.38c and g). Once the SSOs in the line current are damped to less than 2% of current magnitude, the PV power is restored in a ramped manner while keeping the SSO damping function activated in the Partial STATCOM mode, as explained earlier. The PV-STATCOM reactive power Q_{PV} for SSO damping is about 250 Mvar for both P_{WF} = 500 MW and P_{WF} = 300 MW, as seen in Figure 6.38d,h, respectively. It is noted that the corresponding Q_{PV} for SSO damping is relatively smaller (about 200 MW), when the PV system is located at WF terminal.

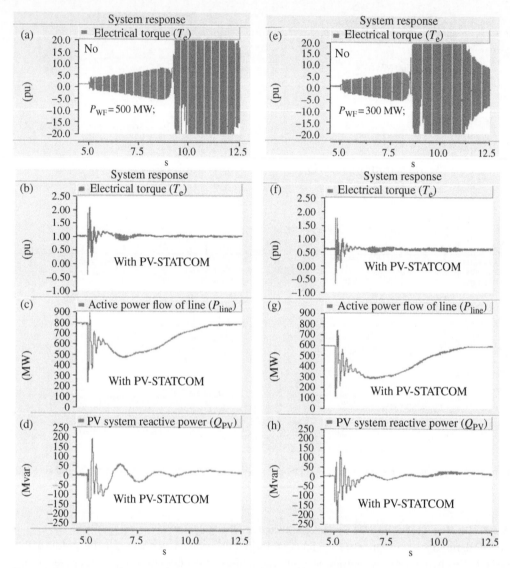

Figure 6.38 System response without and with PV-STATCOM controller, P_{WF} = 500 MW, and 300 MW, PV located at midpoint of the line, and k = 55%. *Source:* Salehi [86].

For both locations of PV solar system (at WF terminal and at line midpoint), SSO damping studies are performed, although not shown here, for different levels of P_{WF}, different levels of series compensation, and different sizes of WF, in all of which cases the PV-STATCOM successfully alleviated SSO.

6.5.3.3 Nighttime Case Study: SSO Alleviation by PV-STATCOM

The nighttime performance of PV-STATCOM for SSO damping is now investigated for a 3LG fault at Bus D (remote from WF terminals) for five cycles at t = 5 s.

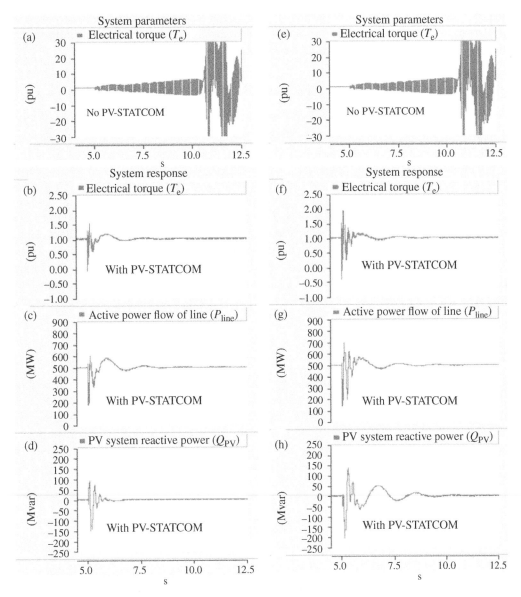

Figure 6.39 System response without and with PV-STATCOM controller, nighttime, PV is connected at bus A and bus C, P_{WF} = 500 MW, and k = 55%. *Source:* Salehi [86].

Figure 6.39a–d and e–h depicts the system variables when PV plant is located at bus A and bus C, respectively. At fault occurrence, the system becomes unstable due to undamped SSO, which get damped with PV-STATCOM. Some initial transients and WF torsional mode oscillations are seen in-line power flow P_{line}. The magnitude of torsional mode oscillations is slightly higher when the PV system is connected at line midpoint instead of at WF terminal.

Comparing Figure 6.39d,h, it is noted that a larger Q_{PV} of 200 Mvar is needed to damp SSO when PV system is located at midline (using local line current as damping signal) than Q_{PV} of 150 Mvar when PV system is connected at WF terminal (utilizing generator speed as damping control signal). In nighttime, the SSO damping process is less than 10 seconds and faster than daytime because no active power curtailment and subsequent restoration is involved at night.

6.5.4 Potential of Utilizing Large Solar Farms for Alleviating SSO in Wind Farms

Nowadays, a large number of PV solar farms rated more than 100 MW (even up to 2200 MW) are already commissioned and increasingly being developed around the world, as described in Section 6.1.

The typical size of STATCOM required for SSR damping in power transmission systems is about 200 Mvar [69, 70, 83, 88]. Hence, given a need, large solar farms can provide PV-STATCOM functionality for mitigation of SSO.

Some other key developments that may be conducive for the above novel application of solar farms are: (i) PV farms are being connected to series compensated lines, e.g. 290 MW Agua Caliente Solar Project in the United States is connected to a series compensated 500 kV line [76], (ii) some IG-based Type 1 WFs are connected to series compensated lines, e.g., 300 MW Stateline WF in the United States is connected to a series and shunt-compensated network in BPA transmission network [89], and (iii) the trend of co-locating WFs and PV farms is also increasing, e.g. in Canada [90] and Australia [91, 92].

6.5.5 Summary

This study presents a new night and day PV-STATCOM application for alleviating instability caused by SSOs in IG-based Type-1 WF connected to series compensated line. A modified IEEE First SSR benchmark including a 500 MW IG-based WF and a 300 MW PV solar farm is employed to study the effectiveness of PV-STATCOM in SSO mitigation through electromagnetic transient studies using PSCAD. Wind generator speed is used as the control signal for damping controller when PV plant is connected at the WF terminal, whereas line current which is a local signal, is utilized when PV plant is located at line midpoint. The following inferences are made (although some results are not shown here:

1) Daytime: PV-STATCOM effectively alleviates SSO for:
   - Different levels of WF power output (P_{WF} = 100, 300, and 500 MW). Higher P_{WF} requires larger PV system reactive power Q_{PV} for SSO mitigation.
   - Various sizes of WFs – 100 and 500 MW. Higher Q_{PV} is needed for stabilizing SSO in larger WFs.
   - Different fault locations – either remote from WF (Bus D) or at terminal of WF (Bus B). Less Q_{PV} is required to mitigate SSO when fault occurs at Bus B.
   - Various series compensation levels (k: 30–80%). Higher k causes more severe SSO which requires larger Q_{PV} for SSO alleviation.
   - Different locations of PV plant – WF terminal (Bus A) or line midpoint (Bus C). Higher Q_{PV} is required for stabilizing SSO when PV system is connected at Bus C than at Bus A.

2) Nighttime: PV-STATCOM mitigates SSO in 500-MW IG-based WF at different PV plant locations: Bus A or Bus C.

The entire process of SSO instability mitigation from the instant of fault to restoration of PV solar farm power to its prefault condition takes place in less than half a minute.

6.6 Mitigation of Fault-Induced Delayed Voltage Recovery (FIDVR) by PV-STATCOM

FIDVR is a postfault phenomenon associated with power systems having a high concentration of induction motor (IM) loads, such as air conditioners [93]. During fault events, IMs consume high reactive current resulting in low voltages for several seconds even after the fault is rapidly cleared [93]. FIDVR is characterized by a post fault steady state voltage recovery of typically more than two seconds [93]. This prolonged low voltage can activate motor protection causing motors to shut down leading to significant load loss and subsequent secondary issues of overvoltages in the system [93–95]. Extended low voltages further affect consumers and lifecycle of utility equipment [93].

FIDVR occurrences are widely described in literature. The FIDVR event in Union City, Metro Atlanta caused a tripping of 1900 MW load and 7 small generating units [93]. A load of 445 MW was lost in Washington due to prolonged low voltage while 1700 MW load was lost in Toronto, Canada, due to FIDVR [94]. In Southern California Edison network alone, more than 50 events have been reported, among which the largest event caused a load loss of 3500 MW [96]. The FIDVR event in Hassayampa, Arizona, caused a loss of 440 MW load and 2600 MW generation [96].

Both customer level and system level solutions are employed for mitigating FIDVR [95]. Customer level solutions aim to reduce FIDVR severity by modifying the motor units to ride through low voltages and adjusting motor protection systems. These solutions are time-consuming and expensive as all motor units need to be modified.

System level solutions include controlled reactive power support, under-voltage load shedding, use of energy-saving devices, and reduced fault clearing time, etc. [93]. A load shedding strategy for FIDVR mitigation is presented in [97]. Dynamic reactive power support is the most widely employed solution at system level [95]. An optimization-based dynamic reactive power support to mitigate FIDVR is presented in [98]. An optimized control of dynamic VAR sources for mitigation of FIDVR is discussed in [99, 100].

FACTS Controllers such as SVC and STATCOM are utilized in transmission systems for providing dynamic reactive power support during system contingencies [4, 101]. SVCs have been employed for FIDVR alleviation in several utility systems as below:

1) In Saudi Arabia, air conditioner motors constitute 80% of the total load in summer months, potentially causing FIDVR. Studies conducted by Saudi Electric Company and ABB revealed that even though the most effective solution for FIDVR mitigation is to connect a large number of small SVCs on 13.8 kV distribution network, this is not practically possible. Therefore, five large SVCs were installed at various locations on 110 kV network, which are at a distance from the FIDVR causing motor loads [102]. This proved to be an efficient solution for FIDVR.

2) In the 138 kV transmission network of Indianapolis, 53 of 83 buses experienced FIDVR. A 300 Mvar SVC is connected in the network at Southwest 138 kV bus to address this issue. This SVC is quite far from the motor loads connected in other parts of the Indianapolis network [103].

3) To address FIDVR caused by motor loads in Metro Atlanta and North East Georgia, an SVC is installed in Barrow County [104]. This SVC is located at a distance from the motor loads connected in the other counties of Metro Atlanta and North East Georgia.

4) A 225 Mvar SVC is connected in 115 kV transmission network at Lower South Eastern Massachusetts, to prevent FIDVR in entire Cape Cod network [105]. This SVC is also at a distance from the motor loads in other regions of Cape Cod network.

It is noted that in all above cases the dynamic reactive power support by SVC alleviates FIDVR even though the SVCs are connected at a distance from the motors. Although SVCs are widely employed for FIDVR mitigation, these Controllers are quite expensive. Similar arguments as above apply to STATCOMs, as well.

Reactive power support by PV plants is also known to improve system voltage stability [18, 106]. Reactive power compensation has been proposed for mitigating voltage sags in IM-rich distribution systems [107]. These control strategies use PV inverter capacity remaining after active power generation and are, therefore, less effective during periods of high PV power generation, i.e. noon hours. Fixed power factor operation of PV plant for voltage support is discussed in [108]. However, the efficacy of this technique is limited during transients as reactive power injected is restricted by both active power output and constant power factor operation. It is noted that all methods in [18, 106–108] are based only on reactive power control.

The potential of smart inverter functions such as volt–var and low voltage ride-through (LVRT) for PV inverters to prevent FIDVR is described in [109]. Dynamic reactive power support by large PV plant per German grid codes [38, 39] to improve short-term voltage stability is presented in [110, 111]. Dynamic reactive current injection during LVRT is described but made optional in IEEE Standard 1547-2018 [112]. It is emphasized that VRT functions according to grid codes [38, 39] and IEEE Standard [112] are based on reactive power and are not defined or prescribed during nighttime (when PV systems do not generate power).

A combined active and reactive power control strategy of injecting active and reactive power in a constant $1:1$ ratio is presented in [113]. The efficacy of this method is, however, limited in transmission systems where $X/R \gg 1$ and also in distribution systems where X/R ratio can vary substantially.

The initial concept of PV-STATCOM control for stabilization of a locally connected motor with PV inverter capacity remaining after active power generation was presented in [114]. However, this PV-STATCOM control did not provide an autonomous capability of STATCOM operation utilizing the full inverter capacity, during any time of the day, for stabilizing remotely connected motors. Moreover, this control was based only on reactive power modulation.

A patented enhanced PV-STATCOM control is described here for solar farms located at a distance from the motor loads, which draws its motivation from the several remote SVC applications for FIDVR mitigation, described above. This study presents the application of the proposed PV-STATCOM control both during night and day for alleviation of FIDVR [115]. The novel features of the proposed PV-STATCOM control are that it:

i) utilizes a combined modulation of active and reactive power during day, and modulation of reactive power at night;
ii) dynamically modulates active and reactive power based on system's existing R/X ratio, i.e. on the sensitivity of PV plant bus voltage to active and reactive power injections;
iii) modulates reactive power to provide voltage control up to the transient over voltage (TOV) limit at PV plant bus for achieving increased stabilization impact;
iv) is more effective than reactive power support required by LVRT specification of German grid code [39];
v) provides voltage recovery similar to a STATCOM, thus reducing the need for expensive STATCOMs.

6.6.1 Study System

The single line diagram of study system is shown in Figure 6.40. This system has been created by incorporating different features of realistic systems in which FIDVR has been experienced [104, 116]. An important factor in considering these systems is that a FACTS Controller such as SVC

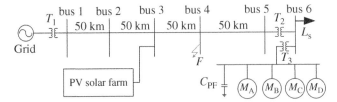

Figure 6.40 Single line diagram of the study system. *Source:* Varma and Mohan [115].

was necessarily required for FIDVR mitigation. The validity of this system was ensured by checking that it provides an FIDVR scenario very similar to the event reported in Barrow County [104].

The study system comprises a 200 km single circuit radial transmission system having four 138 kV line sections of 50 km each. Each line section is represented by an equivalent pi model. This radial transmission system is supplied by a 230 kV transmission system at transformer station T_1. The Thevenin equivalent impedance of the 230-kV network is represented by inductance L_g and resistance R_g.

A 75 MW PV solar farm is connected at bus 3 in the middle of 138 kV transmission line which feeds a total load of 80 MVA through 138/27.6 kV transformer T_2. The PV plant is thus 100 km away from the loads which comprise a static load component of 40 MVA and dynamic motor load component of 40 MVA, both at 0.9 p.f. The voltage-dependent model is used to represent static load L_s. The dynamic load comprises 25% of Type A (compressor motors in commercial cooling and refrigeration systems), 25% of Type B (fan motors in residential and commercial buildings), 25% of Type C (direct connected pump motors in commercial buildings), and 25% of Type D (residential air conditioner) motors [94]. The 460 V rated IMs are connected to T_2 through 27.6/0.46 kV transformer T_3. The capacitor C_{PF} represents the power factor correction capacitor.

6.6.2 Structure of a Large Utility-Scale Solar PV Plant

The single line diagram of a large utility-scale solar plant is depicted in Figure 6.41a. The solar plant comprises multiple inverter units which are connected to point of interconnection (POI) through collector cables. Individual inverters produce power according to locally available solar irradiance. The inverter unit consists of inverters and pad-mount transformers.

In this study, the large-scale 75 MVA solar plant is modeled as a single inverter with inverter controller, plant controller, and plant communication delays [117]. The inverter terminal voltage is 0.6 kV, which is stepped up to 27.6 kV using inverter pad mount transformers. The PV plant power is fed to the network through 27.6/138 kV plant transformer. The delay incurred during the measurement process in PV plant is approximated as 10 ms [117]. The collector cable is modeled by its equivalent π model.

A large PV plant has both power plant control (PPC) and individual inverter level controllers (ILCs). As the plant-level controller is slow, grid supporting functions like LVRT and dynamic reactive power support are performed by ILC [117]. In this study also, since the proposed PV-STATCOM controller is for dynamic voltage support, it is implemented at ILC.

The conventional inverter controller consists of inner current controllers, synchronizing circuit, PLL, and inverter switching circuit [25, 118].

6.6.3 Proposed PV-STATCOM Control

The proposed PV-STATCOM controller is shown in Figure 6.41b. It generates i_{dref} and i_{qref} for providing fast system support during a disturbance. These current references are subsequently

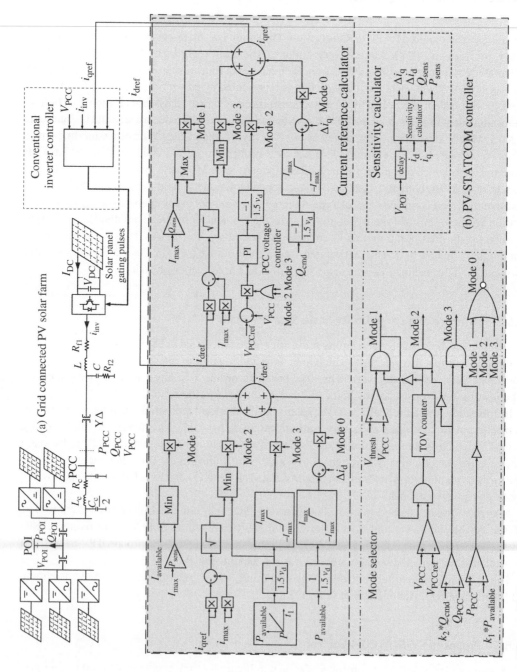

Figure 6.41 Single line diagram of a large PV plant with the proposed PV-STATCOM controller: (a) Grid-connected PV solar farm and (b) PV-STATCOM controller. *Source:* Varma and Mohan [115].

transmitted to the conventional inverter controller. The controller consists of (i) mode selector, (ii) sensitivity calculator, and (iii) current reference calculator.

6.6.3.1 Mode Selector

The typical active/active "*P*" and reactive power exchange capability "*Q*" of a PV solar farm during 24 hours on a sunny day is illustrated in Figure 6.42a. A system disturbance is considered to occur around noon at time $t = t0$. The PV-STATCOM control is immediately initiated and continued till time $t = t3$ to mitigate FIDVR. Thereafter the solar farm returns to normal operation. The PV-STATCOM control operates in three modes: Modes 1, 2, and 3. The enlarged version of the hatched region $t0$–$t3$ where PV-STATCOM control is utilized is shown in Figure 6.42b. This figure illustrates the conceptual *P* and *Q* outputs during the different modes of PV-STATCOM operation. (The actual *P* and *Q* outputs in these different modes are depicted in Figure 6.44.) It is noted that the duration $t0$–$t3$ is generally less than two seconds.

(a)

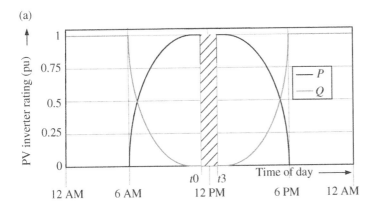

(b)

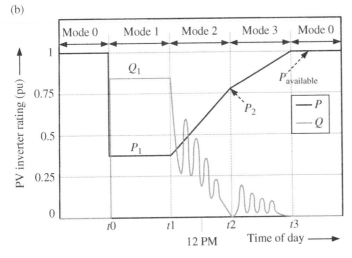

Figure 6.42 (a) Typical active "P" and reactive power "Q" exchange capability of a PV-STATCOM in a sunny day. Hatched region indicates proposed PV-STATCOM controller operation during noontime. (b) Enlarged version of the hatched region of (a) depicting active and reactive power outputs of PV-STATCOM for FIDVR mitigation around noon time. *Source:* Varma and Mohan [115].

The operating modes are decided by mode selector by monitoring V_{PCC}, P_{PCC}, and Q_{PCC}. Mode 0 is the normal prefault/disturbance operating mode of the solar farm.

1) *Mode 0 (controlled by Mode 0 signal):* In this mode, the inverter generates active and reactive power according to active power $P_{available}$ and reactive power Q_{cmd} references provided by PPC. $P_{available}$ corresponds to available solar irradiance. In Figure 6.42, $P_{available} = 1$ pu at noon.

2) *Mode 1 (controlled by Mode 1 signal):* In this mode, PV solar farm injects active power P_1 and reactive power Q_1 in the ratio of effective R/X ratio at POI. If $P_1 < P_{available}$, active power output is curtailed to P_1 (as shown in Figure 6.42) and reactive power is set to Q_1. However, if $P_1 > P_{available}$, active power output is set to $P_{available}$, and the remaining inverter capacity is used for reactive power injection, which may be different than Q_1.

   If, due to a system fault/disturbance, V_{PCC} goes below a threshold voltage V_{thresh}, Mode Selector enables Mode 1 operation by setting Mode 1 signal high. V_{thresh} is the low voltage that results from a fault/disturbance which is sufficient to initiate a FIDVR event leading to shutdown of motors. This voltage is system dependent and determined from off-line simulation studies for the system under study. $V_{thresh} = 0.75$ pu for our study system.

   The PV plant operates in Mode 1 till V_{PCC} increases to just below the utility TOV limit V_{TOVL}. This enhanced voltage support helps in fast voltage recovery. Once TOV limit is reached, Mode 1 operation is disabled, and Mode 2 operation is initiated by setting Mode 1 signal low and Mode 2 signal high. A TOV counter is used to detect TOV limit violation. For instance, a voltage between 1.05 and 1.3 pu is permitted for a period of 166 ms [112]. The counter is activated (output is set high) when V_{PCC} crosses utility's upper steady state voltage limit V_{PCCref} (=1.05 pu, here). The counter is kept activated for 150 ms (to be safely within the utility limit of 166 ms [112]).

3) *Mode 2 (controlled by Mode 2 signal):* In this mode that gives preference to reactive power, the PV-STATCOM control provides reactive power support Q_2 with the objective of maintaining V_{PCC} at utility's upper steady state voltage limit V_{PCCref} (=1.05 pu, here). If active power was curtailed to P_1 in Mode 1, it is increased to P_2 using the inverter capacity remaining after reactive power injection Q_2. It is noted that $P_2 < P_{available}$. Even though V_{PCC} reaches V_{PCCref}, motors still need reactive power support for their speeds to return to their prefault stable operating state, as in Mode 0. Mode 2 operation is, therefore, continued till reactive power support is no longer required for motor stabilization. A practical way of checking is that Q_{PCC} becomes equal to k_2 times predisturbance value Q_{cmd}. In this study $k_2 = 1.1$. Once this condition is satisfied, Mode 2 operation is disabled, and Mode 3 operation is initiated.

4) *Mode 3 (controlled by Mode 3 signal):* In this mode that gives preference to active power, the objective is to ramp active power from P_2 to the predisturbance solar power level $P_{available}$ at a high rate (100 pu/s, here) in a time period t_1. An important aspect of this control is that during power ramp up, voltage control is simultaneously performed to maintain V_{PCC} at V_{PCCref} (=1.05 pu) using inverter capacity remaining after active power ramp up. This unique control prevents onset of any unwanted voltage oscillations at POI while ensuring a fast ramp up. Mode 3 operation is continued till P_{PCC} reaches at least k_1 times $P_{available}$. In this study $k_1 = 0.96$.

6.6.3.2 Sensitivity Calculator

The active and reactive power sensitivities (P_{sens} and Q_{sens}) of POI voltage depend upon the effective R/X ratio. The X and R change due to change in system conditions, e.g. faults, switching of lines/transformers, turning on/off of distributed generators, etc. These active and reactive power sensitivities (P_{sens} and Q_{sens}) of POI voltage are, therefore, computed by monitoring the change in POI voltage by injection/absorption of small amount of active (Δi_d) and reactive (Δi_q) currents, only after the occurrence of a system disturbance which may have impacted X and R in a substantial manner. The occurrences of such events are obtained from the grid/system operator or the utility energy control center. Furthermore, the magnitudes of injected/absorbed active (Δi_d) and reactive (Δi_q) currents are selected such that (i) these do not cause any adverse impact on grid operation and (ii) any errors in the measurement of different system variables associated with the above current injection/absorptions are minimal. These incremental current magnitudes can be computed ahead of time through off-line simulation studies. The approval for performing such X, R determination for providing an effective FIDVR mitigation by the solar farm may be obtained a priori from the grid/system operator.

The motivation for the above technique comes from the real-world applications of SVCs and STATCOMs. Estimation of system strength by injecting a current for short duration is a common practice for both SVCs and STATCOMs [119–124]. SVCs and STATCOMs are equipped with gain optimizer which modifies and optimizes their controller gains by estimating system strength, periodically. The gain optimizer adjusts the gain of voltage regulator of SVC by giving a pulse to the reference voltage of the voltage regulator, or by switching a thyristor switched capacitor (TSC) and subsequently measuring system voltage response for the corresponding reactive power injection. The magnitude and duration of this disturbance are variable (system dependent) and chosen to be small enough to not adversely impact the system performance. This gain optimization is normally performed after a fault or system disturbance which may have affected system strength. Some examples of the use of gain optimizer are provided below:

1) 8 SVCs are connected at different locations in the 1000 km long 735 kV lines to provide dynamic reactive power support for the transmission of nearly 15 000 MW power from James Bay area to Montreal region in Canada [120]. Substantial variations of system strength can potentially occur in James Bay area transmission system. A gain optimizer is used in each SVC to adjust gain according to variation in system strength in real time.

2) In the −265/+300 Mvar SVC located at Oncor Electric Delivery Renner Substation, the gain of the voltage controller is modified automatically after a fault clearance [122]. Here, the magnitude and duration of reactive power injection is adjustable, and are chosen small enough not to disturb the system.

3) The Forbes SVS consists of −110/+149 Mvar SVC and two 300 Mvar mechanically switched capacitors (MSCs). In this SVC, the gain of the voltage regulator is optimized for change in system strength after fault clearance in two ways. In the first method, the system strength is estimated by measuring the change in voltage reference for step variation of susceptance of SVC. In the second method, the gain is optimized by monitoring the voltage response to switching of a 55 Mvar TSC [123].

It is thus evident from above that large SVCs and also STATCOMs routinely inject small amounts of reactive current into the system to estimate system impedance. In all applications, the magnitude, duration, and frequency of the current injection are decided such that it will not negatively impact the normal steady state operation of the system.

The PV-STATCOM is intended to perform similar functions as a FACTS Controller – SVC or STATCOM. Hence, the conventionally utilized techniques in SVCs and STATCOMs are adapted for use in PV-STATCOM. In SVCs and STATCOMs, only an incremental reactive current is injected/absorbed for estimating system strength or alternatively, equivalent system impedance $Z \ (=\sqrt{[X^2 + R^2]})$. However, in PV-STATCOM the above concept is extended to include injection/absorption of both reactive and active currents for estimating individual X and R to obtain X/R ratio.

The sensitivity calculator block calculates the active and reactive power sensitivities (P_{sens}, and Q_{sens}) of POI voltage by monitoring the change in POI voltage on injection/absorption of small amount of active (Δi_d) and reactive (Δi_q) currents, as described above. There is a delay incurred in the measurement and communication of V_{POI} to the ILC, which is considered as 100 ms in study [125]. This delay is considered while computing R and X according to Eqs. (6.13) and (6.14), respectively, for accurate calculation of P_{sens} and Q_{sens}.

Effective R is calculated by varying i_d by a small amount Δi_d and noting the change ΔV_{POI} as below.

$$R = \frac{\Delta V_{POI}}{\Delta i_d} \tag{6.13}$$

Effective X is calculated by injecting a small current Δi_q and noting the change ΔV_{POI} as below.

$$X = \frac{\Delta V_{POI}}{\Delta i_q} \tag{6.14}$$

P_{sens} and Q_{sens} are then calculated as:

$$P_{sens} = \frac{R}{\sqrt{R^2 + X^2}} \tag{6.15}$$

$$Q_{sens} = \frac{X}{R} * P_{sens} \tag{6.16}$$

6.6.3.3 Current Reference Calculator

This block computes i_{dref} and i_{qref} for the inverter based on the mode of operation, as described below.

1) Reference Calculation for Mode 1

   As stated earlier, in this mode the PV solar farm injects active power P_1 and reactive power Q_1 in the ratio of effective R/X ratio at POI. If calculated $P_1 > P_{available}$, the inverter sets i_{dref} to $I_{available}$ ($=P_{available}/[1.5v_d]$). The remaining inverter capacity is utilized to inject reactive power and i_{qref} is set to $\sqrt{(I^2_{max} - i^2_{dref})}$, where I_{max} is the rated inverter current. However, if $P_1 < P_{available}$, active power is curtailed to P_1. Correspondingly i_{dref} is set to $I_{max} * P_{sens}$, and i_{qref} is set to $I_{max} * Q_{sens}$.

$$i_{dref} = \text{minimum}(I_{available}, I_{max} * P_{sens}) \tag{6.17}$$

$$i_{qref} = \text{maximum}\left(\sqrt{I^2_{max} - i^2_{dref}}, I_{max} * Q_{sens} \right) \tag{6.18}$$

2) Reference Calculation for Mode 2

   In this mode, i_{qref} is generated by PCC voltage controller to regulate V_{PCC} to V_{PCCref}. If active power was curtailed during Mode 1 operation to P_1, the inverter will increase active power to P_2 (as described in Mode 2 operation). i_{dref} is obtained by dividing active power by $1.5v_d$.

3) Reference Calculation for Mode 3

   Active power is ramped up from P_2 to $P_{available}$ at a fast rate. The corresponding i_{dref} is obtained by dividing active power by $1.5v_d$. Voltage control is simultaneously performed with reactive power exchange using remaining inverter capacity. Correspondingly, i_{qref} is generated as below:

$$i_{qref} = \text{minimum}\left(I_{VC}, \sqrt{I^2_{max} - i^2_{dref}} \right) \tag{6.19}$$

where I_{VC} is reactive current required by voltage controller to maintain V_{PCC} at V_{PCCref}.

4) Reference Calculation for Mode 0

In this mode, inverter generates current references $I_{available}$ and I_Q from $P_{available}$ and Q_{cmd} signals provided by the power plant controller. i_{dref} and i_{qref} are then calculated as

$$i_{dref} = I_{available} + \Delta i_d \tag{6.20}$$

$$i_{qref} = I_Q + \Delta i_q \tag{6.21}$$

where Δi_d and Δi_q are the incremental currents provided by the sensitivity calculator to calculate P_{sens} and Q_{sens}, respectively; and $I_Q = -Q_{cmd}/(1.5v_d)$.

6.6.4 Design of PV-STATCOM Controllers

The PV-STATCOM PCC voltage controller and current controllers are designed according to frequency domain-based strategy using simplified inverter model [25]. The proposed PV-STATCOM control system involves fast modulation of active and reactive power, hence current controllers are designed with a phase margin of 60° at 500 Hz to ensure fast response. The bandwidth of PCC voltage controller is chosen such that it is at least five times lower than the bandwidth of current controller. The PCC voltage controller is designed with a phase margin of 60° at 100 Hz to ensure fast voltage support in Modes 2 and 3 operation.

6.6.5 Simulation Studies

This section presents PSCAD simulation studies of the performance of proposed PV-STATCOM control for FIDVR mitigation under widely different operating conditions. In all these studies, the PV plant is initially providing its rated power output of 75 MW (1 pu), i.e. at full noon operation. An LLL-G fault is initiated at the middle of transmission line between PV plant and 138/27.6 kV substation (i.e. 50 km away from the motors) and is cleared in 100 ms.

6.6.5.1 Response of IMs for LLL-G Fault with no PV Plant Control

This case is studied to demonstrate the FIDVR phenomenon in absence of any PV system control. The PV solar plant is considered to operate at UPF. The responses of IMs and PV plant are shown in Figure 6.43. Figure 6.43a–d depict PCC voltage V_{PCC} (RMS) and 27.6 kV bus voltage (RMS); apparent power (S_{POI}), active power (P_{POI}), and reactive power (Q_{POI}) output of PV solar farm (pu); total inverter current i, i_d and i_q; and speeds of the type A–D motors (pu), respectively.

The active power from PV farm reduces due to fault-induced drop in voltage. The nonzero Q_{POI} is due to collector cable capacitance. It is seen that in the absence of PV-STATCOM control, the solar farm contributes short circuit current over a substantial period of time (until any corrective action is taken). Also, the voltages at both the POI and 27.6 kV buses continue to remain low even after two seconds. The POI voltage and 27.6 kV bus recover only up to 0.86 and 0.65 pu, respectively, thus causing a FIDVR event. This is due to excessive reactive power drawn by the motors. All motor speeds decline steadily, and motors eventually stall. This demonstrates that the conventional PV solar farm does not provide any improvement in the voltage profile nor prevents FIDVR.

6.6.5.2 Performance of Proposed PV-STATCOM Controller

The performance of the 75 MW PV plant with the proposed PV-STATCOM controller is shown in Figure 6.44. Figure 6.44a–g depicts V_{PCC}, and 27.6 kV bus voltage (RMS), apparent power (S_{POI}), active power (P_{POI}), and reactive power (Q_{POI}) output of PV solar farm, total inverter current (i), direct axis component (i_d) and quadrature axis component (i_q) of inverter current, electromagnetic

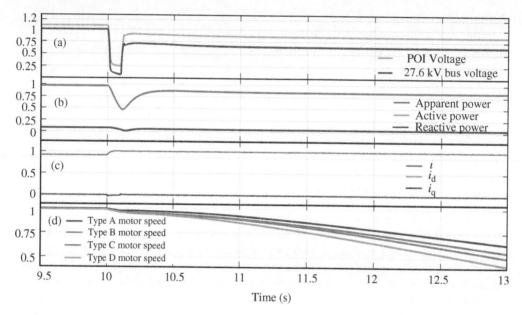

Figure 6.43 Response of IMs with PV plant without any control. (a) PCC voltage (RMS) and 27.6 kV bus voltage (RMS), (b) apparent power (S_{POI}) active power (P_{POI}) and reactive power (Q_{POI}) output of PV solar farm (pu), (c) total inverter current i, i_d and i_q, (d) speeds of the type A–D motors (pu). *Source:* Varma and Mohan [115].

torque (pu), and speed of the type A–D motors (pu), power consumed by loads, and the operating mode signals of the controller, respectively.

$t < 10$ s. The PV plant is operating in Mode 0 by generating 1 pu active power at UPF. V_{PCC} is 1.03 pu and the voltage at LV side of 138/27.6 kV substation is 1 pu. The different motors are operating at steady state with torques between 0.88 and 0.92 pu and slips within 3%. The total load power at 27.6 kV bus is $79 + j12.5$ MVA.

$t = 10$ s. The fault is applied, due to which V_{PCC} and 27.6 kV bus voltages drop to 0.28 and 0.12 pu, respectively. The motor torques become oscillatory and their speeds reduce. The static load consumption also reduces due to drop in voltage. Figure 6.44c demonstrates that the short circuit current contribution from the PV-STATCOM during the fault does not exceed the rated current (1 pu) of the solar farm inverters.

$t = 10.02$ s. The V_{PCC} threshold is violated at 10.01 seconds. As the measurement delay in PV plant is 10 ms, the PV-STATCOM controller detects the drop in voltage at 10.02 seconds and changes the mode of operation to Mode 1. The latest X/R measured at POI is 2.46, so the controller sets $i_{dref} = 0.3765$ pu and $i_{qref} = 0.926$ pu, as explained before.

$t = 10.1$ s. The fault is cleared at 10.1 s. Due to fault clearance, both V_{PCC} and distribution substation voltage increase. Due to P and Q injection by PV-STATCOM, the PCC voltage crosses the steady state limit of 1.05 pu at 10.13 s.

According to IEEE 1547-2018 [112] VRT requirements of each of the Performance Categories I, II, and III, the distributed energy resource (DER) may ride through if the PCC voltage stays between 1.2 and 1.3 pu for 160 ms. As shown in Figure 6.44, the solar farm voltage with PV-STATCOM control reaches 1.22 pu for only 150 ms. This performance of PV-STATCOM conforms with the IEEE 1547-2018 requirements and does not trigger HVRT.

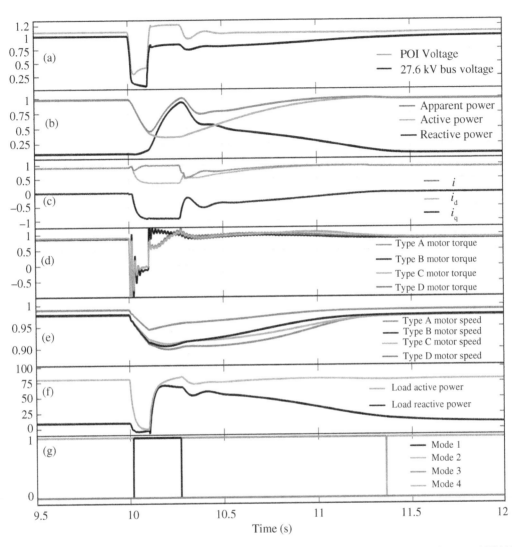

Figure 6.44 Response of the large PV plant with proposed PV-STATCOM control. (a) PCC voltage and 27.6 kV bus voltage (RMS), (b) apparent power (S_{POI}), active power (P_{POI}), and reactive power (Q_{POI}) output of PV solar farm, (c) total inverter current i, direct axis component i_d and quadrature axis component i_q of inverter current, (d) electromagnetic torque (pu) of the type A–D motors (pu), (e) speed of the type A–D motors (pu), (f) power consumed by loads, and (g) operating mode signals of the controller. *Source:* Varma and Mohan [115].

$t = 10.28$ s. The PV plant continues Mode 1 operation till V_{TOVL} is violated. Due to this novel feature of extended voltage support, the IMs start recovering. The PV-STATCOM controller switches to Mode 2 operation at 10.28 seconds. The reactive power output of PV plant is controlled to regulate V_{PCC} at 1.05 pu in this mode. i_{dref} and i_{qref} are limited within ± 0.3765 and ± 0.926 pu, respectively, to provide fast recovery. The PV system injects 0.9 pu reactive power and 0.35 pu active power at 10.28 seconds. As IMs start recovering, their reactive power demand reduces and consequently, the reactive power required to maintain V_{PCC}, also reduces. The active power of the PV plant is also increased.

$t = 11.36$ s. The IMs reach steady state speed at 11.36 s. Due to this, the reactive power Q_{PCC} required to keep V_{PCC} at 1.05 pu falls within k_2 (=1.1) times prefault reactive power output of PV plant, and eventually i_q reduces to zero. The PV-STATCOM then changes the mode of operation to Mode 0. As PV active power has reached the maximum power already (at 10.77 seconds) in Mode 2, Mode 3 operation is skipped. The PV system starts operating at UPF by injecting 1 pu active power to the grid. The IMs resume normal steady state operation and total load consumption reaches its prefault level.

The proposed PV-STATCOM control successfully stabilizes the motors by improving the voltage at the 27.6 kV IM bus to 1 pu within 1.36 seconds. This time frame is within the 2 s limit set by North American Electric Reliability Corporation (NERC) for the event to not be characterized as FIDVR event [93]. Furthermore, the PV plant provides this FIDVR alleviation with an energy curtailment (area of active power dip) of only 12 MWs over 1.36 seconds (equivalent to 3.33 kWh).

6.6.5.3 Advantage of Enhanced Voltage Support up to TOV Limit
The performance of PV-STATCOM controller with and without utilizing the proposed enhanced voltage control up to the TOV limit is compared in terms of time taken for distribution substation voltage recovery, amount of PV power curtailed, and amount of IM loads becoming unstable. The results are presented in Table 6.4 from which the efficacy of proposed enhanced PV-STATCOM control is clearly evident.

This study shows that with the proposed control, even if the inverter experiences an overvoltage for 9–10 cycles, a fast recovery of system voltage and motor stabilization is achieved with minimal PV power curtailment. The high voltage ride-through (HVRT) of IEEE Std. 1547-2018 [112] requires the inverter to remain connected even for a voltage range of 1.05–1.3 pu, and hence all PV inverters will be designed to operate in this voltage range. The proposed enhanced PV-STATCOM control is thus compliant with [112].

6.6.5.4 Comparison of Proposed PV STATCOM Controller and Other Smart Inverter Controls
The performance of the proposed PV-STATCOM controller is compared with other smart inverter technologies viz. (i) dynamic reactive current injection required under German grid code [39] and (ii) reactive power control of PV-STATCOM. According to [39], the PV inverter needs to inject minimum 2% reactive current per 1% voltage drop below 0.9 pu. In this study, the inverter is set to inject 4% reactive current per 1% voltage drop, which is double the minimum requirement set by [39]. In reactive power control of PV-STATCOM, active power is autonomously curtailed to zero during a disturbance and the entire inverter capacity is utilized for providing reactive power support. Figure 6.45 shows the results of this study. Figure 6.45a–e depicts PCC voltage (RMS), 27.6 kV

Table 6.4 Comparison of performance of controller with and without TOV support.

	Time taken for 27.6 kV bus voltage to recover to		Total PV energy lost or curtailed	Total IMs becoming unstable
	0.9 pu	Prefault level		
With TOV control	0.96 s	1.36 s	12 MWs (=3.33 kWh)	0 MVA
Without TOV control	1.21 s	Voltage does not recover	19.45 MWs (=5.4 kWh)	17.5 MVA

Source: Varma and Mohan [115].

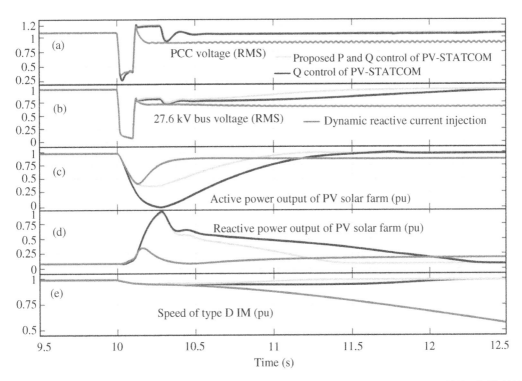

Figure 6.45 Comparison of PV-STATCOM and other smart inverter controls. (a) PCC voltage (RMS), (b) 27.6 kV bus voltage (RMS), (c) active power output of PV solar farm (pu), (d) reactive power output of PV solar farm (pu), and (e) speed of type D IM (pu). *Source:* Varma and Mohan [115].

bus voltage (RMS), active power output of PV solar farm (pu), reactive power output of PV solar farm (pu), speed of type D IM (pu), respectively. Speed of only type D IMs is shown as they have lower inertia and are more prone to instability compared to type A–C motors.

Proposed PV-STATCOM control: The performance is same as already discussed before. The system voltage reaches steady state at 11.58 seconds, and FIDVR is avoided.

Dynamic reactive current injection: V_{PCC} increases to 0.85 pu at 10.5 seconds, and thus i_q is reduced from 1 to 0.2 pu. This reactive support is inadequate, and the IMs eventually stall. The 27.6 kV bus voltage recovers only to 0.676 pu causing FIDVR.

Reactive power control of PV-STATCOM: The PV plant continues providing reactive power support till the TOV limit, and then switches to closed-loop voltage control, i.e. regulating V_{PCC} to V_{PCCref} (=1.05 pu). The system voltage recovers at 12.7 seconds and the motor is stabilized. However, since the voltage recovery takes more than two seconds per NERC [93], FIDVR is technically not prevented.

This study thus shows that among all the investigated smart inverter controls, the proposed P and Q control of PV-STATCOM control is most effective in FIDVR mitigation.

6.6.5.5 Comparison of PV STATCOM Controller and STATCOM

Typically, an SVC or STATCOM is installed close to the distribution substation for FIDVR alleviation, as described earlier. In this section, a performance comparison is presented of: (i) 75 MW PV plant operating with proposed PV-STATCOM controller with no STATCOM connected at the

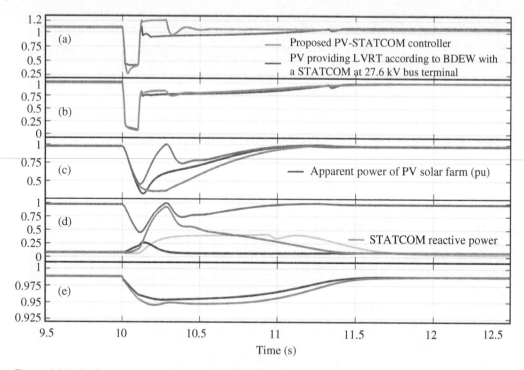

Figure 6.46 Performance comparison of PV-STATCOM and actual STATCOM. (a) PCC voltage (RMS), (b) 27.6 kV bus voltage (RMS), (c) apparent power and active power output of PV solar farm (pu), (d) apparent power and reactive power output of PV solar farm (pu) and STATCOM, and (e) speed of type D IM (pu). *Source:* Varma and Mohan [115].

distribution substation and (ii) 75 MW PV plant operating according to German LVRT require-ments by injecting 4% reactive current per 1% voltage drop for voltage below 0.9 pu [39], with a 34 Mvar STATCOM located at the distribution substation. The results of this study are shown in Figure 6.46. Figure 6.46a–e depicts V_{PCC}, 27.6 kV bus voltage (RMS), apparent power and active power output of PV solar farm (pu), apparent power and reactive power output of PV solar farm (pu), and STATCOM, and the speed of type D IM (pu), respectively. It is clearly seen from Figure 6.46 (and also from Figure 6.44) that with PV-STATCOM control the apparent power S of the solar farm inverter always stays within the limit of 1 pu.

This study shows that a 34 Mvar STATCOM needs to be installed at the distribution substation to mitigate FIDVR, even with 75 MW PV solar providing reactive power support as per [39]. However, the same FIDVR mitigation can be accomplished by the 75 MW solar farm with proposed PV-STATCOM control, even if it is located 100 km away from the distribution substation. This demon-strates that the PV solar farm equipped with the proposed PV-STATCOM controller can reduce or eliminate the need for installing a STATCOM.

6.6.5.6 PV-STATCOM Impact on System Frequency

The impact of PV-STATCOM impact on system frequency, while performing FIDVR mitigation is demonstrated in Figure 6.47. For this study, the power system grid is modeled by a synchronous generator rated 1350 MVA having $H = 3.22$ s. A large variation in system frequency is seen subse-quent to the fault, when the PV plant is operating at UPF. The system frequency variation with PV-STATCOM operation of PV plant is quite similar to the PV plant operating according to German

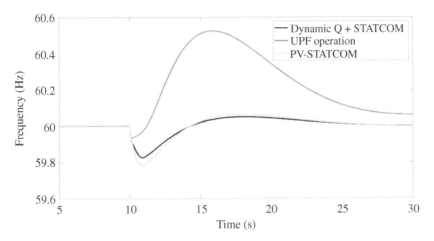

Figure 6.47 Impact on system frequency.

LVRT requirements by injecting 4% reactive current per 1% voltage drop for voltage below 0.9 pu [39], with a 34 Mvar STATCOM located at the distribution substation. PV-STATCOM operation keeps the frequency within the continuous operation capability per IEEE 1547-2018 and NERC Standard. This study demonstrates that the PV-STATCOM successfully alleviates FIDVR without causing any adverse impact on system frequency.

6.6.5.7 Nighttime Performance of PV-STATCOM Controller

Figure 6.48a–e depicts V_{PCC}, 27.6 kV bus voltage (RMS), P_{POI} and Q_{POI} (pu), and the speed of all IMs (pu), respectively for an LLL-G fault at the middle of transmission line between PV plant and 138/27.6 kV substation at night. Performances of PV system both with and without the PV-STATCOM control are illustrated.

As soon as PV-STATCOM controller detects the fault, it initiates dynamic voltage support. Since at night the available active power is zero, the controller uses full inverter capacity to inject reactive power. Once the TOV limit is violated, the controller initiates Mode 2 operation by regulating PCC voltage V_{PCC} to V_{PCCref} (= 1.05 pu). Due to PV-STATCOM support, the 27.6 kV bus voltage recovers and Q_{PCC} falls within k_2 (=1.1) times prefault reactive power output of PV plant. i_q eventually reduces to zero and the controller stops Mode 2 operation. Mode 3 operation is skipped, as the active power is zero. The controller initiates Mode 0 and remains on standby to provide grid support during any next disturbance. Without the support of PV-STATCOM control, neither PCC voltage nor 27.6 kV bus voltage recovers and all motors stall.

The proposed PV-STATCOM controller thus successfully helps voltages to recover and stabilizes motors at night.

6.6.5.8 Compliance with IEEE Standard 1547–2018

In the various studies described in this section (Figures 6.44–6.46), the voltage at POI of the solar farm controlled as PV-STATCOM:

  i) stays below 0.3 pu for 40 ms and
 ii) between 0.3 and 0.65 pu for 60 ms.

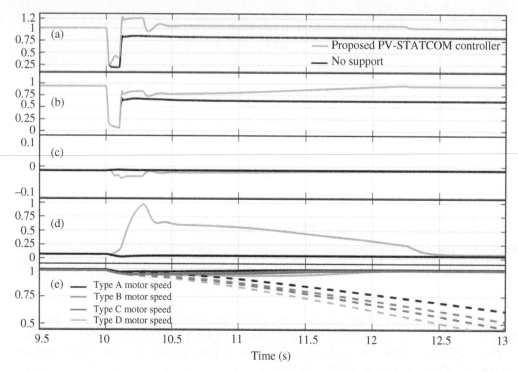

Figure 6.48 Performance of PV-STATCOM at night. (a) PCC voltage (RMS), (b) 27.6 kV bus voltage (RMS), (c) active power output of PV solar farm (pu), (d) reactive power output of PV solar farm (pu), and (e) speed of type D IM (pu). *Source:* Varma and Mohan [115].

The "Category II Voltage Ride Through Requirement of IEEE 1547-2018 based on NERC PRC-024-2" [112] allows the PV solar farm to "may ride through" for case (i) and "permissive operation capability" for case (ii). Hence the PV-STATCOM control of the solar farm does not violate the IEEE Standard 1547-2018. In fact, the PV-STATCOM provides dynamic reactive power control during the VRT period, which is optional under IEEE 1547-2018.

The PV-STATCOM further provides FIDVR alleviation during night. Such nighttime control is not specified either in IEEE 1547-2018 [112] or in the German grid code [39].

6.6.6 Summary

A novel control of large PV solar farm as PV-STATCOM for providing enhanced voltage support utilizing both reactive and active power modulation to alleviate FIDVR is presented. The study system comprises a 138 kV transmission network with a 75 MW PV solar farm installed midline and feeding 40 MVA voltage-dependent static load and 40 MVA IM loads of different types. The PV solar plant injects active and reactive power in proportion to sensitivity of POI voltage to active and reactive power. Extensive PSCAD simulation studies demonstrate that the proposed PV-STATCOM control:

i) mitigates FIDVR even if solar farm is more than 100 km away from the distribution substation with motor loads.

ii) is more effective than (i) reactive current support required by the German grid code, and (ii) PV-STATCOM with only reactive power control.

iii) on the 100-km remote 75 MW solar farm ensures similar FIDVR mitigation as the same PV farm operating according to German grid code along with a 34 Mvar STATCOM connected locally at the distribution substation with motor loads.

iv) successfully alleviates FIDVR for: (i) wide ranging system strengths and (ii) different fault locations.

v) provides voltage recovery and motor stabilization at night. This functionality is beyond the requirements of present-day grid codes.

A PV-STATCOM (i.e. a PV solar plant with the proposed PV-STATCOM controls) is expected to be about 10 times cheaper than an equivalent-sized STATCOM. Even though in this study, a PV-STATCOM is shown to eliminate the need of a half-size STATCOM, the financial savings can be significant. The proposed PV-STATCOM control can, therefore: (i) bring significant saving for utilities by reducing the need for expensive STATCOMs or SVCs for FIDVR mitigation and (ii) opens new revenue-making opportunities for large-scale solar farms for providing night and day FIDVR alleviation service.

6.7 Simultaneous Fast Frequency Control and Power Oscillation Damping by PV-STATCOM

Rapidly increasing deployment of inverter-based generators over the past two decades had led to apprehensions about potential degradation in overall system stability and frequency response [126]. Legacy inverter-based generators affect the system dynamic behavior by (i) reduction of system inertia; (ii) withdrawal of governor response; and (iii) decrease in synchronizing and damping torques [127, 128]. To alleviate these adverse impacts, recent research has focused on the emulation of inertial and governor response of conventional synchronous generators by modulation of active power output of inverter-based generators [129, 130]. The benefits of applying frequency control by solar and wind plants in 16 000-bus North American Eastern Interconnection (EI) are reported in [131]. System operators have initiated measures for making wind and solar generators active players in frequency support by developing new regulations [132]. For instance, German grid code requires PV inverters to ramp down their active power if the system frequency exceeds 50.2 Hz [38]. In Puerto-Rico, PV plants are required to participate in primary frequency control (both up and down-regulation) according to a 5% frequency droop [56]. However, the droop-based response is a slow control since output power changes according to the slowly varying frequency. Recently, frequency control service like FFR has been proposed by ERCOT which can deliver full response in less than 0.5 seconds [133]. It is shown that each 1 MW of FFR can replace 1.4 MW of conventional primary frequency response (obtained by 5% droop) in ERCOT [134]. Similar services are under research and development in the United Kingdom and Australia [135, 136].

In addition to frequency stability, small-signal stability of the power system can also be adversely impacted by high proliferation of PV solar generators. These impacts are potentially due to inertia reduction, path flow alternation, displacement of units equipped with PSSs, and interaction between controls of inverter-based and synchronous generators [137]. The negative effects of inertia reduction on critical oscillatory modes of Western Electricity Coordinating Council (WECC) system is shown by sensitivity analysis [17].

It is known that large system disturbances that excite under/over-frequency events also simultaneously stimulate the oscillatory modes of poorly damped systems and tend to diminish system stability. For instance, the simulation studies reported in [131] show that after a large disturbance (1000 MW generation trip) in EI not only did the frequency fall but also low-frequency power oscillations got initiated. Furthermore, it is reported for the disturbance event of August 4, 2000 [138], while the WECC interconnection frequency started to deviate, low-frequency oscillations occurred simultaneously in system frequency.

Wind power systems successfully provide frequency regulation, however, undergo mechanical stresses on their generator shafts and/or interact with the turbine torsional dynamics [139, 140]. A combined system frequency and POD control using only active power control of Type 4 wind turbines is proposed in [141]. However, the capability of reactive power control has not been investigated for achieving the dual objective of frequency control and POD. The feasibility of providing frequency regulation by PV plants has been demonstrated on a 300 MW PV plant in California [125]. However, simultaneous stabilization of power oscillations and frequency stabilization has not been demonstrated.

This study presents a novel FFR and POD control by large-scale PV plants controlled as PV-STATCOM, to simultaneously enhance frequency regulation and small-signal stability of power systems [142]. Frequency deviations typically occur together with power oscillations in large power systems. The proposed controller comprises (i) POD controller based on reactive power modulation; and (ii) FFR controller based on active power modulation, both of which are applied to the plant level controller of PV-STATCOMs. The PV-STATCOM operating modes are shown in Section 4.7. The proposed composite control is shown to successfully reduce frequency deviations, damp power oscillations, and provide voltage regulation both during over-frequency and under-frequency events. The proposed smart inverter control makes effective utilization of the PV inverter capacity and available solar power to achieve the above objectives. Matlab/Simulink-based simulation studies are conducted on a two-area power system using generic PV plant dynamic models developed by WECC, for a wide range of system operating conditions.

6.7.1 Study System

The study system is a modified version of two-area-four-machine system as depicted in Figure 6.49. This system is selected as it exhibits both inter-machine and inter-area oscillatory modes [1] and has also been used for several frequency regulation studies, e.g. [143]. The generators are represented by

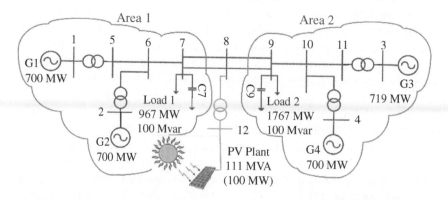

Figure 6.49 Two-area four-machine study system. *Source:* Varma and Akbari [142].

sub-transient models and are equipped with type ST3 exciters [144] and TGOV1 governors [1]. Generators 1 and 3 are also equipped with IEEE PSS1. The loads in each area consist of 20% constant power, 45% constant impedance, and 35% constant current load components. A 111 MVA PV solar system with 100 MW active power rating is added at bus 8. The MVA rating is assumed to be higher than the MW rating because PV inverters are expected to provide reactive power even at the rated active power output, e.g. in IEEE Standard 1547-2018 [112]. The Automatic Generation Control (AGC) control is not modeled, since its response time is much slower than the time frame of interest (<20 seconds) in this study.

6.7.2 System Modeling

6.7.2.1 PV Plant Model

The positive sequence generic dynamic models developed by WECC Renewable Energy Modeling Task Force (REMTF) are utilized for simulation studies [117]. These models can capture dynamics in the frequency range of 0–10 Hz. The DC-bus dynamics are not considered, as they are beyond the normal range of power system stability studies. The irradiance variations and MPPT algorithm are also not modeled [117].

The main modules of PV plant generic model are illustrated in Figure 6.50. *Regc* module emulates the dynamic performance of the inverter interface with the grid. *Reec* module includes the electrical controls of the inverters as well as current limit logic. This module can implement local (inverter level) voltage, reactive, or power-factor control modes. The plant controller model, *Repc*, acts on the POI voltage/reactive power and frequency/active power to emulate volt/var. and active power control at the plant level. The detailed explanations of each module are provided in [145]. The models of plant collector, pad-mounted transformers, and substation transformer according to WECC modeling guidelines are also included in this study system model.

The PV plant model is developed in MATLAB/Simulink with the parameters adopted from [145]. The dynamic model is set to provide plant level voltage control (voltage of bus 8 in Figure 6.49) and plant level active power control.

6.7.2.2 Combined FFR and POD Controller

In this study, the plant controller model (*Repc*) of Figure 6.50 is modified to include the proposed FFR and POD controllers of the PV-STATCOM as illustrated in Figure 6.51.

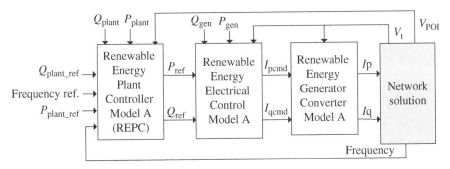

Figure 6.50 Modified WECC generic dynamic model of PV power plant. *Source:* Varma and Akbari [142].

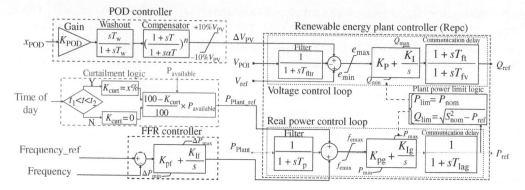

Figure 6.51 Simultaneous FFR and POD control scheme in a PV-STATCOM plant controller. *Source:* Varma and Akbari [142].

Modified Plant Controller (Repc)

The PV plant controller is set to POI voltage regulation mode and plant active power control mode. Therefore, Figure 6.51 only represents the control blocks that are relevant for this study.

In the POI voltage control loop, the measured POI voltage (bus 8 voltage in Figure 6.49) is filtered by a low-pass filter (time constant T_{filt}) and compared with the reference voltage V_{ref}. The voltage error is limited to e_{min}/e_{max} by the limiter block. The PI controller (K_p, K_I) produces reactive power command (Q_{ext}) to individual inverters. The communication delay block models the delay in the communication process [125, 146].

In active power control loop, the plant active power is filtered by a low-pass filter with time constant T_p and compared with reference power P_{ref} that could be the maximum available solar power or a value less than that in the power curtailed mode. The error is limited to f_{min}/f_{max} and passed through the PI controller (K_{pg}, K_{PI}). T_{lag} represents the time delay in sending the active power command (P_{ext}) to inverters.

Plant Power Limit Logic

The active/reactive power limits in the plant level control, *Repc*, are not dynamic in the current WECC generic dynamic models, i.e. the plant level reactive power limits are constant regardless of the available capacity. Since this study proposes a novel plant-level active/reactive power modulation, a dynamic power logic is added to *Repc*. Priority is given to active power for which the limit is set to inverter nominal power as $P_{lim} = P_{nom}$. The reactive power limit, Q_{lim}, is dynamically calculated, i.e. $Q_{lim} = \sqrt{(S_{nom}^2 - P_{ext}^2)}$, where S_{nom} is the plant nominal capacity and P_{ext} is the active power command.

Power Oscillation Damping (POD) Controller

The proposed POD controller consists of a washout filter, phase compensator, and gain [1, 43], as depicted below:

$$\Delta v_{PV} = K_{POD} \frac{sT_W}{1 + sT_W} \frac{1 + sT}{1 + saT} x_{POD} \tag{6.22}$$

where K_{POD} is the controller gain, and T_W is washout filter time constant. T and α are phase compensator parameters. To design the controllers, first, the linearized model of the study system is developed by the MATLAB linearization tool. Utilization of "observability and controllability geometric measures" [147] revealed that POI voltage has higher controllability over the inter-area mode compared to PV plant active power. Therefore, POD controller is added to the POI voltage control loop. The feedback signal, x_{POD}, is selected from a set of potential local and wide-area signals with the highest observability of the inter-area mode. The frequency difference between areas, i.e. frequency difference between buses 7 and 9, described as $\Delta\omega_{7-9} (= \omega_7-\omega_9)$, shows a good observability of the inter-area mode. This signal $\Delta\omega_{7-9}$ is, therefore, selected as the input signal x_{POD}. A delay of 300 ms ($e^{-0.3s}$) is considered to model the measurement and communication delays [125, 146]. The washout filter and phase compensator are designed using residue-based approach [43].

Fast Frequency Response (FFR) Controller

The proposed FFR controller acts on the active power of the PV-plant ΔP_{PV}. The POI frequency is fed to the FFR controller and compared with reference value f_0. A PI controller is used to eliminate the frequency error Δf, with the available PV plant active power capacity:

$$\Delta P_{PV} = \left(K_{pf} + \frac{K_{If}}{s} \right)(f - f_0) \tag{6.23}$$

The FFR *PI* gains are tuned using a trial and error approach. The design objective is to reduce, as fast as possible, the frequency deviation to within ± 36 mHz which is the suggested dead-band of governors [148] to obviate governor's action.

6.7.3 Simulation Scenarios

The performance of the proposed composite FFR + POD controller is evaluated through time-domain simulations in this section. Frequency deviations are classified as under-frequency (generation less than demand) and over-frequency (generation more than demand) events. Over-frequency events can be readily mitigated by rapid reduction of the PV plant active power production. In contrast, under-frequency support is more challenging, since PV plants need to keep a percentage of available power as reserve capacity. As a result, the available support depends on the selected level of curtailment. Figure 6.52 (repeat of Figure 4.19 for ease of reference) shows the conceptual *P* and *Q* capability curves of the study PV-system over the course of a day. All the simulation studies are performed around noon hours when nominal PV power (100 MW) is available. For over-frequency studies, the PV plant is assumed to operate in peak power generation mode without any pre-curtailment. The under-frequency studies are conducted in the curtailed active power mode (PV power is curtailed to 50 MW). This allows 100 Mvar reactive power capacity to be made available for POD.

Five different scenarios are simulated:

1) *No control:* neither FFR nor POD is performed. The PV plant operates only in dynamic voltage control mode.
2) *POD control:* only POD controller is active.
3) *FFR control:* only FFR controller is active.
4) *FFR + POD control:* Both FFR and POD controls are active. This is the main contribution of this study, which demonstrates the best realistic response possible from a PV solar plant.
5) *Ideal control:* Only POD controller is active. It is assumed that the time instant of occurrence and magnitude of the power imbalance are exactly known; and the PV solar farm compensates the imbalance instantaneously. This assumption is not made in 2), 3), or 4), above.

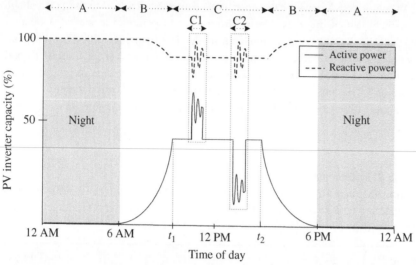

Figure 6.52 Simultaneous modulation of active power and reactive power with pre-existing power curtailment.

6.7.4 Over-Frequency Control

A portion of the load is disconnected to initiate an over-frequency event. The PV plant is considered to operate in normal mode (no power reserve) and generate its nominal power ($\approx$100 MW). The simulations are carried out for load disconnections of 25 and 200 MW to respectively create cases where power imbalance is smaller or larger than the PV plant capacity.

6.7.4.1 25 MW Load Trip in Area 1 ($P_{available}$ = 100 MW, K_{curt} = 0); (Power Imbalance less than PV Plant Capacity)

Figure 6.53 shows the behavior of the system when a 25 MW load is rejected in area 1 while PV plant generates 100 MW. Figure 6.53a–e depicts POI frequency (*Hz*), power transferred from area 1 to 2 (MW), PV plant active power (MW), PV plant reactive power (Mvar), and POI voltage (pu), respectively.

Figure 6.53a shows that in No control case the frequency rises to around 60.06 Hz. When POD controller is activated, the oscillations in frequency are stabilized in less than four seconds, but the frequency deviation is the same as in No control case. With FFR controller the frequency deviation is less than 36 mHz. Nevertheless, inter-area oscillations still exist in the system frequency. The proposed FFR + POD controller stabilizes both frequency excursion and inter-area oscillations, making the response very close to that of the ideal controller. Figure 6.53b shows that with no POD controller, inter-area power oscillations in tie-line power continue for more than 20 seconds. The POD controller damps these oscillations in less than four seconds. The four synchronous generators reduce their generation (through governor action) to reinstate the generation-demand balance. Since all the generators have similar characteristics, each area compensates half of the 25 MW

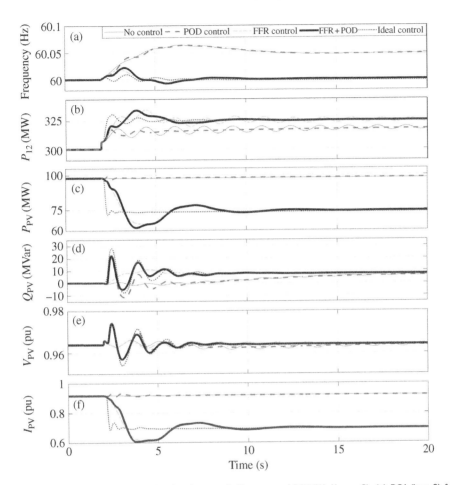

Figure 6.53 25 MW load rejection in area 1 ($P_{available}$ = 100 MW, K_{curt} = 0): (a) POI (bus 8) frequency, (b) Transferred power from area 1 to area 2, (c) PV plant active power, (d) PV plant reactive power, (e) PV-plant voltage, and (f) PV-plant current. *Source:* Varma and Akbari [142].

($\approx$12.5 MW) surplus generation. Thus, the tie-line steady-state power increases by about 12.5 MW as illustrated in Figure 6.53b for no control and POD cases.

With FFR control, the PV plant rapidly reduces its active power generation equal to the lost load, so the synchronous generators do not need to contribute to frequency control. Therefore, the extra 25 MW power of area 1 flows in the tie-line to replace the lost generation from the PV plant. It is evident that the combined FFR + POD controller is most effective among all controllers in damping power oscillations. Figure 6.53c illustrates the PV plant active power. In the No control and POD control cases, PV plant power does not change, whereas in FFR or FFR + POD cases active power decreases and settles at 25 MW in almost eight seconds.

Figure 6.53d depicts the PV plant reactive power. As shown in Figure 6.52, the available reactive power capacity of the simulated PV plant is around 48 Mvar when PV plant is generating its nominal power ($\sqrt{[111^2-100^2]}$). However, the proposed dynamic power limit logic described above automatically increases the reactive power limit based on the amount of active power reduction for over-frequency support. In No control case, the reactive power increases because the increase of power transfer in the tie-line necessitates more capacitive reactive power from the PV plant for

POI voltage regulation in PV-STATCOM mode (in which the PV plant is operating in dynamic voltage control mode). When FFR controller is applied, the reactive power increases, but in a relatively smoother manner as no modulations are needed for damping power oscillations. In POD, Ideal, and FFR + POD cases, not only does the reactive power increase for voltage control, but the POD controller also stabilizes the power oscillations through reactive power modulation. Figure 6.53e illustrates the POI voltage variations. The PV-STATCOM dynamic voltage controller is capable of responding within two cycles, however, due to the presence of plant communication delays and power oscillations, the combined FFR and POD control rapidly stabilizes the voltage in about seven seconds. The PV plant current waveforms are shown in Figure 6.53f. In no control case, the current change is negligible since the PV plant does not react to frequency changes. In POD control case, current oscillates slightly to perform reactive power modulation. With FFR control, current reduces significantly, as the PV plant generation should be reduced. Similarly, in FFR + POD case, the PV plant current is reduced while exhibiting slight oscillations for reactive power modulation. In summary, when imbalance is smaller than PV plant generating power, the combined FFR and POD control can maintain the frequency in the ±36 mHz band, increase the system damping, and provide effective dynamic voltage control.

6.7.4.2 200 MW Load Trip in Area 1 ($P_{available}$ = 100 MW, K_{curt} = 0); (Power Imbalance more than PV Plant Capacity)

Figure 6.54 shows the results of 200 MW load disconnection. The variables depicted are correspondingly similar to Figure 6.53. Since the PV plant is operating at its nominal power (100 MW), it can mitigate only 100 MW of the generation surplus, whereas the synchronous generators compensate the remaining power surplus (100 MW). Figure 6.54a illustrates that with the proposed FFR + POD control, peak of frequency is reduced from 60.37 to 60.2 Hz. Furthermore, the frequency settles at 60.1 Hz compared to 60.27 Hz in No control and POD control cases. As demonstrated in Figure 6.54b, the FFR + POD controller also stabilizes inter-area oscillations in about seven seconds. Figure 6.54c shows that for such a large power imbalance, the FFR controller acts very rapidly (comparable with ideal control) to reduce the PV active power to zero in less than a second. As expected and shown in Figure 6.54d, larger reactive power modulation takes place in the case of 200 MW compared to Figure 6.53d for POD. Similarly, Figure 6.54e demonstrates that the FFR + POD control provides fastest voltage control. This study thus reveals that even when power imbalances are more than the PV plant capacity, the proposed composite FFR and POD controller alleviates frequency deviation, power oscillations, and voltage variation most effectively.

The proposed control is effective in a similar manner for load trips in area 2.

6.7.5 Under Frequency Control

To provide frequency support during under-frequency events, the power output of the PV plants is curtailed as shown in Figure 6.52. The PV solar system can then utilize this reserve capacity to compensate the generation deficiency during under-frequency scenarios. Curtailment of active power releases additional inverter capacity for reactive power exchange which is used for providing effective voltage control and/or POD. For the studies in this section, the PV plant is curtailed by 50%, i.e. K_{curt} = 50%, and simulations are conducted for available PV power of 100 MW.

6.7.5.1 25 MW Load Connection in Area 1 ($P_{available}$ = 100 MW, K_{curt} = 50%); (Power Imbalance less than PV Plant Capacity)

Figure 6.55 demonstrates the results of a 25 MW sudden load increase in area 1, which initiates an under-frequency event. The variables depicted are correspondingly similar to Figure 6.53. As shown

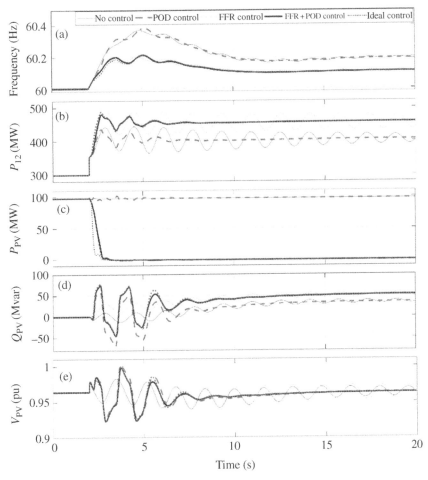

Figure 6.54 200 MW load rejection in area 1 ($P_{available}$ = 60 MW, K_{curt} = 0): (a) POI (bus 8) frequency, (b) Transferred power from area 1 to area 2, (c) PV plant active power, (d) PV plant reactive power, and (e) PV plant voltage. *Source:* Varma and Akbari [142].

in Figure 6.55a, the frequency drops to 59.94 Hz without FFR control. However, this frequency deviation is restricted to only 20 mHz with the proposed FFR control.

Figure 6.55b depicts that inter-area oscillations decay when POD controller is applied. The transferred power from area 1 to 2 is reduced by about 12.5 MW in No control and POD cases because of the equal share of both the areas in compensating generation insufficiency. However, with FFR control, the power export from area 1 is reduced by 25 MW to supply the increased load in area 1. Figure 6.55c also shows that the PV plant rapidly increases its active power output to 75 MW. The remaining inverter capacity makes available 82 Mvar reactive power modulation capability for POD. Figure 6.55d shows that the PV plant absorbs reactive power in steady state to avoid the voltage rise (due to power transfer reduction). Figure 6.55e shows that the POI voltage oscillations settle down within 5% steady state value in eight seconds. Figure 6.55f demonstrates the PV plant current. The PV plant current is initially around 0.47 pu as the PV plant is curtailed by 50%. In no control case, the current remains constant, while it slightly oscillates for reactive power

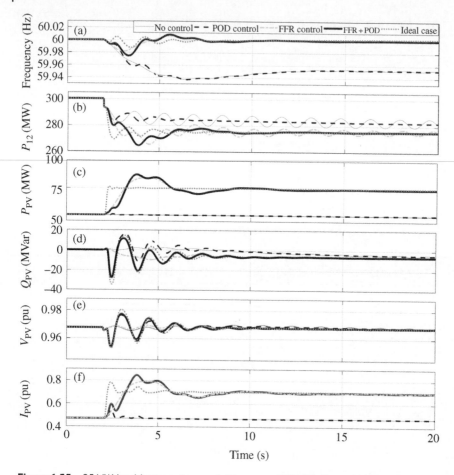

Figure 6.55 25 MW load increase in area 1 ($P_{available}$ = 100 MW, K_{curt} = 50%): (a) POI (bus 8) frequency, (b) transferred power from area 1 to area 2, (c) PV plant active power, (d) PV plant reactive power, (e) PV plant voltage, and (f) PV plant current. *Source:* Varma and Akbari [142].

modulation in POD control case. With FFR control, the PV current reaches 0.7 pu to increase the PV power for frequency support. In conclusion, for this case study, the proposed FFR and POD controller successfully reduces frequency deviations to within ±36 mHz band, damps power oscillations, and regulates voltage in eight seconds.

6.7.5.2 100 MW Load Connection in Area 1($P_{available}$ = 100 MW, K_{curt} = 50%); (Power Imbalance more than PV Plant Capacity)

Figure 6.56 presents the simulation results for a 100 MW sudden increase in load of area 1. The variables illustrated are correspondingly similar to Figure 6.53. Figure 6.56a shows that the FFR controller quickly releases 50 MW reserved power thereby alleviating the frequency nadir from 59.83 to 59.91 Hz. Figure 6.56b demonstrates that power oscillations decay in less than eight seconds with POD controller. Figure 6.56c depicts that the PV plant effectively increases its power output from 50 to 100 MW in less than one second. Figure 6.56d shows that even though the available

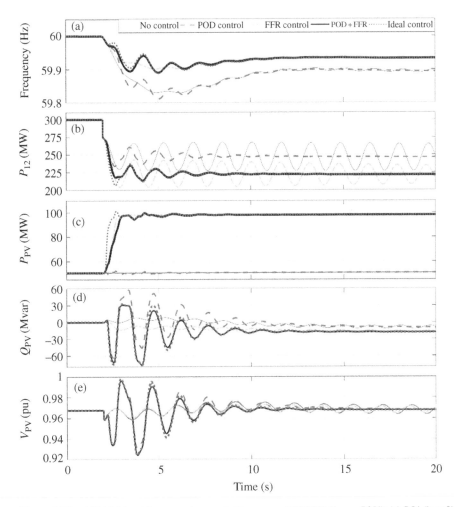

Figure 6.56 100 MW load increase in area 1 ($P_{available}$ = 100 MW, K_{curt} = 50%): (a) POI (bus 8) frequency, (b) transferred power from area 1 to area 2, (c) PV plant active power, (d) PV plant reactive power, and (e) PV plant voltage. *Source:* Varma and Akbari [142].

reactive power capacity is minimum in this case study (as 100 MVA of inverter capacity is used up for active power output), the power oscillations are stabilized in less than eight seconds. Figure 6.56e illustrates effective voltage regulation simultaneously with POD. These results show that even with minimum available reactive power from PV plant, the proposed FFR + POD controller successfully provides improved frequency regulation, POD, and voltage control in less than eight seconds.

The proposed control is effective in a similar manner for load connections in area 2.

Another under-frequency simulation study is also performed for an operating point when it is not noon, i.e. when the power output is less than the rated output (100 MW), i.e. K_{curt} = 50%, $P_{available} \approx$ 60 MW [142]. This is done to investigate the sensitivity to operating conditions which may be more realistic than noontime. The proposed FFR + POD control is still seen to be effective although the nadir is slightly lower as compared to the case shown in Figure 6.56. This study illustrated an interesting feature. While the available PV power and reserve power are reduced, more reactive power capacity is available for POD controller, which makes the overall control quite effective.

6.7.6 Performance Comparison of Proposed FFR + POD Control with Conventional Frequency Control

This study demonstrates the efficacy of the proposed FFR + POD controller as compared to the conventional (5%) droop control which is recommended by NERC for all generators including PV plants [148]. The comparison is made for an under-frequency event during high level of tie-line power flow of 550 MW. The power output of PV plant is curtailed by 50%. A load increase of 100 MW is initiated in area 2. The results are demonstrated in Figure 6.57 which has similar description of variables as in Figure 6.54. Figure 6.57a shows that the conventional droop-based controller has very little impact on the system frequency, whereas the FFR controller noticeably reduces the

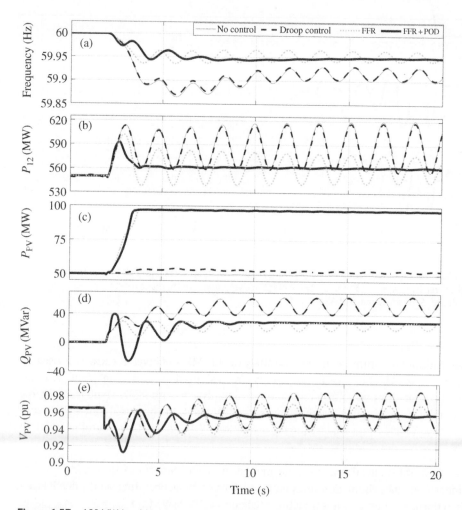

Figure 6.57 100 MW load increase in area 2, PV plant output curtailed by 50%: (a) POI (bus 8) frequency, (b) transferred power from area 1 to area 2, (c) PV plant active power, and (d) PV plant reactive power (e) PV plant voltage. *Source:* Varma and Akbari [142].

frequency deviation. The ineffectiveness of droop control can be understood by observing the PV plant active power variation in Figure 6.57c.

Only a minor active power contribution (~5 MW) is made by the PV plant operating with 5% droop, which mainly corresponds to its size with respect to the conventional generators (100 vs. 2800 MW of synchronous generation). In the no control case, load increase in area 2 leads to a rise in tie-line power which destabilizes the system as shown in Figure 6.57b. The droop control has no positive impact on dynamic stability. The FFR controller preserves system stability although the tie-line power is highly oscillatory.

The proposed FFR + POD controller enhances system damping, so that the power oscillations decay in less than five seconds. It is seen from Figure 6.57d, that when FFR + POD control is activated, the system stability is maintained with lower reactive power injection. The system voltage variations shown in Figure 6.57e, are controlled in a significantly better manner with the proposed combined controller. This study clearly demonstrates that in a situation where the conventional 5% droop control on PV plant is unable to maintain stability, the proposed PV-STATCOM with FFR + POD control not only stabilizes the system but effectively regulates system frequency, power oscillations, and voltage.

6.7.7 Summary

A novel composite FFR and POD control, FFR + POD, of PV solar plants operated as PV-STATCOM is presented. The proposed controller utilizes the fast and flexible response of PV plants to enhance frequency stability and POD concurrently while respecting the capacity limitations of PV inverters. Following inferences are made:

1) The proposed combined FFR + POD controller provides the best performance in regulating frequency and damping of power oscillations as compared to either FFR or POD control acting alone. In the combined control, the individual controls exhibit the following features:
   - If the power imbalance caused by disturbances (e.g. load addition/rejection) is smaller than PV generation or reserve at the time, the FFR controller can rapidly regulate the frequency to within ±36 mHz band to avoid governor's action.
   - If the power imbalance is larger than PV plant capacity, FFR controller still reduces the over/under-frequency deviations substantially.
   - The POD control enhances system stability considerably even when minimum reactive power capacity (48 Mvar) is available during rated power generation from the PV plant.
   - The dynamic voltage controller of PV-STATCOM together with POD control rapidly stabilizes the low-frequency oscillations in the system voltage.

2) The efficacy of the proposed FFR + POD control over conventional droop control is especially evident during high tie-line power flow conditions. For the studied case of a high level of power transfer (550 MW), the conventional droop control has little impact on system frequency and is unable to maintain system stability. However, the proposed control successfully reduces frequency excursion, stabilizes power oscillations, and regulates voltage in a rapid manner.

This study demonstrates that the PV-STATCOM has the unique capability of providing combined dynamic active and reactive power-based services. This is achieved by effective utilization of the PV inverter capacity and available active power production.

6.8 Conclusions

This chapter describes several novel transmission system applications of PV-STATCOM technology. A 100 MW solar PV plant as PV-STATCOM can increase the power transfer by 200 MW in a two-area power system both during night and day, through only reactive power modulation. The PV-STATCOM provides an enhanced POD with a combined modulation of active and reactive power during daytime. A new ramp-up technique is presented, wherein the above damping controls are kept activated while ramping up power from solar PV plants. This results in much faster power restoration times.

A 300 MW solar PV plant as PV-STATCOM located at the terminals of 892 MW synchronous generator connected to a series compensated line in a modified IEEE First SSR Benchmark system can successfully mitigate all four torsional modes at all the four critical levels of series compensation in less than 10s, both during night and day. Furthermore, a 300 MW PV solar plant as PV-STATCOM can alleviate SSOs in a 500 MW IG-based WF connected to a series compensated line in another modified IEEE First SSR Benchmark system over a wide range of series compensation levels on a 24/7 basis. The PV-STATCOM may be connected either at generator terminals or at the line midpoint.

A 75 MW PV-STATCOM located in a 138 kV transmission line can alleviate FIDVR caused by a cluster of 40 MVA motors located 100 km away, both during night (through reactive power modulation) and day (with combined active and reactive power modulation). This PV-STATCOM control obviates the need of a 34 Mvar STATCOM at the terminals of the motor load for providing the same FIDVR mitigation. The PV-STATCOM control does not create any adverse impact on system frequency while providing the above grid support functionalities.

The PV-STATCOM is thus shown to perform grid support functions for which typically expensive SVCs or STATCOMs are employed.

A 100 MW PV-STATCOM is also shown to successfully provide simultaneous FFR (through active power modulation) and POD (with reactive power modulation) in a two-area system both for overfrequency and underfrequency events. These functions can be of great benefit to transmission utilities across the world.

References

1 Kundur, P. (1994). *Power System Stability and Control*. New York: McGraw Hill.

2 Hingorani, N.G. and Gyugyi, L. (1999). *Understanding FACTS*. Piscataway, NJ: IEEE Press.

3 Padiyar, K.R. (2007). *FACTS Controllers in Power Transmission and Distribution*. New Delhi, India: New Age International Publishers.

4 Mathur, R.M. and Varma, R.K. (2002). *Thyristor-Based FACTS Controllers for Electrical Transmission Systems*. New York: Wiley-IEEE Press.

5 Ying, X., Song, Y.H., Chen-Ching, L., and Sun, Y.Z. (2003). Available transfer capability enhancement using FACTS devices. *IEEE Transactions on Power Systems* 18: 305–312.

6 Song, Y.H. and Johns, A.T. (1999). *Flexible AC Transmission Systems (FACTS)*. London, UK: IEE Press.

7 CIGRE (1986). Static var compensators. CIGRE, Paris, France. *Technical Brochure 025*.

8 CIGRE (1999). Static synchronous compensator (STATCOM). CIGRE, Paris, France. *Technical Brochure No. 144*.

9 Litzenberger, W.H. (1994). *An Annotated Bibliography of High-Voltage Direct-Current Transmission and Flexible AC Transmission (FACTS) Devices, 1991–1993*. Portland, OR: Bonneville Power Administration and Western Area Power Administration.

10 Litzenberger, W.H. and Varma, R.K. (1996). *An Annotated Bibliography of High-Voltage Direct-Current Transmission and FACTS Devices, 1994–1995*. Portland, OR: Bonneville Power Administration and U.S. Department of Energy.

11 Litzenberger, W.H., Varma, R.K., and Flanagan, J.D. (1998). *An Annotated Bibliography of High-Voltage Direct-Current Transmission and FACTS Devices, 1996–1997*. Portland, OR: Electric Power Research Institute and Bonneville Power Administration.

12 Rahman, S.A., Maleki, H., Mohan, S., Varma, R.K., and Litzenberger, W.H. (2015). Bibliography of FACTS 2013–2014: IEEE working group report. In *Proc. 2015 IEEE Power & Energy Society General Meeting*, 1–5.

13 Bellini, E. (2020). World's largest solar plant goes online in China. https://www.pv-magazine.com/2020/10/01/worlds-largest-solar-plant-goes-online-in-china

14 Cheng, D., Mather, B.A., Seguin, R. et al. (2016). Photovoltaic (PV) impact assessment for very high penetration levels. *IEEE Journal of Photovoltaics* 6: 295–300.

15 Wolfe, P. (2018). An overview of the world's largest solar power plants. pv magazine. https://www.pv-magazine.com/2019/06/18/an-overview-of-the-worlds-largest-solar-power-plants

16 Shah, R., Mithulananthan, N., Bansal, R., and Ramachandaramurthy, V. (2015). A review of key power system stability challenges for large-scale PV integration. *Renewable and Sustainable Energy Reviews* 41: 1423–1436.

17 Eftekharnejad, S., Vittal, V., Heydt, G.T. et al. (2013). Small signal stability assessment of power systems with increased penetration of photovoltaic generation: a case study. *IEEE Transactions on Sustainable Energy* 4: 960–967.

18 Tamimi, B., Cañizares, C., and Bhattacharya, K. (2013). System stability impact of large-scale and distributed solar photovoltaic generation: the case of Ontario, Canada. *IEEE Transactions on Sustainable Energy* 4: 680–688.

19 Varma, R.K., Rahman, S.A., and Seethapathy, R. (2010). Novel control of grid connected photovoltaic (PV) solar farm for improving transient stability and transmission limits both during night and day. In *Proc. 2010 World Energy Conference*, Montreal, Canada.

20 Varma, R.K., Rahman, S.A., Mahendra, A.C., Seethapathy, R., and Vanderheide, T. (2012). Novel nighttime application of PV solar farms as STATCOM (PV-STATCOM). In *Proc. 2012 IEEE Power & Energy Society General Meeting*, 1–8.

21 Varma, R.K., Rahman, S.A., and Vanderheide, T. (2015). New control of PV solar farm as STATCOM (PV-STATCOM) for increasing grid power transmission limits during night and day. *IEEE Transactions on Power Delivery* 30: 755–763.

22 Rahman, S.A., Varma, R.K., and Vanderheide, T. (2014). Generalised model of a photovoltaic panel. *IET Renewable Power Generation* 8: 217–229.

23 Hussein, K.H., Muta, I., Hoshino, T., and Osakada, M. (1995). Maximum photovoltaic power tracking: an algorithm for rapidly changing atmospheric conditions. *IEE Proceedings – Generation, Transmission and Distribution* 142: 59–64.

24 Chatterjee, K., Fernandes, B.G., and Dubey, G.K. (1999). An instantaneous reactive volt-ampere compensator and harmonic suppressor system. *IEEE Transactions on Power Electronics* 14: 381–392.

25 Yazdani, A. and Iravani, R. (2010). *Voltage-Sourced Converters in Power Systems: Modeling, Control, and Applications*. New York, NY: Wiley.

26 Kim, S.-K., Jeon, J.-H., Cho, C.-H. et al. (2009). Modeling and simulation of a grid-connected PV generation system for electromagnetic transient analysis. *Solar Energy* 83: 664–678.

27 Rashid, M.H. (2007). *Power Electronics Handbook: Devices, Circuits, and Applications*. Elsevier.

28 CIGRE (2000). Impact of interactions among power system controls, CIGRE, Paris, France. *Technical Brochure 166*.

29 Shah, R., Mithulananthan, N., and Lee, K.Y. (2013). Large-scale PV plant with a robust controller considering power oscillation damping. *IEEE Transactions on Energy Conversion* 28: 106–116.

30 Wandhare, R.G. and Agarwal, V. (2014). Novel stability enhancing control strategy for centralized PV-grid systems for smart grid applications. *IEEE Transactions on Smart Grid* 5: 1389–1396.

31 Varma, R.K., Khadkikar, V., and Seethapathy, R. (2009). Nighttime application of PV solar farm as STATCOM to regulate grid voltage. *IEEE Transactions on Energy Conversion (Letters)* 24: 983–985.

32 Varma, R.K. and Maleki, H. (2019). PV solar system control as STATCOM (PV-STATCOM) for power oscillation damping. *IEEE Transactions on Sustainable Energy* 10-4: 1793–1803.

33 Padiyar, K.R. and Varma, R.K. (1991). Damping torque analysis of static var system controllers. *IEEE Transactions on Power Systems* 6: 458–465.

34 Kundur, P. (1994). *Power System Stability and Control*. New York: McGraw-Hill.

35 Shan, J., Annakkage, U.D., and Gole, A.M. (2006). A platform for validation of FACTS models. *IEEE Transactions on Power Delivery* 21: 484–491.

36 Yazdani, A., Fazio, A.R.D., Ghoddami, H. et al. (2011). Modeling guidelines and a benchmark for power system simulation studies of three-phase single-stage photovoltaic systems. *IEEE Transactions on Power Delivery* 26: 1247–1264.

37 Yazdani, A. and Dash, P.P. (2009). A control methodology and characterization of dynamics for a photovoltaic (PV) system interfaced with a distribution network. *IEEE Transactions on Power Delivery* 24: 1538–1551.

38 BDEW (2008). *Technical guideline: generating plants connected to the medium-voltage network*. (Guideline for Generating plants' Connection to and Parallel Operation with the Medium-Voltage Network). BDEW (Bundesverband der Energie- und Wasserwirtschaft e.V.), Berlin, Germany (June 2008, revised January 2013).

39 E. ON Netz (2006). *Grid Code: High and extra high voltage*, E.ON Netz GmbH, Bayreuth, Germany.

40 EPRI (2016). Common functions for smart inverters (4th ed.). EPRI, Palo Alto, CA, USA. *Tech. Rep. 3002008217*.

41 Nelder, J.A. and Mead, R. (1965). A simplex method for function minimization. *The Computer Journal* 7: 308–313.

42 Gole, A.M., Filizadeh, S., Menzies, R.W., and Wilson, P.L. (2005). Optimization-enabled electromagnetic transient simulation. *IEEE Transactions on Power Delivery* 20: 512–518.

43 Gibbard, M., Pourbeik, P., and Vowles, D. (2015). *Small-Signal Stability, Control and Dynamic Performance of Power Systems*. University of Adelaide Press, Australia.

44 Martins, N. and Lima, L.T.G. (1990). Determination of suitable locations for power system stabilizers and static var compensators for damping electromechanical oscillations in large scale power systems. *IEEE Transactions on Power Systems* 5: 1455–1469.

45 Morjaria, M., Anichkov, D., Chadliev, V., and Soni, S. (2014). A grid-friendly plant: the role of utility-scale photovoltaic plants in grid stability and reliability. *IEEE Power and Energy Magazine* 12: 87–95.

46 Haque, M.H. (2004). Improvement of first swing stability limit by utilizing full benefit of shunt FACTS devices. *IEEE Transactions on Power Systems* 19: 1894–1902.

47 Mithulananthan, N., Canizares, C.A., Reeve, J., and Rogers, G.J. (2003). Comparison of PSS, SVC, and STATCOM controllers for damping power system oscillations. *IEEE Transactions on Power Systems* 18: 786–792.

48 Cong, L. and Wang, Y. (2002). Co-ordinated control of generator excitation and STATCOM for rotor angle stability and voltage regulation enhancement of power systems. *IEE Proceedings - Generation, Transmission and Distribution* 149: 659–666.

49 Beza, M. and Bongiorno, M. (2015). An adaptive power oscillation damping controller by STATCOM with energy storage. *IEEE Transactions on Power Systems* 30: 484–493.

50 Zhu, Y., Liu, C., Sun, K. et al. (2019). Optimization of battery energy storage to improve power system oscillation damping. *IEEE Transactions on Sustainable Energy* 10: 1015–1024.

51 Yao, W., Jiang, L., Fang, J. et al. (2016). Adaptive power oscillation damping controller of superconducting magnetic energy storage device for interarea oscillations in power system. *International Journal of Electrical Power and Energy Systems* 78: 555–562.

52 Shintai, T., Miura, Y., and Ise, T. (2014). Oscillation damping of a distributed generator using a virtual synchronous generator. *IEEE Transactions on Power Delivery* 29: 668–676.

53 Maleki, Hesamaldin (2017). *Novel Night and Day Control of a PV Solar System as a STATCOM (PV-STATCOM) for Damping of Power Oscillations*. PhD Thesis, ECE Department, The University of Western Ontario, London, ON, Canada.

54 NERC (2020). *Frequency and voltage protection settings for generating resources*, NERC Standard PRC-024-3.

55 Remon, D., Cantarellas, A.M., and Rodriguez, P. (2016). Equivalent model of large-scale synchronous photovoltaic power plants. *IEEE Transactions on Industry Applications* 52: 5029–5040.

56 Gevorgian, V. and Booth, S. (2013). Review of PREPA technical requirements for interconnecting wind and solar generation. NREL, Golden, CO, USA, *Report NREL/TP-5D00-57089*.

57 (2018). *Rule 14H interconnection of distributed generating facilities with the company's distribution system*, Hawaii Electric Company, Inc.

58 NERC (2018). BPS-Connected Inverter-Based Resource Performance. NERC, Atlanta, GA, USA, *Reliability Guideline*.

59 Walker, D.N., Bowler, C.E.J., Jackson, R.L., and Hodges, D.A. (1975). Results of subsynchronous resonance test at Mohave. *IEEE Transactions on Power Apparatus and Systems* 94: 1878–1889.

60 Anderson, P.M., Agrawal, B.L., and Ness, J.E.V. (1990). *Subsynchronous Resonance in Power Systems*. Piscataway, NJ: IEEE Press.

61 Padiyar, K.R. (1999). *Analysis of Subsynchronous Resonance in Power Systems*. Norwell, MA: Kluwer.

62 (1992). Reader's guide to subsynchronous resonance. *IEEE Transactions on Power Systems* 7: 150–157.

63 Bahrman, M., Larsen, E.V., Piwko, R.J., and Patel, H.S. (1980). Experience with HVDC - turbine-generator torsional interaction at Square Butte. *IEEE Transactions on Power Apparatus and Systems* PAS-99: 966–975.

64 Mortensen, K., Larsen, E.V., and Piwko, R.J. (1981). Field tests and analysis of torsional interaction between the Coal Creek turbine-generators and the CU HVdc system. *IEEE Transactions on Power Apparatus and Systems* PAS-100: 336–344.

65 IEEE Subsynchronous Resonance Working Group of the System Dynamic Performance Subcommittee, Power System Engineering Committee (1980). Countermeasures to subsynchronous resonance problems. *IEEE Transactions on Power Apparatus and Systems* PAS-99: 1810–1818.

66 Abi-Samra, N.C., Smith, R.F., McDermott, T.E., and Chidester, M.B. (1985). Analysis of thyristor-controlled shunt SSR countermeasures. *IEEE Transactions on Power Apparatus and Systems* PAS-104: 583–597.

67 Ramey, D.G., Kimmel, D.S., Dorney, J.W., and Kroening, F.H. (1981). Dynamic stabilizer verification tests at the San Juan Station. *IEEE Transactions on Power Apparatus and Systems* PAS-100: 5011–5019.

68 Hammad, A.E. and El-Sadek, M. (1984). Application of a thyristor controlled var compensator for damping subsynchronous oscillations in power systems. *IEEE Transactions on Power Apparatus and Systems* PAS-103: 198–212.

69 Patil, K.V., Senthil, J., Jiang, J., and Mathur, R.M. (1998). Application of STATCOM for damping torsional oscillations in series compensated AC systems. *IEEE Transactions on Energy Conversion* 13: 237–243.

70 Padiyar, K.R. and Prabhu, N. (2006). Design and performance evaluation of subsynchronous damping controller with STATCOM. *IEEE Transactions on Power Delivery* 21: 1398–1405.

71 Varma, R.K. and Salehi, R. (2017). SSR mitigation with a new control of PV solar farm as STATCOM (PV-STATCOM). *IEEE Transactions on Sustainable Energy* 8: 1473–1483.

72 (1977). First benchmark model for computer simulation of subsynchronous resonance. *IEEE Committee Report, IEEE Transactions on Power Apparatus and Systems* 96: 1565–1572.

73 Chen, C., Wasynezuk, O., and Anwah, N.A. (1986). Stabilizing subsynchronous resonance using transmission current feedback. *IEEE Transactions on Power Systems* 1: 34–41.

74 Thirumalaivasan, R., Janaki, M., and Prabhu, N. (2013). Damping of SSR using subsynchronous current suppressor with SSSC. *IEEE Transactions on Power Systems* 28: 64–74.

75 (2015). *ERCOT Nodal Operating Guides*. Electric Reliability Council of Texas.

76 Agua Caliente Solar Project. http://www.firstsolar.com/en/Resources/Projects/Agua-Caliente-Solar-Project (accessed 10 Sep. 2021).

77 Varma, R.K., Auddy, S., and Semsedini, Y. (2008). Mitigation of subsynchronous resonance in a series-compensated wind farm using FACTS controllers. *IEEE Transactions on Power Delivery* 23: 1645–1654.

78 Varma, R.K. and Moharana, A. (2013). SSR in double-cage induction generator-based wind farm connected to series-compensated transmission line. *IEEE Transactions on Power Systems* 28: 2573–2583.

79 Leon, A.E. and Solsona, J.A. (2015). Sub-synchronous interaction damping control for DFIG wind turbines. *IEEE Transactions on Power Systems* 30: 419–428.

80 Adams, J., Carter, C., and Huang, S. H. (2012). ERCOT experience with sub-synchronous control interaction and proposed remediation. In *Proc. 2012 IEEE/PES Transmission and Distribution Conference & Exposition (T&D)*, 1–5.

81 Wang, L., Xie, X., Jiang, Q. et al. (2015). Investigation of SSR in practical DFIG-based wind farms connected to a series-compensated power system. *IEEE Transactions on Power Systems* 30: 2772–2779.

82 Wu, B., Lang, Y., Zargari, N., and Kouro, S. (2011). *Power Conversion and Control of Wind Energy Systems*. New York, USA: Wiley-IEEE Press.

83 Moharana, A., Varma, R.K., and Seethapathy, R. (2014). SSR alleviation by STATCOM in induction-generator-based wind farm connected to series compensated line. *IEEE Transactions on Sustainable Energy* 5: 947–957.

84 The Wind Power: Wind energy market intelligence. https://www.thewindpower.net (accessed 10 April 2021).

85 Cheng, Y., Huang S.F., Rose J., Pappu V.A., and Conto J. (2016), "ERCOT subsynchronous resonance topology and frequency scan tool development," in *Proc. 2016 IEEE Power & Energy Society General Meeting*.

86 Sharafdarkolaee, R. Salehi. (2019). *Novel PV Solar Farm Control as STATCOM (PV-STATCOM) for SSR Mitigation in Synchronous Generators and Wind Farms*. PhD Thesis, ECE Department, The University of Western Ontario, London, ON, Canada.

87 NERC (2017). *Reliability standards for the bulk electric systems of North America*, Atlanta, GA, USA.

88 El-Moursi, M.S., Bak-Jensen, B., and Abdel-Rahman, M.H. (2010). Novel STATCOM controller for mitigating SSR and damping power system oscillations in a series compensated wind park. *IEEE Transactions on Power Electronics* 25: 429–441.

89 Stateline Wind Project. https://www.crpud.net/wp-content/uploads/stateline.pdf (accessed 9 Sep. 2021).

90 Grand Renewable Energy Park. Haldimand County, Ontario. https://www.power-technology.com/projects/grand-renewable-energy-park-haldimand-county-ontario (accessed 9 Sep. 2021).

91 Iberdrola builds hybrid wind energy and solar power plant in Australia. https://www.evwind.es/2020/10/09/iberdrola-builds-hybrid-wind-energy-and-solar-power-plant-in-australia/77656 (accessed 9 Sep. 2021).

92 Australia's first hybrid wind and solar farm officially launched. https://www.ecogeneration.com.au/australias-first-hybrid-wind-and-solar-farm-officially-launched/ (accessed 9 Sep. 2021).

93 NERC (2009). A Technical Reference Paper Fault-Induced Delayed Voltage Recovery Version 1.2, NERC, Princeton, NJ, USA. *Report*.

94 Ito, J. (2015). *U.S. DOE-NERC Workshop on Fault-Induced Delayed Voltage Recovery (FIDVR) & Dynamic Load Modeling*. US Department of Energy and NERC, Alexandria, VA, USA.

95 Lawrence Berkeley National Laboratory (2010). Final Project Report Load Modeling Transmission Research, Lawrence Berkeley National Laboratory, Berkeley, CA, USA. *Report*.

96 US Department of Energy (2008), *Workshop on the Role of Residential AC in Contributing to Fault-Induced Delayed Voltage Recovery (FIDVR)*, US DOE, Dallas, TX, USA.

97 Bai, H. and Ajjarapu, V. (2011). A novel online load shedding strategy for mitigating fault-induced delayed voltage recovery. *IEEE Transactions on Power Systems* 26: 294–304.

98 Paramasivam, M., Salloum, A., Ajjarapu, V. et al. (2013). Dynamic optimization based reactive power planning to mitigate slow voltage recovery and short term voltage instability. *IEEE Transactions on Power Systems* 28: 3865–3873.

99 Tiwari, A. and Ajjarapu, V. (2016). Addressing short term voltage stability problem – Part I: Challenges and plausible solution directions. In *Proc. 2016 IEEE/PES Transmission and Distribution Conference & Exposition (T&D)*, 1–5.

100 Tiwari, A. and Ajjarapu, V. (2016). Addressing short term voltage stability problem – Part II: A case study. In *Proc. 2016 IEEE/PES Transmission & Distribution Conference and Exposition (T&D)*, 1–5.

101 Varma, R.K. and Paserba, J. (2012). Flexible AC transmission systems (FACTS). In: *Electric Power Engineering Handbook on Power System Stability and Control*, vol. 3 (ed. P. Kundur). USA: CRC Press/Taylor & Francis.

102 Al-Mubarak, A.H., Bamsak, S.M., Thorvaldsson, B., Halonen, M., and Grunbaum, R. (2009). Preventing voltage collapse by large SVCs at power system faults. In *Proc. 2009 IEEE/PES Power Systems Conference and Exposition*, 1–9.

103 Reed, G., Grainger, B., Kempker, M. et al. (2016). Technical requirements and design of the Indianapolis Power & Light 138 kV Southwest static var compensator. In *Proc. 2016 IEEE/PES Transmission and Distribution Conference & Exposition (T&D)*, 1–5.

104 Sullivan, D., Pape, R., Birsa, J. et al. (2009). Managing fault-induced delayed voltage recovery in Metro Atlanta with the Barrow County SVC. In *Proc. 2009 IEEE/PES Power Systems Conference and Exposition*, 1–6.

105 Boström, A., Grunbaum, R., Dahlblom, M., and Oheim, H. V. (2013). SVC for reliability improvement in the NSTAR 115 kV Cape Cod transmission system. In *Proc. 2013 IEEE Power & Energy Society General Meeting*, 1–5.

106 Tamimi, B., Cañizares, C., and Bhattacharya, K. (2011). Modeling and performance analysis of large solar photo-voltaic generation on voltage stability and inter-area oscillations. In *Proc. 2011 IEEE Power & Energy Society General Meeting*, 1–6.

107 Gutiérrez, H. A. V. and Molina, M. G. (2017). Analysis of voltage sags due to induction motors in distribution systems with high PV penetration. In *Proc. 2017 IEEE PES Innovative Smart Grid Technologies Conference - Latin America (ISGT Latin America)*, 1–6.

108 Abdel-Aziz, E. Z., Ishaq, J., Al-Khulayfi, A. M., and Fawzy, Y. T. (2016). Voltage stability improvement in transmission network embedded with photovoltaic systems. In *Proc. 2016 IEEE International Energy Conference (ENERGYCON)*, 1–7.

109 Bravo, R. J. (2015). DER volt-var and voltage ride-through needs to contain the spread of FIDVR events. In *Proc. 2015 IEEE Power & Energy Society General Meeting*, 1–3.

110 Lammert, G., Boemer, J. C., Premm, D. et al. (2017). Impact of fault ride-through and dynamic reactive power support of photovoltaic systems on short-term voltage stability. In *Proc. 2017 IEEE Manchester PowerTech*, 1–6.

111 Lammert, G., Premm, D., Ospina, L.D.P. et al. (2019). Control of photovoltaic systems for enhanced short-term voltage stability and recovery. *IEEE Transactions on Energy Conversion* 34: 243–254.

112 IEEE (2018). *IEEE standard for interconnection and interoperability of distributed energy resources with associated electric power systems interfaces.* IEEE Std 1547-2018 (Revision of IEEE Std 1547-2003).

113 Kawabe, K., Ota, Y., Yokoyama, A., and Tanaka, K. (2017). Novel dynamic voltage support capability of photovoltaic systems for improvement of short-term voltage stability in power systems. *IEEE Transactions on Power Systems* 32: 1796–1804.

114 Varma, R.K., Rahman, S.A., Atodaria, V. et al. (2016). Technique for fast detection of short circuit current in PV distributed generator. *IEEE Power and Energy Technology Systems Journal* 3: 155–165.

115 Varma, R.K. and Mohan, S. (2020). Mitigation of fault induced delayed voltage recovery (FIDVR) by PV-STATCOM. *IEEE Transactions on Power Systems* 35-6: 4251–4262.

116 Wang, W. V., Ming, Z., and Wah, S. (2015). Challenges to supplying large induction motor loads in a long radial transmission system. In *Proc. 2015 IEEE Power & Energy Society General Meeting*, 1–4.

117 Western Electricity Coordinating Council Renewable Energy Modeling Task Force (2014). *WECC PV Power Plant Dynamic Modeling Guide.* WECC, Salt Lake City, UT, USA.

118 Varma, R.K. and Siavashi, E.M. (2018). PV-STATCOM: A new smart inverter for voltage control in distribution systems. *IEEE Transactions on Sustainable Energy* 9: 1681–1691.

119 CIGRE (1999). Coordination of controls of multiple FACTS/HVDC links in the same system. CIGRE, Paris, France, *Technical Brochure No. 149.*

120 ABB (2011). SVC to stabilize a large 735 kV transmission system in Canada. ABB, Vasteras, Sweden.

121 Belanger, J., Scott, G., Anderson, T., and Torseng, S. (1984). Gain supervisor for thyristor-controlled shunt compensators. In *Proc. CIGRE Conference*, Paris, France.

122 Reed, G. F., Larsson, D., Rasmussen, J., Rosenberger, T., and Fakir, R. E. (2011). Advanced control methods and strategies for the Oncor Electric Delivery Renner SVC. In *Proc. 2011 IEEE/PES Power Systems Conference and Exposition*, 1–9.

123 Sybille, G., Giroux, P., Dellwo, S. et al. (1996). Simulator and field testing of Forbes SVS. *IEEE Transactions on Power Delivery* 11: 1507–1514.

124 IEEE (2019). *IEEE guide for specification of transmission static synchronous compensator (STATCOM) systems.* IEEE Std 1052-2018, 1–115.

125 C. Loutan, M. Morjaria, V. Gevorgian et al. (2017). Demonstration of essential reliability services by a 300-MW solar photovoltaic power plant. National Renewable Energy Laboratory, Golden, CO, USA. *Technical Report NREL/TP-5D00–67799.*

126 Miller, N.W., Shao, M., Pajic, S., and D'Aquila, R. (2014). Western wind and solar integration study phase 3—Frequency response and transient stability: Executive Summary. NREL, Golden, CO, USA, *Report NREL/SR-5D00–62906-ES.*

127 Miller, N. W., Shao, M., Pajic, S., and D'Aquila, R. (2013). Eastern Frequency Response Study. NREL, Golden, CO, USA, *Report No. NREL/SR-5500-58077.*

128 Miller, N. W., Shao, M., Venkataraman, S., Loutan, C., and Rothleder, M. (2012). Frequency response of California and WECC under high wind and solar conditions. In *Proc. 2012 IEEE Power & Energy Society General Meeting*, 1–8.

129 Crăciun, B., Kerekes, T., Séra, D., and Teodorescu, R. (2014). Frequency support functions in large PV power plants with active power reserves. *IEEE Journal of Emerging and Selected Topics in Power Electronics* 2: 849–858.

130 Alatrash, H., Mensah, A., Mark, E. et al. (2012). Generator emulation controls for photovoltaic inverters. *IEEE Transactions on Smart Grid* 3: 996–1011.

131 Liu, Y., Gracia, J. R., Hadley, S. W., and Liu, Y. (2013). Wind/PV generation for frequency regulation and oscillation damping in the Eastern Interconnection (EI). Oak Ridge National Laboratory, Oak Ridge, TN, USA. *Technical report ORNL/TM-2013/587.*

132 García-Gracia, M., Halabi, N.E., Ajami, H., and Comech, M.P. (2012). Integrated control technique for compliance of solar photovoltaic installation grid codes. *IEEE Transactions on Energy Conversion* 27: 792–798.

133 Matevosyan, J., Sharma, S., Huang, S. et al. (2015). Proposed future ancillary services in electric reliability council of Texas. In *Proc. 2015 IEEE Eindhoven PowerTech*, 1–6.

134 Li, W., Du, P., and Lu, N. (2018). Design of a new primary frequency control market Market for hosting frequency response reserve offers from both generators and loads. *IEEE Transactions on Smart Grid* 9: 4883–4892.

135 NationalGridESO. (2018). Enhanced frequency response (EFR). https://www.nationalgrideso.com/balancing-services/frequency-response-services/enhanced-frequency-response-efr

136 Miller, N., Lew, D., Piwko, R. et al. (2017). Technology capabilities for fast frequency response. Report prepared by GE Energy Consulting for Australian Energy Market Operator, Schenectady, NY, USA. *Report.*

137 Quintero, J., Vittal, V., Heydt, G.T., and Zhang, H. (2014). The impact of increased penetration of converter control-based generators on power system modes of oscillation. *IEEE Transactions on Power Systems* 29: 2248–2256.

138 NERC (2017). Forced Oscillation Monitoring & Mitigation. NERC, Atlanta, GA, USA, *Reliability Guideline.*

139 Arani, M.F.M. and Mohamed, Y.A.I. (2016). Analysis and mitigation of undesirable impacts of implementing frequency support controllers in wind power generation. *IEEE Transactions on Energy Conversion* 31: 174–186.

140 Fan, L., Yin, H., and Miao, Z. (2011). On active/reactive power modulation of DFIG-based wind generation for interarea oscillation damping. *IEEE Transactions on Energy Conversion* 26: 513–521.

141 Rimorov, D., Joós, G., and Kamwa, I. (2016). Design and implementation of combined frequency/oscillation damping controller for type 4 wind turbines. In *Proc. 2016 Power Systems Computation Conference (PSCC)*, 1–7.

142 Varma, R.K. and Akbari, M. (2020). Simultaneous fast frequency control and power oscillation damping by utilizing PV solar system as PV-STATCOM. *IEEE Transactions on Sustainable Energy* 11-1: 415–425.

143 Wilches-Bernal, F., Chow, J.H., and Sanchez-Gasca, J.J. (2016). A fundamental study of applying wind turbines for power system frequency control. *IEEE Transactions on Power Systems* 31: 1496–1505.

144 IEEE (2016). *IEEE recommended practice for excitation system models for power system stability studies.* IEEE Std 421.5-2016 (Revision of IEEE Std 421.5-2005), 1–207.

145 WECC (2015). Central Station Photovoltaic Power Plant Model Validation Guideline. Western Electricity Coordination Council, Salt Lake City, UT, USA, *Guideline.*

146 Pourbeik, P. (2015). Model user guide for generic renewable energy system models. EPRI, Palo Alto, CA, USA, *Technical Report 3002006525.*

147 Heniche, A. and Kamwa, I. (2002). Control loops selection to damp inter-area oscillations of electrical networks. *IEEE Transactions on Power Systems* 17: 378–384.

148 NERC (2012). Frequency Response Initiative Report. North American Electric Reliability Corporation, Atlanta, GA, USA, *Report.*

7

INCREASING HOSTING CAPACITY BY SMART INVERTERS – CONCEPTS AND APPLICATIONS

This chapter describes hosting capacity for solar photovoltaic (PV) systems and its enhancement in distribution networks. The concept of hosting capacity is introduced and the factors impacting hosting capacity are elucidated. Different non smart inverter-based methods for increasing hosting capacity are presented. The characteristics of different smart inverter (SI) functions and their effectiveness in improving hosting capacity are discussed. The methodologies and guidelines for selecting the settings of different smart inverter functions are explained. Several simulation studies of increasing hosting capacity in utility networks are described [1]. Finally, different worldwide field implementations of smart inverters in enhancing hosting capacity are presented [1]. Although the focus of this chapter is on hosting capacity of solar PV systems, the concepts presented apply equally to other distributed energy resources (DERs) (e.g. wind power systems).

7.1 Hosting Capacity of Distribution Feeders

Hosting capacity is defined as "the amount of PV that can be accommodated without impacting power quality or reliability under existing control and infrastructure configurations" [2]. This definition, although provided for PV systems, is also applicable for other types of DERs such as wind power systems, energy storage systems, electric vehicle (EV) charging systems, etc. Hosting capacity limits the amount of DERs that can be connected on any given distribution feeder without incurring any expensive upgrades such as construction of lines with higher capacity and/or higher rated transformers.

Hosting capacity is the maximum amount of DERs being connected on a distribution feeder, beyond which adverse impacts start occurring on either one or several of the following [2]:

- Voltage
- Protection
- Unintentional islanding
- Thermal overloading
- Reverse power flow
- Harmonics
- Short circuit currents

Every distribution feeder has its own hosting capacity which may not be a single value but can vary with different factors. Moreover, the hosting capacity of two feeders with similar characteristics can be different due to location of the DERs to be connected.

Smart Solar PV Inverters with Advanced Grid Support Functionalities, First Edition. Rajiv K. Varma.
© 2022 The Institute of Electrical and Electronics Engineers, Inc. Published 2022 by John Wiley & Sons, Inc.

The above impacts have already been described in Chapter 1. This chapter focuses on the following two considerations that can limit hosting capacity.

7.1.1   Voltage

Increase in DER penetration leads to injection of higher amounts of active power on the feeder which tends to increase the steady-state voltages beyond the operating limits specified by the utilities. Variability of power injection from DERs may also cause voltage fluctuations above or below operating limits which may cause increased transformer tap operations or capacitor switchings.

If the utility has implemented conservation voltage reduction (CVR), the amount of DERs that can be hosted will become lower due to reduced voltage limits. High DER penetration causes voltage to rise. With CVR the utility acceptable upper voltage limit will be lower than under normal operation. Hence integration of lesser number of DERs will cause the voltage to rise to the upper voltage limit required under CVR. Without CVR, since the upper voltage limit is higher, more number of DERs can be connected on the feeder.

In some cases, the transient overvoltages (TOVs) during faults may restrict the connectivity of DERs, although not considered here. Voltage unbalance may also become a limiting factor for hosting capacity if the amount of single-phase DERs is large and not balanced appropriately among phases possibly due to the specific set of loadings on different phases of the distribution lines.

7.1.2   Thermal Overloading

The maximum amount of DERs connected on a given feeder should not inject currents which exceed the thermal limit of the distribution line or substation transformers. In many cases, DERs can offset local loads and therefore have a lesser impact on the thermal overloading. If DERs need to be connected on thermally constrained lines, power curtailment, or grid reinforcements (e.g. upgrades to conductors, substations, etc.) will be required.

7.2   Hosting Capacity Based on Voltage Violations

Overvoltage is one of the main factors which restricts the hosting capacity in a distribution feeder. The concept of voltage-based hosting capacity and its increase by the volt–var SI function on a PV system is demonstrated in [1, 3, 4]. The hosting capacity in a distribution feeder with multiple PV systems operating with unity power factor (PF) is shown in Figure 7.1a [4]. This figure depicts a plot of the maximum feeder voltage with increasing amount of PV system penetration. Every point on the plot represents the maximum feeder voltage for a specific size and location of the PV system on the feeder. These voltages are compared with the ANSI maximum voltage limit of 1.05 pu. Three regions of PV penetration are indicated. The first region from left depicts the case where no overvoltage occurs beyond 1.05 pu, irrespective of the size and location of PV systems on the feeder. The upper boundary of this first region is considered to be the minimum hosting capacity for that feeder. This limit represents the minimum number of PV systems beyond which any addition will begin causing a negative voltage impact on the feeder.

The middle region shows that larger sizes of PV systems may be connected in the feeder but only at specific locations so as not to violate the voltage limit. The upper limit of this region is considered to be the maximum hosting capacity for that feeder. Beyond this limit, any additional PV system regardless of size or location in the feeder will cause a voltage violation.

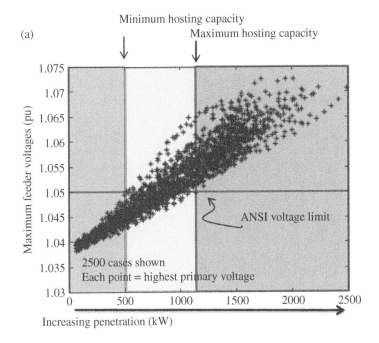

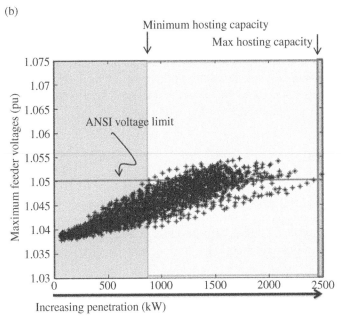

Figure 7.1 Hosting capacity of a distribution feeder. (a) PV systems operating at unity power factor and (b) PV systems operating with volt-var control. *Source*: Reprinted with permission from EPRI [4].

This case study demonstrates that the hosting capacity of a distribution feeder for PV systems is not a specific number but lies between two limits [2]. However, conservatively, the minimum hosting capacity may be used for planning purposes.

The plot of maximum feeder voltage with increasing PV penetration when PV systems are operating with volt–var control (VVC) is shown in Figure 7.1b. Both the minimum and maximum hosting capacity limits hosting capacity are seen to increase significantly. The enhancement in minimum hosting capacity is almost 160%.

This clearly demonstrates how smart PV inverter functions can effectively increase the hosting capacity of distribution grids. The above estimates of the hosting capacity limits are obtained using the metric of "maximum feeder voltage." Hosting capacity limits may also be determined based on other voltage metrics [5, 6]. These metrics are described later in this chapter.

Variants of Hosting Capacity

Further insights into the hosting capacity are provided by the following variants of hosting capacity limits [6]:

i) *Theoretical hosting capacity*: This is the hosting capacity of a distribution feeder evaluated after removing (in simulation studies) all the existing PV systems on the feeder.
ii) *Remaining hosting capacity*: This represents the additional amount of PV systems (at unity PF) that can be connected on the feeder while keeping the existing PV systems at their locations in the feeder.
iii) *Hosting capacity with new smart inverters but no retrofit*: This corresponds to the increase in remaining hosting capacity when new PV systems with SI functions are connected on the feeder, without retrofitting the existing PV systems with SI functions.

   In this context, a term "smart inverter fraction ($Frac_{SI}$)" is defined as the ratio of the capacity of PV that has SI functionality to the total capacity of PV on the feeder [7],

$$Frac_{SI} = \frac{PV_{kVASI}}{PV_{kVAtotal}} \tag{7.1}$$

   where PV_{kVASI} is the kVA rating of PV systems with SI functionality and $PV_{kVAtotal}$ is the total kVA rating of all the PV systems connected on the feeder.
iv) *Advanced inverter impact (full-retrofit)*: This relates to the increase in remaining hosting capacity when all the PV system on the feeder are required to have SI functionality.

Different techniques are employed to increase the above hosting capacity, which are discussed in subsequent sections.

7.3   Increasing Hosting Capacity with Non Smart Inverter Techniques

This section provides a brief introduction to the techniques for increasing hosting capacity in distribution networks, that are not based on SIs. The objective here is to present the capabilities of these techniques so that they can be used in conjunction with SIs to provide enhanced benefits to utilities in increasing hosting capacity. Several international case studies of high penetration of PV solar systems and methods used to increase the hosting capacity are presented in [8]. Some of these main techniques for increasing hosting capacity are as follows [1, 9]:

i) Active power curtailment (APC)
ii) Change in orientation of PV panels
iii) Correlation between load and PV systems
iv) Demand-side management (DSM)
v) On load tap changer (OLTC) transformers, voltage regulators, and switched capacitors
vi) Application of decentralized energy storage systems

7.3.1 Active Power Curtailment (APC)

The concept of APC is that if the active power output of a PV power system is curtailed the amount of installed PV power can be correspondingly increased.

An autonomous curtailment strategy may involve a partial reduction in the power output from the PV systems when an overvoltage is experienced. This is compared with a strategy of complete shutdown of PV systems due to operation of overvoltage protection systems, which represents a 100% loss in power from the PV systems. The application of the above strategy is studied on a 400 kVA, 400 V/230 V typical Dutch low voltage residential feeder having a high penetration of PV systems [10]. It is demonstrated that the PV energy can be increased by 77.4% with implementation of the proposed curtailment strategy as compared to the application of overvoltage protection. Alternatively stated, a total of 550 kWh energy was saved by curtailment which is enough to cover the needs of a household for a month.

This strategy, however, results in different levels of power curtailment for PV systems at different locations, resulting in different monetary losses for each of them. To balance curtailment for all participating PV systems, a communication-based coordination of curtailment at different buses based on local voltages, is recommended.

Another strategy for increasing PV hosting capacity in radial low-voltage distribution lines based on droop-based APC technique in described in [11]. Droop control relating active power with frequency is an established method for power sharing among generators connected in parallel. In low voltage systems, the voltage is more strongly related to active power than reactive power, due to low X/R ratios [12]. Considering this feature, it is proposed that the power injected by the inverter be a function of the bus voltage.

Studies were conducted on a typical 75-kVA 14.4 kV–120/240 V, Canadian suburban residential feeder with several PV systems [11]. If APC is implemented on all the PV systems with same droop coefficients, the PV hosting capacity can be increased from 63% (without APC) to 100% of substation transformer capacity (with APC). However, the curtailment required from each PV inverter for overvoltage prevention was different. The downstream PV inverters were required to curtail more power than the upstream ones, affecting their revenues.

An alternative curtailment approach using different droop coefficients that results in approximately equal sharing of the curtailed active power among all inverters was proposed. However, this feature resulted in increased power losses as compared to the basic APC scheme. It is noted that while the APC strategies are effective in increasing hosting capacity, they cause different levels of curtailments in differently located PV systems, which may not be desirable.

It was recommended that the selection of either technique be based on: (i) need for sharing the curtailment costs equally among all households, or (ii) increasing the net PV power exported from the feeder.

7.3.2 Change in Orientation of PV Panels

The daily peak of irradiance typically occurs at midday if a PV power plant is installed horizontally or elevated southwards. If PV panels are mounted facing different directions, the peak power generation can be shifted to a different time. Since the hosting capacity is limited by the peak power, a shifting of the peak can increase the hosting capacity.

In one study, if the PV panels are oriented eastwards or westwards, the peak can be shifted into the morning or afternoon [9]. However, the tilt angle also plays an important role in determining the peak power production.

A different study has reported that west-facing PV systems were able to reduce the peak load at the substation by 60% of the nameplate rating of the PV systems, as compared to only 20% reduction provided by generally south-facing PV systems [6]. The west-facing PV systems also successfully shifted the feeder peak loading to a later time of the day [6].

7.3.3 Correlation between Load and PV Systems

If the feeder load is high during periods of high PV power generation, the net PV power generated will be lower. This will cause lesser voltage rise thereby allowing more PV systems to be connected on the same feeder.

7.3.4 Demand Side Management (DSM)

DSM provides similar end-result as APC. Part of the power output from PV plants can be utilized locally by appropriate switching or management of flexible local loads through DSM. This reduces the net PV generation causing a lower voltage rise, thus helping in the increase of hosting capacity.

7.3.5 On Load Tap Changer (OLTC) Transformers, Voltage Regulators, and Switched Capacitors

Since voltage rise is a key reason for limiting hosting capacity, lowering the feeder voltage through OLTC, voltage regulators or operating switched capacitors can increase the hosting capacity on the feeder. Utilities routinely utilize these devices for voltage control on their feeders. These devices can be effectively combined with other methods of voltage control (i.e. SIs) for increasing hosting capacity.

Several utility case studies are available where hosting capacity could be increased by modifying the tap settings of OLTC to lower values, and/or switching of capacitors [8, 13]. In long cable feeders, the voltage rise becomes a limiting factor for hosting capacity. In this case switched inductors are utilized to reduce the voltage.

7.3.6 Application of Decentralized Energy Storage Systems

Decentralized energy storage systems provide a similar effect as power curtailment on PV hosting capacity. A portion of excess PV power generation can be stored in local energy storage system, thus reducing the net power injected into the feeder. This allows more PV systems to be added, while not losing energy. The excess PV power can also be utilized in locally available EV charging systems producing a similar effect as energy storage. Lower PV power injection into the feeder results in reduced voltage rise, which helps increase the hosting capacity for more PV systems.

7.3.7 Energy Storage Requirements for Achieving 50% Solar PV Energy Penetration in California

Energy storage can help achieve significant increases in hosting capacity of PV systems on a large scale. The energy storage needs to increase the PV penetration in California up to 50% (with renewable penetration over 66%) are analyzed in a report published in 2016 [14]. The studies are performed using National Renewable Energy Laboratory's Renewable Energy Flexibility (REFlex) model, a reduced-form dispatch model that calculates the supply/demand balance of an electricity system. Some key findings of this study are summarized in this section [1].

A major limiting factor in deploying PV systems is the PV energy that needs to be curtailed in order to maintain the supply-demand balance. Energy storage is, therefore, a major enabling technology which helps in reducing curtailment to acceptable levels and increase PV penetration. The energy storage requirement is also related to the cost of PV systems. Levelized cost of electricity (LCOE) is considered to be the average price that the generating asset must receive in a market to break even over its lifetime. In the context of PV systems, the net LCOE is defined as the cost of PV energy that can be used by the grid after considering curtailment and energy storage losses. If the prices of PV systems can get lowered significantly, even with substantial curtailments an LCOE of 3 cents/kWh may be achieved by 2030. However, a reasonable expectation of LCOE is 7 cents/kWh. This cost is comparable to projected variable costs of combined-cycle gas generators in California. Another mechanism of reducing curtailment is through charging of EVs during hours of high PV power production.

The energy storage needs for California to achieve different levels of PV penetration for various system flexibility cases are depicted in Figure 7.2 [14]. The energy storage systems are assumed to have eight-hours discharge capacity with round-trip efficiency of 80%. The top bar corresponds to the high flexibility case with low-cost PV (@ 3 cents/kWh) with 25% of California light-duty vehicle fleet being EVs, i.e. equivalent to about 6.4 million vehicles. This study showed that the expected deployment (i.e. 4 GW) of energy storage systems up to 2020 will be sufficient to support 40% PV penetration. An additional 15 GW of energy storage will be needed to increase the PV penetration from 40 to 50%.

Energy storage requirements are then computed for cases when the aggressive flexibility assumptions are lowered. The second bar represents reduced EV penetration from 25% to 5% (or reaching a total EV fleet of 1.3 million vehicles in California by 2030). The third bar represents a case with the base PV LCOE increased to 5 cents/kWh. This assumes that the reductions in PV costs may not be as much as expected by 2020. The last bar depicts a combination of both the above-reduced flexibility scenarios. This demonstrates that if California can substantially increase grid operational flexibility but not achieve either large-scale deployment of EVs, or a reasonably reduced PV cost, still about 10 GW of new energy storage systems would be required to achieve 40% PV penetration. This will increase to about 28 GW if a PV penetration of 50% needs to be achieved. Since the costs of energy storage systems are declining such deployment of storage systems are quite accomplishable.

7.3.8 Comparative Evaluation of Different Techniques for Increasing Hosting Capacity

A comparative evaluation of seven techniques for increasing hosting capacity is presented in [1, 9]. These include (i) correlation between load and PV systems, (ii) use of reactive power control (RPC), (iii) APC, (iv) orientation change of PV power systems, (v) application of decentralized storage, (vi) DSM, and (vii) on-load tap-changing transformers. While these techniques are based on certain conditions and assumptions which may be different in different utility feeders, they offer valuable insight into the benefits of each technique.

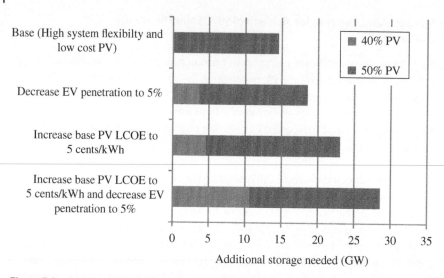

Figure 7.2 Additional energy storage needed to achieve a marginal PV net LCOE of 7 cents/kWh for the high flexibility case and three reduced flexibility cases. *Source*: Reprinted with permission from the National Renewable Energy Laboratory [14].

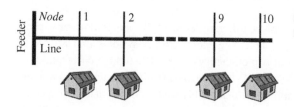

Figure 7.3 Study system. *Source*: Reprinted with permission from EU PVSEC [9].

A radial 10-node feeder with varying number of houses and varying sizes of PV systems is considered as the study system which is shown in Figure 7.3 [9]. The PV systems and loads are equally distributed at each bus. The feeder impedance is chosen to provide a 6% voltage drop during no generation from PV plants for all study cases. Realistic household loading and solar irradiance data are utilized in the studies.

7.3.8.1 Active Power Curtailment

The usefulness of APC depends on amount of energy curtailed. If the output of PV systems is reduced the number of PV systems installed in the system can be correspondingly increased. This is, however, associated with loss of energy and corresponding loss of revenues to the PV systems. Figure 7.4 [9] depicts the increase in PV hosting capacity as a function of the annual energy lost. It is observed that for a loss of 5% in the annual energy yield, an increase in PV hosting capacity by about 50% can be achieved. The energy loss will be reduced if: (i) large number of households are connected (i.e. energy can be consumed by more loads), or (ii) energy storage systems are used at customer locations.

7.3.8.2 Different PV System Orientations

Figure 7.5 [9] shows the additional PV hosting capacity depending on the tilt angle of a PV power system. It clearly shows the benefit of east-west-oriented systems, but only if they have a large tilt angle. The peak of 30% additional PV hosting capacity is reached at a tilt angle of 70°.

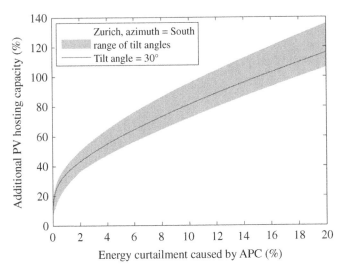

Figure 7.4 Additional PV hosting capacity using APC. *Source*: Reprinted with permission from EU PVSEC [9].

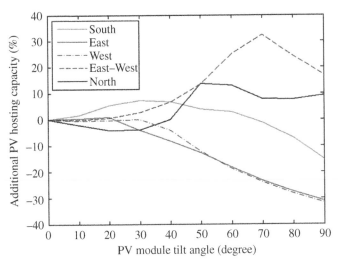

Figure 7.5 Additional PV hosting capacity using different module orientations. *Source*: Reprinted with permission from EU PVSEC [9].

7.3.8.3 Correlation with Load

Realistic load patterns with a high temporal resolution are utilized instead of selecting all loads to be zero, for evaluating the hosting capacity of distributed generators (DGs). The "Technical Rules for the Assessment of Network Disturbances" published by the associations of electric utilities of Germany, Austria, Switzerland, and the Czech Republic ("DACHCZ") require calculations of voltage rise be done for all loads at zero [15]. However, if the correlation with load is considered, this can result in 6–47% increase in PV hosting capacity, as demonstrated in Figure 7.6 [9]. This technique is found to be more effective if there are a large number of household loads.

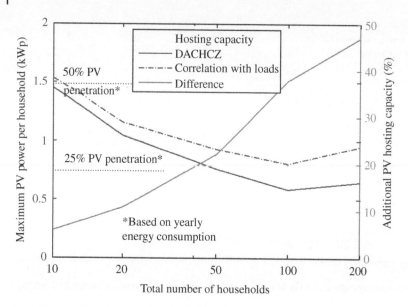

Figure 7.6 PV hosting capacity of the 10 node test feeder in accordance with DACHCZ (solid dark line), and respecting the correlation between PV and load (dashed line). *Source*: Reprinted with permission from EU PVSEC [9].

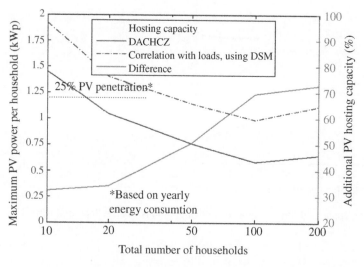

Figure 7.7 Additional PV hosting capacity using DSM. *Source*: Reprinted with permission from EU PVSEC [9].

7.3.8.4 Demand Side Management (DSM) Approach

In this case study, DSM can help increasing PV hosting capacity by 30–70%. It should be utilized in combination with a smart APC algorithm. If no load is available to switch on a day with high solar irradiance, the PV plant should curtail its power output accordingly. The enhancement of PV hosting capacity by DSM is demonstrated in Figure 7.7 [9].

7.3.8.5 On Load Tap Changer Transformer (OLTC)

An OLTC transformer can typically be switched in several steps and change the voltage in a range of ±5%. If PV hosting capacity is limited only by voltage constraints, the hosting capacity can be increased substantially through OLTC based voltage control. It is shown that reducing the LV grid voltage by 5% given a voltage tolerance of 3% (per DACHCZ) increases the tolerance to a total of 8%, which corresponds to an additional PV hosting capacity of 167% [9].

7.3.8.6 Storage

The required rating of energy storage for providing enhanced hosting capacity is depicted in Figure 7.8 [9]. This relationship is almost linear up to four hours of storage, beyond which it declines. This is because the stored energy of four hours can be discharged into the grid during nighttime, even between two consecutive sunny days. Higher stored energy will take days to discharge.

7.3.8.7 Reactive Power Control (RPC)

The effectiveness of RPC method depends on R/X ratio of the distribution line. Smaller R/X increases the effectiveness of RPC. Figure 7.9 [9] shows the impact of RPC at different PFs in increasing the PV hosting capacity. It is noted that RPC can be very effective in case of long and thin cables, but not as much in short cables.

7.3.9 Summary

Several techniques for increasing PV systems hosting capacity of distribution systems are presented in this section. These techniques can provide even higher hosting capacities if appropriately coordinated with SIs.

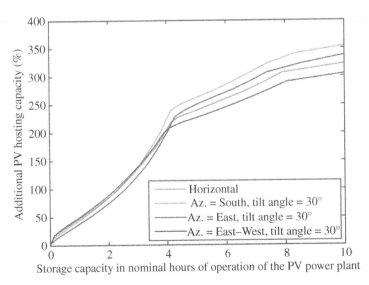

Figure 7.8 Additional PV hosting capacity using distributed storage systems. *Source*: Reprinted with permission from EU PVSEC [9].

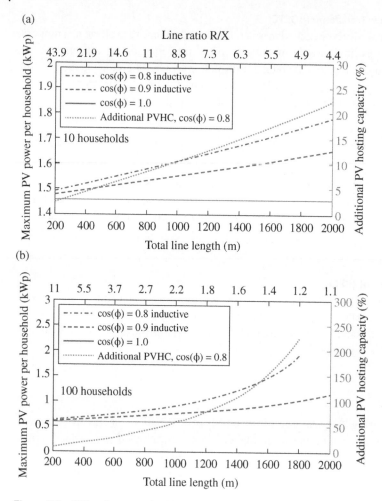

Figure 7.9 PV hosting capacity (absolute and additional) using RPC. *Source*: Reprinted with permission from EU PVSEC [9].

7.4 Characteristics of Different Smart Inverter Functions

This section presents the characteristics of different SI functions and their role in increasing the hosting capacity of distribution feeders [6, 16, 17, 18].

7.4.1 Constant Power Factor

The PV system with this SI function modulates its reactive power output to maintain the specified power factor (PF) at all levels of active power production.

7.4.1.1 Advantages

This is a simple function to configure and implement. It is effective for:

i) Mitigation of voltage issues (typically overvoltages)
ii) PF control in the feeder

Both the above objectives can be pursued either independently or in an inter-related manner.

This reactive power-based function is more effective in feeders with high X/R ratios. It is commonly used for alleviation of voltage issues.

7.4.1.2 Potential Issues

This function may exchange reactive power with the feeder even when it may not be needed. For instance, an inductive PF setting will result in large reactive power absorption even during high system loading conditions when the voltage is already low, thereby potentially causing an under-voltage situation. Furthermore, a capacitive (injection of reactive power) PF setting will result in a large reactive power injection even during light loading conditions when the voltage is generally high, likely causing an overvoltage condition.

This function causes a continuous flow of reactive power (absorption or injection) in the feeder. In case of an inductive PF setting, this function may result in an increased flow of reactive power at the feeder head or at the substation resulting in a reduced PF. This reactive power drawn by the smart PV inverters needs to be supplied from the grid or from some other means such as switched capacitors. Moreover, an increased reactive power flow causes a higher current flow in the feeder thereby increasing line losses and heating in series-connected distribution equipment.

Coordination of the constant PF function with other existing voltage regulating equipment may also present challenges.

In the normal "watt-precedence/priority" mode, at high levels of PV active power production, there may not be enough inverter capacity left for exchanging the required amount of reactive power. This requires APC and operation of the PV inverter in "var precedence/priority" mode. Even though curtailment may involve loss of usable active power or revenue, the var priority mode provides improved voltage control. The comparative performance of different operating modes of a PV inverter from a simulation study is illustrated in Figure 7.10 [18]

Normal unity PF mode of operation of a PV inverter results in an unacceptable voltage rise to 1.09 pu around noon time. If the inverter is operated in the watt precedence mode with a constant PF setting of 0.9 inductive, the voltage does not rise due to reactive power absorption by the PV inverter. Around noontime, the entire inverter capacity is used up for producing active power, and no inverter capacity is left for providing reactive power absorption. Hence the voltage rises to 1.09 pu as with unity PF mode of operation. However, in the var precedence mode, the PV inverter can operate at 0.9 PF even at noontime, thereby ensuring that the voltage does not rise beyond 1.045 pu.

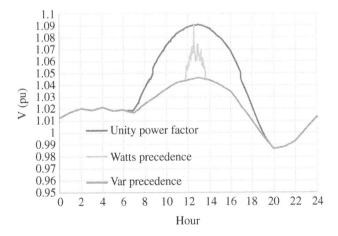

Figure 7.10 Voltage comparison with different PV inverter controls. *Source*: Reprinted with permission from EPRI [17].

7.4.2 Volt–Var Function

This function enables a PV inverter to absorb or inject reactive power during overvoltage or under-voltage conditions, respectively. It also does not exchange reactive power when the voltage is normal. In other words, this function exchanges reactive power with the feeder only when it is required, not continuously as in constant PF function.

The volt–var function can be implemented with either watt (active power) priority or var (reactive power) priority.

7.4.2.1 Advantages

This function provides the following advantages:

i) The volt–var function (with deadband) exchanges reactive power with the feeder only when it is required, not continuously as in constant PF function.
ii) The amount of reactive power exchanged is proportional to the magnitude of voltage deviation from its nominal value.

For the above reasons, the volt–var function is widely used for voltage control especially in feeders with high penetration of PV systems. It can also be indirectly utilized for PF control in the feeder.

This reactive power-based function is more effective in feeders with high X/R ratios.

7.4.2.2 Potential Issues

During high power production, there may not be enough capacity left in the inverter for reactive power exchange in the "watt priority" mode of operation. This may require operating the PV inverter in the "var priority" mode in which active power is curtailed to release enough capacity for the required reactive power exchange.

Default volt–var settings (as specified in IEEE 1547–2018 [19], CA Rule 21 [20], HI Rule 14 [21]) are generally not aggressive and in some cases may not take full advantage of the entire reactive power capability of DERs. The effectiveness of default volt–var settings decreases in systems with lower X/R ratios. To make an appreciable impact on voltage during periods of rated output of PV systems, the needed reactive power may become more than the 44% capability specified in IEEE 1547–2018 [19, 22, 23].

If the settings of the volt–var function are not selected appropriately, an adverse impact may occur on the operation of conventional voltage regulating equipment, and feeder PF. Moreover, undesirable oscillations may also potentially occur between adjacent smart PV inverters.

7.4.3 Volt–Watt Control

A smart PV inverter operating with volt–watt control function curtails its active power output to reduce overvoltage at its point of common coupling (PCC).

7.4.3.1 Advantages

This function offers the following advantages:

i) The volt–watt function provides effective voltage control, especially if the PV penetration level is high.

ii) The volt–watt function may be used in combination with other SI functions for voltage control such as constant PF or volt–var function. This will prevent any curtailment of active power during reactive power exchange with the distribution system.

iii) It keeps the DER online even during high voltages caused by temporary circuit reconfigurations.

This active power-based function is more effective in feeders with low X/R ratios.

7.4.3.2 Potential Issues

This function curtails active power generation, which may also lead to a loss of revenue for the PV solar system.

7.4.4 Active Power Limit

This SI function restricts the maximum active power output of a PV inverter to a specified value.

7.4.4.1 Advantages

This function helps to:

i) Reduce overvoltages in distribution feeders
ii) Limit the power output from PV systems during periods of overgeneration in the power system
iii) Prevent reverse power flow
iv) Avoid overloading of distribution lines

This active power-based function is also more effective in feeders with low X/R ratios.

7.4.4.2 Potential Issues

This function also leads to a loss of active power generation and potential revenues.

This section has presented the benefits and associated challenges for different SI functions. An understanding of these unique characteristics of each SI function will help in selecting the appropriate function for increasing hosting capacity in a given distribution feeder.

7.5 Factors Affecting Hosting Capacity of Distribution Feeders

Hosting capacity of a distribution system for accommodating DERs is dependent on the following factors:

i) Size and location of DER
ii) Physical characteristics of the distribution system
iii) DER technology

7.5.1 Size and Location of DER

The impact of PV active power injection on PCC voltage is dependent upon the system strength and X/R ratio of the distribution feeder as viewed from the PCC [12]. Therefore, a PV system connected close to the substation end of the distribution feeder will have a different voltage impact than a same size PV system connected downstream at the far end of the distribution line. Also, the voltage impact of customer-owned PV systems widely dispersed along a distribution feeder will be different from an equivalent size centralized utility-class PV system at a specific location on the feeder.

7.5.2 Physical Characteristics of Distribution System

A distribution feeder is characterized by the voltage class of the overhead line, type of conductor, layout of conductors, feeder topology, length of feeder, etc. Similar considerations also apply for underground cables. In addition, number of existing voltage control equipment such voltage regulators on the line, switched capacitors, etc. will also impact the hosting capacity of a distribution line.

The location of loads on the distribution feeder and variability of loading levels are other important factors to be considered in determining the hosting capacity of the feeder. The voltage impact will be different from loads that are coincident with PV generation than those which are not [24].

7.5.3 DER Technology

The voltage influence of variable generating sources such PV power systems based on inverters is different from that of relatively constant power-producing generators based on rotating machines. The most important factor here is the ability of such inverter-based generating systems to provide dynamic control of active and reactive power which can obviate the adverse impacts on feeder voltage.

It is, therefore, evident that hosting capacity for PV power systems will be substantially different when they are equipped with SI functions as described in Chapter 2 [25, 26], than if they are not. Furthermore, different SI functions may help achieve higher and different hosting capacity limits.

7.5.4 PV Hosting Capacity Estimation

A sensitivity analysis of PV hosting capacity with respect to distribution feeder characteristics and PV systems (without and with SI functions) presented in [1, 27] is described below.

7.5.4.1 Impact of Feeder Characteristics

PV hosting capacities for seventeen distribution feeders from five different utilities are computed using a stochastic analysis approach. All feeders are modeled from the substation transformer to individual customers, using open-source distribution system simulator (OpenDSS) software.

Different characteristics of these feeders including voltage class, peak load power, total number of customers, feeder length, maximum impedance for all primary buses, weighted average magnitude for all primary buses, maximum load, total number of capacitors, total capacitor kvar, and total number of voltage regulators are depicted in Figure 7.11 [27]. This figure displays the wide diversity of characteristics of different feeders.

Figure 7.12 [27] shows PV hosting capacity for the 17 feeders, expressed as a percentage of peak load. These values range approximately from 0.2% to 70%. The hosting capacity values are correlated with different characteristics of the feeders as shown in Figure 7.13 [27]. This figure also illustrates the second-order polynomial which best fits the correlation in each subplot. The average error between the fitting curve and actual plots is indicated on top right. A monotonically decreasing fitting curve indicates a decline of hosting capacity with respect to the concerned variable.

It is concluded that higher PV hosting capacity is obtained for feeders with lower impedance (strong feeders), shorter length, less loads, and less capacitive reactive power (kvar) support. Higher

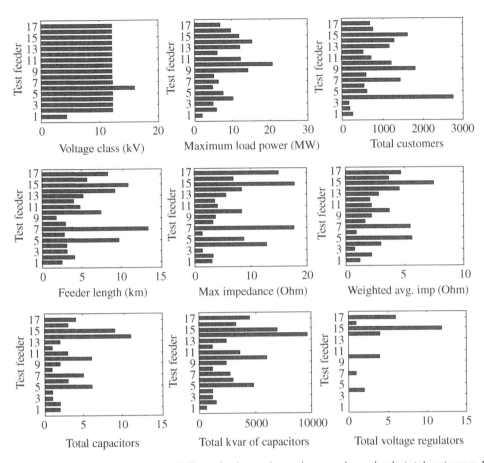

Figure 7.11 Key characteristics of 17 test feeders: voltage class, maximum loads, total customers, feeder length, maximum impedance, weighted average impedance, total capacitors, total capacitor kvar, total voltage regulators. *Source*: Ding and Mather [27].

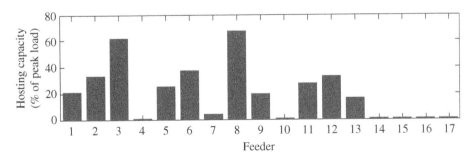

Figure 7.12 PV hosting capacity results of 17 test feeders. *Source*: Ding and Mather [27].

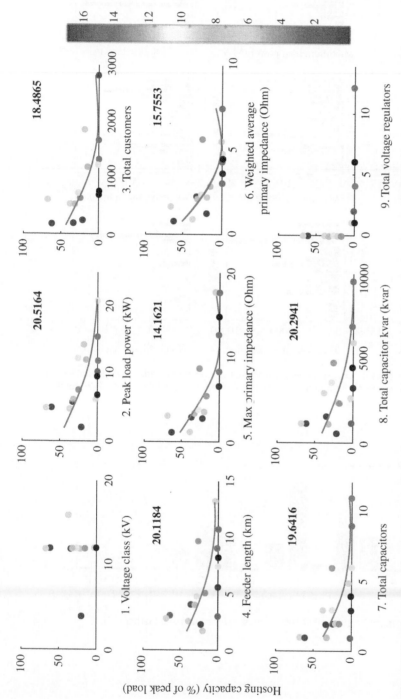

Figure 7.13 Correlation between PV hosting capacity and feeder characteristic: each shaded dot represents a feeder; x-axis is the studied feeder characteristic and y-axis is the PV hosting capacity; the solid line is the second-order polynomial fitting curve; the top right number at the subplot is the average error between the fitting curve and actual values. *Source:* Ding and Mather [27].

voltage class feeders typically have larger PV hosting capacity. Moreover, feeders with voltage regulators tend to have lower hosting capacity than those without voltage regulators.

7.5.4.2 Impact of Smart Inverter Functions

The effect of constant PF and volt–var SI functions on PV hosting capacity is then investigated [27]. These SI functions provide voltage control and thereby help increase hosting capacity. In this study, all PV inverters are modified to operate at 0.98, 0.95, or 0.9 lagging PFs. They are further equipped with the volt-var control curve, shown in Figure 7.14 [27]. The available Volt–Amp Reactives (vars) are computed from the inverter kVA capacity S and the present kW output P as Q (= $\sqrt{(S^2 - P^2)}$). This volt-var curve does not have a deadband.

The PV hosting capacity of each test feeder is recomputed and compared with the existing hosting capacity as shown in Figure 7.15 [27]. It is noted that the PV hosting capacities of feeders 2, 3, 6, and 8 are unchanged as they were already at their maximum. PV hosting capacities of feeders 1, 4, 5, 7, 9, 11, 12, 13, 14, and 16 increase significantly when the PV system operate at off-unity PF, as compared to unity PF operation. Since no voltage violation occurs for all PV scenarios, the hosting capacity of feeders 5, 9, 11, 12, 13, 16 stays constant for all three PFs. However, when voltage violations do

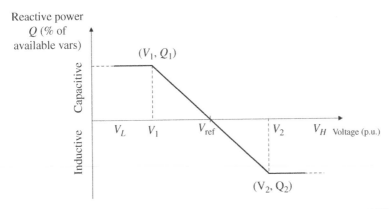

Figure 7.14 Volt-var function; V_1 = 0.95 pu, Q_1 = 100%, V_2 = 1.05 pu, Q_2 = 100%, V_{ref} = 1.0 pu.

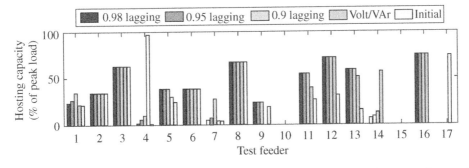

Figure 7.15 PV hosting capacity results of 17 test feeders after applying various smart inverter functions: 0.98, 0.95, 0.9 lagging power factors, volt-var control, compared with the initial result. *Source*: Ding and Mather [27].

occur, the hosting capacities of feeders 1, 4, 7, and 14 increase gradually with decreasing lagging PF. Volt-var control significantly increases PV hosting capacity of feeders 1, 4, 5, 7, 11, 12, 13, 14, 16, and 17.

It is concluded that SI functions can generally increase PV hosting capacity by 1.5–3.5 times. The lowest lagging PF provides the highest PV hosting capacity among different PFs. This is because it absorbs the highest amount of reactive power thus reducing voltage most effectively. Volt-var control increases PV hosting capacity substantially. Depending on feeder characteristics, either volt-var function or constant PF function can be more effective.

In the absence of a voltage regulator, PV hosting capacity can be increased by locating PV systems close to substation and utilizing SI functions. However, the presence of voltage regulators can either have both a positive or negative impact on hosting capacity.

This study provides a broad understanding of the impact of feeder characteristics, PV system locations, and SI functions on hosting capacity. It is, however, emphasized that the specific impacts are system dependent and must be evaluated for each study system.

7.6 Determination of Settings of Constant Power Factor Function

SI functions are very helpful in controlling distribution feeder voltages. However, it is important to select the appropriate settings of the SI function being employed to provide effective voltage control for different power generation, network conditions, and loading levels.

This section describes three techniques for determining default "constant (off-unity) PF settings for smart PV inverters [1, 18].

7.6.1 Single DER System

In case of a single PV-based DER connected to a grid, the change in voltage ΔV caused by injection of active and reactive power can be expressed as

$$\Delta V \cong \frac{R*P + X*Q}{V} \ (pu) \tag{7.2}$$

where V = base voltage; X, R = equivalent reactance and resistance at the point of interconnection (POI) primary side; P, Q = DER active and reactive power.

To mitigate the voltage deviation, $\Delta V = 0$, i.e.

$$\frac{X}{R} \cong \frac{-P}{Q} \tag{7.3}$$

For this condition, the required PF is given by

$$\text{Power factor} \cong \frac{\frac{X}{R}}{\sqrt{\left(\frac{X}{R}\right)^2 + 1}} \tag{7.4}$$

It is noted from above that a feeder with lower X/R ratio requires a lower PF, i.e. higher amount of reactive power to mitigate voltage rise. For instance, if the bus where DER is located has an X/R of 1.833, a DER PF setting of 0.9 absorbing will be needed to obviate the voltage rise. Furthermore, DERs located at buses having low X/R requiring PF settings less than 0.9 are generally not effective in mitigating voltage.

It is known that the short circuit impedance increases and X/R ratio decreases as the distance increases from the substation (feeder head) toward the feeder end. Hence PF setting should be high for DERs located near substations and low for those located further downstream.

7.6.2 Multiple DERs

The procedures for determining default PF settings in feeders having multiple DERs are described in this section [18]. The PF settings can be calculated from three methods as below:

7.6.2.1 Median Feeder X/R Ratio
A median X/R for the given distribution feeder is calculated and the default PF settings for all DERs in the feeder are selected as

$$\text{Power Factor} \cong \frac{\left(\frac{X}{R}\right)_{\text{median}}}{\sqrt{\left\{\left(\frac{X}{R}\right)_{\text{median}}\right\}^2 + 1}} \tag{7.5}$$

If the calculated PF setting is less than 0.9, the PF is set to 0.9. At buses close to substation where X/R is high and where voltage deviation is less than 1%, the default PF is set to unity.

This method provides a single PF setting for all DERs in a given feeder.

7.6.2.2 Weighted Average X/R Ratio
In this method an average X/R weighted by the DER size is calculated as below:

$$\left(\frac{X}{R}\right)_{\text{Weighted average}} = \frac{\sum\limits_{i=1}^{N} S_i \left(\frac{X}{R}\right)_i}{\sum\limits_{i=1}^{N} S_i} \tag{7.6}$$

where S_i = MVA rating of ith DER; $(X/R)_i = X/R$ at the ith DER POI.

The default PF setting for all the DERs is then calculated by substituting weighted average X/R in Eq. (7.4) for single DER.

If the calculated PF setting is less than 0.9, the PF is set to 0.9. At buses close to substation where X/R is high and where voltage deviation is less than 1%, the default PF is set to unity.

This method provides a single PF setting for all DERs in a given feeder but incorporates the effect of size and location of individual DER in the feeder.

7.6.2.3 Sensitivity Analysis Based Technique
This method provides customized PF setting for each DER in a given feeder. In this method, the aggregated voltage deviation at any ith bus in an N-bus distribution system is calculated as:

$$\Delta V_i = \sum_{j=1}^{N} \left(SP_{ij}P_j + SQ_{ij}Q_j\right) \tag{7.7}$$

where ΔV_i = aggregated voltage deviation at ith bus; P_j = active power injected at jth bus, $j = 1, 2, 3, ..., N$; Q_j = reactive power absorbed at jth bus, $j = 1, 2, 3, ..., N$; SP_{ij} = sensitivity of voltage at ith bus due to active power injection at jth bus, $j = 1, 2, 3, ..., N$; SQ_{ij} = sensitivity of voltage at ith bus due to reactive power absorbed at jth bus, $j = 1, 2, 3, ..., N$.

These sensitivities are obtained from load flow studies.

The voltage deviation at each bus is minimized as follows:

$$\text{Minimize} \sum_{i=1}^{N} \Delta V_i^2 \tag{7.8}$$

Subject to constraint:

$$\left| \frac{Q_i}{P_i} \right| \leq 0.4843, i = 1, 2, 3, ..., N \tag{7.9}$$

This constraint is another way of expressing the constraint $PF_i \in [1, \pm 0.9]$

The optimization provides the reactive power Q_i at each bus and thereby the ratio (Q_i/P_i). The PF setting at the ith DER is then determined from Eqs. (7.3) and (7.4).

7.6.2.4 Performance Comparison of the Three Methods

Studies performed on seven study feeders have shown the following characteristics of each technique:

Median Feeder X/R Ratio

This technique is based on simple calculations and provides limited improvement based on actual DER site. In systems with high DER penetration, the single PF setting may cause over-mitigation of voltage rise and even cause voltage drop.

Weighted Average X/R Ratio

This technique does not require complex calculations and is found to be effective in feeders with single DER. It provides limited improvement in feeders with multiple DERs but more than the previous method. Some over-mitigation or under-mitigation of voltage may occur.

Sensitivity Analysis Based Technique

This is the most effective among all techniques in feeders both with single or multiple DERs, although it requires complex computations.

In general, DER locations having high X/R ratio are most responsive to reactive power, whereas locations having low X/R ratios may need to use alternative methods (such as volt–watt, etc.). Constant PF settings based on X/R ratios are beneficial but generally require high amount of reactive power. Also, such calculated settings are less effective in weak systems having low short circuit levels [23, 22].

7.7 Impact of DER Interconnection Transformer

DERs provide SI functions at their terminal, i.e. at the point of connection (PoC) [19]. However, the impact of the SI function needs to be implemented at the PCC on the distribution feeder. The PoC is typically interfaced with the PCC through the interconnection transformer. Therefore, the impact of the SI function at PoC needs to be adjusted to reflect its behavior at the PCC [1, 17].

Let X and R be the network equivalent resistance and reactance, respectively, as seen from the PCC of the DER. These parameters are obtained from short circuit studies. Also, let R_T and X_T be the resistance and reactance, respectively, of the interconnection transformer.

The adjusted X/R at the PoC of the DER is then expressed as [17]:

$$\left(\frac{X}{R}\right)_{\text{adjusted}} = \left(\frac{X}{R}\right) + \left[\left(R_T + \left(\frac{X}{R}\right)X_T\right]\sqrt{1 + \left(\frac{X}{R}\right)^2} \tag{7.10}$$

The effective PF of the PV site with the adjusted X/R is given by Eq. (7.4) as

$$\text{Power factor} \cong \frac{\left(\frac{X}{R}\right)_{\text{adjusted}}}{\sqrt{\left\{\left(\frac{X}{R}\right)_{\text{adjusted}}\right\}^2 + 1}} \tag{7.11}$$

In a similar manner, the volt–var SI settings seen at the PCC need to be modified to reflect the volt–var settings at the PoC due to the impact of the interconnection transformer. The new set of volt–var points are obtained as [17]:

$$V_{\text{new}} = V + \frac{P_{\text{gen}}}{V} * R_T + \frac{Q_{\text{gen}}}{V} * X_T \tag{7.12}$$

where P_{gen} and Q_{gen} are the active and reactive power generation of the DER.

7.8 Determination of Smart Inverter Function Settings from Quasi-Static Time-Series (QSTS) Analysis

It is important to select the appropriate parameters and settings of the SI function being employed, to accomplish the desired voltage control in a distribution feeder [28]. To understand the need for determining the most appropriate settings, consider the example of a PV solar system DER operating with volt–var function. Figure 7.16 [29] depicts the feeder voltage as impacted by a single DER with a multitude of volt–var settings.

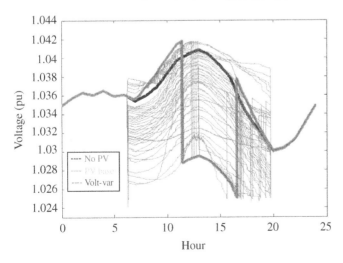

Figure 7.16 Voltage responses obtained with different volt–var settings. *Source*: Reprinted with permission from CIGRE [29] © 2014.

Some settings are seen to improve the voltage over the unity PF operation of DER while some others worsen the voltage. Evidently, it is difficult to determine the best volt–var setting from this figure. A detailed procedure is, therefore, required to determine the best setting of this volt–var function.

Quasi-static time-series (QSTS) analysis-based methodologies are generally utilized for performing hosting capacity analysis and determining optimal settings for SIs connected on secondary systems [28, 30–32]. QSTS analysis performs a series of sequential steady state power flow solutions where the converged state of each iteration is used as the starting state of the next. This captures the impact of time-varying parameters such as load, and time-dependent discrete controls in the system such as capacitor switchings, regulator tap positions, etc. [32].

It is noted that QSTS simulations use simple models of inverter controls and generally do not have the capability of representing system dynamic behavior and transient conditions [33]. While QSTS simulations are effective in determining SI settings as described in this chapter, they are not suitable for examining controller interactions among SIs described in the next chapter.

These methodologies can be implemented on different softwares, e.g. OpenDSS software developed by EPRI. It is noted that the optimal settings are determined to meet specified objectives and may be different for different objectives in the same distribution system.

This methodology comprises the following four major steps, which are graphically depicted in Figure 7.17 [1, 30].

Step 1: Development of detailed feeder model
Step 2: Simulation of quasi-static time-series model
Step 3: Analysis of results
Step 4: Selection of the appropriate setting

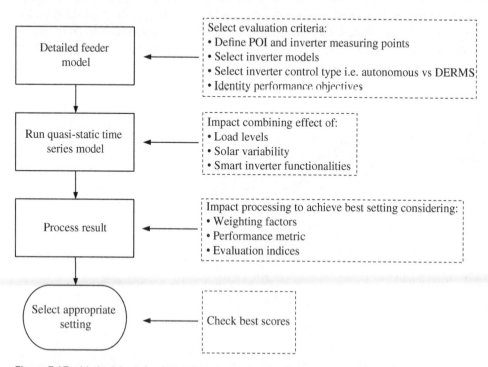

Figure 7.17 Methodology for determining optimal smart inverter controller parameters.
Source: Bello et al. [30].

7.8.1 Development of Detailed Feeder Model

This step includes modeling of the following aspects:

7.8.1.1 Distribution System

The distribution system model is derived from Geographic Information System (GIS) data provided by the utility and subsequently implemented in the OpenDSS software. This dynamic model is generally more detailed than the models used by utilities for system planning. The model includes distribution transformers with their tap changers, capacitor banks, voltage regulators together with their control systems, etc. The secondary service drops and individual customer loads are represented in the model. Validation of the model is performed at different levels to ensure accuracy. The above steady state validation is further extended to a time series validation.

7.8.1.2 Conventional Voltage Regulation Equipment

The feeder model should incorporate detailed representation of all control modes, e.g.

- Switched capacitors with their setpoints, delays, etc.
- OLTC transformers with their controls and delays; tap positions from the nominal and the bandwidth of each tap position
- Line drop compensation
- Voltage regulators

7.8.1.3 PV Systems with Smart Inverter Controls

A detailed range of PV solar conditions is modeled. The power outputs for all the PV systems over a year are represented.

Various SI functions can be modeled, which include the following:

- Constant PF control
- Volt–var control
- Volt–watt control

Each control can also be modeled with different characteristics. For instance, the constant PF function can be modeled with inductive PFs from 0.9 to 1.0 in increments of 0.01.

The volt–var function can be represented by the following cases: (i) no deadband, (ii) with deadband, or (iii) inductive operation only. Volt–var functions can be studied with different slopes and deadbands with different widths. Nonlinear characteristics beyond the endpoints of slopes may need to be incorporated, if needed. Different volt–var curves with deadband are depicted in Figure 7.18.

It needs to be clarified whether all PV systems have SIs with reactive power capabilities at all times – including periods when PV is not generating active power, i.e. nighttime or only daytime. It is understood that during daytime, the PV inverters have a minimum reactive power capability of 44% of inverter nominal rating as specified in IEEE Standard 1547-2018 [19].

This study involves consideration of only those SI functions that will have a direct impact on distribution system performance. SI functions that impact the bulk power system, such as frequency-watt, low/high voltage ride-through, low/high frequency ride-through are not included in this analysis.

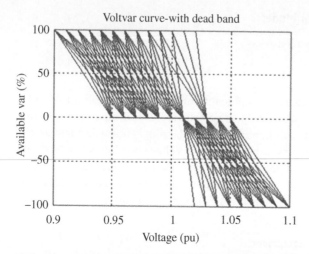

Figure 7.18 Volt–var control curves with deadband. *Source*: Bello et al. [30].

7.8.1.4 Type of Smart Inverter Control

Smart PV inverters can be controlled either autonomously or through an integrated Distributed Energy Resource Management System (DERMS) control. In the autonomous control mode, each PV inverter output is based on the primary node voltage at its POI. In the DERMS-based control, all the smart PV inverters connected to a distribution feeder are controlled and coordinated in an integrated manner.

7.8.1.5 Performance Criteria

The feeder performance with respect to SI settings is analyzed in terms of certain performance metrics and the ability to meet voltage constraints. The performance of each SI setting is usually evaluated in terms of impacts observed at [28, 30]:

- POI on the feeder
- Start of feeder
- End of feeder
- Overall feeder

The criteria/metrics for evaluating the performance of SIs are defined. The different criteria relating to impact on the considered distribution feeder, that may possibly be considered, are listed below [30, 31].

1) Minimum feeder head PF
2) Tap operations
3) Maximum and minimum tap positions
4) Capacitor switchings
5) Average PF at feeder head
6) Feeder losses (kWh)
7) Feeder consumption (MWh)
8) Max feeder-wide voltage (pu)
9) Time above ANSI max voltage (s)
10) Min feeder-wide voltage (pu)

11) Time below ANSI min voltage (s)
12) Difference between feeder-wide max and min voltages
13) Voltage drop across the feeder
14) Max feeder head voltage (pu)
15) Mean feeder head voltage (pu)
16) Min feeder head voltage (pu)
17) Max feeder end voltage (pu)
18) Mean feeder end voltage (pu)
19) Min feeder end voltage (pu)
20) Max PCC voltage (pu)
21) Mean PCC voltage (pu)
22) Min PCC voltage (pu)
23) Voltage variability index at inverter terminals

It is emphasized that not all criteria need to be used in a study: only those which are relevant to a utility in a given application may be utilized. Furthermore, weights are assigned to different criteria/metric depending on the objective of the study.

7.8.2 Simulation of Quasi-Static Time-Series Model

The performance of smart PV inverters is strongly dependent on solar characteristics and load levels. Different SI characteristics may be more appropriate for different days of solar irradiance, i.e. some characteristics that are suitable for clear sunny days may not be appropriate for cloudy days. Furthermore, characteristics that are suitable for heavy load conditions may not be suitable for light load conditions. It is, therefore, important to model all major solar categories and load conditions over all the 365 days in a year at the location of the PV inverters.

7.8.2.1 Characterization of Solar Conditions
The solar condition on any day is expressed in terms of two indices: (i) variability index and (ii) clearness index.

i) *Variability index (VI)*: It is defined as [30]:

$$\text{VI} = \frac{\sum_{k=2}^{n} \sqrt{(\text{GHI}_k - \text{GHI}_{k-1})^2 + \Delta t^2}}{\sum_{k=2}^{n} \sqrt{(\text{CSI}_k - \text{CSI}_{k-1})^2 + \Delta t^2}} \tag{7.13}$$

where global horizontal irradiance (GHI) and clear sky irradiance (CSI) are two sequences of irradiance values that are sampled at an incremental time interval Δt.

ii) *Clearness index*: It is defined as the ratio of actual solar energy measured on a given surface to the theoretical maximum energy on that same surface during a clear sky day.

Based on these indices, typically three types of days are identified for simulation studies:

i) Clear sunny day
ii) Highly variable day
iii) Overcast (cloudy) day

7.8.2.2 Characterization of Load Conditions

The time series load profile is developed based on the customer meter recording including billing details. Information on customer-owned PV systems is integrated in the above recordings. This load profile is validated from energy management systems of the utility. Simulations are conducted to obtain the voltage profile for the different loading scenarios including peak loads. These voltage profiles and expected short circuit currents at different locations in the distribution network are also correlated with values already obtained by the utility during planning studies to validate the load models.

Loads are broadly classified into two categories:

i) Peak/maximum loads
ii) Minimum loads

Therefore, the different solar and load categories result in six broad combinations:

i) Variable day – Peak load
ii) Variable day – Offpeak load
iii) Overcast day – Peak load
iv) Overcast day – Offpeak load
v) Clear day – Peak load
vi) Clear day – Offpeak load

The probability weightings of each of the above combinations expressed in number of days in a year are then calculated. This helps in evaluating the impact of relevant SI characteristics over different time horizons relating to the probability of those scenarios.

Any additional load or solar irradiance profiles may be helpful, if time and efforts permit.

7.8.2.3 Simulation Studies

A detailed time series analysis is performed where the distribution system is simulated in sequential time intervals and transferring the results of one simulation step to the next. This also helps in tracking controller interactions in the time frame of simulation time-step.

For each characteristic of a given SI function, simulations are run at 1 minute interval for 24 hours (i.e. 1440 times) per day for each of the six combinations of solar variability – load variability days as described in previous subsection, and the results are stored. Simulations are then repeated for the next characteristic of the same SI function till all characteristics are examined. Studies are then initiated for the next SI function.

Although simulation of 1 minute is generally used for study of SI function impacts, a simulation time step of less than 20 seconds is recommended for the analysis of load tap changer (LTC) and capacitor switchings [32]. In principle, the simulation time step in a QSTS simulation should be smaller than the fastest delay in any device with discrete controls on the feeder to ensure modeling accuracy of that device. The minimum requirements for QSTS simulation time step and the length of time for each type of QSTS analysis are described in [32]. It is recommended that in order to capture all distribution system analysis metrics together accurately, a time step of less than a five seconds and a time horizon of one year may be selected.

7.8.3 Analysis of Results

Each SI controller setting is ranked based on the performance criteria/metric selected by the utility for this study.

Weights are assigned for each performance criteria, as well as for each solar-load condition, as explained before. The weighting based on performance criteria relates to the impact of the specific SI function characteristic/setting on a distribution feeder. This weighting helps analyze the impacts on several feeders at the same time. Meanwhile, the weighting based on the solar-load condition brings an understanding of the time horizon of its impact.

7.8.4 Selection of Appropriate Setting

The best SI function setting is determined based on the weighted performance criteria and weighted solar-load condition. If there are several settings which meet the above criteria, the one that satisfies voltage constraints in the best possible manner is chosen.

As an example, Figure 7.19 illustrates the bus voltage response for a SI setting with one performance metric optimized (base case). Since this is seen to violate the voltage constraint, another less-optimal setting is selected which does not violate the voltage constraint [29].

It is noted that the best settings may vary depending on the selected/weighted performance criteria. For instance, the best SI controller setting from the perspective of reducing the voltage difference between maximum and minimum voltages across a feeder may be different from that which minimizes line losses, or that which reduces the time duration of voltages below the ANSI undervoltage limit. In one study [34], the best volt–var curve for optimizing substation (feeder-head) PF on a clear day with high loading has a voltage setpoint of 0.99 pu and a slope of 16% reactive power per volt. On the other hand, the best volt–var curve for minimizing line losses for a similar day has a setpoint of 1.02 pu and a slope of 26% reactive power per volt.

The size (MVA rating) of the PV inverter also has an important role in the determination of the optimal setting [30].

It is noted that the above procedure is fairly detailed and exhaustive. A simplified version may be developed by a utility along the above lines, to meet its specific needs [35].

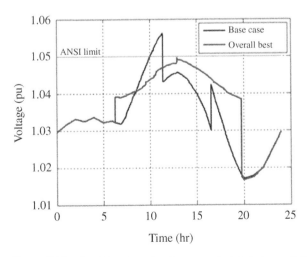

Figure 7.19 Consideration of voltage constraint during optimization of performance metrics. *Source*: Reprinted with permission from CIGRE [29] © 2014.

7.9 Guidelines for Selection of Smart Inverter Settings

SI settings may be selected by procedures that have different levels of complexity [1].

7.9.1 Autonomous Default Settings

Recommendations on SI settings are provided in [16]. These recommendations are based on experiences gained from different SI installations in smaller systems on distribution systems. An assumption is made that in such systems small signal stability and control interactions may not be an issue.

This report [16] recommends that the default settings for volt–var with reactive power priority as specified in IEEE Standard 1547-2018 [19] be adopted together with Category B for DER inverters. This is because utility experience has shown that volt–var function can control system voltages very effectively. This function reduces the need for changing the settings and hence delays the use of communications to modify the settings in future. It is effective in networks having both high or low penetrations of DERs. Moreover, volt–var function does not create any major challenges with existing conventional voltage regulation equipment.

The report recommends that DER developers and owners may consider oversizing their inverters to provide desired reactive power without affecting active power production [16]. This is anyway needed under the new IEEE Standard 1547–2018 to provide reactive power at a PF of 0.9 even at rated active power production.

Volt–watt function may also be enabled which starts to reduce/curtail power only if the voltage starts increasing beyond 1.06 pu. The volt–watt function provides an effective backstop (support) to volt–var in increasing hosting capacity. Actual implementation in Hawaii has shown that this function has led to curtailment is less than 1% of all installed DERs. Hence there is no real concern of losing power due to use of volt–watt functions.

The utilities should, however, address overvoltage issues beyond ANSI C84.1 limits through other means such as network reconfiguration or use of conventional voltage regulating equipment.

This report, however, clarifies that these settings are not universally applicable for all distribution systems. While the individual needs of different distribution networks may vary and corresponding system studies may be necessary, adoption of the default settings have largely worked for most distribution systems. The default settings in IEEE 1547-2018 will also cause less misunderstanding among manufacturers, developers, and utilities; and will make the interconnection process more harmonious.

These recommended SI settings are expected to be beneficial to manufacturers and utilities for the following reasons, as they will (i) harmonize the SI settings in various jurisdictions, as far as possible, and (ii) be autonomous; and take care of present and future system requirements without requiring costly communications to modify the settings.

7.9.2 Non-optimized Settings

Methods for determining the best settings of SIs for increasing hosting capacity based on utility operating experience are described in [17]. These methods vary in complexity and are described in Table 7.1 [17]. The complexity is expressed in terms of the data required and the computational resources available to the utility. The methods with low complexity need very little or no feeder data and use only spreadsheet tools. However, methods with higher degree of complexity need more detailed feeder data and commercial software tools for computations.

Table 7.1 Different methods of determining settings of smart inverter functions.

Level of complexity	Fixed power factor	Volt–var control
Low	Dependent on X/R ratio of distribution feeder	Generic setting
Medium	Dependent on model of feeder and location of PV system	Dependent on model of feeder and location of PV system
High	Dependent on model of feeder, location of PV system, and impedance of service transformer	Dependent on model of feeder, location of PV system, and impedance of service transformer

Source: Based on [17]. Used with permission from EPRI.

The approach adopted here is not to optimize the SI settings for specific day types and loading conditions, which will need to be updated regularly through advanced controls and communications.

Instead, an intermediate approach is utilized which improves the overall feeder performance without optimizing it. The SI settings are determined to perform satisfactorily over longer durations, i.e. either seasonally or annually. It is understood that in this method the PV sites will be operating at conditions that are different from those which were used to obtain the SI settings. Nevertheless, the inverters will still be operating within the original design conditions [17].

7.9.3 Optimized Settings

The maximal benefits from smart PV inverter functions in increasing hosting capacity are obtained from SI settings that are optimized for the specific feeder characteristics, size, and location of PV inverters; and solar-load conditions. Determination of these settings in a utility network can be a challenging task. Furthermore, these settings need to be regularly communicated to all the PV systems in the grid in an accurate manner. Streamlined methods for determining optimal settings which also do not cause adverse control interactions are still being researched.

However, the following high-level guidance is provided for selection of SI settings based on size of distribution systems [36]:

1) Small scale system with highly distributed PV solar systems: Constant PF function with absorbing PF if overvoltage issues are experienced.
2) Medium-scale system with secondary voltage POI and PV systems less than 500 kW: Volt–var function which may be set to default or customized if appropriate interconnection studies are performed.
3) Large scale system with primary voltage POI: both the type of SI function and their settings need to be determined from appropriate interconnection studies.

7.10 Determination of Sites for Implementing DERs with Smart Inverter Functions

It is important to determine the sites where implementation of DER SI functions will help in enhancing the hosting capacity of a feeder. The impacts of both active power and reactive power from a PV-based DER may be examined with respect to some criteria relevant for the utility feeder. If the hosting capacity is being considered with respect to voltage considerations, these criteria could include [1, 17]:

1) Primary overvoltage
2) Primary voltage deviation
3) Regulator voltage deviation
4) Primary undervoltage

A study describing the hosting capacity evaluation at specific PV sites for each of the above criteria for three scenarios is described in [17]:

Scenario 1: No PV based DER
Scenario 2: All DERs on-line; study site PV system operating with unity PF
Scenario 3: All DERs on-line; study site PV system operating with 0.95 PF (absorbing reactive power)

For each PV site, the impacts of active power and reactive power with respect to each criteria are evaluated as below:

$$\text{Active power impact} = \text{Hosting capacity for Scenario 1} - \text{Hosting capacity for Scenario 2}$$

$$\text{Reactive power impact} = \text{Hosting capacity for Scenario 3} - \text{Hosting capacity for Scenario 2}$$

The impact of DER active power injection is thus measured by the reduction in hosting capacity of a feeder from a case with no DER to a case with the specific DER operating at unity PF. Meanwhile, the impact of reactive power is quantified by the increase in hosting capacity from the case with DER operating at unity PF to the case with specific DER operating with reactive power-based SI functions.

A hosting capacity impact index is then computed for each study site PV system, as depicted in Table 7.2 [17] with some example values.

Table 7.2 Calculation of PV site impact factor.

"i"	Criterion "i"	Active power impact: hosting capacity reduction "a_i"(MW)	Reactive power impact: hosting capacity increase "b_i"(MW)	Minimum Impact Min (a_i, b_i) (MW)	Impact factor = Average of all minimum impacts
1	Primary overvoltage	1.2	1.8	1.2	0.7
2	Primary voltage deviation	0	0.2	0	
3	Regulator voltage deviation	0.8	1.2	0.8	
4	Primary undervoltage	0.8	1.0	0.8	

Source: Based on [17]. Used with permission from EPRI.

In a similar manner, impact factors for all the potential PV sites and with different SI functions are evaluated. The sites having high indices denote the locations of PV systems which will have a substantial impact on the hosting capacity of the feeder if reactive power-based SI functions are implemented there.

7.11 Mitigation Methods for Increasing Hosting Capacity

The impacts of DERs on the hosting capacity are system dependent. Their impact on one distribution system will be different from that on the other. Consequently, the methods to mitigate their adverse impacts will also be unique and will depend on [1, 37]:

- Size and location of the DER
- Impacted power system criteria
- Specific distribution system design and operating parameters
- Specific DER characteristics

Furthermore, a single mitigation option may not solve all power system criteria issues and hence a suite of mitigation options needs to be considered [35, 37]. A methodology for automating the analysis of different technology-based mitigation options with the objective of enhancing hosting capacity in distribution networks is described in [35, 37]. This methodology consists of the following steps:

Step 1: Assessment of Base Feeder

a) Get the model of feeder
b) Apply the targeted DER size at the location specified with SI options
c) Technical issues (i.e. thermal overloading, voltage) are checked
d) If there are no technical issues, done
e) If any technical issues are found, continue to the Integration Solution Assessment

Step 2: Integration Solution Assessment

a) Specify the suite of settings and options
    i) Volt–var
    ii) Volt–watt
    iii) Watt-power factor
    iv) Fixed PF

b) Run Assessment for each option

The settings of the SI functions are specified by the user and may vary anywhere between the default settings of different Standards to customized aggressive settings.

This technique may be integrated in the existing planning software of utilities and utilized to determine suitable mitigation options that can help enhance hosting capacity. It is further noted that a mitigation option can be viewed from different perspectives. It can be viewed in terms of its outcome, i.e. the overall improvement in hosting capacity. Alternatively, it can be evaluated financially, i.e. as the lowest cost per change in hosting capacity [35].

7.12 Increasing Hosting Capacity in Thermally Constrained Distribution Networks

In general, SIs can increase hosting capacity in distribution feeders but not in those feeders which are constrained by thermal limits [34]. However, if the magnitude of feeder current reaches the thermal limit of conductors or transformer, the utilities should check the PF of the current. It is quite possible that the PF may not be unity. If so, the volt–var curves of DERs installed on the feeder may be optimized to achieve "Average Power Factor at Feeder Head" as unity. If the PF is found to be unity when the thermal limits are violated for certain hours, APC may be considered through volt–watt function, during those hours. Management of active power output from PV systems can help alleviate time-dependent grid constraints and avoid expensive network upgrades [23].

If the PF is unity and thermal limits are violated only sporadically, the PV-STATCOM technology [38] may be considered for reducing the active power for those short durations and provide dynamic reactive power compensation for voltage control or PF correction to unity.

In other words, before any expensive network upgrades are contemplated, all cost-effective alternatives must be examined to operate the feeder at unity PF for as long as possible so that the hosting capacity of the feeder can be increased up to the thermal limit.

7.13 Utility Simulation Studies of Smart Inverters for Increasing Hosting Capacity

7.13.1 Voltage Control in a Distribution Feeder with High PV Penetration

A case study for demonstrating the benefit of volt–var SI function on the steady state voltages of a distribution feeder is reported in [3, 39]. This study is conducted on an actual 12 kV three-phase 15 miles long distribution feeder which has 1800 customers having a total load of 10 MW, as depicted in Figure 7.20 [3]. The study is conducted using Open DSS software developed by EPRI.

A high PV penetration scenario is created by selecting about 450 customers who have rooftop solar systems (shown by small circles in Figure 7.20). Each PV solar system is rated at 4 kW but has an overrated inverter of 4.8 kVA. The inverter is overrated to leave inverter capacity for providing reactive power support even during full irradiance conditions, i.e. while generating rated amount of active power. This constitutes about 20% PV penetration on the feeder. A typical solar irradiance profile for the day and a realistic daily load profile are chosen and illustrated in Figure 7.21 [3]. Autonomous volt-var control (VVC) having coordinates shown in Table 7.3 is implemented on all the solar systems.

Three studies are performed for the following cases:

1) No PV system
2) 20% PV systems
3) 20% PV systems with VVC

The voltages for all the three cases are plotted in Figure 7.22 [3]. A wide variation in voltage is observed for the case of no PV connection (due to the daily load changes). An addition of 20% PV systems causes the voltage to not only rise but undergo a high degree of fluctuation as a result of cloud passages during the day. The reactive power support provided by the VVC on the PV solar farms successfully reduces the voltage fluctuations. It further decreases the overall voltage variation due to load changes during the day, thereby improving the overall voltage profile on the feeder.

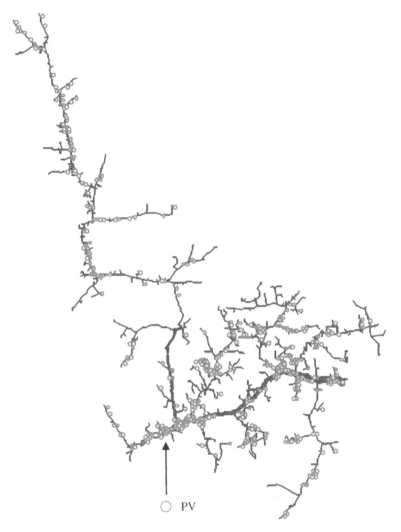

○ PV

Figure 7.20 Study distribution system. *Source*: Smith et al. [3].

Key Takeaways
VVC provides effective voltage control and mitigation of voltage fluctuations.

7.13.2 Smart Inverter Functions for Increasing Hosting Capacity in New York Distribution Systems

The effectiveness of different SI functions in increasing the hosting capacity of both small-scale (typically rooftop systems) and large-scale PV systems (rated more than 500 kW) in four distribution feeders in New York state is investigated in [40]. The SI functions considered are (i) constant PF function, (ii) volt–var function, and (iii) volt–watt function. All PV inverters are assumed to have an inverter rating which is 20% higher than the maximum DC rating of the PV panels. The voltage-based hosting capacity is considered to be the average of minimum and maximum hosting capacities as depicted in Figure 7.1. Further, the hosting capacity is evaluated for the case when all

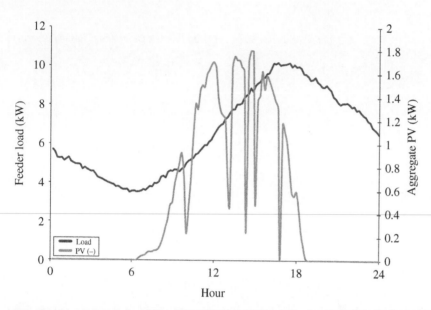

Figure 7.21 Aggregate solar and customer load profile over a day. *Source*: Smith et al. [3].

Table 7.3 Coordinates of volt–var function.

Voltage V_x (V)	Reactive power Q_x (% of rated kVA)
114.0	100.0
119.0	0.0
121.0	0.0
126.0	−100

Source: Smith et al. [3]

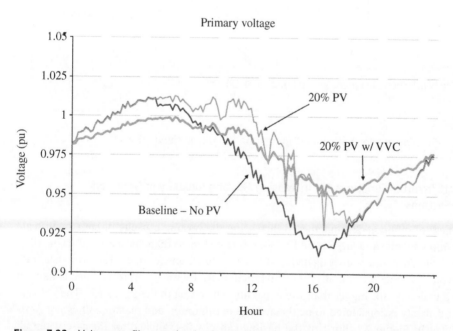

Figure 7.22 Voltage profile over the day for the three study cases. *Source*: Smith et al. [3].

PV systems are equipped with SIs and compared with the hosting capacity obtained for unity PF operation of all PV systems.

Following is the summary of hosting capacity results when several small scale PV systems are connected on the study feeders:

Almost all SI functions increase hosting capacity. Both constant PF and volt–var functions are effective in increasing the Minimum hosting capacity, i.e. at lower penetration levels. Constant PF function is successful in increasing Maximum hosting capacity, i.e. at higher penetration levels.

Constant PF control results in the largest increase in hosting capacity varying between 37 and 150%. PF setting of 0.98 absorbing is most effective. VVC with appropriate settings can increase the hosting capacity in the range 42–123%. However, improper settings of volt–var function can actually reduce the hosting capacity. Volt–watt control increases the hosting capacity by a relatively small amount 11–82%.

Following is the summary of hosting capacity results when large scale PV systems are connected at five locations on the study feeders:

Almost all SI functions increase hosting capacity. VVC is effective in increasing both the minimum and maximum hosting capacity, i.e. it is effective at both low and high penetration levels.

Constant PF control (at 0.98 absorbing) results in the highest increase in hosting capacity with the maximum increase to be 260%. VVC also increases hosting capacity in the range 130–247%. There is no perceptible increase in hosting capacity with volt–watt function.

Key Takeaways

In the studied feeders, constant PF function is more effective than volt–var function in increasing hosting capacity. The volt–var function with deadband is more effective than without it.

7.13.3 Impact of Different Smart Inverter Functions in Increasing Hosting Capacity in Hawaii

National Renewable Energy Laboratory (NREL) has performed extensive studies to demonstrate that SI technology can help increase the PV system hosting capacity of distribution feeders through effective voltage control [1, 41]. These studies were conducted on a power-hardware-in-the-loop (PHIL) platform with real-time simulation models of Hawaiian feeders and hardware models of different inverters dynamically connected to the simulation. This PHIL simulation facilitated actual inverters to be connected to Hawaiian feeders in real time. The following conclusions were drawn from the studies:

Distribution circuits have a combination of legacy inverters (without any SI functions) and SIs with grid support functions. If the number of legacy inverters is lower than SIs, constant PF at 0.95 inductive (absorbing reactive power) operation was sufficient to reduce the voltage to acceptable limits without resorting to use of volt–watt function which would have required curtailment of active power. However, if the legacy inverters were larger in number, volt–watt function had to be invoked more frequently for control of voltages.

There needs to be a minimum set of SIs to be able to perform voltage control without APC. The size of this minimum set depends upon the number of legacy inverters present in the distribution circuit, loads, circuit impedance, X/R ratio, circuit topology, and above all the type and parameters of the SI function under consideration.

A subsequent collaborative project was undertaken between NREL and Hawaiian Electric, titled "Voltage Regulation Operational Strategies (VROS)" [42]. The primary objective of this project was to determine most appropriate voltage control strategies for enhancement of hosting capacity of PV systems in Hawaiian distribution grids while creating minimum impact on active power generation of the PV systems. Extensive quasi-static time series simulation studies yielded the following conclusions:

Voltage profiles improve substantially with the implementation of SI functions (grid support functions) on PV systems. Also, SI operation did not result in any increase in the operation of utility voltage regulation equipment, e.g. LTCs.

VVC is more effective than constant PF of 0.95 inductive for voltage control during periods of high active power production by PV systems (typically 10 a.m. to 2 p.m.). The VVC, therefore, reduces the need for APC from PV systems for voltage control.

VVC also results in lower reactive power absorption compared to the constant PF of 0.95 inductive operation. Since distribution voltages generally lie within the slope of the volt–var function, the reactive power absorbed is proportional to the voltage. However, in constant PF operation, reactive power is absorbed continuously, irrespective of voltage. The volt–var function thus results in less reactive power flow (import) at the start of the distribution circuit, causing a smaller impact on the upstream high voltage network.

Although VVCs require less import of reactive power from the distribution–transmission interface, the net reactive power drawn from the bulk transmission system may become large if very high penetration of PV systems with volt–var functions are installed in the distribution system. This aspect needs to be investigated further.

Volt–var function with reactive power priority provides effective control of momentary overvoltages during hours of peak PV power generation. On the contrary, the conventional volt–var function with active power priority will not have any reactive power capacity during peak active power production (based on the studies performed in [42]). It is, however, noted that the new IEEE Standard 1547-2018 requires that DER inverters be capable of providing reactive power at 0.9 PF even at rated active power generation. Hence, DER inverters compliant with the new IEEE Standard 1547-2018 will be better able to control momentary overvoltages as described above.

A combination of volt–var function with volt–watt function results in lesser curtailment of active power generation from PV systems. A combination of volt–var function with volt–watt function, per Hawaiian DG Interconnection Rule 14H, is depicted in Figure 7.23 [42].

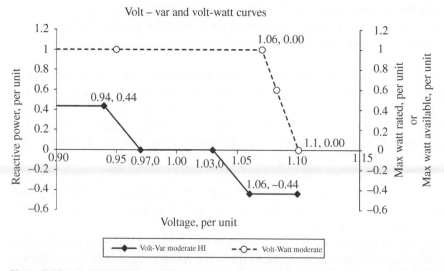

Figure 7.23 Combined volt–var and volt–watt smart inverter functions. *Source*: Reprinted with permission from the National Renewable Energy Laboratory [42].

The settings of the volt–var and volt–watt function are chosen such that as the PCC voltage increases due to active power production from the PV system, the reactive power capacity is first utilized through volt–var function for voltage control. Only when the entire available reactive power capacity is exhausted at $V = 1.06$ pu, and the voltage continues to rise, the volt–watt function is initiated. Active power is subsequently reduced to regulate the voltage. If the PCC voltage still rises beyond 1.1 pu, the entire PV production is curtailed. This combined control was seen to cause less than 1% annual energy curtailment for more than 90% of the customers in distribution circuits having high PV penetration levels [42]. The significance of this combined volt–var and volt–watt control is that more PV systems can be connected to the distribution circuits.

Further simulation studies with more accurate modeling and a field pilot study [43] were done on a 12 kV feeder in Hawaii to validate the conclusions made in VROS project [42]. The studies showed that volt–var function is very effective in reducing voltages during PV active power generation periods. The need to activate volt–watt function is not high.

Volt–watt is an effective SI function to prevent large voltages, in case the voltage at a PV location becomes high due to temporary situation in the grid, i.e. feeder reconfiguration. If volt–watt function causes significant active power reduction for a customer, it may be that the regular voltages at that bus are anyway higher than those specified by the Hawaiian Electric Rule 2 and ANSI Standard C84.1. In such a situation, it is the responsibility of the utility to resolve the voltage issue by appropriate circuit upgrades or other means, instead of the customer needing to curtail power at its location.

With combined volt–var and volt–watt SI functions, there is minimal curtailment of PV active power production, as long as the voltage is within the ANSI Standard C84.1.

Key Takeaways

Reactive power-based SI functions such as constant PF and volt–var functions may be used before the use of volt–watt function. The volt–var function offers better voltage control than constant PF, and especially if it operates in var priority mode. The combination of volt–var and volt–watt SI functions is very effective in increasing the hosting capacity of PV systems, especially when the PV connectivity is limited by voltage constraints. Reactive power import across the distribution/transmission interface may become an issue if the number of PV systems with reactive power-based SI controls is large.

7.13.4 Smart Inverter Impacts on California Distribution Feeders with Increasing PV Penetration

The impacts of volt–var SI function in a real 11.7 km long rural feeder in California with very high levels of PV penetration are presented in [1, 7]. The impacts of this function are examined on voltage profile, minimum and maximum voltages, tap operations, line losses, and voltage variability along the feeder. This study considers PV penetrations up to 200% and also accounts for the fact that not all PV systems may be equipped with SI functions. PV penetration in a feeder is defined as the ratio between the installed amount of PV systems and peak load in that feeder. A new term smart inverter fraction (Frac_{SI}) is defined as the ratio of the capacity of PV that has SI functionality to the total capacity of PV systems on the feeder, as given in Eq. (7.1).

The volt–var function is considered both with and without deadband as shown in Figure 7.24 [7]. The deadband is of 0.04 pu between 0.98 and 1.02 pu. This implies that no reactive power exchange occurs for voltages within the deadband.

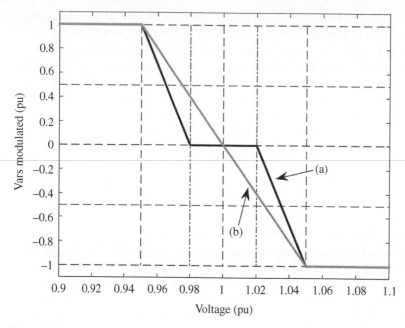

Figure 7.24 Volt–var control curves: (a) with deadband and (b) without deadband. *Source*: Pecenak et al. [7].

7.13.4.1 Methodology

A QSTS simulation study was performed using OpenDSS software and MATLAB. The simulation period extended over 95 days (10 December 2014–15 March 2015). The load data and PV data were obtained at 30 seconds intervals. Simulation studies were performed with a total of 107 PV systems with two of them being larger than 0.5 MW [7]. The PV inverters were oversized by 5% (i.e. having AC/DC ratio = 1.05), and operated on active power priority. It was noted that undersized inverters will not be able to provide adequate reactive power compensation for voltage control.

7.13.4.2 Voltage Profile Along Feeder

The voltage profile along the feeder for PV penetrations of 5–200% are shown in Figure 7.25a,b, respectively [7]. Results are presented for different values of $Frac_{SI}$ and the two volt–var curves (a) and (b) depicted in Figure 7.24.

The voltage decreases along the length of the feeder for 5% PV penetration but increases significantly with 200% penetration due to active power generation. The effectiveness of volt–var function in controlling the voltage increases with higher $Frac_{SI}$ and furthermore when deadband is absent. In fact, the volt–var function without deadband is able to increase voltages at the sending end and reduce voltage rise at the feeder end, thus creating a flatter voltage profile, as compared to the volt–var function with deadband. This is because the volt–var function with deadband does not begin to provide voltage control until the voltage has exceeded the upper limit of the deadband voltage, i.e. until the PV penetration has increased to 75%.

7.13.4.3 Maximum Voltage

The maximum voltage along the feeder increases with increase in PV penetration. In general, the volt–var curve without deadband (Figure 7.24 curve (b)) is able to reduce this voltage more effectively than the function with deadband. However, the capability of volt–var curve to reduce overvoltages at

(a)

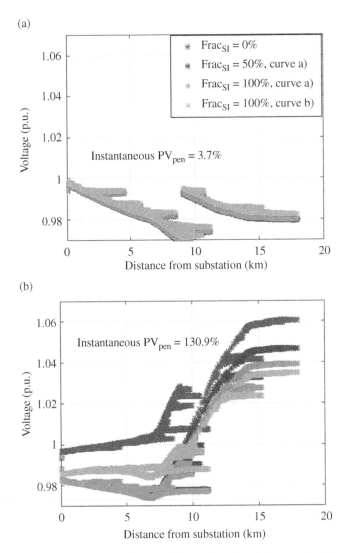

(b)

Figure 7.25 Feeder voltage profile as a function of distance for 1 March 2015 12 a.m. PST. (a) 5% nominal PV penetration. (b) 200% nominal PV penetration. The curves in each plot represent different SI fractions and VVC curves simulated. *Source*: Pecenak et al. [7].

very high PV penetrations becomes limited due to the active power priority considered for volt–var function. Such priority does not leave adequate reactive power exchange capability at peak power production periods.

7.13.4.4 Minimum Voltage

The minimum voltage occurs on the feeder when PV output is zero and load is maximum. As PV penetration and $Frac_{SI}$ increase, the efficacy of SIs in mitigating under-voltage increases. With increasing PV penetration more reactive power capability becomes available for controlling the voltage. It is noted that a high percentage of $Frac_{SI}$ with no deadband results in higher minimum feeder voltages.

7.13.4.5 Tap Operations

The number of tap operations increases when more PV systems not having SI functions are added. The volt–var function reduces tap operations, substantially. The reduction in number of tap operations is more when $Frac_{SI}$ is high and deadband is removed from the volt–var curve. Oversizing of PV inverters is expected to reduce tap operations further.

7.13.4.6 Line Losses

The normalized power losses in the feeder with increasing PV penetration are illustrated in Figure 7.26 [7].

An initial increase in PV penetration up to 25% decreases line losses. However, higher increases in PV penetration with larger $Frac_{SI}$ result in an increase in line losses. The volt–var function with deadband results in more losses than without deadband. In general, the more reactive power a SI function exchanges with the feeder in providing voltage control, more will be the current flow in the line, resulting in higher I^2R line losses.

7.13.4.7 Voltage Fluctuations

A statistical voltage variability score (VS) is proposed to measure voltage fluctuations in the feeder [7]. For all cases of PV penetration considered, the voltage variability decreases with the implementation of volt–var SI function. This variability reduces further when the deadband is removed from the volt–var function. While the SI function reduces the voltage variability up to a PV penetration of 125%, its effectiveness decreases with higher PV penetrations.

Key Takeaways

The efficacy of SI VVC increases with an increase in $Frac_{SI}$. The voltage control capability can be increased by oversizing the inverter, i.e. by having a higher AC/DC ratio in PV inverters. The VVC without deadband is more effective than VVC with deadband. Higher penetration of SIs VVC mitigates overvoltage but causes an increase in line losses.

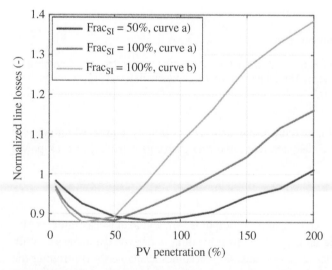

Figure 7.26 Normalized power losses with respect to feeder without SI in feeder lines as a function of PV penetration. *Source*: Pecenak et al. [7].

7.13.5 Mitigation Methods to Increase Feeder Hosting Capacity

A methodology for evaluating different SI-based technological options to increase the hosting capacity of PV systems in distribution networks is described in [1, 37]. A case study is presented to demonstrate the application of above assessment methodology in a distribution feeder depicted in Figure 7.27 [37]. The different options considered in this case study are as follows:

i) Volt–watt function with different settings
ii) Volt–var function with different settings
iii) Watt-power factor
iv) Fixed PF with different settings
v) Power priority of PV inverters involving active power priority or reactive power priority
vi) Oversized inverter wherein the inverter size is increased by 10% to allow 0.9 PF reactive power exchange capability even at rated active power output.

7.13.5.1 Assessment of Base Feeder

The PV system under consideration with the user-defined settings for the PV inverter is described below [37]:

- *Location*: The location shown in the triangle is selected for examining the potential connection of a PV system.
- *kWp capacity*: A 3.6 MW PV is desired to be connected at the selected location.
- *Power priority*: Active power priority is selected.
- *Oversized inverter*: The inverter is oversized by 10%.

A three-phase transformer connects the three-phase PV system to the distribution network. The connection of this 3.6 MW PV system causes overvoltage issues shown with small circles in Figure 7.27.

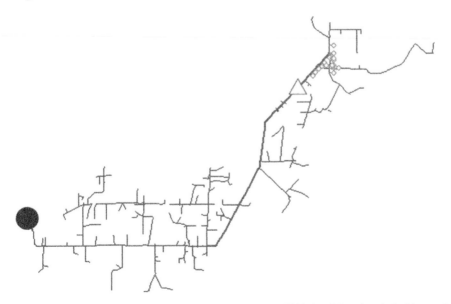

Figure 7.27 Case study feeder with the location of new PV (triangle) and technical issues (circles).
Source: Reprinted with permission from EPRI [37].

7.13.5.2 Integration Solution Assessment

Time-series analyses and static power flow studies are used to examine the different mitigation options. The mitigation options and their settings in this case study utilize the default settings of various standards across the world. Different international standards have different voltage limits such as 0.9–1.1 pu in Europe, 0.94–1.1 pu in Australia and 0.95–1.05 pu in the United States. All these sets of default settings are normalized using the limits considered in U.S. International settings of volt–watt function are compiled in Table 7.4 [37] and plotted in Figure 7.28 [37].

Aggressive settings are also considered to provide an increased active power response to changes in PCC voltage. In the custom aggressive setting of volt–watt the APC starts before the voltage reaches the limit. These settings do not allow any power generation beyond the upper voltage limit.

Normalized settings used in this case study for volt–var function with their visual representation are depicted in Table 7.5 [37] and Figure 7.29 [37], respectively. In case of custom aggressive setting of volt–var function, the reactive power is exchanged (absorbed/injected) as soon as the voltage deviates from 1 pu. Also, the maximum amount of reactive power is exchanged when the voltage reaches the statutory limits.

The different settings of watt–PF function and fixed PF function are compiled in Table 7.6 [37] and Table 7.7 [37], respectively.

7.13.5.3 Volt–Watt Function

The daily profiles of active power and voltage for different volt–watt functions and without the volt–watt function (w/o solution) are depicted in Figure 7.30 [37]. Without any solution, the voltage rises above the limit of 1.05 pu. With utilization of different volt–watt functions, the rise in voltage is

Table 7.4 International default settings of volt-watt function normalized to ANSI voltage limits

Volt-Watt Settings	IEEE/California/Hawaii	Australia	Europe	Custom Aggressive
V_1	1.06	1.038	1.05	1.04
P_1	100%	100%	100%	100%
V_2	1.1	1.1	1.07	1.05
P_2	0%	20%	0%	0%

Source: Based on [37]. Used with permission from EPRI.

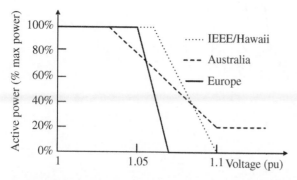

Figure 7.28 Visual representation of international default settings of volt–watt function normalized to ANSI voltage limits. *Source*: Based on [37]. Used with permission from EPRI.

Table 7.5 International default settings of volt-var function normalized to ANSI voltage limits.

Volt-Var Settings	California	Hawaii	IEEE – Category A	IEEE – Category B	Australia	Europe	Custom Aggressive
V_1	0.92	0.97	0.9	0.92	0.91	0.95	0.95
Q_1	+30%	+44%	+25%	+44%	+30%	+43.6%	+44%
V_2	0.967	0.97	1	0.98	0.966	0.97	1.0
Q_2	0	0	0	0	0	0	0
V_3	1.033	1.03	1	1.02	1.038	1.031	1.0
Q_3	0	0	0	0	0	0	0
V_4	1.07	1.06	1.1	1.08	1.1	1.05	1.05
Q_4	-30%	-44%	-25%	-44%	-30%	-43.6%	-44%

Source: Based on [37]. Used with permission from EPRI.

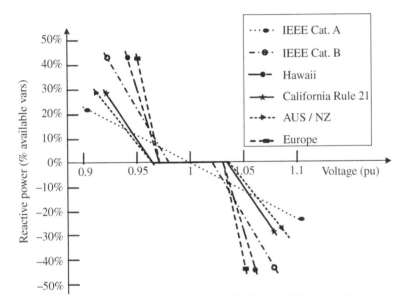

Figure 7.29 Visual representation of international settings of volt–var function normalized to ANSI voltage limits. *Source*: Based on [37]. Used with permission from EPRI.

Table 7.6 Settings of watt-power factor.

cos φ (P) settings	Australia	Europe
cos φ_1	1	1
P_1 (%)	50	50
cos φ_2	0.95 (lag)	0.9 (lag)
P_2 (%)	100	100

Source: Based on [37]. Used with permission from EPRI.

Table 7.7 Settings of fixed power factor.

Fixed cos φ settings	Set 1	Set 2	Set 3	Set 4
cos φ	0.95	0.90	0.85	0.80

Source: Based on [37]. Used with permission from EPRI.

(a)   (b)

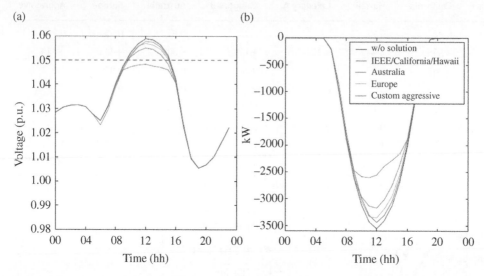

Figure 7.30 Analysis of volt–watt settings (a) voltage and (b) active power. *Source*: Reprinted with permission from EPRI [37].

lower, however, with different levels of power curtailment based on the setting adopted. Only the aggressive setting keeps the voltage within limits. This is because APC starts when voltage reaches 1.04 pu and power generation is completely stopped when the voltage reaches 1.05 pu. This does not happen with other settings where power generation is curtailed only after the voltage reaches 1.05 pu. It is noted that the aggressive settings entail higher power curtailment.

7.13.5.4   Volt–Var Function
The daily profiles of voltage, reactive power, and active power of the PV system for each of the examined volt–var settings are shown in Figure 7.31 [37].

Only the Hawaiian, European, and Custom Aggressive settings of volt–var function are able to prevent voltage rise beyond the acceptable limit. This is due to the substantially high amount of reactive power absorption when the voltage is above or at the statutory limit. The settings of IEEE, California, and Australia help slightly in reducing the overvoltage issues. In this case study, the inverter is considered to be oversized by 10%.

It is noted that PV inverter with active power priority has limited capability of absorbing reactive power during the periods of peak power generation. Therefore, the reduction in voltage is lesser. On the other hand, PV inverter with reactive power priority has larger reactive power absorption capability during the times of peak power generation and hence substantially reduces the voltage rise. This is, however, accompanied by curtailment of active power.

7.13.5.5   Watt-Power Factor Function
The daily profiles of voltage, reactive power, and active power of the PV system for each of the examined Watt–PF settings are shown in Figure 7.32 [37].

Both the examined settings help in lowering the voltages substantially as shown in Figure 7.32 [37]. This is because the inverter is oversized by 10% in this case with active power priority. This allows a larger reactive power absorption capability during peak power production. However, similar results were not observed if the inverter was not oversized.

Figure 7.31 Analysis of volt–var settings (a) voltage, (b) reactive power, and (c) active power. *Source*: Reprinted with permission from EPRI [37].

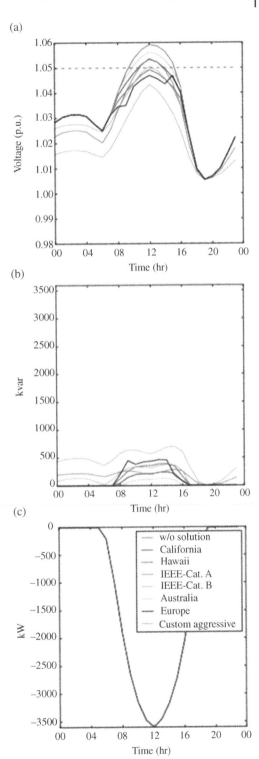

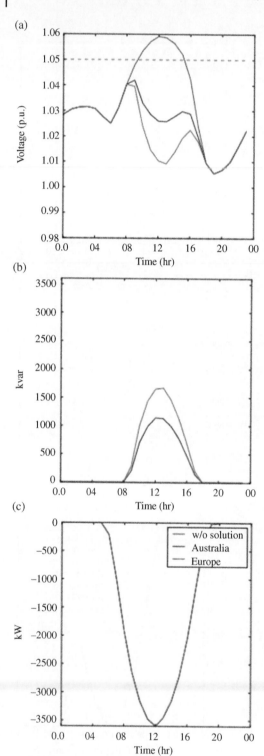

(a)

(b)

(c)

Figure 7.32 Analysis of watt–PF settings (a) voltage, (b) reactive power, and (c) active power. *Source*: Reprinted with permission from EPRI [37].

There is larger absorption of reactive power with the watt–PF function than with the volt–var function. This results in an increase in the current flow in the feeder, which can likely cause thermal congestion issues.

7.13.5.6   Fixed Power Factor Function

The daily profiles of voltage, reactive power, and active power of the PV system for each of the considered fixed PF settings are shown in Figure 7.33 [37].

This function results in the lowest voltage profiles. Reactive power absorption increases as the PF is decreased. A PF lower than 0.9 results in significant reduction of voltage. If the PF is reduced further, the voltage may fall below the acceptable lower limit. The fixed PF causes a larger absorption of reactive power which increases the current flow in the system. This increases the risk of thermal overloading.

Due to 10% oversized inverter, full reactive power absorption capability during the peak power generation is available for PFs of 0.95 and 0.9. However, this reactive power absorption capability reduces for PFs of 0.85 and 0.8.

Based on this assessment study, mitigation options are identified that can increase the feeder hosting capacity to accommodate connection of the 3.6 MW PV system without causing any voltage or thermal issues.

Key Takeaways

Aggressive settings of volt–watt function are more effective in this case study although they cause a higher level of power curtailment. If voltage rise is large, aggressive volt–var settings are needed for voltage control as they provide higher amount of reactive power absorption. The volt–var settings are more effective in the var priority mode. Watt–PF function and constant PF functions are effective in voltage control, especially if an overrated inverter is considered. Lower inductive settings of the constant PF function can cause undervoltage problems. Both Watt–PF function and constant PF functions draw large reactive power and result in a high amount of current flow in the lines. This can potentially create a thermal overloading issue.

7.13.6   Impact of Smart Inverter Functions in Increasing Hosting Capacity in Five California Distribution Feeders

Simulation studies for enhancing hosting capacity by 25% in five real California distribution feeders utilizing different SI functions are described in [1, 44]. These functions are covered under the California Rule 21 Phase I and Phase III SI functions.

QSTS simulations are first conducted to determine the existing baseline hosting capacity in these feeders by considering the following cases:

1) Three types of solar production days: clear, partly cloudy, and overcast
2) Two characteristic load days: low loading and peak loading
3) Three PV locations: instead of studying the impact of multitude of combinations of small PV systems all along the feeders, the impact of aggregated utility-scale PV systems at three specific locations in the feeder termed, "front," "middle," and "end" are studied.

While these consolidated locations are expected to provide a representative impact of the distributed PV systems, this strategy may not be entirely suitable for studying the impact of substantially distributed residential PV systems [44].

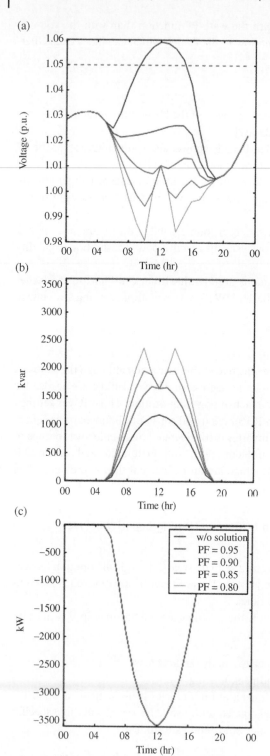

(a)

(b)

(c)

Figure 7.33 Analysis of fixed power factor settings. *Source*: Reprinted with permission from EPRI [37].

The AC:DC rating (ratio of AC inverter rating to the DC power rating) of the PV inverters is considered to be 1 : 1.2.

A total of 18 scenarios per feeder were studied with 24-hour QSTS simulations, for all the five feeders. Hosting capacity was obtained by incrementing the PV DC power rating by 1 kW each time until any of the following two violations were experienced:

- Voltage variation outside the 0.95–1.05 pu (per ANSI Standard C84.1 Range A for voltages)
- Current flow in excess of conductor rating or power flow higher than the transformer rating.

The entire process of determining the baseline hosting capacity is depicted in Figure 7.34 [44].

Extensive studies were performed with the objective of increasing hosting capacity by 25% on all feeders by the following SI functions:

- Volt–var function with watt priority
- Volt–var function with var priority
- Volt–watt
- Limit maximum active power output

QSTS simulations were conducted for a wide range of settings of the above functions to determine the "compliant" settings which did not cause either voltage or thermal violations. These "compliant" settings were then ranked based on the important metrics for the distribution system.

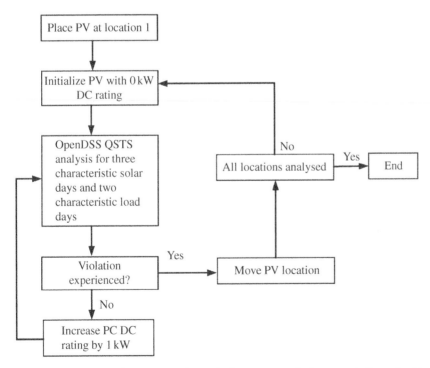

Figure 7.34 Baseline hosting capacity analysis process applied separately for each of the three locations on each of the five feeders. *Source*: Haghi et al. [44].

7.13.6.1 Volt–Var Control

VVC can be implemented with either watt (active power) priority or var (reactive power) priority. A comparison of the two types of volt–var function is illustrated in Figure 7.35 [44].

Although there is no power curtailment involved, the watt priority is unable to provide any voltage regulation. On the contrary, the var priority provides substantial reduction in voltage from 1.062 to 1.039 pu. There is, however, a slight APC involved with the var priority, in order to release enough inverter capacity for providing reactive power compensation.

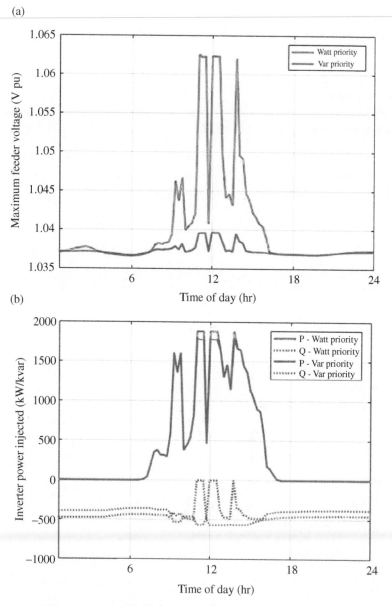

Figure 7.35 Comparison of watt and var priority for a specific setting on a feeder when the PV is at the "end" of the feeder for the partly cloudy day with minimum feeder loading. (a) Maximum voltage obtained from the voltage time series. (b) Active and reactive power output from the inverter. *Source*: Haghi et al. [44].

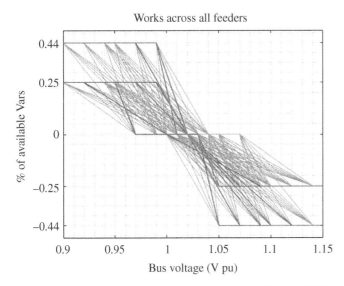

Figure 7.36 Volt–var with var priority curves that achieved a 25% increase in hosting capacity for all feeders and PV locations. *Source*: Haghi et al. [44].

No setting of volt–var function with watt priority was able to increase the hosting capacity by 25%. However, in general, volt–var function with var priority was able to increase the hosting capacity by 25% in all the feeders. The volt–var curves which could accomplish the 25% increase in hosting capacity are plotted in Figure 7.36 [44].

It is noted that in general, volt–var curves which had a higher voltage V_3 setpoint (the voltage level at which the droop starts downwards in inductive range) were less effective in increasing hosting capacity. Also, the curves which had a higher voltage V_4 setpoint (the voltage level at which the downward droop ended in inductive range) were found to be effective only when 44% (instead of 25%) reactive power limit was selected. Stated alternatively, reduced deadband and steeper slopes of the volt–var curves were seen to be more beneficial in this study.

7.13.6.2 Volt–Watt Control
The volt–watt control was not effective in enhancing the hosting capacity due to unbalance in feeders and due to the specific choice of reference voltage for triggering the power curtailment. Three-phase PV inverters are controlled based on phase-average voltage. However, due to feeder unbalance, the phase-maximum voltage is higher than the phase-average voltage. A volt–watt function based on phase-average voltage is, therefore, not able to prevent the phase-maximum voltage violating the upper ANSI limit. An appropriate choice of reference voltage is, therefore, essential for successful operation of the volt–watt control.

Such an issue does not arise in single-phase feeders, where only one (phase) voltage is available for control which can be effectively utilized for volt–watt control.

7.13.6.3 Limiting Maximum Real Power (LMRP) Output Function
This function could increase the hosting capacity by 25% only when the real power was reduced by 70%. This result is considered trivial because 70% (LMRP setting) of 125% (PV penetration compared to baseline) equals 88%. This implies a net increase in hosting capacity only by 88%, which is even lower than the existing hosting capacity.

7.13.6.4 Selection of the Smart Inverter Settings

From the QSTS studies, a range of settings for a specific SI function may be obtained which do not cause either voltage or thermal violation. However, not all may satisfy different performance metrics considered important for the distribution feeder. The most appropriate setting may be determined by assigning weights to the different criteria/metric on a case-to-case basis, depending on the objective of the study.

Key Takeaways

Only volt–var function with var priority was able to increase the hosting capacity by 25% in all the feeders. The volt–var settings which achieved the above objective had reduced deadband and steeper slopes. The volt–watt control was not effective in enhancing the hosting capacity due to the unbalance in three-phase feeders and the specific choice of reference voltage for triggering the power curtailment. In such cases, the volt–watt function should be made to operate when phase-maximum voltage (and not phase-average voltage) starts to violate the upper ANSI limit.

7.13.7 Hosting Capacity Experience in Utah

Extensive simulation studies and laboratory testing has provided following insights into hosting capacity issues in several feeders in PacifiCorp, UT, USA [1, 34, 45]:

SIs can increase hosting capacity but not in lines which are constrained by thermal limits. Increase in hosting capacity is strongly dependent on the location of SI deployment. A larger increase in hosting capacity is obtained for SI functions implemented on central PV inverter systems than on distributed PV systems. This is because the location of central inverter systems can be controlled more easily than distributed systems. Also, buses near feeder-head, i.e. substation transformers, that see a lower grid impedance looking into the source have a larger increase in hosting capacity compared to buses that are farther downstream.

7.13.7.1 Constant Power Factor Function

Fixed PF function is a simple function to implement as it requires specification of only two parameters: (i) magnitude of PF (typically 0.9 lagging to 0.9 leading), and (ii) direction of reactive power flow (i.e. (absorbing or injecting).

Fixed PF function with gradually decreasing PF (absorbing reactive power) tends to increase hosting capacity especially at the substation and moving downstream. However, as PF decreases, the factor restricting hosting capacity changes from overvoltage to thermal limits due to the increased flow of reactive current in the feeder.

7.13.7.2 Volt–Var Function

The parameters are much more in number with volt–var functions: (i) maximum and minimum reactive power outputs (typically in the range 0.5 to −0.5 pu), (ii) droop or slope (typically in the range 4–40% Q/V), (iii) voltage setpoint (typically 0.96–1.04 pu), and (iv) deadband.

Volt–var functions with optimized settings were more effective than fixed PF functions in improving hosting capacity. However, nonoptimized volt–var functions resulted in an even worse performance. Incorrect settings of SI functions can cause adverse impacts on voltage levels and voltage variability in addition to increasing the number of operations of regulating equipment (e.g. transformer taps, capacitor switchings, etc.).

Well-selected SI functions can potentially increase hosting capacity, reduce line losses and improve PF, in a simultaneous manner. However, inappropriate settings can worsen either one or more of these metrics [45]. Both volt–var curves and fixed PF settings, if not optimized, are likely to result in aggravated line losses. Similarly, "good" volt–var settings can improve substation PF whereas a "bad" selection of settings will worsen the feeder PF substantially. In some feeders, the best volt–var characteristics for optimizing substation PF (close to unity) may be quite different from the best volt–var characteristics for minimizing line losses for same type of day (clear/overcast/variable) and similar loading condition.

In some feeders, the best volt–var settings for achieving various objectives under different solar outputs and load conditions may not be quite different. This means that the best settings once determined from system studies, may be maintained continuously.

Furthermore, the best volt–var settings may be more strongly influenced by load level rather than by solar irradiance. This implies that best volt–var settings need not be changed from day to day based on solar irradiance but may need to be altered when loading level changes from season to season.

Even though IEEE 1547-2018 requires DERs to have a communication capability over an open protocol, utilities are still developing a common approach for interfacing with these DERs.

Key Takeaways

Both volt–var and constant PF settings need to be optimized to obtain the desired performance metric, as nonoptimized settings may cause system performance issues. The best settings to achieve one performance metric may be different from those needed to achieve another metric. There may be feeders for which it may not be necessary to optimize settings for each and every system operating condition and performance metric, as some optimized settings may serve different objectives. This can save efforts in both computing the settings and communicating them frequently to a large number of PV systems. Finally, SIs cannot increase hosting capacity beyond thermal limits.

7.14 Field Implementation of Smart Inverters for Increasing Hosting Capacity

The operating experience of actual distribution feeder systems worldwide, where improvements in the hosting capacity were accomplished through smart inverter (SI) controllers are presented in this section [1, 26].

7.14.1 Voltage Control by Constant Power Factor Function in a Distribution Feeder in Fontana, USA

A field demonstration of advanced inverter (i.e., smart inverter) functionality was performed in Fontana, California, USA, in Fall 2013 under a collaboration between NREL and Southern California Edison (SCE) [46].

A successful demonstration of the fixed PF (off-unity) SI function was performed on a feeder system that had 4.5 MW of grid-connected rooftop PV solar systems. A nonunity PF value (0.95 inductive implying absorption of reactive power by the PV inverter) was determined from simulation studies, which could reduce the undesirable voltage variation impacts of the PV solar farms operating at unity PF. The off-unity PF was successfully implemented on a 2 MW solar farm comprising 4×500 kW PV inverters.

Key Takeaways

An unusual situation occurred in this demonstration project. The PF at the PV system was actually set to +0.95 (capacitive) while the intention was to set it to −0.95 (inductive). This was an inadvertent error. Utility engineers and line crew persons are more accustomed to operating utility equipment with capacitive reactive power, and not so much with inductive reactive power. A lesson learned was that in future installations this should be clarified in advance with better communication.

The PF settings should also be stored in appropriate nonvolatile memories so that they do not get erased during soft or hard restarts of the PV inverters.

Further, the voltage measurements should be as close to the PCC of the PV inverters as possible, to ascertain the effectiveness of the SI function implemented.

7.14.2 Smart Inverter Demonstration in Porterville Feeder in California

An SI field demonstration was conducted in a distribution network close to Porterville, California over two weeks starting on 23 November 2014. The objective of this field test was to demonstrate the capability of fixed power (off-unity PF) smart PV inverter function in reducing voltage deviations on the distribution circuit and thereby increase the hosting capacity.

This test was conducted on a 12 kV distribution circuit having a 5 MW fixed-tilt ground-mounted PV solar farm connected about 9 miles from the substation as shown in Figure 7.37 [47]. The PV system comprised 10 inverters. A fixed inductive PF of −0.95 (involving reactive power absorption) was implemented on each inverter of the PV system.

The voltage at the POI of the PV system and the reactive power exchanged by the PV system with unity PF operation on 23 November 2014, and with −0.95 PF on 27 November 2014 are illustrated in Figure 7.38 [47]. It is seen that the inductive PF operation of the PV system successfully mitigates the voltage rise during peak power production around noon time.

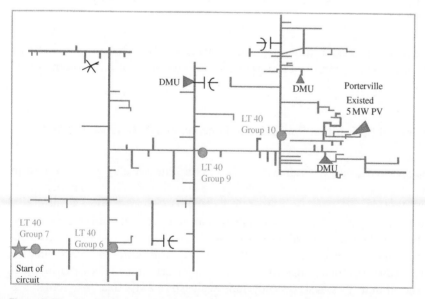

Figure 7.37 Distribution circuit of Porterville, CA [47]. *Source*: Kadam et al. [48].

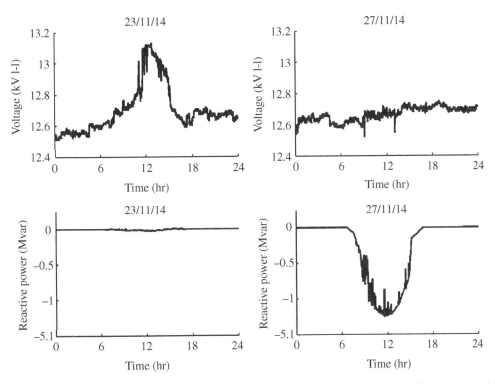

Figure 7.38 PV system POI voltage and reactive power exchange on 23 November 2014 (prior to field demonstration) and on 27 November 2014 (during field demonstration) [47]. *Source*: Kadam et al. [48].

A notable aspect of lagging PF operation of the PV solar system or any reactive power absorption-based SI operation of the PV system is that the reactive (inductive) current (and consequently the reactive power) being absorbed by the PV system needs to be supplied from an external means in the distribution system. This reactive current may be supplied either from switched capacitors in the network or from the distribution substation which in turn draws this current from the upstream high voltage system.

The reactive current absorption by the smart PV inverter leads to an increase in current flow in the distribution line, especially during peak PV active power generation. Since the voltage on the network is fairly well regulated by the smart PV inverter, no capacitor switching occurs. Therefore, the entire reactive current absorbed by the PV inverter is fed from the substation, causing an increase in the substation current. Even though a relatively large amount of reactive current may be absorbed by the PV system, it combines in quadrature with the active current flow in the circuit to cause only a modest increase in the total magnitude of current in the distribution circuit. For this study, at the start of the circuit, the average current is seen to increase from 42 A (before field demonstration) to 58.8 A (during demonstration), while the maximum current increases from 174.6 A (before field demonstration) to 201.3 A (during demonstration).

This is further evident from Figure 7.39 [47] which depicts the start of the circuit (substation) current in phase A and the reactive power absorbed by the PV system. The maximum increase in the current is observed during maximum reactive power absorption by the PV system.

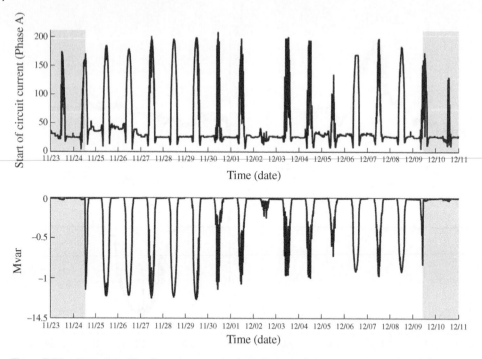

Figure 7.39 Start of the circuit current magnitude in Phase A and reactive power absorbed by the PV system during normal days and during field demonstration period [47]. *Source*: Kadam et al. [48].

Key Takeaways

This project demonstrated that off-unity (−0.95) PF SI function was successful in decreasing the PV system POI voltage by 0.03 pu and also reducing the variability of the voltage during hours of peak power production.

7.14.3 Improvement in Feeder Hosting Capacity Through Smart Inverter Controls in Upper Austria

This nine-month field trial project was carried out under the MOREPV2GRID project in Austria to demonstrate that SI controls can be deployed to increase the hosting capacity of feeders, which was restricted due to the voltage rise from high PV penetration [48].

The study feeder was more than 800 m long with five single-phase PV installations totaling 15.3 kW. The control parameters of the inverters could be remotely transmitted. A voltage rise of 4.9% was experienced with normal operation of PV inverters.

Different SI controls were implemented and tested. These included fixed PF control (0.85 PF), VVC with and without deadband, and volt–watt control. Simulation and laboratory tests were carried out before the field implementation. The VVC not only reduced the voltage during high power generation but also increased the voltage with reactive power injection during low voltage conditions. During high power production, the volt–watt control was seen to be effective in reducing the voltage rise. Reactive power compensation during symmetrical power injection into the grid from the single-phase PV inverters was found to be more effective.

Key Takeaways

The effectiveness of volt–var and volt–watt functions for voltage control was demonstrated. In the most conservative testing scenario, the voltage rise was reduced by 20–30% thus leading to an increase in hosting capacity by 20–30%.

7.14.4 Demonstration of Smart Inverter Controls under the META PV Project Funded by the European Commission

Field testing was performed over two years around 2013 in the Limburg province in Belgium on several low and medium voltage lines to demonstrate the increase in hosting capacity through SI controls on PV systems [49, 50]. The field testing was done in LV (230 V) distribution system having 128 customers each having 4 kW installed PV systems totaling 512 kW. Meanwhile, the MV (10 kV) test feeder system had 31 customers each having about 200 kW PV solar system installed totaling about 6.2 MW.

The performances of various SI functions were demonstrated on PV systems including

  i) RPC through Fixed PF control and volt–var function
 ii) Volt–watt functions
iii) Coordination of distribution system controls with SI controls
iv) Low/high voltage ride-through function

It was noted that voltage limit was the most important constraint in increasing the hosting capacity of PV systems. The thermal loading limit of feeders and transformers were, however, less of a concern.

RPC through VVC with deadband and fixed PF control was most effective in reducing voltage rise. The effectiveness of RPC was seen to be strongly related to the X/R ratio of network impedance. Cable networks have low X/R ratio as compared to overhead lines, hence RPC was more effective in overhead line networks than cable-based feeders.

Hosting capacity with unbalanced installations of PV systems over the different phases is much lower (about six times for the study system) than with symmetrical installation of the single-phase PV systems.

It was shown that the capacity of the distribution systems to connect PV solar systems could be increased by 50%. Also, the cost incurred in doing so would be only 10% of the cost of conventional compensating equipment to achieve the same task.

Key Takeaways

The effectiveness of reactive power-based SI functions is strongly dependent on the X/R ratio. Hosting capacity increase with SI functions is lower in unbalanced installations as compared to symmetrical ones. It was recommended that the voltage control responsibility should be shared by as many PV systems (equipped with smart functions) as possible, to be fair to all inverters. Further coordinated control of multiple smart PV systems operated from a utility control center is preferred.

7.14.5 Arizona Public Service Solar Partner Program

During 2016–2017, Arizona Public Service (APS) deployed about 1600 utility owned, residential PV systems, each rated 6 kW on an average, and equipped with SI functionalities. A significant collaborative activity between APS and EPRI was undertaken to address various challenges related to field implementation of SIs, energy storage systems, and control systems. The outcomes of studies on six utility feeders and overall project recommendations are described in [6].

The following SI functions were investigated in the field tests:

- Fixed PF (at values of 0.97, 0.95, and 0.90 absorbing)
- Volt–var (at slopes of 3%, 6%, and 9% of nameplate kVA rating per volt [120-V Base])
- Active power limit (at values of 95%, 85%, and 75% of nameplate rating)
- Fixed reactive power (at values of 10%, 20%, and 30% of the inverter's kVA rating)

Feeder circuits were evaluated under four different conditions with the corresponding study questions defined as below:

i) *Theoretical hosting capacity* – removing all of the PV currently connected, how much PV can the circuit ideally accommodate?

ii) *Remaining hosting capacity* – with the currently connected PV in place, how much additional PV can be accommodated?

iii) *Advanced inverter impact (no-retrofit)* – If new PV systems were added with advanced inverters (any existing PV would be left at unity PF), how would that increase the remaining hosting capacity?

iv) *Advanced inverter impact (full-retrofit)* – What would be the remaining hosting capacity if all of the inverters on the feeder were required to be advanced?

Some of the results of this project are summarized below:

Reactive power absorption by multiple distributed smart PV inverters as in rooftop systems has a less perceptible impact as compared to a large centralized smart PV inverter on the feeder.

Volt–var was found to be the best SI function for controlling the voltage in APS research feeders, given the high variability of summertime temperatures and feeder loads. Aggressive settings (e.g. steep slopes and narrow deadbands in volt–var functions) are seen to be effective in controlling the voltage to increase hosting capacity. If the number of SIs in a distribution feeder is not high, aggressive settings of SIs can be used without causing interference with the operation of other voltage regulating equipment in the feeder. Aggressive settings can create a significant impact only if the number of SIs with such settings is low. On the other hand, if the number of SIs having aggressive settings is large, they may adversely impact the operation of conventional voltage regulation equipment.

SIs have the potential to substantially increase hosting capacity limitations (increasing hosting capacity by more than three times in one simulated case). This can occur only if all the PV inverters are retrofitted with the SI functionality. If there are a large number of pre-existing PV systems which cannot be retrofitted with SI functions the improvement in hosting capacity will be somewhat lower.

SIs generally always followed the settings communicated to them, which was once a day. In the event of communication failure, the inverters should remain in the same state as before losing communication, or move to a default "safe state." Commands for changing settings may have to be repeated several times to ensure that any communication outages will not hinder the implementation of new settings. In future, if the settings need to be changed more frequently in real time, and especially on a large number of SIs, communication latency can become an issue. Integration of more SIs will require additional efforts to standardize interoperability among these SI devices and their subsystems.

In this field-testing, no adverse control interaction was observed among SIs themselves, or between SIs and other voltage control equipment. However, the project plans to investigate in future the potential of control interactions among SIs, battery storage systems, and other voltage management systems.

Key Takeaways

A good performance comparison of different SI functions can only be made if the number of participating PV systems is large.

Aggressive volt–var settings (e.g. steep slopes and narrow deadbands) are seen to be effective when there are small number of smart PV inverters in the distribution systems. However, aggressive settings may have adverse impacts if the number of smart PV systems is high.

Due to the variability of wireless communication and especially when the SIs are large in number, flawless communication with SIs may not be practical and should also not be expected.

It is recommended that future research should address automating the process of determining controller settings and implementing the updated settings on the SIs.

As configured, the inverters did not operate at night due to manufacturer-imposed limitations. Joint industry efforts should discuss the merits of advanced inverters performing functions (such as VAR support) continuously (not just during the day) [6].

7.14.6 Increasing Renewables Hosting Capacity in the Czech Republic

CEZ Distribuce, the largest distribution system operator (DSO) in the Czech Republic have implemented several SI-based technical solutions for enhancing the connectivity of DERs and EVs, under the Horizon 2020 Interflex Project (2017–2019) [51]. Autonomous SI functions on PV systems are proposed and implemented to increase the voltage-based hosting capacity. The total installed capacity of PV in the Czech Republic is approximately 2.1 GWp which is about 10% of the total capacity of generators installed in the country (22 GW). It is anticipated that in 2040, the amount of PV installed in low voltage grids will increase to 6 GWp [51].

The Czech demonstration project WP6 was conducted in various areas in the Czech Republic in the territory of CEZ Distribuce. Two of the four use cases related to PV systems as below:

7.14.6.1 Use Case 1: Increasing DER Hosting Capacity of LV Distribution Networks

The PV inverters are equipped with a combination of two autonomous SI functions: (i) volt–var $Q(V)$ function, and (ii) volt–watt $P(V)$ function similar to that shown in Figure 7.23, however, with a slight difference. As the voltage rises beyond 1.05 pu, volt–var function reduces the voltage by absorbing reactive power. At $V = 1.08$ pu, maximum reactive power is absorbed. APC by volt–watt function begins only after the voltage has risen beyond $V = 1.09$ pu. Active power is completely curtailed at $V = 1.11$ pu. It is thus noted that the APC does not begin at the same voltage at which maximum reactive power is absorbed, as is the case in Figure 7.23.

Initial results for some distribution feeders with very limited hosting capacity (long feeders with thin cross-sections) are presented in [51], which are described below. The daily behavior of a 10 kW PV system is depicted in Figure 7.40 [51]. The shown voltages are 1 minute average values. The nominal voltage of 230 V is denoted by V_n (i.e. U_n).

Significant increase in voltage is observed with PV active power production. The volt–var function starts absorbing reactive power if the average voltage (middle dark trace) exceeds 1.05 V_n or 241.5 V. Maximum reactive power is absorbed if the average voltage exceeds 1.08 V_n or 248 V. APC is commenced only if maximal phase-to-neutral voltage exceeds 1.09 V_n or 250.7 V. This strategy results in effective voltage control with only few minor curtailments of active power around noon time.

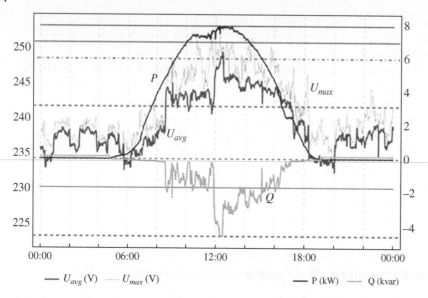

Figure 7.40 Daily behavior of smart PV inverter with Q (V) and P (V) functions. *Source*: Hes et al. [51].

7.14.6.2 Use Case 2: Increasing DER Hosting Capacity in MV Networks

In this use case, VVC is implemented on several DERs connected to medium voltage grids. The field test results for a 1.1 MW PV system at Zamberk for a sunny day and a cloudy day are illustrated in Figure 7.41 [51] and Figure 7.42 [51], respectively. The VVCs on both days show accurate changes between the reactive power absorption and reactive power injection modes when the voltage goes above and below the voltage tolerance band (indicated by dashed lines), respectively. A statistical compilation of the PV system behavior over a three-month period confirmed reliable and adequate reactive power response of the VVC in stabilizing the voltage in MV grid.

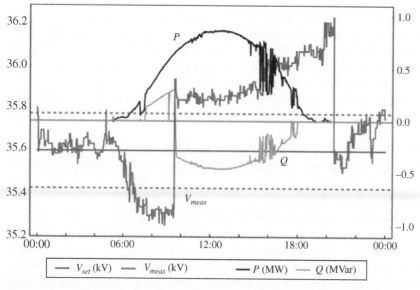

Figure 7.41 Demonstration of volt–var control on a 1.1 MW PV system on a sunny day. *Source*: Hes et al. [51].

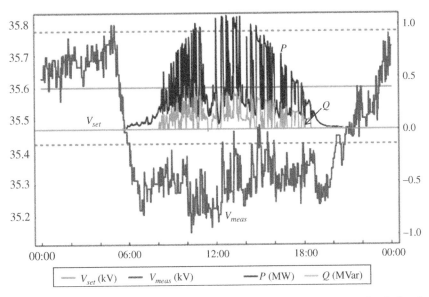

Figure 7.42 Demonstration of volt–var control on a 1.1 MW PV system on a cloudy day. *Source*: Hes et al. [51].

Key Takeaways

This project anticipates that VVC system on DER connected to MV level could significantly enhance DER hosting capacity by 20–100%, depending on the grid topology [51]. It is also expected that the autonomous $Q(V)$ and $P(V)$ functions together with smart energy storage concept should significantly reduce the number of regions with insufficient DER hosting capacity in LV grids and thus reduce costs for DER integration (costs for grid reinforcement). An increase between 30% and 50% in hosting capacity of DERs is foreseen [51].

7.14.7 Hosting Capacity Experience in Salt River Project

Field implementation studies were performed under a joint project between Salt River Project (SRP) and EPRI to examine the benefits of operating smart PV inverters alone and in combination with conventional voltage control equipment controlled from a distribution management system (DMS) [23]. Smart PV inverters were tested with the following three types of controls on more than 775 PV generation sites:

1) SIs with autonomous settings that were not changed (i.e. "set and forget"). These settings served as baseline for comparing the other two cases 2) and 3)
2) SIs with settings that were changed seasonally
3) Real-time control of SI settings and conventional voltage control equipment (i.e. LTCs and capacitor banks) on a continuous basis from a DMS

Following outcomes were reported:

Conventional voltage control equipment are effective in decreasing voltage violations and maintaining a desirable voltage profile along the feeder. The most effective voltage control is obtained from conventional equipment having high reactive power rating or from those which control a large segment of distribution feeders, i.e. LTCs. Modifying the settings of existing voltage control equipment can provide a low-cost and effective solution for increasing hosting capacity, especially in systems with low penetrations of SIs. In this study, voltage control through operation of traditional voltage controller equipment by DMS increased the number of LTC operations by 2–7 per day and total capacitor banks by 0.5–1 per day, on an average.

Voltage control from conventional voltage control equipment can be complemented by SIs of large sizes utilizing DMS to provide enhanced benefits. However, the costs of communication and control of a large number of small size residential smart PV inverters may not be justified by the relatively small benefits of such implementation.

SIs can increase hosting capacity when feeders are experiencing voltage constraints. If the voltage is at the upper limit SI functions that absorb reactive power (i.e. constant PF) are more effective. If the feeder voltage hovers near both the upper and lower limit, SI functions that both absorb and inject reactive power (i.e. volt–var function) will be more appropriate. The beneficial impact of SI increases with the size of PV inverters, as larger reactive power support will have a bigger impact on feeder voltages.

Reactive power-based SI functions did not have any adverse impact (i.e. need for curtailment) on active power generation of PV systems. Maximum grid benefits are obtained when SI functions are customized or optimized based on feeder characteristics; and size and location of smart PV inverters. However, methods for determining optimal settings under diverse conditions are still under research and development.

In some cases, SI functions were unable to increase hosting capacity in a substantial manner as the increased current flow due to reactive power absorption caused thermal limits to be violated.

Key Takeaways

Conventional voltage control equipment can be coordinated with SIs through DMS to provide an enhanced voltage control.

In commissioning of smart PV inverters a better terminology to indicate the direction of reactive power flow is to specify PV inverters as either "injecting" or "absorbing" reactive power, instead of "leading" or "lagging."

It is desirable to request SI capabilities on PV inverters even if they are not needed at the time of installation. This is because the cost of providing these functions is very small compared to the benefits they can provide in future, if and when needed.

The benefits of controlling large smart PV inverters through DMS are substantial. However, controlling smaller smart PV inverters increases the cost of communication and control, which may not be justifiable from the benefits obtained. The DMS industry capabilities required for communicating and controlling distribution system equipment are still evolving.

7.15 Conclusions

This chapter presents a review of various technologies with case studies for enhancing the PV hosting capacity of distribution networks. These technologies include conventional equipment such as OLTC, voltage regulators and switched capacitors; energy storage systems; and SIs.

Hosting capacity in distribution systems is limited by several factors among which the most important is overvoltage caused by active power injection from PV systems. The hosting capacity is influenced by the size and location of DERs, electrical characteristics of the distribution networks, the type of DERs (without or with SI functions), and the specific SI function employed. Higher PV hosting capacity is obtained for feeders with higher voltage class, lower impedance (strong feeders), shorter length, less loads, and less capacitive reactive power (kvar) support. PV hosting capacity can be further increased by locating PV systems close to substation.

Several case studies around the world have demonstrated that SI functions are very effective in enhancing PV hosting capacity. Lower constant PF is better for improving hosting capacity. For instance, 0.9 inductive PF provides superior performance than 0.95 inductive PF. However, constant PF causes thermal violations in some feeder segments due to large absorption of reactive power from substation.

VVC is generally more effective than constant PF in improving hosting capacity. VVC results in lower reactive power absorption compared to constant PF operation. Hence a substantial amount of volt–var-based SI controls can be implemented in distribution networks with only a moderate increase in the total circuit current. This increased reactive current may be supplied by capacitor banks installed in the distribution circuit (e.g. at substation) or may be imported from the upstream high voltage level sub-transmission circuit. This aspect, however, needs examination on a case by case basis, so as not to cause any adverse impact on the bulk transmission system.

Volt–var function with reactive power priority is more effective than watt priority. In several studies, it was noted that volt–var curve without deadband is more effective in increasing hosting capacity than with deadband. Volt–var combined with volt–watt is most effective in mitigating high overvoltages, with minimal power curtailment.

At the simplest level, default settings for SI functions as specified in various Standards, i.e. IEEE 1547-2018, Rule 14H, or California Rule 21, may be used. These settings are not aggressive. Autonomous operation of SIs with default settings may avoid costly communications to modify the settings. In an intermediate level, good settings may be determined for different SI functions that generally perform well under a wide range of system conditions. In the most detailed level, maximal grid benefits are obtained when SI functions are customized or optimized based on feeder characteristics; and size and location of smart PV inverters; and solar and load conditions. The optimal settings for SIs for achieving different performance criteria can be obtained by QSTS studies.

DMS control provides enhanced control of both smart PV inverters and conventional voltage control equipment. The benefits of controlling large smart PV inverters through DMS are substantial. However, controlling smaller smart PV inverters increases the cost of communication and control, which may not be justifiable from the benefits obtained. The DMS industry capabilities required for communicating and controlling distribution system equipment including SIs, are still evolving.

It is recommended that utilities should ask for SI capabilities in PV inverters at the time of installation, even if they are not required immediately. These may be substantially beneficial in future.

References

1 CEATI International Inc. (2020). Enhancing Connectivity of DGs by Control Coordination of Smart Inverters. CEATI International Inc. Montreal, QC, Canada, *Rep. No. T184700 #5178*.

2 EPRI (2015). Distribution Feeder Hosting Capacity: What Matters when Planning for DER? EPRI, Palo Alto, CA, USA, *White Paper 3002004777*.

3 Smith, J.W., Sunderman, W., Dugan, R., and Seal, B. (2011). Smart inverter volt/var control functions for high penetration of PV on distribution systems. In *Proc. 2011 IEEE/PES Power Systems Conference and Exposition*, 1–6.

4 Key, T. (2018). Solar realities and grid capacity for hosting with smart inverters. IEEE Tutorial on 'Smart Inverters for Distributed Generators', In *Proc. 2018 IEEE PES General Meeting*.

5 Jain A.K., Horowitz K., Ding F., Gensollen N. et.al. (2019). Quasi-Static Time Series PV Hosting Capacity Methodology and Metrics: Preprint. National Renewable Energy Laboratory, Golden, CO, USA, *NREL/CP-5D00-72284*.

6 EPRI (2017). Arizona Public Service Solar Partner Program - Advanced Inverter Demonstration Results. EPRI, Palo Alto, CA, USA, *Rep. 3002011316*.

7 Pecenak, Z.K., Kleissl, J., and Disfani, V.R. (2017). Smart inverter impacts on California distribution feeders with increasing PV penetration: a case study. In *Proc. 2017 IEEE Power & Energy Society General Meeting*, 1–5.

8 IEA (2014). High Penetration of PV in Local Distribution Grids: Subtask 2: Case Study Collection. International Energy Agency PVPS Program *Rep. IEA PVPS T14-02*.

9 Bucher, C., Andersson, G., and Küng, L. (2013). Increasing the PV hosting capacity of distribution power grids – a comparison of seven methods. In *Proc. 28th European PV Solar Energy Conference and Exhibition (EU PVSEC)*.

10 Gagrica, O., Nguyen, P.H., Kling, W.L., and Uhl, T. (2015). Microinverter curtailment strategy for increasing photovoltaic penetration in low-voltage networks. *IEEE Transactions on Sustainable Energy* 6: 369–379.

11 Tonkoski, R., Lopes, L.A.C., and El-Fouly, T.H.M. (2011). Coordinated active power curtailment of grid connected PV inverters for overvoltage prevention. *IEEE Transactions on Sustainable Energy* 2: 139–147.

12 Stetz, T., Marten, F., and Braun, M. (2013). Improved low voltage grid-integration of photovoltaic systems in Germany. *IEEE Transactions on Sustainable Energy* 4: 534–542.

13 Varela, J., Hatziargyriou, N., Puglisi, L.J. et al. (2017). The IGREENGrid project: increasing hosting capacity in distribution grids. *IEEE Power and Energy Magazine* 15: 30–40.

14 NREL (2016). Energy Storage Requirements for Achieving 50% Solar Photovoltaic Energy Penetration in California. National Renewable Energy Laboratory, Golden, CO, USA, *Report NREL/TP-6A20-66595*. https://www.nrel.gov/docs/fy16osti/66595.pdf, (accessed November 16th, 2020).

15 Bartak, G., Holenstein, H., and Meyer, J. (2007). *Technical Rules for the Assessment of Network Disturbances*, 2e. Aarau: Verband Schweizerischer Elektrizitätsunternehmen.

16 O'Connell, R., Volkmann, C., and Brucke, P. (2019). Regulating Voltage: Recommendations for Smart Inverters. GridLab. Berkeley, CA, USA, *Report*.

17 EPRI (2018). Recommended Smart Inverter Settings for Grid Support and Test Plan. Interim Report. EPRI, Palo Alto, CA, USA, *Report 3002012594*.

18 Smith, J. (2015). Analysis to Inform CA Grid Integration: Methods and Default Settings to Effectively Use Advanced Inverter Functions in the Distribution System. EPRI, Palo Alto, CA, USA, *Report 3002007139*.

19 IEEE (2018). *IEEE standard for interconnection and interoperability of distributed energy resources with associated electric power systems interfaces*. IEEE Std 1547-2018 (Revision of IEEE Std 1547-2003).

20 CPUC (2019). *California Electric Tariff Rule 21*. https://www.cpuc.ca.gov/Rule21/ (accessed 21 July 2019).

21 Hawaii Electric Company, Inc. (2018). *Rule 14H Interconnection of Distributed Generating Facilities with the Company's Distribution System*. Hawaii Electric Company, Inc.

22 EPRI (2016). Analysis to Inform California Grid Integration Rules for Photovoltaics: Final Results on Inverter Settings for Transmission and Distribution System Performance. EPRI, Palo Alto, CA, USA, *Rep. No. 300200830*.

23 EPRI (2019). SRP Advanced Inverter Project - Research Findings. EPRI, Palo Alto, CA, USA, *Rep. 3002016625*.

24 Rylander, M. (2014). Computing solar PV hosting capacity of distribution feeders. In *Proc. 2014 IEEE/PES Transmission and Distribution Conference & Exposition (T&D)*.

25 EPRI (2016). Common Functions for Smart Inverters: 4th Edition. EPRI, Palo Alto, CA, USA, *Tech. Rep. 3002008217*.

26 CEATI International Inc. (2016). Investigation of Smart Inverters. CEATI International Inc. Montreal, QC, Canada, *Rep. No. T154700 #50/128*.

27 Ding, F. and Mather, B. (2017). On distributed PV hosting capacity estimation, sensitivity study, and improvement. *IEEE Transactions on Sustainable Energy* 8: 1010–1020.

28 Rylander, M., Smith, J., and Li, H. (2014). Determination of smart inverter control settings to improve distribution system performance. In *Proc. CIGRE US National Committee 2014 Grid of the Future Symposium*, 1–6.

29 Li, H., Smith, J., and Rylander, M. (2014). Multi-inverter interaction with advanced grid support function. In *Proc. CIGRE US National Committee 2014 Grid of the Future Symposium*.

30 Bello, M., Montenegro-Martinez, D., York, B., and Smith, J. (2018). Optimal settings for multiple groups of smart inverters on secondary systems using autonomous control. *IEEE Transactions on Industry Applications* 54: 1218–1223.

31 Kraiczy, M., York, B., Bello, M. et al. (2018). Coordinating smart inverters with advanced distribution voltage control strategies. In *Proc. 2018 IEEE Power & Energy Society General Meeting*, 1–5.

32 Reno, M.J., Deboever, J., and Mather, B. (2017). Motivation and requirements for quasi-static time series (QSTS) for distribution system analysis. In *Proc. 2017 IEEE Power & Energy Society General Meeting*, 1–5.

33 IEEE (2018). Impact of IEEE 1547 Standard on Smart Inverters. IEEE Power & Energy Society New York, NY, USA, *Techn. Rep. PES-TR67*.

34 EPRI (2019). Advancing Smart Inverter Integration in Utah - Research Highlights. EPRI, Palo Alto, CA, USA, *Tech. Rep. 3002015319*.

35 New York State Energy Research and Development Authority (NYSERDA). (2019). Mitigation Methods to Increase Feeder Hosting Capacity, NYSERDA *Report Number 19-45*. Prepared by EPRI, Palo Alto, CA, USA.

36 EPRI (2018). Smart Inverter Functions and Settings for Distribution Applications. EPRI, Palo Alto, CA, USA, *Rep. 3002013380*.

37 EPRI (2018). Mitigation Methods to Increase Feeder Hosting Capacity. EPRI, Palo Alto, CA, USA, *Rep. 3002013382*.

38 Varma, R.K. and Siavashi, E.M. (2018). PV-STATCOM: A new smart inverter for voltage control in distribution systems. *IEEE Transactions on Sustainable Energy* 9: 1681–1691.

39 CEATI International Inc. (2017). Electrical Energy Storage in Distribution Systems for Mitigation of Power Quality Issues. CEATI International Inc. Montreal, QC, Canada, *Rep. No. T164700 #5173*.

40 EPRI (2015). Smart Grid Inverters to Support Photovoltaic in New York Distribution Systems. EPRI, Palo Alto, CA, USA, *Rep. No. 3002006278*.

41 Hoke, A., Giraldez, J., Symko-Davies, M. et al. (2018). Integrating More Solar with Smart Inverters: Preprint. National Renewable Energy Laboratory, Golden, CO, USA, *Rep. NREL/CP-5D00-71766*.

42 Giraldez, J., Nagarajan, A., Gotseff, P. et al. (2017). Simulation of Hawaiian Electric Companies Feeder Operations with Advanced Inverters and Analysis of Annual Photovoltaic Energy Curtailment. National Renewable Energy Laboratory, Golden, CO, USA, *Techn. Rep. NREL/TP-5D00-68681*. https://www.nrel.gov/docs/fy17osti/68681.pdf, (accessed November 16th, 2020).

43 Giraldez, J., Hoke, A., Gotseff, P. et al. (2018). Advanced Inverter Voltage Controls: Simulation and Field Pilot Findings. National Renewable Energy Laboratory, Golden, CO, USA, *Techn. Rep. NREL/TP-5D00-72298*.

44 Haghi, H.V., Pecenak, Z., Kleissl, J. et al. (2019). Feeder impact assessment of smart inverter settings to support high PV penetration in California. In *Proc. 2019 IEEE Power & Energy Society General Meeting*, 1–5.

45 EPRI (2019). Advancing Smart Inverter Integration in Utah - Final Report. EPRI, Palo Alto, CA, USA, *Rep. No. 3002015334*.

46 Mather, B. (2014). NREL/SCE High-Penetration PV Integration Project: Report on Field Demonstration of Advanced Inverter Functionality in Fontana, CA. National Renewable Energy Laboratory, Golden, CO, USA, *Rep. NREL/TP-5D00-62483*.

47 Mather, B. and Gebeyehu, A. (2015). Field demonstration of using advanced PV inverter functionality to mitigate the impacts of high-penetration PV grid integration on the distribution system. In *Proc. 2015 IEEE 42nd Photovoltaic Specialist Conference (PVSC)*, 1–6.

48 Kadam, S., Bletterie, B., Lauss, G. et al. (2014). Evaluation of voltage control algorithms in smart grids: results of the project: morePV2grid. In *Proc. 29th European Photovoltaic Solar Energy Conference and Exhibition*.

49 Dierckxsens, C., Bletterie, B., Deprez, W. et al. (2015). Cost-effective integration of photovoltaics in existing distribution grids: results and recommendations. Meta PV Project, Europe, *Report*.

50 Dierckxsens, C., Bletterie, B., Lemmens, J. et al. (2015). Meta PV, Technical Project Overview and Results. Meta PV Project, Europe, *Report*.

51 Hes, S., Kula, J., and Svec, J. (2018). Increasing of renewables hosting capacity in the Czech Republic in terms of European Project InterFlex (Case Study). In *Proc. 2018 International Conference and Utility Exhibition on Green Energy for Sustainable Development (ICUE)*, 1–7.

8

CONTROL COORDINATION OF SMART PV INVERTERS

Multiple smart inverter-based Distributed Energy Resources (DERs) are being increasingly connected in close vicinity on distribution feeders. Since these are power electronic converters performing similar functions as reactive power control (RPC) or voltage control at similar speeds of response, their controllers can potentially interact with each other. The concept of coordinating multiple control devices is explained in this chapter, using examples from Flexible AC Transmission System (FACTS) technology. FACTS Controllers and smart inverters both perform voltage control at a rapid rate. Therefore, the challenges already experienced with coordinating FACTS Controllers and their methods of resolution will help in understanding the coordination issues amongst smart inverters and determining their remedial measures.

This chapter presents the issues related to the control coordination of smart photovoltaic (PV) inverters with conventional voltage control equipment. Control coordination issues of smart inverter functions within a PV system are first discussed. Case studies of control interactions between same and different smart inverter functions among neighboring smart inverters are described [1]. A detailed small-signal study of the various factors causing control interaction between two smart inverters in a distribution feeder, validated by electromagnetic transients simulations, is presented. A comprehensive control coordination study of 100 MW PV-STATCOM and 100 MW Doubly fed induction generator (DFIG)-based wind farm connected to series compensated line for mitigating subsynchronous oscillations (SSO) is described.

8.1 Concepts of Control Coordination

8.1.1 Need for Coordination

A simple analogy is provided here to explain the need for control coordination. We all know that TV news channels present multiple pieces of information simultaneously. For instance, there is a presentation of news by an individual or a panel conversation which takes place in real time; one moving strip carrying breaking news; another moving strip containing stocks and shares prices; another containing sports news; and a flashing screen depicting weather or highway conditions. Have we ever wondered why such diverse array of information through various modes of delivery does not cause any confusion in our minds? This is because all the different data are being delivered at different speeds; or in the language of control systems, with different time constants. If all the various

Smart Solar PV Inverters with Advanced Grid Support Functionalities, First Edition. Rajiv K. Varma.
© 2022 The Institute of Electrical and Electronics Engineers, Inc. Published 2022 by John Wiley & Sons, Inc.

strips of news and on-screen information were moving at the same speed, the human mind would very likely get confused. The same principle also applies to devices involved in controlling power system quantities. If multiple devices are operating with similar speeds of control, and controlling the same quantity, e.g. voltage, control-conflicts are likely to arise.

In essence, potential control interactions can occur among devices which operate at similar speeds, regardless of whether the speed is slow or fast.

8.1.2 Frequency Range of Control Interactions

This concept is explained from the realm of FACTS Controllers. The frequency ranges in which control interactions of FACTS Controllers and High Voltage DC (HVDC) systems occur are enumerated below [2, 3]. These are also the frequency ranges in which control interactions are expected to occur with smart inverters, both for solar PV and wind power systems.

- 0 Hz for steady-state interactions
- 0–3/5 Hz for electromechanical oscillation interactions
- 2–15 Hz for control system interactions
- 5–50/60 Hz for subsynchronous oscillation (SSO) interactions
- >15 Hz for electromagnetic transients, high-frequency resonance or harmonic resonance interactions, and network-resonance interactions.

Smart inverters may also experience similar controller interactions as they also provide voltage control and may possibly, in near future, provide other grid support functions similar to FACTS Controllers.

8.1.2.1 Steady-State Interactions

These interactions relate to the system level controls of the FACTS Controllers and are not linked to controller dynamics. For instance, the interactions correspond to the slopes of the $V–I$ characteristics of shunt connected FACTS Controllers, the frequency droop characteristics of battery energy storage systems or wind power systems, etc.

Load flow software with the appropriate models of FACTS Controllers is generally utilized to examine such control interactions. Centralized controls or a combination of local and centralized controls of participating devices are recommended for achieving the desired coordinated control performance.

8.1.2.2 Electromechanical Oscillation Interactions

Electromechanical oscillation interactions typically occur among FACTS controls, synchronous generator controls involving power system stabilizers (PSS), HVDC power oscillation damping controls, synchronous condensers, etc. [4]. These oscillatory modes can be classified into two categories: *local area mode* (associated with single generator or a small group of generators), typically in the range of 0.7–2.5 Hz; and *interarea modes* (associated with large group of machines across the system), generally in the range of 0.1–1.2 Hz[5].

Different FACTS Controllers can be assigned to damp different oscillatory modes, in which case there will be no control interaction. Conversely, coordinated controls of different FACTS Controllers are utilized to simultaneously damp the same oscillatory modes.

Power oscillation damping can be provided by Type 3 (DFIG) wind plants [6, 7] and Type 4 (full converter based) wind plants [8]. Inverter-based wind plants are currently used to provide power oscillation damping [8]. Recently, novel controls of solar PV systems as STATCOM (PV-STAT-COM) have been proposed to provide damping for electromechanical modes [9–12] in addition to other solar PV farm controls for damping power oscillations [13]. These are subject to the same principles of control coordination as FACTS Controllers.

Eigenvalue analysis programs are utilized in the power industry to determine the oscillatory modes in a given power system and design the coordinated controls for participating devices. These studies are subsequently validated by transient stability simulation studies.

8.1.2.3 Control System Interactions

Interactions may occur among the control systems of different voltage control devices such as FACTS Controllers and/or HVDC converters. These result in the onset of oscillations in the range of 2–15 Hz (which can extend up to 30 Hz in some cases). These oscillations have been reported to occur among the voltage controllers of multiple static var compensators (SVCs) [2, 14] and also due to the resonance between shunt reactors and series capacitors [2, 15]. The interactions are highly influenced by system strength and are resolved either by slowing the voltage controller gains (not a preferred method) or through coordinated design of controller gains of different devices.

These control interactions are likely to be experienced with smart inverters since they also engage in fast voltage control of nearby buses on a distribution feeder.

The relatively higher-frequency oscillations can be investigated by eigenvalue analysis programs with modeling capabilities extended to analyze higher-frequency modes. The performance of coordinated controls is validated by electromagnetic-transient programs (EMTP) and/or real-time digital simulators (e.g. RTDS).

8.1.2.4 Subsynchronous Oscillation (SSO) Interactions

Subsynchronous oscillations can be caused by interaction between the generator torsional system and series-compensated-transmission lines [16, 17], HVDC converter controls [18, 19], generator excitation controls, or even the SVCs [20, 21]. Subsynchronous interactions have been reported with Type 3 wind turbine generators and series compensated lines [22, 23]. These interactions are initiated in the frequency range of 5–60 Hz and can potentially damage generator shafts, impair control circuits due to overvoltages, etc.

Frequency scanning is typically utilized as an initial screening tool to identify the resonant modes [24]. Subsequently, eigenvalue analysis is used to identify the interactions and design coordinated controls. It is important to note that eigenvalue programs must be enhanced to include adequate network dynamics. Finally, electromagnetic transient studies are used to validate the performance of the designed coordinated controls.

8.1.2.5 High-frequency Interactions

High-frequency oscillations in excess of 15 Hz are triggered by large nonlinear disturbances, such as switching of capacitors, reactors, or transformers. Network resonant modes resulting from interactions between line/generator/transformer inductances and shunt capacitances, contributed by bus capacitors, line charging capacitances, cable capacitances, etc., constitute a major cause of controller interaction among FACTS Controllers [2]. These resonant modes can potentially lead to controller instabilities [2]. Harmonic instabilities are also known to occur from synchronization or voltage-measurement systems, transformer energization, or transformer saturation caused by geomagnetically induced currents (GICs) [2]. FACTS Controllers must be coordinated to minimize such interactions.

8.1.3   Principle of Coordination

Controls for individual devices, such as PSS, FACTS Controllers, HVDC systems, etc., are usually optimally tuned under the assumption that the remaining power system control devices are passive. However, in reality, the dynamics of the other devices are very much active. Therefore, individually optimized controllers are not optimal when all devices are performing control actions simultaneously. The adverse control interactions can only be avoided through control coordination. Control coordination is thus a process that involves simultaneous tuning of the controllers of various active devices, resulting in an overall improvement of the different control schemes.

8.2   Coordination of Smart Inverters with Conventional Voltage Controllers

Smart inverter controls such as volt-var, volt-watt are fast acting (with a response time of few seconds). They still need to work in unison with the relatively slow-acting voltage control equipment such as transformer tap changers, voltage regulators, and capacitor banks, etc. Studies are ongoing to understand these challenges and develop their solutions. The experiences gained so far are described below.

8.2.1   European IGREENGrid Project

Integrating Renewables in the EuropEaN electricity Grid (IGREENGrid) is a major European initiative involving several countries for studying methods for increasing the grid integration of renewables [1, 25]. Some of the major results from this project are summarized here.

Hosting capacity under various system conditions was evaluated using optimum power flow in which the objective function (OF) was designed to maximize the injection of active power from different generating sources without violating either voltage or loading limits. This was done for distribution systems in different European countries. The effectiveness of different smart grid (SG) solutions in increasing hosting capacity (HC) is compared in Table 8.1 [25]. It was noted that centralized solutions were, in general, more effective than decentralized approaches. Use of individual solutions (STATCOM, on-load tap changers [OLTC], etc.) or several local solutions operating individually were not found to be as effective. This is because the effectiveness of distributed solutions depends on the location of distributed generators (DGs). The most effective solutions were found to be a coordination of OLTC and other mechanisms such as STATCOM, RPC of distributed generators (e.g. smart inverter functions) and intelligent deployment of field measurements. The centralized solution utilized optimal power flow (OPF) studies to determine the best settings for OLTCs as well as the RPCs on DGs for coordinated operation. This approach provided significantly higher hosting capacity than those achieved with individual controls.

A small number of sensors placed at well-chosen locations to capture voltages of most sensitive buses were seen to be more effective than installing a large number of sensors extensively throughout the distribution systems.

Voltage and line loading are two major constraints that limit hosting capacity in distribution feeders. For instance, rural feeders are mainly voltage constrained, whereas urban feeders are typically loading constrained. In the two major distribution grids of different distribution system operators (DSO), studied in this project, one distribution grid had 90% voltage constrained feeders, whereas the other had 77% voltage constrained lines.

Table 8.1 Increase in hosting capacity obtained from different smart grid (SG) solutions.

Centralized smart grid solutions			
		HC increase	
Demo	**MV centralized voltage control with** ···	**Average value (%)**	**Relative standard deviation**
Germany	Field measurements – OLTC	67.00	0.57
Spain	OLTC + STATCOM control	64.53	0.67
Austria	Field measurements – OLTC + DG reactive power control	62.99	1.38
Austria	Field measurements – OLTC control	53.36	1.27
Spain	STATCOM control	20.02	0.19

Distributed SG solutions			
		HC increase	
Demo	**MV centralized voltage control with** ...	**Average value (%)**	**Relative standard deviation**
Germany	AVR	37.75	2.87
France	DG reactive power control (droop Q-V control)	37.13	1.62
France	DG reactive power control (fixed $\tan(\pi)$)	34.81	3.10
Italy	OLTC + DG reactive power control	29.72	3.53
Greece	DG curtailment + DG reactive power control	13.89	2.25
Greece	DG curtailment	10.83	0.39

Source: Modified from Varela et al. [25].

From studies of various networks of different countries, it was seen that voltage control provided by DGs through volt–var controls (VVCs) could successfully increase hosting capacity in about 60% of feeders examined. Coordinated control of OLTC and DGs resulted in highest increase in hosting capacity. The coordination is achieved through OPF which provided the best settings for significantly increasing the hosting capacity of networks.

In systems exhibiting contrasting voltage issues, a combined use of OLTC at both high voltage (HV)/medium voltage (MV) and MV/low voltage (LV) transformers enhanced the flexibility of voltage control, increasing hosting capacity by as much as 179% in the systems investigated. Insertion of autotransformers in middle of critical MV lines was also found quite helpful in enhancing hosting capacity. Active demand management and installation of energy storage at customer locations further contributed to voltage control of distribution lines.

8.2.2 Interaction of Smart Inverters with Load Tap Changing Transformers

A detailed interaction study of PV inverters with different RPC based smart inverter functions such as constant (nonunity) power factor (PF), watt–power factor PF(P), volt–var control $Q(V)$, and transformers with OLTC is presented in [1, 26]. It is shown that smart PV inverters may cause an increased variation of reactive power flow across the substation transformer causing voltage fluctuations, that may lead to increased transformer tap-change operations.

Studies are conducted on the CIGRE MV benchmark grid [27], which is representative of a German MV grid as shown in Figure 8.1. The 20 kV network connects to the transmission network through a 110/20 kV OLTC transformer. An 800 kW solar system is connected at each MV bus. A medium-level PV penetration is considered with the total PV capacity being 36% of the peak load. RMS simulations are performed using PowerFactory software using 1 second simulation time-step.

According to grid code EN 50160, the maximum allowable voltage magnitude at the German MV and LV levels is 110% of V_N (=20 kV in this case). The OLTC regulates the voltage at T1 to 104% of V_N with a 1% deadband. The maximum allowable voltage in the MV level is set to 107% of V_N while a voltage rise of 3% of V_N is allocated for the LV network.

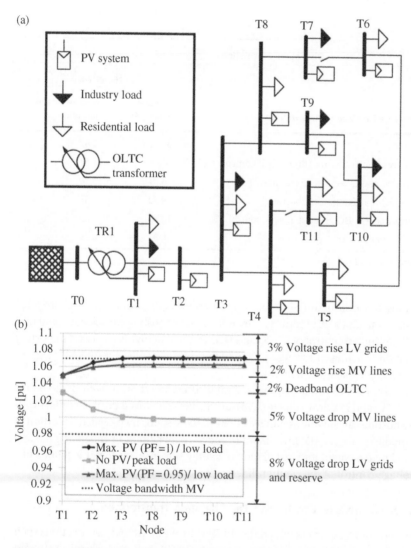

Figure 8.1 MV grid based on the CIGRE MV benchmark grid. (a) results of the worst-case analysis for the critical feeder T1–T11. (b) the applied allocation of the voltage bandwidth for a German MV and LV grid (right). (PF = power factor). *Source:* Kraiczy et al. [26].

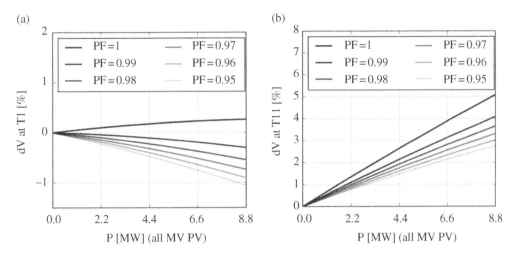

Figure 8.2 Voltage variations dV caused by all PV systems. (a) at node T1, (b) at node T11. *Source:* Kraiczy et al. [26].

It is noted from Figure 8.1 that the maximum voltage at bus T11 during low load conditions is 1.073 pu which violates the stipulated voltage limits. This makes a case for the use of reactive power based smart inverter functions on PV systems.

The voltage variations at buses T1 and T11 are depicted in Figure 8.2. It is assumed that the slack bus voltage at bus T0 is 1.0 pu. To begin with, OLTC control is deactivated.

Increase in the active power output from all solar PV systems causes a large increase in voltage at bus T11, but a relatively much lower voltage rise at Bus T1. The implementation of constant PF smart inverter function, especially at PF = 0.95 absorption, reduces the voltage rise at T11. However, an increased reactive power absorption by all the PV systems causes the voltage at Bus T1 to drop substantially, which will likely initiate OLTC switching. On the other hand, volt–var $Q(V)$ control is sensitive to bus voltages. Hence, if volt–var $Q(V)$ control is implemented on the solar PV systems, less reactive power is absorbed during LVs at T1. This will not reduce voltage at Bus T1 to levels that may require OLTC switching.

In this study, the OLTC controls the voltage of a strong node (having high short circuit level and a high X/R ratio), and therefore its voltage is more sensitive to reactive power flow across it (caused by reactive power-based controls of smart PV inverters).

8.2.2.1 System Modeling

Interactions between smart PV inverters (with RPCs) and OLTC transformers are studied with RMS based simulations using PowerFactory software using 1 second simulation time-step.

It is known that the variability of solar power output decreases substantially as the size of PV plants increases. Thus, the variation in solar power output from a single residential PV system is much more than an aggregation of several residential solar plants over a geographical area, or from a utility size PV system. In this study, the conglomeration of PV systems was modeled by a point sensor measurement with an adjustable low pass filter (of first order), to represent the smoothing effect of solar variability in the overall MV network. This approach is especially useful when comprehensive measurement data (e.g. solar irradiation data at different grid locations with high temporal resolution) is not available for all solar PV systems.

Loads are comprised of industrial and residential loads. These are modeled using ZIP (constant impedance, constant current, and constant impedance) models with appropriately selected coefficients for active power and reactive power to represent actual load behavior [28]. The ZIP load model can be represented by the equations [29]:

$$P = P_0\left[a_1\left(\frac{V}{V_o}\right)^2 + b_1\left(\frac{V}{V_o}\right) + c_1\right] \tag{8.1}$$

$$Q = Q_0\left[a_2\left(\frac{V}{V_o}\right)^2 + b_2\left(\frac{V}{V_o}\right) + c_2\right] \tag{8.2}$$

where, P denotes the active power at a voltage V, P_0 represents load active power at voltage V_o, and a_1, b_1, and c_1 are coefficients of constant impedance, constant current, and constant power loads, respectively. Similarly, Q, Q_0, a_2, b_2, and c_2 represents similar quantities with respect to reactive power. The sum of the three coefficients must be equal to 1.

The OLTC transformer is modeled with a voltage deadband of ±1% around the voltage setpoint (e.g. 104% of nominal voltage) and a delay function. This delay prevents OLTC switching during short duration voltage variations.

8.2.2.2 Simulation Studies

Simulation studies are performed using a Monte Carlo simulation of external grid voltage, i.e. the voltage at upstream HV terminal of the OLTC transformer. It is assumed that the external grid voltage (i.e. the slack bus voltage) may itself be varying due to network conditions. Multiple Monte Carlo simulation studies are performed to avoid basing conclusions on a single external bus voltage condition. For each external bus voltage conditions, all the four RPC techniques are evaluated. Almost 100 Monte Carlo iterations are performed for each RPC mode [26]. The different RPC techniques are as follows:

i) RPC Mode 1: No RPC (PF = 1);
ii) RPC Mode 2: Fixed PF (PF = 0.95);
iii) RPC Mode 3: PF(P) control (watt - power factor control);
iv) RPC Mode 4: $Q(V)$ control (volt-var control).

The above RPC strategies are evaluated in terms of

a) Maximum voltage magnitude at different buses
b) Additional OLTC switching operations, and
c) Provision of reactive energy

An effectiveness ratio EFF_Q is introduced to quantify the benefit of different RPC techniques. This effectiveness ratio is defined as the reduction in maximum grid voltage with the additional reactive power absorption by PV systems. It is noted that additional reactive power support from PV systems will cause increased losses in the PV inverters, enhanced reactive power flows across the network leading to overloading of grid assets, line losses, and increased costs of balancing reactive power flows in the network. Hence, a high effectiveness ratio is desired to be achieved in all cases.

The following conclusions are drawn based on the studies performed:

1) The unintended switching of OLTC transformers adversely impacts the coordination of transformers and smart inverters with different RPC strategies for voltage control.

2) RPC strategies of smart PV inverters are less likely to cause undesirable OLTC switching if the transformer is located at a weak bus, i.e. having low short circuit level.

3) RPC strategies at PV inverters are more likely to cause unintended switching operations of OLTC transformers connected at strong buses with high X/R than those connected at weak buses with low X/R.

4) RPC strategies can potentially cause an increase in unintended OLTC switchings, but the impact is different for different RPC strategies.

5) The constant PF and watt–power factor PF(P) strategies are more likely to cause an increase in unintended OLTC switchings with a decreased effectiveness ratio.

6) The volt–var $Q(V)$ control can either avoid or decrease the risk of unintended OLTC switching operations, thereby resulting in a high effectiveness ratio.

7) The OLTC switching operations can be decreased by increasing the deadband and also by extending the delay time of the OLTC transformer, by a slight amount.

Even though coordinated control of PV systems and voltage regulators are desirable, autonomous controls will still be important due to rapid response, no requirement of communication links, and simple methods of determining control parameters.

8.2.3 Coordination of Smart Inverters with Distribution Voltage Control Strategies

Optimal settings of volt–var smart inverter function are determined in [1, 30] to obviate or minimize unintentional OLTC switching operations over a wide range of system operating conditions. Inappropriate settings of smart inverter functions can lead to adverse interaction and increase in number of OLTC switching operations. Studies are performed on the IEEE 13 bus feeder system [31] with minor modifications. The 4.16 kV distribution feeder has three single-phase feeder head voltage regulating device i.e. load tap change (LTC) transformer and two capacitor banks installed at bus 611 and 675, respectively, as depicted in Figure 8.3 [30].

8.2.3.1 System Modeling
Load
The quasi-static time-series (QSTS) simulations are performed by applying the feeder loads to a time series profile of typical peak and light load scenario. The IEEE 13 bus feeder has an unbalanced load distribution with a total peak load of 3.47 MW. The nominal load power is the active power consumption described in [31]. Also, the loads operate with a fixed PF based on the active and reactive power consumption of different loads as mentioned in [31].

Load Tap Changer
Three single-phase voltage regulating devices – LTCs are connected at the start of the feeder, each of which controls the voltage in each phase of the feeder. The LTC regulates the voltage at the feeder end (Node 680) to a reference voltage of 1.0 pu with a bandwidth of $\pm 1.63\%$ (2 V/122 V). The LTC tap range is ± 16 tap positions from the nominal setting. Two capacitor banks as indicated in [31] are considered in this study, although are disabled during low load conditions.

PV Systems
The PV systems are distributed across the network. A PV system is connected at each load bus. The rating of PV systems can be varied to simulate four different levels of PV penetration which are as follows:

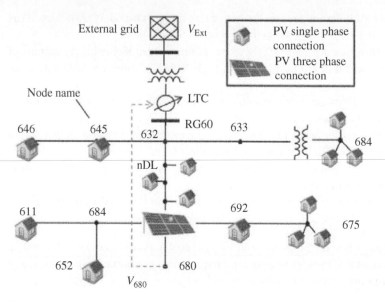

Figure 8.3 Study feeder system. *Source:* Kraiczy et al. [30].

i) PV25: PV generation is 25% of total load, i.e. 0.87 MW
ii) PV50: PV generation is 50% of total load, i.e. 1.73 MW
iii) PV75: PV generation is 75% of total load, i.e. 2.6 MW
iv) PV100: PV generation is 100% of total load, i.e. 3.47 MW

Three PV generation profiles are considered, which correspond to a PV clear day, PV variable day, and PV overcast day.

Volt–Var Smart Inverter Functions

Nine volt–var characteristics (C1–C9) of smart PV inverters are considered for simulation studies as shown in Figure 8.4. Each characteristic has a different voltage reference setpoint. For instance, curve C5 has a voltage reference setpoint of 1.0 pu Curves C0–C4 have lower voltage setpoints whereas curves C6–C8 have higher voltage reference setpoints. Each PV inverter can provide 44% reactive power even while generating rated active power [32].

8.2.3.2 Simulation Studies

Extensive QSTS simulation studies are performed using DIgSILENT PowerFactory software for low loading level and widely varying system conditions involving [31]:

i) Four PV penetration scenarios, as described above
ii) Eleven configurations of LTC controls and smart inverter volt–var characteristics as below:
   - Base case: Regulator (LTC) control inactive (tap position = 0) and PV systems with unity power factor (UPF)
   - LTC: Regulator (LTC) control is active, PV systems with UPF
   - LTC + C0, ..., LTC + C8: LTC control is active, PV inverters are considered with nine different volt–var curves (Figure 8.4)

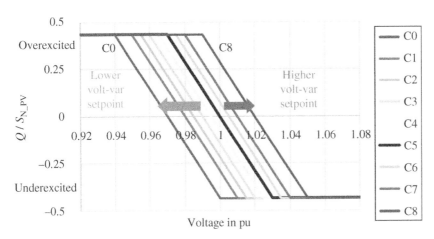

Figure 8.4 Different volt–var curves considered for smart PV inverters. *Source:* Kraiczy et al. [30].

The following impacts of volt–var smart inverter functions are considered in the simulation studies:

- Voltage range and compliance with ANSI voltage limits
- Maximum voltage drop/rise along the feeder
- Total LTC tap operations
- Maximum and minimum LTC tap position
- Total feeder losses
- Average PF at start of the feeder

Figure 8.5 presents the simulation results of number of LTC operations corresponding to the four PV penetration scenarios (along x-axis) and eleven LTC and volt–var settings (along y-axis), during a clear day (left) and a variable day (right). The impact is expressed using a bar with numbers on the right hand side of the figure. Smaller numbers on the bar imply good performance whereas larger numbers indicate weak performance for the specific criterion under consideration.

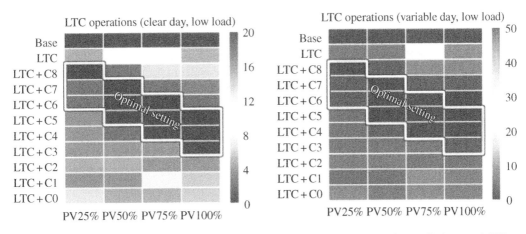

Figure 8.5 LTC tap operations during a clear day (left) and variable day (right). *Source:* Kraiczy et al. [30].

The optimal volt–var settings resulting in best performance are highlighted in Figure 8.5. It is seen that LTC with optimal volt–var settings of PV inverters significantly reduces the number of LTC tap operations as compared to the case with LTC alone both during clear and variable days. In other words, the entire voltage control is performed mainly by the smart PV inverter controls while needing LTC operations only sporadically.

The voltage impact studies demonstrate that the optimal volt–var settings can successfully alleviate the maximum voltage drop and also the maximum voltage rise along the feeder compared to the base case and LTC alone, scenarios. For low PV penetration, volt–var curves with a voltage reference higher than nominal are suggested. However, for high PV penetration, lower voltage references are more appropriate to counteract large voltage rises.

Smart inverter controls should not drive the taps to either maximum or minimum positions in order to leave room for further voltage control. It is observed that with recommended volt–var settings the required range of taps is similar or lower than the LTC case alone. Furthermore, the recommended volt–var settings do not adversely impact feeder losses and the average PF at the start of the feeder.

It is recommended that sensitivity analysis be conducted to determine the applicability and robustness of the recommended volt–var settings for:

i) variation in upstream (external) grid voltages,
ii) limitations of reactive power capability of inverters especially during peak active power production, and
iii) variation in simulation step sizes from the recommended step of 20 seconds [33] to 1 minute as considered in [30].

For all the above variations in the study system, the optimal volt–var settings were not affected in any significant manner. Further studies are being planned to investigate volt–var curves with deadbands, active control of capacitor banks, etc., using real and detailed distribution feeder models.

8.2.4 Coordination of Transformer On-Load Tap Changer and PV Smart Inverters for Voltage Control of Distribution Feeders

An effective methodology for voltage control in distribution feeders having a large penetration of solar PV systems through a coordinated use of smart inverter functions and OLTC is described in [1, 34]. The SCADA system is utilized to collect the voltages at the ends of each distribution feeder connected to the main transformer, along with the power generation profile of all the PV systems on those feeders. Based on this information as well as the changes in feeder load demand, the hourly tap settings of the main transformer are determined by the system operator. Subsequently, smart inverter functions on the PV inverters are used to provide autonomous control of the point of common coupling (PCC) voltage of each PV system to the desired level. Additionally, conservation voltage regulation (CVR) is implemented by regulating the voltage at the end point of feeders to 0.95 pu. The effectiveness of this strategy is demonstrated through simulation studies on a system comprising a 69/11 kV main transformer supplying six distribution feeders of the Taiwan Power Company (Taipower). A total of 114 PV systems with a total installed capacity of 21.42 MW are installed in the above six feeders. While coarse voltage control is provided by the transformer tap changer, finely tuned voltage control is enabled by smart inverters.

The proposed autonomous smart inverter control in PV systems is shown in Figure 8.6 [34]. If an overvoltage problem occurs, the PF of the PV inverter is first reduced and subsequently power is curtailed, if needed. The PCC voltage is measured every 100 ms by the inverter. If the voltage rises

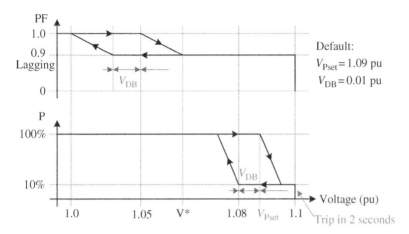

Figure 8.6 Control scheme of power factor and output power of smart inverters (autonomous control with voltage setting). *Source:* Ku et al. [34].

beyond 1.05 pu, the PF of the inverter is reduced in decrements of 1.0% in Mode 1 operation. This reduction is continued till either the overvoltage issue is mitigated or the PF has decreased to a limit of 0.9. The PCC voltage when the PF reaches 0.9 is denoted by V^*. If V_{PCC} decreases below 1.05 pu by the deadband voltage V_{DB} the PF is increased in increments of 1% every 100 ms in Mode 2 operation.

However, if the PF has been reduced to 0.9 and V_{PCC} is still higher than the setpoint voltage V_{Pset} the power output of the PV system is curtailed by 10% every 100 ms. Power curtailment continues till the PV system power goes less than 10% of its rated capacity. If V_{PCC} decreases below V_{Pset} by the deadband voltage V_{DB}, the PV power output is increased by 10% every 100 ms. However, even after power curtailment to 10%, if V_{PCC} is more than 1.1 pu, the smart inverter is tripped with a time delay of 2 seconds due to overvoltage.

This strategy exhausts the capability of reactive power compensation and power curtailment before tripping the PV inverter, thereby maximizing the harvest of PV energy.

The performance of the test solar PV farm without and with the proposed smart inverter control is depicted in Figure 8.7 [34] and Figure 8.8 [34], respectively. In the former case, voltage control is achieved through power curtailment alone, whereas in the latter case, the proposed smart inverter control regulates the voltage. It is evident that the proposed strategy significantly reduces the amount of power curtailment for performing voltage control.

The proposed coordinated control of OLTC and smart inverters is thus an effective technique for integrating a large number of PV power systems, while not putting the burden of voltage control only on smart inverters.

8.3 Control Interactions – Lessons Learned from Coordination of FACTS Controllers for Voltage Control

FACTS Controllers such as SVCs and STATCOMs have been utilized in power systems worldwide for several decades for dynamic voltage control (within few cycles). Several FACTS Controllers have experienced stability issues when they were not appropriately coordinated with neighboring FACTS Controllers [2, 3, 35]. For instance, the voltage controllers of several SVCs in Hydro-Quebec

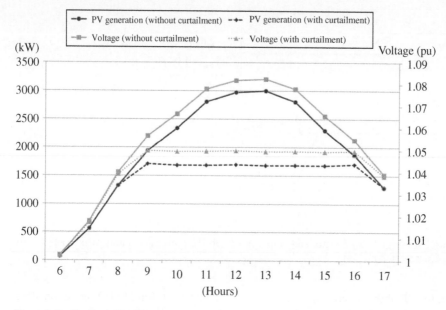

Figure 8.7 Profiles of active power generation and bus voltage of test PV farm (using active power curtailment). *Source:* Ku et al. [34].

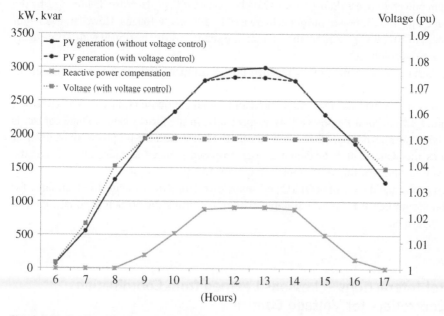

Figure 8.8 Profiles of active power generation, reactive power compensation, and bus voltage of test PV farm (using smart inverter). *Source:* Ku et al. [34].

had to be coordinated to ensure system stability instead of being slowed down. Numerous studies have been done to identify the controller modes that get destabilized in the absence of controller coordination.

Smart inverters are similar to FACTS Controllers as they also provide voltage regulation with fast speed of response. It is not only expected but also observed from several system studies that smart inverters need to be control-coordinated else they may become unstable. This section presents the main lessons learned from the realm of control coordination of FACTS Controllers which can be helpful for control coordination of smart inverters.

8.3.1 Controller Interaction Among Static Var Compensators in a Test System

Figure 8.9 shows a transmission network with two SVCs, which is used to study the controller interaction between the SVCs [2]. Impact of two factors are considered as follows:

1) System strength, and
2) Distance between the two SVCs

If the power system to which both SVCs are connected is strong, even if the two SVCs are located close to each other no controller interaction will occur between them. This implies that if the controller gain of SVC 1 is increased it will not affect the eigenvalues of SVC 2 in any substantial manner. This is because the interlinking variable between the two SVCs is voltage which stays fairly

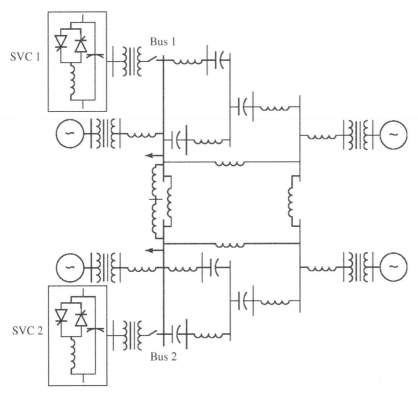

Figure 8.9 SVC interaction analysis network. *Source:* Mathur and Varma [2].

constant in a strong system. Thus the controls of both SVCs can be independently designed and optimized.

However, if the power system is weak, varying the controller gain of SVC 1 will strongly influence the eigenvalues associated with SVC 2. It is, therefore, imperative that a coordinated control design be undertaken for both SVCs.

These controller interaction results for SVCs were reported around 1997–1998. It is interesting to note that the controller interaction between two volt–var controllers of smart PV inverters studied in 2014 is also similar.

8.3.2 Controller Interaction Among Multiple Static Var Compensators in Hydro-Quebec System

A Hydro-Quebec summertime test power system with multiple SVCs depicted in Figure 8.10 [2] is studied in [2, 14]. Control interaction between SVCs is investigated for the summer transmission system which was revealed to be more crucial than that of winter transmission system not only because of reduced short-circuit levels but due to lower loads as well. It is noted that loads contribute to the damping of power system oscillations.

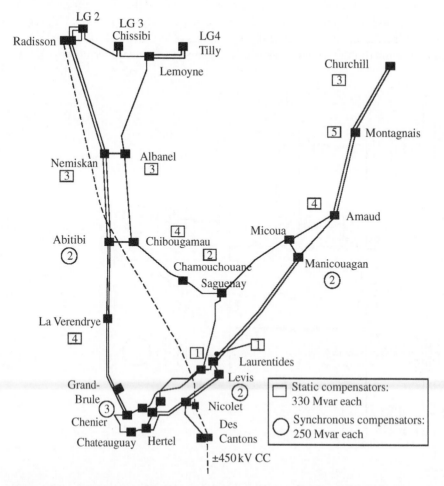

Figure 8.10 A proposed Hydro-Quebec summertime study system with shunt compensation only. *Source:* Gerin-Lajoie et al. [14].

A critical contingency seen to influence the SVC interactions is the loss of two lines south of La Verendrye, called the La Verendrye contingency. The performance of SVCs is examined over a range of operating conditions spanning zero output to the maximum capacitive output. The SVC voltage regulator is modeled by the gain–time-constant representation: the gain is the inverse of the SVC slope, whereas the time constant is the response rate. On the AC side of the SVC measurement systems are notch filters (80 and 96 Hz) to counteract network resonances; on the DC side are low-pass and harmonic-notch filters to obtain the pure DC equivalent of the SVC bus voltage. The interaction between SVC controllers is examined through both eigenvalue analysis and the simulation of transients by using EMTP.

The response to small-reactor switching for the La Verendrye contingency is depicted in Figure 8.11a [2]. The initial conditions are selected as those which are likely to occur 30 cycles after this contingency. Increasing oscillations of 16-Hz frequency are noticed in the transient response because of the adverse interaction between the fast SVC controllers. An increase in the regulator time constant (the slowing down of the SVC) from 0.133 to 0.5 seconds stabilizes the response as depicted in Figure 8.11b [2].

This is also revealed from the root loci of critical modes with varying controller-response rates, T_r, as shown in Figure 8.12 [2]. In this figure, the 16-Hz mode is seen to be unstable for $T_r = 0.133$ s (symbol 2) but stable for $T_r = 0.5$ s (symbol 4).

Another study presented in [2, 14] shows that when multiple SVCs are connected on the same line, the controller mode of the SVC with the lower effective short-circuit ratio (ESCR) becomes susceptible to instability. The difference in ESCR, as viewed from the thyristor controlled reactors (TCR) terminals of each SVC, is caused by the unequal rating of the SVCs; hence a coordination in controller parameters of the different SVCs is necessary. The interactions between voltage controllers become more prominent with high controller gains.

The use of voltage signal on the HV-side instead of LV-side as the feedback signal effectively makes the SVCs become closer electrically and more interlinked.

The learned lessons from SVC control interaction studies which will be helpful for investigation of interaction among smart PV inverters with voltage control are as follows:

- Controller interaction studies should be performed for weak system conditions and light loading scenarios.
- Adverse interactions are typically observed between voltage controllers.
- SVC voltage controllers connected to a weak network are more susceptible to instability.
- Higher controller gains lead to controller instability. Slowing down the voltage controllers prevents the adverse controller interaction and stabilizes the overall system.
- More the electrical closeness between SVCs, more they are likely to interact adversely.

Similar considerations will be helpful in the study and mitigation of adverse controller interactions among smart inverters.

8.4 Control Interactions Among Smart PV Inverters and their Mitigation

The need for control coordination in smart inverter applications arises from several considerations:

- Different smart inverter controls may be operative simultaneously in a smart inverter
- Different smart inverter outputs may be impacting same power system variable, e.g.
  - Q control for voltage (volt-var function, dynamic current injection)
  - P control for voltage (volt-watt function)

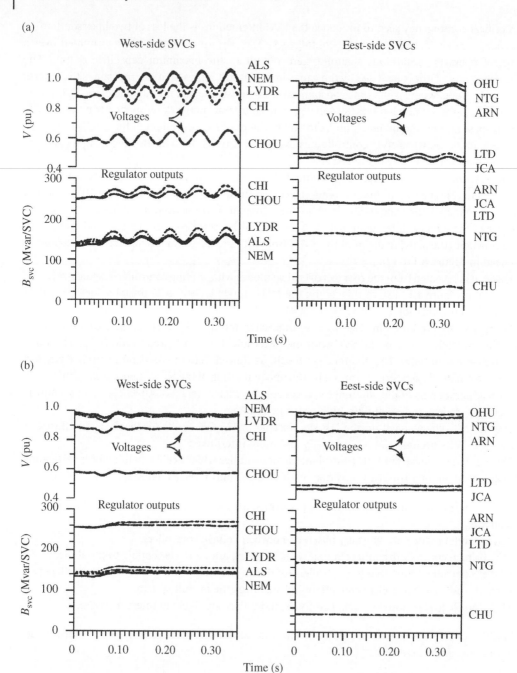

Figure 8.11 SVC transient behavior in La Verendrye system due to "snapshot" reactor switching at Abitibi. (a) The existing SVC response rate ($T_r = 0.133$ s) and (b) the reduced SVC response rate ($T_r = 0.5$ s). *Source:* Gerin-Lajoie et al. [14].

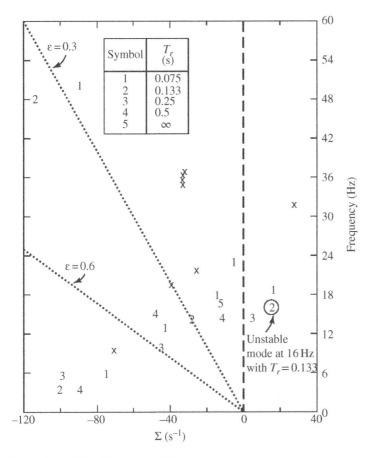

Figure 8.12 The effect of the SVC response rate on system eigenvalues in La Verendrye system stimulated by reactor switching at Abitibi. *Source:* Gerin-Lajoie et al. [14].

- Same smart inverter output may be impacting different power system variables
  - *P* control for voltage (volt-watt function)
  - *P* control for frequency (frequency-watt function)
- Different smart inverters on a distribution feeder may be performing same functions or multiple functions.

This section describes some of the above-described controller interactions. Various case studies are presented where uncoordinated controls of smart PV inverters result in inverter instability, voltage oscillations and high circulating current, or hunting [11].

8.4.1 Concepts of Control Stability with Volt–Var Control in a Single Smart PV Inverter

The impact of different system variables on the stability of a single inverter-based smart inverter is presented through a linearized system analysis in [1, 36], which is summarized below.

A simple model of a grid-connected smart PV inverter operating on VVC is depicted in Figure 8.13a. The inverter receives DC power from the solar panels, transforms into AC, and after appropriate filtering feeds it into the grid through an interfacing transformer. The corresponding

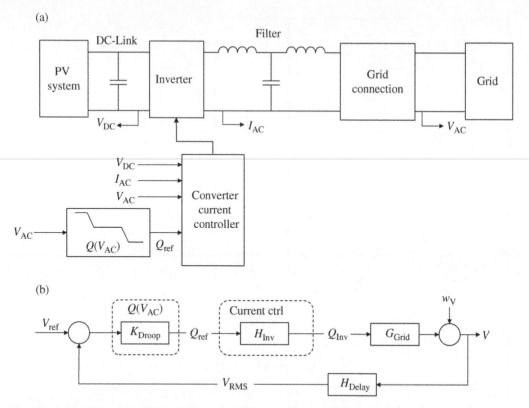

Figure 8.13 (a) Simplified grid-connected model of the smart PV inverter, (b) block diagram of the overall control system. *Source:* Modified from Andrén et al. [36].

equivalent control system model is illustrated in Figure 8.13b. Since this analysis is for voltage control, only the voltage controller part of the overall inverter control system is included in the control system. The reactive power output of the filter is not considered here as it can be rapidly compensated through the inverter control within a few cycles. The different control system blocks in Figure 8.13b are explained below.

K_{Droop}: This block represents the volt–var characteristic of the smart inverter. The overall nonlinear characteristic comprises three segments – the linear droop, deadband, and the reactive power limits.

H_{Inv}: This block is a simple model of the reactive current controller of the inverter for voltage regulation. The controller can be represented by any of the following:
  i) first-order transfer function with an adjustable time constant, or a proportional-integral controller
  ii) rate limiter with an adjustable maximal reactive power gradient

G_{Grid}: This block represents the grid which comprises a constant voltage source behind an impedance $R + j X$

H_{Delay}: This block represents the entire time delay from the instant of voltage measurement at the PCC to the instant it is fed to the input of the voltage controller of the inverter. This delay comprises:
  i) delay due to voltage RMS computation, typically half a cycle
  ii) delay due to filtering of the transduced voltage signal, signal processing, and moving average computation, etc.,

iii) delays due to communication of the voltage signal from the measurement location to the input of the specific inverter by the overall solar PV plant controller, as well as any communication delay between the controllers of the specific inverter

w_V: This represents the disturbance in the PCC voltage caused by any system disturbance or change in irradiance due to clouds, etc.

In an ideal condition, in the absence of any delay, the inverter control system is always stable since only one pole is contributed by the first order reactive current controller or rate limiter.

The considered system exhibits the following nonlinearities in addition to those in the network:

i) nonlinearity in the volt–var curve (due to deadband and limits)
ii) nonlinearities in inverter control system due to limits
iii) nonlinearity due to the delay

The stability of the overall control system is analyzed by linearizing the system about an operating point. With respect to the VVC block, an operating point either in the deadband ($Q = 0$) or the regions with Q limits ($Q = \pm Q_{MAX}$) is stable since there will be no change in reactive power output with changing voltage in those regions. The system stability is, therefore, examined for operation in the droop region of the volt–var curve. It is assumed only for this analysis that inverter can exchange a large amount of reactive power even when it is producing rated active power output. In reality, this is partially true as some standards [32] and grid codes require the inverters to be capable of operating with PF of 0.9 or 0.95 even at rated active power output.

The transfer functions of the different control blocks are expressed as:

$$K_{Droop} = k_1 = \frac{Q_{MAX}}{\Delta V_{Droop}} = \frac{P_{N1} \tan \theta_{MAX}}{\Delta V_{Droop}} \tag{8.3}$$

$$H_{Inv} = \frac{1}{1 + s\tau_1} \tag{8.4}$$

$$G_{Grid} = \frac{X_1}{V_N^2} \tag{8.5}$$

where, s is the Laplace variable, k_1 is the droop of the volt–var function at the operating point, P_{N1} is the nominal active power output of the inverter, θ_{MAX} is the maximum operating PF at the nominal power output, X_1 is the network reactance at the PCC, V_N is the nominal voltage at the PCC, and ΔV_{Droop} is the total voltage range of the droop region over which the inverter reactive power output varies from zero to Q_{MAX}.

The open-loop transfer function of the control system is expressed as

$$G_{ol} = \frac{K_{ol}}{1 + s\tau_1} \tag{8.6}$$

where, K_{ol} is the open-loop gain

$$K_{ol} = \frac{k_1 X_1}{V_N^2} \tag{8.7}$$

The change in the PCC voltage ΔV_{PV} caused by the injection of active power P_{PV} and reactive power Q_{PV} from the PV inverter is given by

$$\Delta V_{PV} \approx \frac{RP_{PV} + XQ_{PV}}{V_N} \tag{8.8}$$

Manipulating the above equations we get

$$K_{o1} = \frac{\Delta V_{PV} \tan \theta_{MAX}}{\Delta V_{Droop}} \frac{X}{R} \tag{8.9}$$

where, ΔV_{PV} is the voltage rise in the grid caused by the injection of nominal power output from the inverter. X/R is the ratio of the reactance and resistance of the equivalent grid network.

Padé equivalent [37] is generally used to represent the transfer of the time delay T by the nth order transfer function

$$H_{Padé} = \frac{1 - k_1 s + k_2 s^2 - \dots \pm k_n s^n}{1 + k_1 s + k_2 s^2 + \dots + k_n s^n} \tag{8.10}$$

Using the above Padé equivalent, the delay block H_{Delay} is modeled by a transfer function of second order

$$H_{Delay} = \frac{1 - k_1 s + k_2 s^2}{1 + k_1 s + k_2 s^2} \tag{8.11}$$

where, $k_1 = T/2$, and $k_2 = T^2/12$

The overall system transfer function between w_V and PCC voltage V (with V_{ref} being constant) is obtained as

$$G_{w_V V} = \frac{1}{1 + K_{Droop} H_{Inv} G_{Grid} H_{Delay}} \tag{8.12}$$

This is a third-order transfer function with three poles. The stability of this PV inverter grid system is determined from this transfer function.

A study system with a single smart PV inverter shown in Figure 8.14 is considered for stability analysis. An inverter with rating S_{N1} feeds power into a network.

The impact of the following three factors is examined through a detailed root loci studies for the study system:

- Open-loop gain K_{o1}
- Time constant τ of the inverter reactive current control, and
- Time delay T

The following behavior is observed for the system:

i) An increase in the time delay beyond a limit causes instability. This limit becomes lower if higher damping ratios of the most critical pole need to be achieved.
ii) An increase in the open-loop gain K_{o1} beyond a limit also leads to system instability.

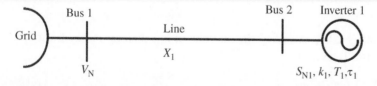

Figure 8.14 Single smart inverter connected to a simple network. *Source:* Modified from Andrén et al. [36].

iii) It is further demonstrated that for ensuring system stability, the ratio of time delay T and time constant τ must satisfy the following criterion:

$$\frac{T}{\tau} \leq \frac{1}{f(K_{o1})} \tag{8.13}$$

where,

$$f(K_{o1}) = a_\xi K_{o1} + b_\xi \tag{8.14}$$

a_ξ and b_ξ are constants that can be obtained from the parametric analysis of T/τ and root loci studies. These will be different for achieving different damping ratios.

Substituting Eqs. (8.9) and (8.14) in Eq. (8.13), the stability criterion is rewritten as

$$\frac{T}{\tau} \leq \frac{1}{a_\xi \dfrac{\Delta V_{PV} \tan \theta_{MAX}}{\Delta V_{Droop}} \dfrac{X}{R} + b_\xi} \tag{8.15}$$

The above criterion can be interpreted differently from different perspectives I or II.

I) In a system where the time delay T and inverter time constant τ are fixed, the smart inverter system can become unstable for the following conditions:
   - The PV inverter causes a large voltage impact ΔV_{PV} on the grid. This occurs when
     - The network is weak, i.e. has a low short circuit level
     - The active power injection from the PV system is high
   - The network has a high X/R ratio, i.e. the network is more reactive than resistive.
   - ΔV_{Droop} is small, i.e. the slope of the volt–var curve is high
   - $\tan \theta_{MAX}$ is high, i.e. PF is low
II) If the above conditions exist in the network, the system stability can only be ensured if the ratio T/τ is kept high.

The maximum delay for an example case is obtained in [36]. A worst-case scenario is considered with the following parameters:

   Maximum voltage rise with very high PV penetration: $\Delta V_{PV} = 6\%$
   High slope of the volt–var function: $\Delta V_{Droop} = 1\%$
   Reactive network: $X/R = 1$
   Low PF of the PV system: $\cos \theta = 0.9$

The root loci for such a system yields the open-loop gain $K_{o1} = 2.91$ and $f(K_{o1}) = 1.57$. According to the stability criterion in Eq. (8.15)

$$T \leq \frac{\tau}{f(K_{o1})} = 0.64\tau \tag{8.16}$$

For a specified VVC response of 10 seconds, the time constant is about 3 seconds (since $3\,\tau < 10$ seconds for reaching 95% steady state value). Hence the delay T should be less than 1.9 seconds. If a damping ratio of 10% is desired, the delay T should be less than 1.4 seconds.

If, however, the VVC response is specified to be 1 second, a stable performance with a damping ratio of 10% will be achieved for the delay T less than 160 ms. Such a time delay can be easily accomplished for inverters which have in-built measurement system and volt–var function. However, in case of a large solar PV system with a central plant controller controlling multiple inverters, the communication delay can become an issue.

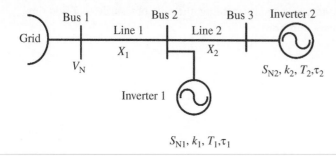

Figure 8.15 Distribution line with two smart PV inverters. *Source:* Modified from Andrén et al. [36].

8.4.2 Concepts of Control Stability with Volt–Var Control with Multiple Smart PV Inverters

The above concept is extended to two smart PV inverters with VVC as illustrated in Figure 8.15 as well as for multiple smart inverters in [36]. It is shown in [36,38] that the X/R ratio has a strong influence on the stability of a PV system. A higher resistance does not mean more damping in this case. A high settling time constant will have a positive contribution to the system stability, being direct proportional with the stability limit. In this context, a detailed case study is presented in Sec. 8.6.

8.4.3 Control Interactions Within a Smart Inverter

The PCC voltage control is impacted both by active power exchange and reactive power exchange by the smart PV inverter. Hence, an interaction can potentially occur between the PCC voltage control (RPC) with DC-link voltage control (active power control).

A study of impact of filters of grid voltage feed-forward signals on the coupling between DC-link voltage and AC voltage controllers in smart PV solar systems for voltage control in distribution systems is presented in [39]. A comprehensive linearized state-space model of a PV system including the dynamics of grid voltage feed-forward filters, DC-link voltage, and AC voltage controllers is developed. This model is validated by electromagnetic transients software PSCAD. This model is then used to identify the coupling between DC-link voltage and AC voltage controllers by eigenvalue and participation factor analyses. The impact of time constant of feed-forward filters on this coupling is studied by eigenvalue sensitivity analysis for PV systems with proportional and proportional integral (PI) type AC voltage controllers. Smart inverter functions such as volt-var control and dynamic reactive current injection typically have a proportional controller type PCC voltage regulator. On the other hand, the voltage controller can also be implemented using a PI controller with reactive current droop.

The range of time constants over which the filters interact with controllers and create instability for systems with different strengths and X/R ratios is identified. Insights are provided on the choice of time constants of grid voltage feed-forward filters for stable system design. These are summarized below:

1) The value of filter time constant τ should be kept as small as possible, for both systems with proportional and PI type AC voltage controllers. A certain range of high time constant may provide stable response for system with proportional type AC voltage controller, which can be determined from eigenvalue studies.

2) For a given time constant τ, systems with lower X/R ratios (more resistive networks) are more likely to cause instability.
3) Strong distribution systems with proportional type PCC voltage controllers are stable over a wider range of τ, however, this range is limited for PI type voltage controllers. In weak distribution systems, the PI controller enables a wider range of time constants for stable operation.
4) An appropriate time constant value can be chosen by eigenvalue analysis to ensure system stability.
5) However, if any available filter time constant results in system instability, and if this time constant cannot be changed, a reduction in the gain crossover frequency (or speed of response) of either the proportional or the PI type PCC voltage controller can stabilize the system.

8.4.4   Smart Inverters with Volt–Var Functions

A case study of control interaction between smart inverters with VVCs is reported in [1, 40]. Two 50 kW PV inverters are connected at the same PCC of a distribution feeder which has Thevenin's equivalent resistance $R = 0.05\,\Omega$ and inductance $= 0.1$ mH. This results in an X/R ratio of 0.75. The PV inverters are modeled with IGBT switches and an L-C filter at their output. A simple d-q axis-based current controller is modeled for the PV inverters. Both inverters are equipped with volt–var smart inverter controls. The VVC is implemented through a look-up Table. This controller provides reactive power Q at the PCC in response to the PCC voltage V based on the look-up table. The parameters of the volt–var curve such as slope, deadband, delay in responding to voltage change, and response time of the inverter, are settable by the user. Three simulation studies using Simulink/SimPowerSystems software are described below.

Study 1: Normal Slope of Q-V Characteristic of Both Inverters
Slope of Q-V curve $= 10$ pu (pu var/pu voltage). Delay time $= 1$ ms, and response time $= 1$ ms.
Figure 8.16a–d illustrates the inverter volt–var characteristic, average line-neutral voltage at PCC, active and reactive power of inverter 1 and active and reactive power of inverter 2, respectively. Inverter 1 is started with UPF at $t = 0.05$ s [40]. The VVC is activated at $t = 0.2$ s. Inverter 2 is started at $t = 0.4$ s, and the VVC is activated at $t = 0.6$ s.
In this case, PCC voltage increases beyond acceptable levels when the inverters operate at UPF, i.e. with no reactive power exchange. However, when the volt–var function is enabled on both inverters with the parameters as specified above, the voltage is regulated within stipulated limits.

Study 2: High Slope of Q-V Characteristic of Both Inverters
Slope of Q-V curve $= 30$ pu (pu var/pu voltage). Delay time $= 1$ ms, and response time $= 1$ ms.
Figure 8.17a–d illustrates the inverter volt–var characteristic, average line-neutral voltage at PCC, reactive power exchange by the two inverters, and the reference reactive power (depicting the selected delay time), respectively [40]. Inverter 1 is activated at $t = 0.05$ s, whereas inverter 2 is started at $t = 0.4$ s. As soon as inverter 2 is activated, an adverse interaction between the two VVCs occurs and results in voltage and reactive power oscillations at a frequency of about 46 Hz. Even though these oscillations do not grow in time but are detrimental to the operation of power system equipment including the inverters.

Study 3: Lower Slope of Q-V Characteristic and Slower Response of Both Inverters
Slope of Q-V curve $= 10$ pu (pu var/pu voltage). Delay time $= 0.05$ ms, and response time $= 0.05$ ms.
The implication of these settings is that the inverters will exchange less reactive power with grid for the same PCC voltage, and will also respond after a larger delay. Figure 3.10e–h shows the inverter volt–var characteristic, average line-neutral voltage at PCC, reactive power exchange by

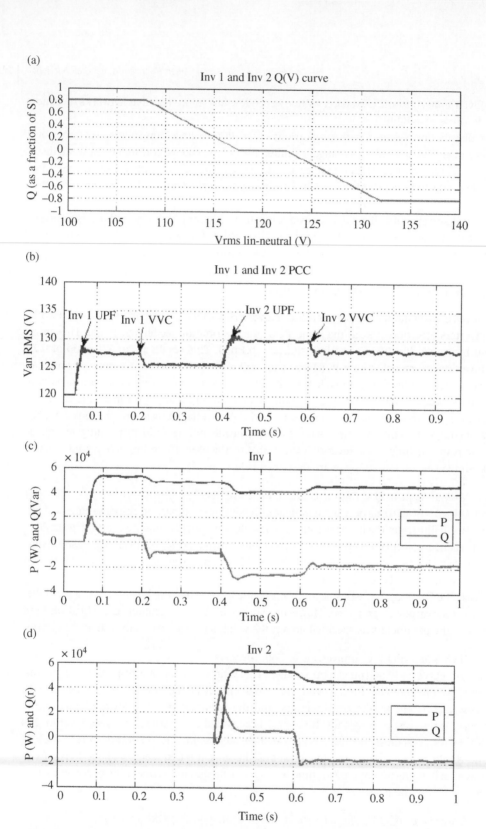

Figure 8.16 Normal operation of two PV inverters with volt–var control. (a) Volt-var curves of inverters 1 and 2, (b) PCC phase a voltage rms, (c) active and reactive power of inverter 1, (d) active and reactive power of inverter 2. *Source:* Chakraborty et al. [40].

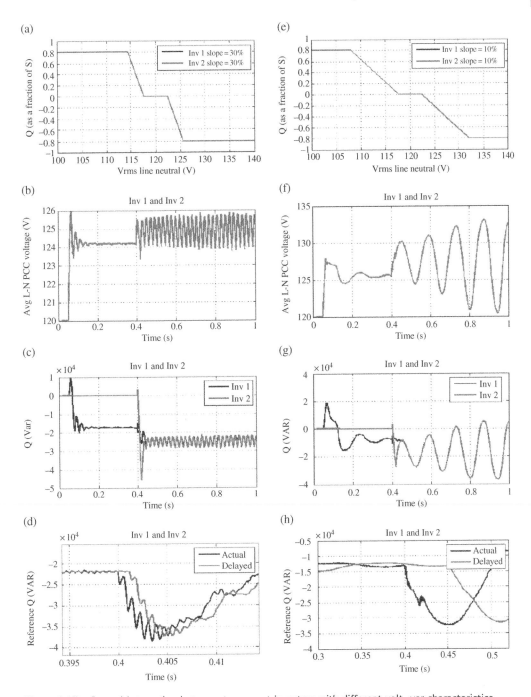

Figure 8.17 Control interaction between two smart inverters with different volt–var characteristics. For inverters 1 and 2: (a) volt-var curves with 30% slope, (b) average L–N PCC voltage, (c) reactive power, (d) reference reactive power, (e) volt-var curves with 10% slope, (f) average L–N PCC voltage, (g) reactive power, (h) reference reactive power. *Source:* Chakraborty et al. [40].

the two inverters, and the reference reactive power, respectively. As soon as the second inverter is activated the adverse interaction between the two VVCs causes the onset of growing oscillations of 7 Hz. The overall system becomes unstable. This instability, however, disappears if the VVC is disabled on the second inverter (although not shown in the figure above).

It is concluded from this simulation study that undesirable voltage oscillations or instability may result with the following:

1) High slope of VVC
2) Large delay in the response (or slower response) of inverters
3) High grid impedance (although not shown here)

The above control interactions observed with only two 50 kW three-phase inverters was also observed with a large number (more than 10) of single-phase inverters of 3–5 kW PV inverters [40]. It is, therefore, concluded that even a large number of small size inverters can create a similar adverse impact as a few large size inverters.

8.4.5 Control Interaction Between Volt–Var Controls of Smart Inverters

A case study of multi-inverter interaction with volt–var smart inverter functions is presented in [1, 41]. The study system comprises a simplified distribution system model with a load at far end. Two three-phase PV inverters are connected to the feeder at a distance from each other. A capacitor bank is connected at the source end. Both inverters are equipped with VVC. A disturbance is created by switching the capacitor banks. The control interaction of the two smart inverters while performing voltage control is investigated with MATLAB/Simulink simulation studies. The inverter kVA is considered to be more than the kW rating to allow for reactive power exchange at all times during the day. The PV inverters provide both reactive and active power control using PI controllers based on Q_{ref} and P_{ref} signals. The inverter operates on V, α control instead of the d-q control. The functioning of the voltage control is depicted in Figure 8.18. The PCC instantaneous voltage V is processed through an averaging window to generate the average voltage. A look-up table representing the volt–var function is then used to produce the reactive power reference Q_{ref} based on the average voltage. The inverter reactive power Q is compared with Q_{ref} and the error signal is processed through a PI controller to generate the appropriate signal for determining inverter switching functions.

The following studies are performed:

A capacitor switching is performed at $t = 10$ s and the response of the inverters is obtained for two volt–var characteristics. Figure 8.19a, b portray the volt–var characteristics for the base case with slope = 100 pu (pu var/pu volts), and the compared case with slope = 20 pu. The PI controller parameters are $K_P = 0.3$, and $K_I = 3.0$. The averaging window for voltage measurement is 0.05 seconds.

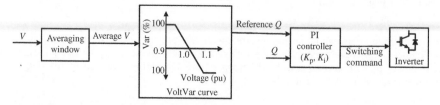

Figure 8.18 Voltage control through volt–var function. *Source*: Used with permission from CIGRE [41] © 2014.

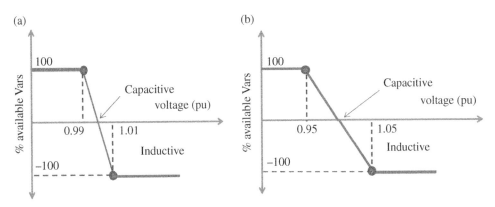

Figure 8.19 (a) Volt–var function for base case, (b) volt–var function for compared case. *Source*: Reprinted with permission from CIGRE [41] © 2014.

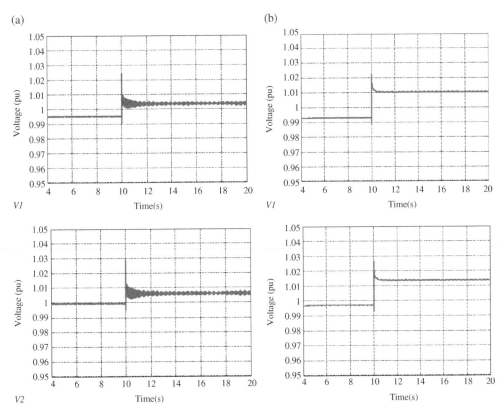

Figure 8.20 Voltage response of both inverters with $K_P = 0.3$ and $K_I = 3.0$, voltage averaging window = 0.05 s; (a) slope of volt–var curve = 100, (b) slope of volt–var curve = 20. *Source*: Reprinted with permission from CIGRE [41] © 2014.

The voltage response for both inverters is depicted in Figure 8.20a for the base case with larger slope (=100) of the volt–var curve, and in Figure 8.20b for the compared case with lower slope (=20) of the volt–var curve. It is evident that a higher slope or higher sensitivity of the reactive power exchanged with respect to voltage, causes voltage oscillations and consequent instability.

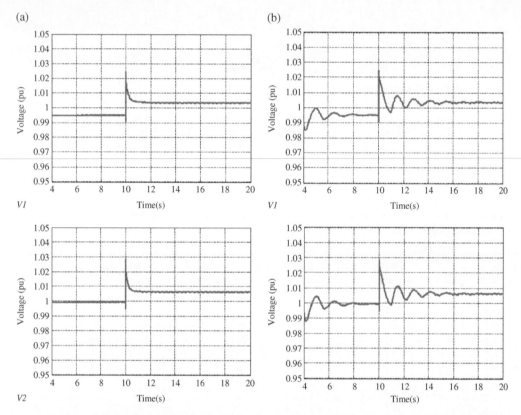

Figure 8.21 Voltage response of both inverters with slope of volt–var curve = 100; (a) $K_P = 0.1$ and $K_I = 1.0$, voltage averaging window = 0.05 s, (b) $K_P = 0.3$ and $K_I = 3.0$, voltage averaging window = 1 s. *Source*: Reprinted with permission from CIGRE [41] © 2014.

The responses of both inverters with controller gains reduced by three times, i.e. $K_P = 0.1$ and $K_I = 1.0$, and with increased value of moving averaging window for voltage measurement to 1 second are depicted in Figure 8.21a, b, respectively. Comparing Figures 8.20a and 8.21a demonstrate that voltage instability can be prevented by decreasing the controller gains, i.e. by making the voltage controller respond slower. Further comparison between Figures 8.20a and 8.21b illustrate that increasing the averaging window although increases the initial oscillations in voltage but damps them subsequently.

This study demonstrates that the choice of slope of volt–var function, controller gains, and the width of averaging window can strongly impact the interactions among smart inverters.

8.4.6 Oscillations Due to Voltage Control by Smart Inverter

A study was conducted on an actual segment of the distribution network of Southern California Edison [1, 42]. The study system comprised a 50 kVA single phase transformer feeding 18 residential loads rated at 3.32 kVA each. Every house has a small scale 5 kVA PV system installed. This results in a PV penetration of 150% with respect to the total circuit load. A QSTS analysis was performed by

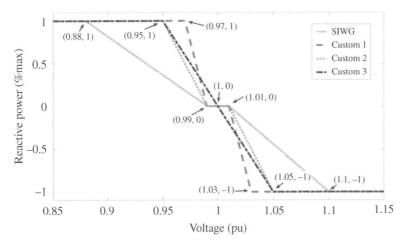

Figure 8.22 Different volt–var characteristics implemented on PV systems. *Source:* Schoene et al. [42].

open-source distribution system simulator (OpenDSS) which was also validated by Transient simulations with EMTP-RV software. In the system studies, all PV systems were equipped with four volt–var characteristics. One was the characteristic proposed by Smart Inverter Working Group (SIWG) which has conservative slopes. The other three custom volt–var characteristics had different higher slopes. All these characteristics are depicted in Figure 8.22.

System studies were conducted for a constant irradiance of 0.4 pu, thereby leaving enough inverter capacity for providing reactive power support. A worst-case overvoltage scenario was simulated where the voltage resulted in PV system operation beyond the corner point of the slope, i.e. 1.104, 1.035, 1.055, and 1.055 pu for the SIWG, Custom 1, Custom 2, Custom 3 volt–var curves, respectively. Studies were conducted for the cases: (i) without Var Ramp Rate Limit and (ii) with Var Ramp Rate Limit to 50% of maximum available reactive power (var) per second on all PV systems. A high initial overvoltage scenario was created and the response measured at one of the PV systems without Var Ramp Rate Limit is depicted in Figure 8.23.

About 30 Hz oscillations are observed in both bus voltage and reactive power output of the PV inverters. These oscillations decay within about 10 cycles and 40 cycles for PV systems with Custom 3 and Custom 2 volt–var curves, respectively. However, for Custom 1 volt–var curve which has the highest slope, the oscillations continue indefinitely. These oscillations cause both power quality and system stability issues.

The system response with Var Ramp Rate Limit implemented is illustrated in Figure 8.24. Since the reactive power exchanged is restricted by the limit, no oscillations are observed. However, bus voltage regulation is achieved in about 100 cycles whereas the this was accomplished in only a few cycles when the VAR limit was not imposed.

Following conclusions were made as a result of these studies:

1) Oscillations in voltage and reactive power result if Var Ramp Rate limit is not employed for all the volt–var characteristics considered. These oscillations decay rapidly if the slope of the volt–var curve is small, but take longer to settle if the slope increases. If the slope is quite high, oscillations may persist without decay. The occurrence of oscillations also depends upon the remaining inverter capacity available for reactive power support.

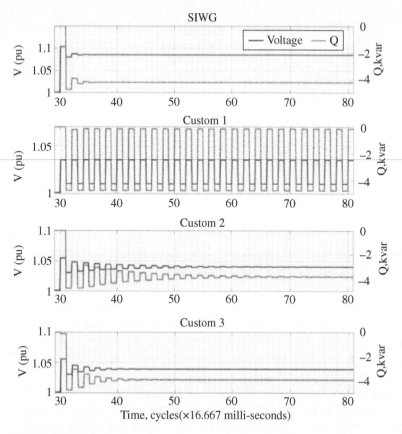

Figure 8.23 Bus voltage and reactive power output of a PV system, without var ramp rate limit. *Source:* Schoene et al. [42].

2) Although Var Ramp Rate limit helps in stable operation of the PV systems, they delay the voltage regulation process which may not be acceptable in some situations.
3) The deadband considered in this study did not have any effect on the onset of oscillations. However, if the deadband is made larger, oscillations may occur since increased deadbands will imply higher slopes of the volt–var curves.
4) It is further recommended that smart inverters be utilized for voltage support during nighttime to provide greater grid support.

8.4.7 Control Interactions of Volt–Var Controllers

The stability issues related to distribution system voltage control by multiple PV systems equipped with volt–var controllers have been studied both analytically (linearized system eigenvalue analysis) and through real-time hardware in loop simulations using RTDS [1, 43]. The study system shown in Figure 8.25 comprises two PV inverters connected to a grid having short circuit impedance Z_{Th}. P_{refi} and Q_{refi} represent the active and reactive power references of the ith PV inverter, whereas P_{PVi} and Q_{PVi} denote their actual outputs. Z_f represents the equivalent filter and transformer impedance for the PV system. A cable of impedance Z_{cable} separates the two PV inverters.

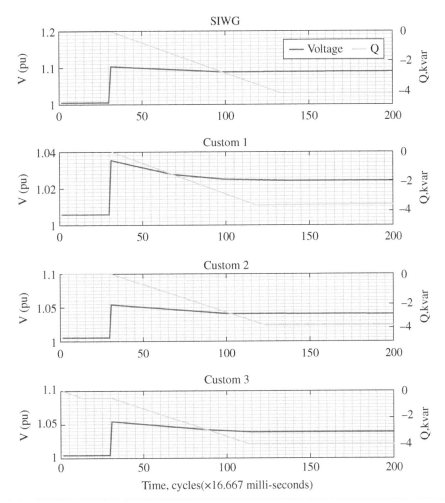

Figure 8.24 Bus voltage and reactive power output of the PV system, with var ramp rate limit. *Source:* Schoene et al. [42].

The control system of the two PV inverters is shown in Figure 8.26 [43]. The active power and reactive power controllers are modeled as PI regulators with gains K_p and K_i, respectively. The delay block represents the time delay in voltage measurement. The active power P_{PVi} and reactive power Q_{PVi} of the PV system are controlled by the phase angle and the magnitude of the PV system terminal voltage V_{PCC}, using d–q reference frame control. Active power control is used for control of DC-link voltage and maximum power point tracking (MPPT), whereas RPC is employed for VVC. The impact of variations in grid impedances, slope of the volt–var curve, PI controller parameters, response time, and measurement delay are studied. The impact of PI controller parameters is expressed in terms of the response time of the PI controller.

Following outcomes have been obtained that can be helpful in design of volt–var controllers:

1) A weak grid (high system equivalent impedance) typically results in a less stable system.
2) The system becomes unstable for higher values of slope of volt–var curve and for larger time delays. This is especially so for weak grids.

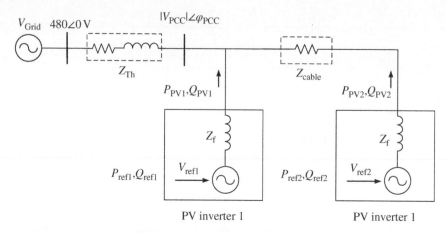

Figure 8.25 Study system for smart PV inverter controller interactions. *Source:* Modified from Kashani et al. [43].

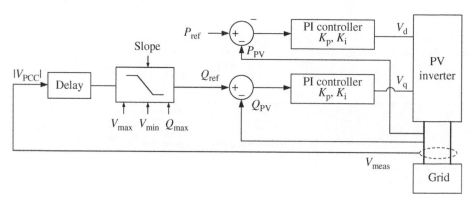

Figure 8.26 Control system of the smart PV inverters. *Source:* Modified from Kashani et al. [43].

3) Making the controllers slower i.e. by increasing the response times of the PI controllers through adjustment of the two gains K_p and K_i, helps improve system stability.
4) If the grid X/R ratio is decreased by keeping X same while increasing R, system becomes unstable for high slope values. Also, high PV penetration is difficult to achieve in highly resistive distribution feeders.
5) The amount of PV that can be connected becomes less as the distance from the source to PV system increases.

8.4.8 Controller Interaction Between Volt–Var and Volt–Watt Controllers

A detailed mathematical analysis of the controller interaction between volt–var and volt–watt inverter functions in a DER inverter is presented in [1, 44]. An adverse interaction is shown to occur between the two smart inverter functions when both the functions operate simultaneously in the watt-preference mode.

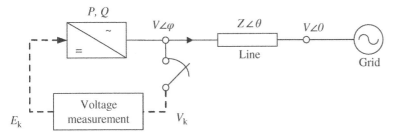

Figure 8.27 DER inverter connected to infinite bus. *Source:* Modified from Braslavsky et al. [44].

The study system is depicted in Figure 8.27 [44]. A DER inverter is connected to the infinite bus through a line with impedance $Z\angle\theta$. P, Q are the active and reactive power outputs of the DER, respectively. The voltage at point of connection (PoC) of the DER is $V\angle\varphi$.

The voltage V_k is sampled at time k by a voltage measurement system providing outputs E_k. All the delays and the dynamics associated with the voltage measurement and generation of P and Q outputs of the DER are lumped together in a low pass filter model as below:

$$E_{k+1} = aE_k + (1-a)V_k \tag{8.17}$$

where, "a" = parameter representing filter bandwidth; $0 \leq a \leq 1.0$.

The volt–var and volt–watt smart inverter functions that are simultaneously operative in the DER are shown in Figure 8.28 [44].

A realistic system scenario is simulated with the following parameters:

$$E_A = 240\,\text{V}, \ E_B = 250\,\text{V}, \ V = 230\,\text{V}$$
$$R = 0.606\,\Omega, \ X = 1.251\,\Omega, \ S = 10\,\text{kVA}$$

Simulation studies are conducted with PYPOWER, an open-source port of MATPOWER developed in the Python programming language.

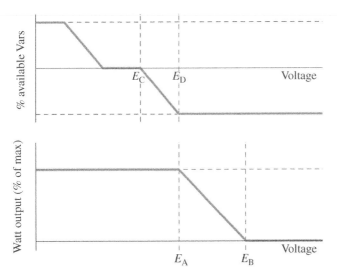

Figure 8.28 Volt–var function (a) and volt–watt function (b) operative in the DER. *Source:* Braslavsky et al. [44].

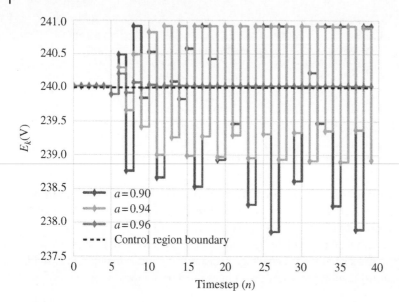

Figure 8.29 Time-domain response of voltage E_k at the point of connection of the inverter to the line. *Source:* Braslavsky et al. [44].

The time-domain response of voltage E_k is portrayed in Figure 8.29 [44]. It is seen that the response is unstable for $a = 0.90$ and $a = 0.94$. However, the system is exponentially stable for $a = 0.96$. Stability studies are performed for varying X/R ratios of the network and filter bandwidth parameter "a." The system is noted to be stable for $1.4 \leq X/R \leq 2.1$ for $a > 0.96$. For low X/R (<1.5), system stability is attained for generally lower values of "a" (i.e. 0.9).

The explanation for this instability is as follows. Reactive power support is initiated when DER inverter voltage $E > E_C$. However, the amount of reactive power is constrained due to the already high active power output of the DER which is operating in watt-precedence mode. If the DER voltage rises to E_A, active power is curtailed enabling a high amount of reactive power to be exchanged with the grid. This results in a high rate of change of voltage and active/reactive power due to a quadratic coupling between the active and reactive power outputs, eventually causing system instability. This adverse interaction between the volt–var and volt–watt controls can be avoided by selecting var-precedence mode of operation, if both volt–var and volt–watt smart inverter functions are used simultaneously.

8.5 Study of Smart Inverter Controller Interactions

Study of smart controller interactions of multiple PV systems requires detailed modeling of network dynamics, solar PV system smart inverter controllers, inner current controllers, and loads. The system components that are modeled for such studies include [1, 3, 41, 45, 46]:

Network Dynamics:

- System short circuit impedance
- Line resistances and reactances
- Cable resistances, reactances, and capacitances

- Feeder capacitors and PF correction capacitors
- Transformers with tap changers

PV System Dynamics

- Measurement systems with filters
- Phase-locked loop
- *d-q* axis current controllers
- DC-link voltage controllers
- PCC voltage controller
- Smart inverter functions
- Pulse width modulation (PWM) units
- PV plant controller
- Communication delays in plant-level control

Load Dynamics:

- Static loads
- Dynamic loads such as induction motor loads

Controller interactions can be studies with linearized eigenvalue analysis using root loci techniques. However, high-frequency interactions can only be investigated with detailed electromagnetic transients studies.

- Small signal state-space analysis

Differential equations representing the dynamics of each component are developed and linearized around an operating point. These are combined to develop the state-space model for the complete system. Eigenvalue analysis is performed to determine the critical oscillatory modes. Loci of these critical roots (eigenvalues) are plotted to observe their sensitivity to different network conditions and controller parameters. System parameters and controller parameters that provide the maximum damping of these critical modes are determined to result in an overall stable system.

- Electromagnetic transient (EMT) studies

These studies are performed with time step of typically 1–50 ms to capture the oscillatory modes resulting from smart inverter controller interaction and the electromagnetic transient behavior of the overall power system. The electromagnetic transient (EMT) studies can predict the behavior of network modes which are generally not modeled in small-signal analysis. In several cases, these network modes are responsible for controller interactions [2] as well as harmonic interactions [47].

A case study of determination of coordinated controllers of solar PV systems with smart inverter functions utilizing the above techniques is presented below.

8.6 Case Study of Controller Coordination of Smart Inverters in a Realistic Distribution System

A comprehensive study of control interactions between two PV plants while providing VVC using a detailed system model considering PV plant dynamics, load dynamics, and network dynamics, is described in [48]. This study also bridges the different perspectives reported in the literature on the impact of "measurement and communication delays" and "inverter response time" on smart inverter controller interactions.

For instance, studies reported by authors from [41] suggest that a slow response of voltage controller and large voltage averaging time window can prevent adverse oscillatory interaction between two smart PV inverters with VVC.

However, studies performed by authors from [40] have shown that slow response (large delay and large inverter response time) on two smart PV inverters causes their controllers to become unstable. It is further stated, "This particular observation is contradictory to some common belief that a delay in volt–var control (VVC) response will be better to reduce interaction between inverters and other voltage regulation assets (e.g. regulator, LTCs, capacitor banks). From this case study, it is evident that depending on inverter controller implementations, slower VVC response may cause control instability and needs to be evaluated.... More studies need to be done to determine the effect of VVC delay on the closed-loop response of the grid-inverter systems [40]."

Studies presented in [43] also demonstrate that increasing delay time causes controller instability between two smart PV inverters with VVC. Furthermore, slower inverter response time reduces controller oscillations.

This comprehensive study [48] is, therefore, performed is to examine and explain the apparently different conclusions of the above papers [40, 41, 43].

A detailed linearized state-space model for the study system is developed. Root loci studies are conducted to study the impacts of different system parameters. These small signal studies are subsequently validated by time-domain simulations using MATLAB Simulink.

8.6.1 Study System

The study system is a modified version of a real utility feeder in Ontario [48] and depicted in Figure 8.30. The 27.6 kV distribution feeder is 45 km long and is supplied by 115 kV transmission network through a 115/27.6 kV transformer T_1 at the distribution substation. Two PV plants rated 6 MW each are assumed to be connected to the feeder at Bus 2. These are at a distance of 35 km from the distribution substation. A total static load of 5.3 MVA at 0.9 PF is connected at the end of the feeder.

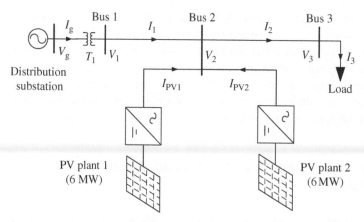

Figure 8.30 Single line diagram of study system. *Source:* Reprinted with permission from CIGRE [48] © 2020.

8.6.2 Small Signal Modeling of Study System

The impact of different variables is first studied through small-signal eigenvalue analysis. The small-signal model of the overall study system is developed in the common *d-q* reference frame which is synchronously rotating at the network frequency. The study system is separated into different subsystems. State-space models are developed for each subsystem based on their individual rotating *d-q* reference frames. For instance, each individual PV inverter is modeled on a *d-q* frame whose rotational frequency is set by its phase-locked loop (PLL). Subsequently, these subsystem models are transformed from their individual reference frames to the common network *d-q* reference frame and integrated to obtain the state-space model of the complete study system.

8.6.2.1 Network

The network model consists of the 115 kV transmission network, transformer T_1, distribution feeder and the static load. The 115 kV transmission network is modeled as a voltage source behind Thevenin equivalent impedance. The Thevenin's equivalent voltage, resistance, and inductance are denoted by V_g, R_g, and L_g, respectively. The resistance and reactance of the transformer T_1 referred to HV side are R_T and L_T, respectively. The grid current is denoted as I_g.

The distribution feeder segments are represented by their lumped π equivalent models comprising lumped resistance, inductance, and capacitances of the feeder segment. The two feeder segments connecting Bus 1 and Bus 2, and Bus 2 and Bus 3, are modeled in a similar manner. The static load is modeled as constant impedance resistive–inductive (R-L) load.

V_j denotes the voltage of the *j*th bus. I_j is the current flowing through the feeder connecting *j*th and (*j*+1)th. I_{PVj} denotes the output current of *j*th solar PV farm. The load current is given by I_3.

The network model is developed in *abc* frame and then transformed to the synchronously rotating *d-q* frame.

8.6.2.2 PV Plant

The general modeling features described below apply to both PV plant 1 and PV plant 2.

A PV plant (PV plant 1 or PV plant 2) with VVC connected to the distribution feeder is depicted in Figure 8.31. The PV plant consists of two subsystems (i) PV power circuit and (ii) PV control system. The PV power circuit comprises the voltage source converter (VSC) also referred as the PV inverter, filter, pad-mount transformer, and collector cables. The control circuit includes PLL, current controllers, measurement filter, volt–var controller, and active power controller. The description of the components is provided below.

PV Power Circuit

The PV power circuit consists of the PV panels, DC-link capacitor, LC filter, and pad-mount transformer. As the objective is to study control interactions with VVC which has response time of few seconds, the slow variation in PV panel output due to cloud passage, that is typically over a few minutes, is neglected.

V_{DC}, I_{DC}, I_{inv}, V_{Cf}, I_{PPCC}, V_{PCC}, and m represent the DC-link voltage, PV panel output current, VSC output current, filter capacitor terminal voltage, current measured at VSC side of pad mount transformer, PCC voltage (measured at HV side of pad mount transformer) and modulation index, respectively. C_{DC} is the capacitance of DC-link voltage. The resistance R_{f1} represents the ON

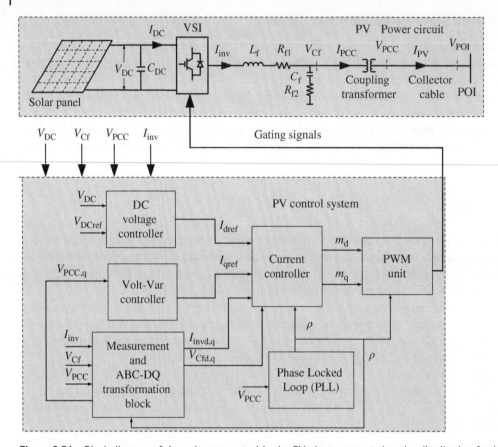

Figure 8.31 Block diagram of the volt–var control in the PV plant connected to the distribution feeder.

resistance of IGBT and the resistance of L_f. C_f is the filter capacitance, and R_{f2} is the resistance of the damping resistor. L_{PVT} and R_{PVT} are the inductance and resistance of the pad mount transformer, respectively. V_{POI}, V_{PCC}, I_{PCC}, and I_{PV} are the point of interconnection (POI) and PCC voltage, current measured at VSC side of pad mount transformer, and current flowing in cable, respectively. The collector cable is modeled by its lumped π equivalent model including its inductance, resistance, and capacitance.

The state-space model of the PV power circuit is developed in *abc* frame and transformed to its individual *d-q* frame.

PV Control System
The components of control system of the smart inverter for developing the state-space model are described below [46, 48, 49].

a) Measurement Filter

A measurement filter is employed to obtain the fundamental component of the AC system voltage and to attenuate any sideband harmonics in the measured signal. The measurement system however incurs delays due to signal processing, computation, and communication of the measured signal to the controller. A time constant τ_1 of the measurement filter is incorporated to represent the delays incurred in measurement process.

V_{PCCdm} and V_{PCCqm} are the filtered components of V_{PCCd} and V_{PCCq}, respectively. Also, V_{Cfdm} and V_{Cfqm} are the filtered components of filter capacitor terminal voltage V_{Cfd} and V_{Cfq}, respectively.

b) *Phase-Locked Loop*

The PLL is used to extract the voltage angle of PCC voltage.

c) *Current Controller*

This controller provides decoupled (independent) control of active power and reactive power. The active and reactive power is controlled by controlling d axis current (I_{invd}) and q axis current (I_{invq}). The active and reactive power controllers generate d axis current reference (I_{dref}) and q axis current reference (I_{qref}) according to the active and reactive power required to be injected into the grid. The current controller ensures that the VSC currents I_{invd} and I_{invq} track the references I_{dref} and I_{qref} in a rapid manner. Proportional integral controllers are used as the compensator for current controller.

d) *DC Voltage Controller*

The DC-link voltage controller regulates the DC-link voltage (V_{DC}) at the reference voltage (V_{DCref}) to extract the corresponding power from the solar panels. In the MPPT strategy, the MPPT controller determines V_{DCref}. A corresponding active current reference I_{dref} is generated to enable maximum power output from of the solar panels. V_{DC1ref} and V_{DC2ref} are the DC-link reference voltage for PV plant 1 and 2, respectively. In this study, they are assumed to be same.

e) *Volt–Var Controller*

A typical volt–var curve for a smart inverter is shown in Figure 8.32.

The structure of the volt–var controller considered in this study is depicted in Figure 8.33. The blocks corresponding to the current controller and Inverter are equivalents of the detailed model shown in Figure 8.31. The volt–var controller generates the reactive power reference signal based on the moving fixed-width time window average value of PCC terminal voltage. The moving average time window duration represents the delay. The different parameters in the volt–var controller including the delay, deadband, and slope of volt–var curve can be varied to study their impact.

$Slope_{\text{ind}}$ and $Slope_{\text{cap}}$ are the droops of inductive and capacitive operating regions of the volt–var curve. τ_d is the time constant of voltage averaging window. The calculation of I_{qref} is performed differently for the inductive (voltage swell), and capacitive (voltage sag) operating modes of the VVC.

Figure 8.32 Typical volt–var curve of a smart inverter.

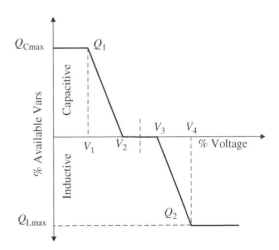

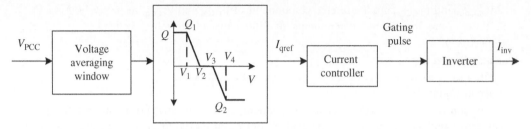

Figure 8.33 Structure of a volt–var controller. *Source*: Reprinted with permission from CIGRE [48] © 2020.

f) Overall Model of PV plant

The overall model of PV plant is obtained by combining the state-space models of the PV power circuit and the different constituents of the PV control system.

8.6.2.3 Overall Study System Model

The state-space equations developed for the generic PV plant with VVC in Section 8.6.2.2 are customized to PV plant 1 and PV plant 2, and written separately.

Both the PV plant 1 and PV plant 2 are connected to Bus 2 of the feeder. The d-q components of the input voltage to both the PV plant models are V_{2d} and V_{2q}, respectively. The d-q components of output current of PV plant 1 are I_{PV1d} and I_{PV1q}, respectively. The d-q components of output current of PV plant 2 are expressed by I_{PV2d} and I_{PV2q}, respectively.

The linearized state-space equations of the network, PV plant 1 and PV plant 2 are systematically combined to result in the overall state-space model for the study system, as below:

$$\dot{\widetilde{X}}_{sys} = A_{sys}\widetilde{X}_{sys} + B_{sys}\widetilde{U}_{sys} \tag{8.18}$$

The above system model has 56 states and 4 inputs.

The detailed derivation of the above system model is given in [48]. This model is general and can be used for performing impact studies of different system variables on control interactions among smart PV inverters.

8.6.3 Small Signal Studies

The impacts of various system variables on smart PV inverter controller interactions under diverse operating conditions are described in [48]. However, only results relating to the impact of the delay time and PV inverter response time on smart PV inverter controller interactions are described below. The objective here is to explain the apparently different conclusions of the papers [40, 41, 43].

Both PV plants are assumed to be generating 0.9 pu active power and operating in inductive mode of VVC.

8.6.3.1 Impact of Delay

Time delay T_d represents the delays incurred in the measurement system due to signal processing and filtering, and the delay incurred in communication of the measured signal to the controller. The impact of time delay on control interaction is investigated by varying the delay between 0.001 and 1 second. The response time and slope are set at 0.007 seconds, and 15 seconds,

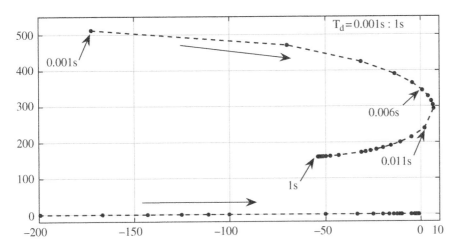

Figure 8.34 Variation of dominant poles in ($\sigma + j\omega$) plane, for varying delay. *Source*: Reprinted with permission from CIGRE [48] © 2020.

respectively. The X/R of feeder and short circuit ratio at POI are 2.5 and 3.5, respectively. The locus of dominant poles for the variation of delay is illustrated in Figure 8.34. With increase in delay, a dominant pole shifts to right, and eventually makes system unstable. However further increase in delay shifts the pole to left thus making the system stable. This observation shows that increase in delay can have either a positive or negative impact on control interactions.

8.6.3.2 Impact of Response Time

This study is performed to investigate the impact of response time T_r of the q-axis current controller of the smart inverter which primarily provides the reactive power response to variations in PCC voltage.

The loci of dominant poles for variation of response time is studied by varying response time between 0.001 and 1 seconds, and the results are depicted in Figure 8.35. The delay and slope are 0.005 seconds, and 15 seconds, respectively. The X/R of feeder and short circuit ratio at POI are 2.5 and 3.5, respectively. For the variation of response time, dominant pole shows similar behavior as in the case of variation of delay. The control interaction is minimal for the fastest response (small response time). The frequency of oscillation is high for the smallest response time (fastest control), but the oscillations damp fast due to large damping ratio. The frequency and damping ratio of the sensitive mode decreases with increase in response time, and the mode eventually becomes unstable. With further increase in response time, the dominant mode shifts towards left and becomes stable, again.

The above studies show that response time and delay have similar impact on control interactions. The increase in delay and response time can both positively or negatively damp the control interactions. Thus, based on the initial operating point and range of variation of these factors considered, different results can be obtained. This clarifies the different perspectives reported in literature on impact of delay and response time.

8.6.4 Time Domain Simulation Studies

Time-domain simulation studies are done on PSCAD software to validate the results of small signal studies. In the following simulation studies, both PV plants are initially generating 0.9 pu active

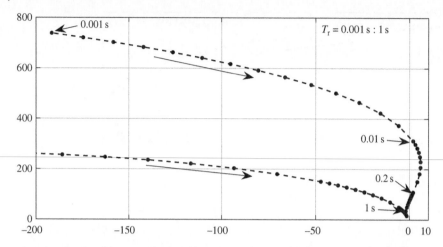

Figure 8.35 Variation of dominant poles in ($\sigma + j\omega$) plane, for varying inverter response time. *Source*: Reprinted with permission from CIGRE [48] © 2020.

power at UPF. The VVC of PV plant 1 (PV1) and PV plant 2 (PV2) are enabled at 1.5 and 2.5 seconds, respectively. The system conditions are identical to small signal studies, with X/R, SCR, and slope as 2.5, 3.5, and 15, respectively.

8.6.4.1 Impact of Delay

Figure 8.36 illustrates the reactive power output (Q) of both PV plants for three different time delays. Figure 8.36a–c depicts response for delays of 0.003, 0.009, and 0.05 seconds, respectively. The inverter response time is 0.007 seconds for all the cases. The system is stable for small value of delay (0.003 seconds), and a large value of delay (0.05 seconds), whereas it is unstable for an intermediate value of delay (0.009 seconds). These results correlate well with the small-signal analysis. This study clearly explains the conclusions in earlier publications that both increasing and decreasing delay time makes the controllers unstable. It depends on the starting value of T_d from where the delay is either increased or decreased.

8.6.4.2 Impact of Response Time

Figure 8.37 shows the reactive power output (Q) of both PV plants for three different response times. Figure 8.37a–c depicts reactive power for response times of 0.001, 0.015, and 0.5 seconds, respectively. The delay is 0.005 seconds. Stable response is observed for small response time (0.001 seconds), and large response time (0.5 seconds). But the system is unstable for intermediate value of response time (0.015 seconds). This result also matches with the small-signal results.

This study demonstrates that the inverter is stable in both cases of fast response and sluggish response. It further provides an explanation of the seemingly disparate conclusions in the published literature that slowing down of the inverter makes the inverter stable and also unstable. It indeed depends upon the starting value of the inverter response time, whether it is in the stable region or unstable region.

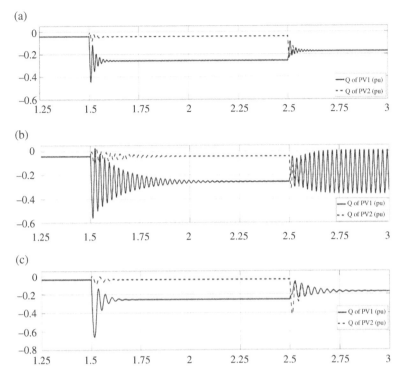

Figure 8.36 Reactive power output (*Q*) of both PV plants for delays of: (a) 0.003 seconds, (b) 0.009 seconds, and (c) 0.05 seconds. *Source*: Reprinted with permission from CIGRE [48] © 2020.

8.6.5 Summary

This detailed study demonstrates that controller instability can be caused by either increasing the time delay or decreasing the time delay. However, a small delay provides fast response with very little oscillations. The study further shows that the controller can become either stable and unstable by increasing the inverter response time. However, a fast response time is preferred to provide effective VVC in the distribution grid. Thus, the impact of delay and response time on system stability should be analyzed on system by system basis to ensure stable operation.

8.7 Control Coordination of PV-STATCOM and DFIG Wind Farm for Mitigation of Subsynchronous Oscillations

Series compensation is a widely utilized method to increase power transmission limit of transmission lines. However, series compensation results in SSO which, if uncontrolled, can potentially result in damage to synchronous generators connected to such series compensated lines [5, 17, 50, 51]. SSO issues have been studied in series compensated induction generator based wind farms [52, 53] and DFIG based wind farms [22, 23]. The adverse interaction between DFIG controller and series compensated electrical network results in subsynchronous control interaction (SSCI). This has caused severe overvoltages issues and damage to DFIG generator control circuits in 2007 in Minnesota (USA) [54], in 2009 in Texas [55], and in 2012 in China [56].

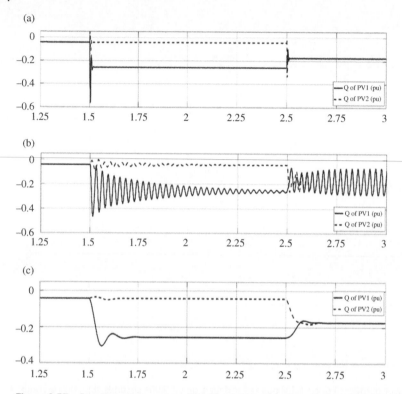

Figure 8.37 Reactive power output (Q) of both PV plants for response time of: (a) 0.001 seconds, (b) 0.015 seconds, and (c) 0.5 seconds. *Source*: Reprinted with permission from CIGRE [48] © 2020.

With the rapid growth of wind and solar PV plants worldwide, it is quite likely that wind farms and solar PV farms may be located in close vicinity and be connected to series compensated lines. For instance, wind plants and PV plants have been co-located in several countries including Canada [57], UK [58], Australia [59], and India [60].

Subsynchronous damping controllers (SSDC) have been developed for DFIG-based wind farms [61, 62] and solar PV farms [63] to damp SSO when either of these plants are individually connected to series compensated lines. Since the SSDCs on both are based on power electronic converters and operate at similar response speeds, there is a potential for them to interact in an adverse manner, if both the wind plant and PV plant are located in close vicinity and connected to a series compensated line.

This section describes a coordinated control of PV-STATCOM and DFIG converter to damp SSO when the co-located DFIG-based wind farm and PV plant are connected to a series compensated transmission line [64, 65].

8.7.1 Study System

Figure 8.38 depicts the study system which is derived from the IEEE First SSR Benchmark system [61, 66]. The DFIG-based wind farm, shown in Figure 8.38 is rated 100 MW which is an aggregation of 50 identical wind turbine units, each with a power rating of 2 MW [61, 67]. The wind farm is connected to infinite bus through a 161-kV series compensated transmission line. A solar PV farm rated 111 MVA with a rated power output of 100 MW, is also connected to the series compensated

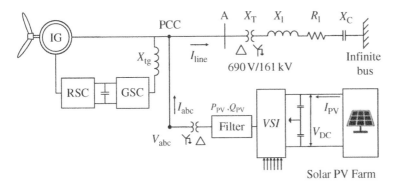

Figure 8.38 Study system: DFIG-based wind farm and solar PV farm connected to a series compensated transmission line. *Source:* Modified from Varma and Salehi [64].

network at the wind farm terminal (Bus A). The series compensation level (k) of transmission is considered to be 75% as in [61].

The DFIG-based wind farm model consists of a wound-rotor induction generator, rotor-side converter (RSC), and grid-side converter (GSC). The converters of the DFIG system are approximately rated 30% of the nominal power of the induction generator [68]. The reactance of the GSC transformer is denoted by X_{tg} in Figure 8.38. An aggregated two-mass wind turbine is considered to model the dynamics of the shaft system. In this study, the wind speed is considered 7 m/s which corresponds to a rotor speed $\omega_r = 0.75$ pu, mechanical power $P_m = 0.32$ pu, and a shaft torque $T_m = 0.43$ pu [61, 67].

The solar PV farm model comprises a voltage source inverter (VSI) which is connected to a capacitor and a controlled current source on its DC side. On its AC side, the VSI is linked to the transmission line through a filter and a coupling transformer. The DC source follows the *I-V* characteristic of PV panels of the solar farm. To extract maximum power from PV panels, the PV system employs MPPT system, although the SSO damping control utilizes a non-MPPT control. The entire system is modeled in MATLAB/Simulink according to the data provided in [61, 63, 65].

8.7.2 Control System of DFIG

Figure 8.39 depicts the configuration of GSC controller, i.e. *d*-axis control loop, and SSDC considered for DFIG converter. All the controllers are modeled in *dq*-reference frame. The voltage vector is aligned with quadrature axis and V_d equals zero. Thus, the reactive powers of the converters are controlled with the *d*-axis control loop whereas their active powers are controlled by the *q*-axis loop

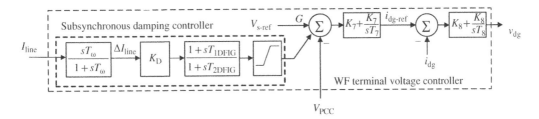

Figure 8.39 Subsynchronous damping controller (SSDC) of DFIG converter. *Source:* Salehi [65].

(Note that this convention is different than used in [69] although it does not impact the study results). Various signals such as output active power of wind farm (P_{DFIG}), line active power (P_{line}), voltage across the series capacitor (V_C), and line current (I_{line}) are used to damp SSO in DFIG-based wind farm [61, 62, 70, 71]. In this study, I_{line} is considered as a local control signal for SSDCs. The configuration of SSDC is depicted in Figure 8.39, which includes a washout filter block, controller gain block, phase compensator (lead-lag compensator), and a limiter block. This controller is connected on the GSC side, i.e. at wind farm terminal voltage (V_{PCC}) controller [61, 67]. The output of SSDC is added to the point G of GSC controller to modulate the reactive power of GSC (Q_g) and mitigate SSO in the system. The full model of the DFIG control system is described in [65].

8.7.3 Control System of PV-STATCOM

The PV-STATCOM operates in the reactive power priority mode wherein active power is curtailed during a system disturbance (Section 4.7.3.2) to make inverter capacity available for providing grid support as a STATCOM. Figure 8.40 depicts the PV inverter controller augmented by PV-STATCOM controller [64, 65] which consists of an SSDC, DC voltage controller with power point tracking (PPT) system, and the conventional PV current controller [69]. The current control block controls the output current of the PV inverter based on the references provided by the DC voltage controller and the SSDC. The DC voltage controller including the PPT block adjusts the active power output of the solar PV system P_{PV} by providing $i_{\text{q-ref}}$ for current controller, as described in [63, 64]. The SSDC consists of a washout filter block, controller gain block, phase compensator (lead-lag compensator), and a limiter block. Similar to the SSDC of DFIG converter, the proposed SSDC for PV system utilizes line current as the control signal. The output of the SSDC, $i_{\text{d-ref}}$, is then passed to the current controller to modulate the reactive power output of the solar PV system Q_{PV} and mitigate SSO in the system.

During a system disturbance which initiates SSO, the PPT system curtails the power from maximum power point P_{MAX} to a lower level P_{Curt} by increasing the voltage from V_{MP} to V_{Curt}. The SSDC modulates Q_{PV} to alleviate SSO in the system by providing $i_{\text{d-ref}}$ for current controller. After the SSO are damped, the PPT system decreases the voltage from V_{Curt} to V_{MP} in a ramped manner over the time period T_{ramp}. Reactive power modulation is continued during the ramp-up in Partial STATCOM mode to mitigate any resurgence of SSO during ramp-up.

8.7.4 Optimization of Subsynchronous Damping Controllers

The optimized parameters of SSDC for either the DFIG converter or PV-STATCOM individually or both of them in coordination, are obtained by employing the optimization toolbox of MATLAB software, based on Genetic Algorithm (GA). The OF which has to be minimized to find the optimal parameters of the controller is defined in Eq. (8.19) where ξ_i is the damping ratio of the ith eigenvalue.

$$\text{OF} = \sum_{j=1}^{n} (1 - |\xi_i|), \xi_i > 0 \tag{8.19}$$

subject to constraints that system eigenvalues are stable and have positive damping ratio.

8.7.5 System Response with No Subsynchronous Damping Controls

A state-space model of the complete system is developed and its eigenvalues are computed. The critical eigenvalues corresponding to subsynchronous and super-synchronous network modes, electromechanical mode, and the wind generator shaft mode are compiled in Table 8.2. It is seen that the system is unstable due to its subsynchronous mode.

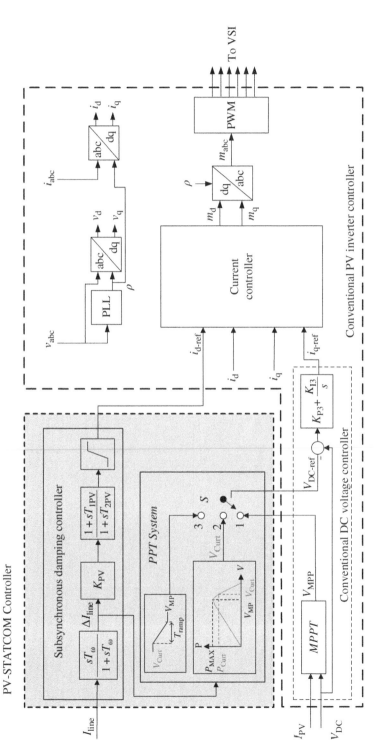

Figure 8.40 Control system of PV-STATCOM. *Source:* Based on Varma and Salehi [64].

Table 8.2 System eigenvalues for different cases.

Modes	No damping control	SSDC in DFIG converter	SSDC in PV-STATCOM	Uncoordinated SSDCs on DFIG converter and PV-STATCOM	Coordinated SSDCs on DFIG converter and PV-STATCOM
Network-1 (Sub)	$4.737 \pm 122.18i$	$-9.808 \pm 121.54i$	$-11.93 \pm 134.72i$	$-33.021 \pm 146.74i$	$-16.196 \pm 131.67i$
Network-2 (Super)	$-7.659 \pm 630.90i$	$-2.380 \pm 626.89i$	$-6.30 \pm 625.92i$	$-0.207 \pm 620.14i$	$-3.956 \pm 625.18i$
Electro-mechanical	$-14.007 \pm 98.67i$	$-13.668 \pm 100.79i$	$-11.879 \pm 99.75i$	$-10.72 \pm 99.34i$	$-11.90 \pm 100.16i$
Shaft mode	$-1.349 \pm 5.88i$	$-1.356 \pm 5.90i$	$-1.36 \pm 5.91i$	$-1.366 \pm 5.94i$	$-1.365 \pm 5.91i$

Source: Based on Varma and Salehi [64].

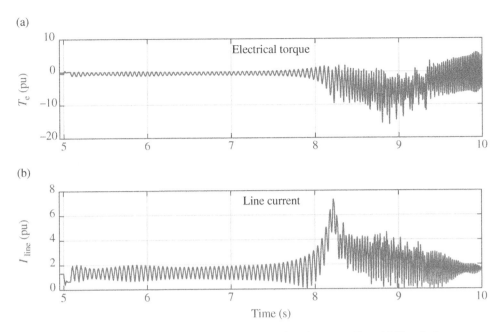

Figure 8.41 System response without subsynchronous damping controllers. (a) Electrical torque, (b) line current. *Source:* Varma and Salehi [64].

A three-line-to-ground (3LG) fault for 5 cycles is initiated at the end of the transmission line at $t = 5$ s. Figure 8.41a, b depicts the post-fault electrical torque (T_e) and line current (I_{line}), respectively. The system is seen to be unstable due to growing SSO in system which validates the results of first column of Table 8.2.

8.7.6 Independently Optimized SSDC of DFIG Converter

The DFIG converter SSDC parameters are optimized based on the procedure given in Section 8.7.4. The critical system eigenvalues utilizing the optimized SSDC are shown in second column of Table 8.2. It is seen that the SSDC successfully stabilizes the subsynchronous mode and thus the overall system. Comparison of column 1 and column 2 of Table 8.2 shows that while SSDC stabilizes the subsynchronous mode, it decreases the damping of supersynchronous mode and electromechanical mode. However, the SSDC does not affect the shaft mode.

8.7.7 Independently Optimized SSDC of PV-STATCOM

The PV-STATCOM SSDC parameters are also optimized based on the procedure given in Section 8.7.4. The critical system eigenvalues utilizing the optimized SSDC are listed in the third column of Table 8.2. Comparison of column 2 and column 3 of Table 8.2 shows that the optimized PV-STATCOM SSDC stabilizes the entire system. It substantially improves the damping of subsynchronous mode, even more than the SSDC of DFIG converter. It slightly decreases the damping of supersynchronous mode and the electromechanical mode as compared to the no damping controller case. It also does not impact the shaft mode.

8.7.8 Uncoordinated SSDCs of PV-STATCOM and DFIG Converter

In this case, the independently optimized SSDCs of the PV-STATCOM and DFIG converter are combined but not coordinated. The corresponding critical eigenvalues are listed in column 4 of Table 8.2. It is seen that the two uncoordinated controllers: (i) increase the damping of subsynchronous mode, but (ii) decrease the damping of supersynchronous mode, as compared to either of the two controllers implemented alone.

The SSO damping by DFIG converter alone, PV-STATCOM alone, and both of them combined (but uncoordinated) are now compared. Figure 8.42a, b show the line current (I_{line}) and electrical torque (T_e) for the cases with (i) only DFIG SSDC, (ii) only PV-STATCOM SSDC, and (iii) combined but uncoordinated SSDCs of DFIG converter and PV-STATCOM.

Figure 8.42 demonstrates that:

1) PV-STATCOM damps SSO faster than DFIG-GSC. However, the combined controllers damp SSO faster than each of them acting individually.
2) Supersynchronous mode oscillations both in-line current and electrical torque persist for a long period of time.

The above observations correlate with the results of the eigenanalysis of the damping controllers presented in Table 8.2.

8.7.9 Coordinated SSDCs of PV-STATCOM and DFIG Converter

In this study, both the SSDCs of the PV-STATCOM and DFIG converter are optimized together i.e. in a coordinated manner, using the procedure described in Section 8.7.4. The corresponding sensitive eigenvalues of the overall system are compiled in column 5 of Table 8.2. It is seen that the coordinated SSDCs: (i) increase the damping of the subsynchronous mode more than either controller acting alone, and (ii) improve the damping of supersynchronous mode much more than uncoordinated controllers. The supersynchronous mode damping is now brought to a level higher than with only DFIG SSDC but slightly lower than with only PV-STATCOM SSDC.

Time-domain simulation studies are conducted for this system with optimized and coordinated SSDCs. A 3LG fault for 5 cycles is initiated at the end of the transmission line at $t = 5$ s. Figure 8.43a–c show the reactive power of GSC (Q_g), reactive power generation of PV system (Q_{PV}), and active power generation of PV system (P_{PV}), respectively.

As seen from Figure 8.43b, as soon as the fault occurs, the DC voltage controller of the PV-STATCOM curtails active power from 100 to 90 MW, although with a slight delay based on the controller time constant, to make enough inverter capacity available for SSDC. The SSDC utilizes this released capacity for modulation of reactive power Q_{PV} in Full-STATCOM mode to mitigates SSO as seen from Figure 8.43c. When oscillations in-line current are alleviated successfully to within 2% within 3 seconds (i.e. at $t = 7.92$ s), the PV system starts increasing P_{PV} from 90 MW to 100 MW by decreasing V_{DC} from V_{Curt} to V_{MPP} in ramped manner over 5 seconds. The PV-STATCOM operates in Partial-STATCOM mode during the power ramp-up, utilizing the remaining capacity of PV inverter after active power generation, to prevent the recurrence of SSO. The PV plant restores its pre-fault active power generation at about $t = 13$ s, i.e. almost 8 seconds after fault occurrence. It is noted from Figure 8.43b, that the maximum reactive power from PV-STATCOM required to damp SSO is about 47 Mvar, which is much less than the total inverter capacity of 111 MVA. This demonstrates that the PV system does not need to fully curtail its active power generation for mitigation of SSO in this study.

(a)

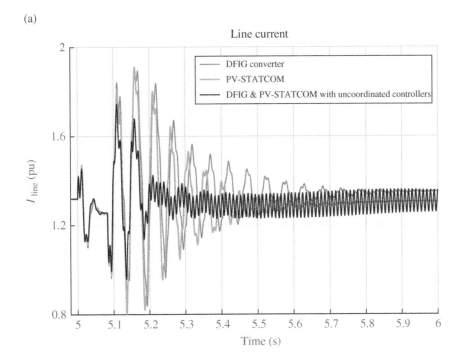

(b)

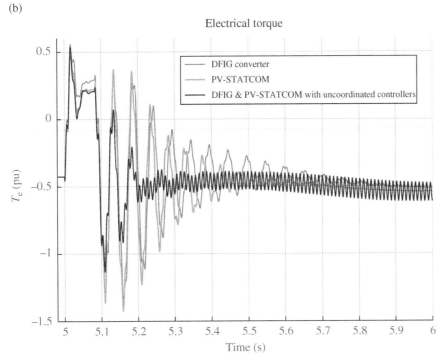

Figure 8.42 Performance of uncoordinated SSDCs of PV-STATCOM and DFIG: (a) line current, (b) electrical torque (T_e). *Source:* Varma and Salehi [64].

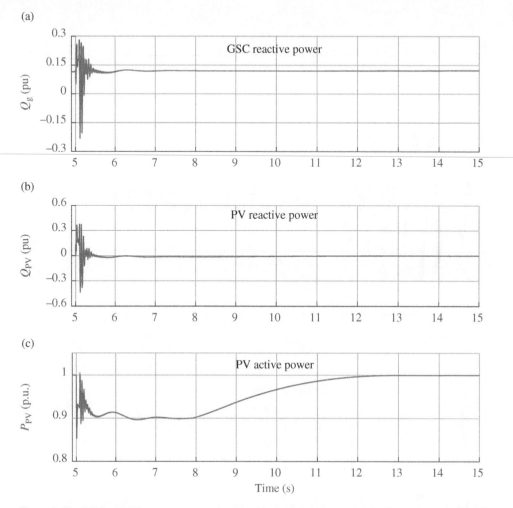

Figure 8.43 DFIG and PV system responses with coordinated SSDCs of PV-STATCOM and DFIG. (a) Reactive power of grid side converter, (b) reactive power of PV-STATCOM, (c) active power of PV-STATCOM. *Source:* Varma and Salehi [64].

Figure 8.44a, b portrays the line current (I_{line}) and electrical torque (T_e) for the cases with (i) only DFIG SSDC, (ii) only PV-STATCOM SSDC, (iii) coordinated SSDCs of DFIG converter and PV-STATCOM.

It is seen that:

- PV-STATCOM damps SSO faster than DFIG-GSC. However, the combined controllers damp SSO faster than each of them acting individually.
- The combined controllers damp both the subsynchronous and supersynchronous modes faster than each of them acting individually.
- The time taken to damp SSO with coordinated controllers is about 3 seconds as compared to 11 seconds in the case with uncoordinated controllers (result not shown).
- The total time taken for the PV system to return to its pre-disturbance power generation level is less than 8 seconds in the coordinated case as compared to 17 seconds in the uncoordinated case (result not shown).

(a)

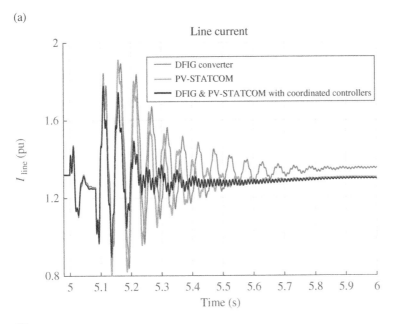

(b)

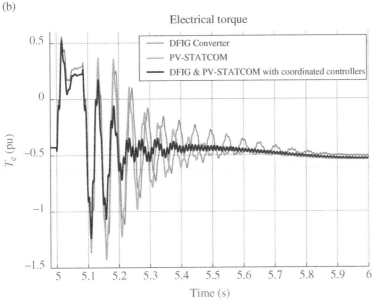

Figure 8.44 Performance of coordinated SSDCs of PV-STATCOM and DFIG: (a) line current, (b) electrical torque (T_e). *Source:* Varma and Salehi [64].

These observations correlate with the eigenanalysis results presented in Table 8.2.

This case study demonstrates that adverse control interactions can occur between a smart wind-turbine generator (DFIG) and a smart PV inverter (PV-STATCOM) when both are performing SSO damping in a series compensated line. A control coordination of the subsynchronous damping controllers of both the DFIG and the PV system results in a much more effective overall performance.

8.8 Control Interactions Among Plants of Inverter Based Resources and FACTS/HVDC Controllers

With the spiraling growth of PV systems, it is quite likely that smart PV inverters will get located in close vicinity of inverter-based wind generators, FACTS Controllers, and HVDC converter stations. This will necessitate a study of control interactions between these dynamic controllers.

It is important that coordination issues should be evaluated in the interconnection process prior to connection of these resources to the BPS [72]. Controller interactions are strongly dependent on system strength. They get aggravated during weak grid conditions. Hence mitigation solutions are site dependent. However, some general solutions are provided as below [72, 73]:

- Utilization of reactive current or reactive power based droop function to facilitate sharing of reactive power output among different IBRs (PV-PV, PV-wind, etc.)
- Optimization of controller gains, deadbands, inverter response times to ensure stable responses
- Specific control or operational strategies
- Master slave or other control systems for electrically close IBRs.

The importance of different filters in the measurement systems of SVCs in preventing various instabilities is discussed in [2]. Recommendations for prevention of controller interactions among FACTS Controllers and HVDC systems will also be useful in obviating controller interactions among IBRs [2, 3, 35].

8.9 Conclusions

This chapter describes the principles of control coordination of different devices which operate at similar speeds. These devices include FACTS Controllers such as SVCs and smart inverters.

The lessons learned from control coordination of SVCs in transmission systems are presented. The voltage controllers of SVCs can adversely interact if the SVCs are in close vicinity, and especially if the system is weak. Higher controller gains lead to controller instability. Slowing down the voltage controllers prevents the adverse controller interaction and stabilizes the overall system. Controller interaction studies are recommended to be performed for weak system conditions and light loading scenarios.

Smart Inverters need to be coordinated with conventional voltage controllers. Coordinated control of OLTC and smart inverters is an effective technique for integrating a large number of smart PV power systems, while not putting the burden of voltage control only on smart inverters.

RPC strategies can potentially cause an increase in unintended OLTC switchings, but the impact is different for different RPC strategies. The constant PF and watt-power factor strategies are more likely to cause an increase in unintended OLTC switchings. However, optimized VVC can either avoid or decrease the risk of unintended OLTC switching operations.

Within a smart inverter, an adverse interaction can occur between volt–var and volt–watt smart inverter functions when both functions operate simultaneously in the watt-preference mode. This adverse interaction between the volt–var and volt–watt controls can be avoided by selecting var-precedence mode of operation.

Multiple smart inverters connected in close vicinity in a distribution system and operating on VVC can experience adverse control interactions if: (i) the slopes of volt–var curve are high, (ii) the system is weak, (iii) and the system has a high X/R ratio. The control interactions are strongly dependent on measurement delays and inverter response times. Both low and high values of delay and response time result in stable responses, while intermediate values may cause instability.

It is demonstrated through a case study that adverse control interactions can occur between a smart wind-turbine generator (DFIG) and a smart PV inverter (PV-STATCOM) when both are used for damping SSO in a series compensated line. A control coordination of the subsynchronous damping controllers of both the DFIG and the PV system results in a much more effective overall performance.

The control coordination principles described in this chapter will likely help in investigating and mitigating adverse controller interactions among smart inverters.

References

1 CEATI International Inc. (2020). Enhancing Connectivity of DGs by Control Coordination of Smart Inverters. CEATI International Inc., Montreal, QC, Canada, *Report No. T184700 #5178*.

2 Mathur, R.M. and Varma, R.K. (2002). *Thyristor-Based FACTS Controllers for Electrical Transmission Systems*. New York: Wiley-IEEE Press.

3 CIGRE (1999). Coordination of Controls of Multiple FACTS/HVDC Links in the Same System. CIGRE, Paris, France, *Technical Brochure No. 149*.

4 Kundur, P. (1994). *Power System Stability and Control*. New York: McGraw Hill.

5 IEEE (2020). Stability Definitions and Characterization of Dynamic Behavior in Systems with High Penetration of Power Electronic Interfaced Technologies. IEEE PES, New York, USA, *Technical Report PES-TR77*.

6 Miao, Z., Fan, L., Osborn, D., and Yuvarajan, S. (2009). Control of DFIG-based wind generation to improve interarea oscillation damping. *IEEE Transactions on Energy Conversion* 24: 415–422.

7 Singh, M., Allen, A.J., Muljadi, E. et al. (2015). Interarea oscillation damping controls for wind power plants. *IEEE Transactions on Sustainable Energy* 6: 967–975.

8 Wang, Y., Meng, J., Zhang, X., and Xu, L. (2015). Control of PMSG-based wind turbines for system inertial response and power oscillation damping. *IEEE Transactions on Sustainable Energy* 6: 565–574.

9 Varma, R.K., Rahman, S.A., and Seethapathy, R. (2010). Novel Control of Grid Connected Photovoltaic (PV) Solar Farm for Improving Transient Stability and Transmission Limits Both During Night and Day. In *Proc. 2010 World Energy Conference*, Montreal, Canada.

10 Varma, R.K., Rahman, S.A., Mahendra, A.C. et al. (2012). Novel nighttime application of PV solar farms as STATCOM (PV-STATCOM). In *Proc. 2012 IEEE Power & Energy Society General Meeting*, 1–8.

11 Varma, R.K., Rahman, S.A., and Vanderheide, T. (2015). New control of PV solar farm as STATCOM (PV-STATCOM) for increasing grid power transmission limits during night and day. *IEEE Transactions on Power Delivery* 30: 755–763.

12 Varma, R.K. and Maleki, H. (2019). PV solar system control as STATCOM (PV-STATCOM) for power oscillation damping. *IEEE Transactions on Sustainable Energy* 10-4: 1793–1803.

13 Shah, R., Mithulananthan, N., and Lee, K.Y. (2013). Large-scale PV plant with a robust controller considering power oscillation damping. *IEEE Transactions on Energy Conversion* 28: 106–116.

14 Gerin-Lajoie, L., Scott, G., Breault, S. et al. (1990). Hydro-Quebec multiple SVC application control stability study. *IEEE Transactions on Power Delivery* 5: 1543–1551.

15 Larsen, E.V., Baker, D.H., Imece, A.F. et al. (1990). Basic aspects of applying SVC's to series-compensated AC transmission lines. *IEEE Transactions on Power Delivery* 5: 1466–1473.

16 Anderson, P.M., Agrawal, B.L., and Ness, J.E.V. (1990). *Subsynchronous Resonance in Power Systems*. Piscataway: IEEE Press.

17 IEEE Subsynchronous Resonance Working Group of the System Dynamic Performance Subcommittee Power System Engineering Committee (1980). Countermeasures to Subsynchronous Resonance Problems. *IEEE Transactions on Power Apparatus and Systems* PAS-99: 1810–1818.

18 Bahrman, M., Larsen, E.V., Piwko, R.J., and Patel, H.S. (1980). Experience with HVDC – turbine-generator torsional interaction at Square Butte. *IEEE Transactions on Power Apparatus and Systems* PAS-99: 966–975.

19 Mortensen, K., Larsen, E.V., and Piwko, R.J. (1981). Field tests and analysis of torsional interaction between the coal creek turbine-generators and the CU HVdc system. *IEEE Transactions on Power Apparatus and Systems* PAS-100: 336–344.

20 Rostamkolai, N., Piwko, R.J., Larsen, E.V. et al. (1990). Subsynchronous interactions with static var compensators - concepts and practical implications. *IEEE Transactions on Power Systems* 5: 1324–1332.

21 Rostamkolai, N., Piwko, R.J., Larsen, E.V. et al. (1991). Subsynchronous torsional interactions with static var compensators-influence of HVDC. *IEEE Transactions on Power Systems* 6: 255–261.

22 Fan, L., Kavasseri, R., Miao, Z.L., and Zhu, C. (2010). Modeling of DFIG-based wind farms for SSR analysis. *IEEE Transactions on Power Delivery* 25: 2073–2082.

23 Irwin, G.D., Isaacs, A., and Woodford, D. (2012). Simulation requirements for analysis and mitigation of SSCI phenomena in wind farms. In *Proc. 2012 IEEE/PES Transmission and Distribution Conference & Exposition (T&D)*: 1–4.

24 Cheng, Y., Huang, S., Rose, J. et al. (2016). ERCOT subsynchronous resonance topology and frequency scan tool development. In *Proc. 2016 IEEE Power & Energy Society General Meeting*, 1–5.

25 Varela, J., Hatziargyriou, N., Puglisi, L.J. et al. (2017). The IGREENGrid project: increasing hosting capacity in distribution grids. *IEEE Power and Energy Magazine* 15: 30–40.

26 Kraiczy, M., Stetz, T., and Braun, M. (2018). Parallel operation of transformers with on load tap changer and photovoltaic systems with reactive power control. *IEEE Transactions on Smart Grid* 9: 6419–6428.

27 Rudion, K., Orths, A., Styczynski, Z.A. et al. (2006). Design of benchmark of medium voltage distribution network for investigation of DG integration. In *Proc. 2006 IEEE Power Engineering Society General Meeting*, 6.

28 Milanovic, J.V., Yamashita, K., Villanueva, S.M. et al. (2013). International industry practice on power system load modeling. *IEEE Transactions on Power Systems* 28: 3038–3046.

29 Bokhari, A., Alkan, A., Dogan, R. et al. (2014). Experimental determination of the ZIP coefficients for modern residential, commercial, and industrial loads. *IEEE Transactions on Power Delivery* 29: 1372–1381.

30 Kraiczy, M., York, B., Bello, M. et al. (2018). Coordinating Smart Inverters with Advanced Distribution Voltage Control Strategies. In *Proc. 2018 IEEE Power & Energy Society General Meeting*, 1–5.

31 Kersting, W.H. (2001). Radial distribution test feeders. In *Proc. 2001 IEEE Power Engineering Society Winter Meeting*, 908–912.

32 IEEE (2018). *IEEE standard for interconnection and interoperability of distributed energy resources with associated electric power systems interfaces*. IEEE Std 1547-2018 (Revision of IEEE Std 1547-2003).

33 Reno, M.J., Deboever, J., and Mather, B. (2017). Motivation and requirements for quasi-static time series (QSTS) for distribution system analysis. In *Proc. 2017 IEEE Power & Energy Society General Meeting*, 1–5.

34 Ku, T., Lin, C., Chen, C., and Hsu, C. (2019). Coordination of transformer on-load tap changer and PV smart inverters for voltage control of distribution feeders. *IEEE Transactions on Industry Applications* 55: 256–264.

35 CIGRE (2000). Impact of Interactions Among Power System Controls. CIGRE, Paris, France, *Technical Brochure No. 166.*

36 Andrén, F., Bletterie, B., Kadam, S. et al. (2015). On the stability of local voltage control in distribution networks with a high penetration of inverter-based generation. *IEEE Transactions on Industrial Electronics* 62: 2519–2529.

37 Liu, G., Zinober, A., and Shtessel, Y.B. (2009). Second-order SM approach to SISO time-delay system output tracking. *IEEE Transactions on Industrial Electronics* 56: 3638–3645.

38 Constantin, A. and Lazar, R.D. (2012). Open loop Q(U) stability investigation in case of PV power plants. In *Proc. 27th European Photovoltaic Solar Energy Conference and Exhibition.* 3745–3749.

39 Subramanian, S.B., Varma, R.K., and Vanderheide, T. (2020). Impact of grid voltage feed-forward filters on coupling between DC-link voltage and AC voltage controllers in smart PV solar systems. *IEEE Transactions on Sustainable Energy* 11: 415–425.

40 Chakraborty, S., Hoke, A., and Lundstrom, B. (2015). Evaluation of multiple inverter volt–var control interactions with realistic grid impedances. In *Proc. 2015 IEEE Power & Energy Society General Meeting,* 1–5.

41 Li, H., Smith, J., and Rylander, M. (2014). Multi-inverter interaction with advanced grid support function. In *Proc. CIGRE US National Committee Grid of the Future Symposium.*

42 Schoene, J., Zheglov, V., Humayun, M. et al. (2017). Investigation of oscillations caused by voltage control from smart PV on a secondary system. In *Proc. 2017 IEEE Power & Energy Society General Meeting,* 1–5.

43 Kashani, M.G., Cho, Y., and Bhattacharya, S. (2016). Design consideration of volt–var controllers in distribution systems with multiple PV inverters. In *Proc. 2016 IEEE Energy Conversion Congress and Exposition (ECCE).* 1–7.

44 Braslavsky, J.H., Collins, L.D., and Ward, J.K. (2019). Voltage stability in a grid-connected inverter with automatic volt-watt and volt-var functions. *IEEE Transactions on Smart Grid* 10: 84–94.

45 Yazdani, A., Fazio, A.R.D., Ghoddami, H. et al. (2011). Modeling guidelines and a benchmark for power system simulation studies of three-phase single-stage photovoltaic systems. *IEEE Transactions on Power Delivery* 26: 1247–1264.

46 Varma, R.K. and Siavashi, E.M. (2018). PV-STATCOM: A new smart inverter for voltage control in distribution systems. *IEEE Transactions on Sustainable Energy* 9: 1681–1691.

47 Enslin, J.H.R. and Heskes, P.J.M. (2004). Harmonic interaction between a large number of distributed power inverters and the distribution network. *IEEE Transactions on Power Electronics* 19: 1586–1593.

48 Mohan, S. and Varma, R.K. (2020). Control interaction of two PV power plants with Volt-Var control. In *Proc. 2020 CIGRE Canada Conference.*

49 Varma, R.K., Siavashi, E.M., Mohan, S., and Vanderheide, T. (2019). First in Canada, night and day field demonstration of a new photovoltaic solar based Flexible AC Transmission System (FACTS) device PV-STATCOM for stabilizing critical induction motor. *IEEE Access* 7: 149479–149492.

50 Subsynchronous Resonance Working Group of the System Dynamic Performance Subcommittee (1992). Reader's guide to subsynchronous resonance. *IEEE Transactions on Power Systems* 7: 150–157.

51 Padiyar, K.R. (1999). *Analysis of Subsynchronous Resonance in Power Systems.* Norwell: Kluwer.

52 Pourbeik, P., Koessler, R.J., Dickmander, D.L. et al. (2003). Integration of large wind farms into utility grids (Part 2 - performance issues). In *Proc. 2003 IEEE Power Engineering Society General Meeting,* 1520–1525.

53 Varma, R.K., Auddy, S., and Semsedini, Y. (2008). Mitigation of subsynchronous resonance in a series-compensated wind farm using FACTS controllers. *IEEE Transactions on Power Delivery* 23: 1645–1654.

54 Narendra, K., Fedirchuk, D., Midence, R. et al. (2011). New microprocessor based relay to monitor and protect power systems against sub-harmonics. In *Proc. 2011 IEEE Electrical Power and Energy Conference*, 438–443.

55 Adams, J., Carter, C., and Huang, S. (2012). ERCOT experience with sub-synchronous control interaction and proposed remediation. In *Proc. 2012 IEEE/PES Transmission and Distribution Conference and Exposition (T&D)*, 1–5.

56 Wang, L., Xie, X., Jiang, Q. et al. (2015). Investigation of SSR in practical DFIG-based wind farms connected to a series-compensated power system. *IEEE Transactions on Power Systems* 30: 2772–2779.

57 (2020). Grand Renewable Energy Park, Haldimand County, Ontario. https://www.power-technology.com/projects/grand-renewable-energy-park-haldimand-county-ontario (accessed 17 June 2020).

58 Madeleine Cuff (2016). Vattenfall starts work on UK hybrid wind and solar farm. https://www.theguardian.com/environment/2016/jan/12/vattenfall-starts-work-on-uk-hybrid-wind-and-solar-farm (accessed 17 June 2020).

59 (2018). Australia's first hybrid wind and solar farm officially launched. https://www.ecogeneration.com.au/australias-first-hybrid-wind-and-solar-farm-officially-launched/. (accessed 9 Sep. 2021).

60 Fred Guteri (2018). The rise of the hybrids: this plant combines wind and solar power to keep renewable electricity flowing. https://www.ge.com/news/reports/rise-hybrids-power-plant-combines-wind-solar-keep-renewable-electricity-flowing (accessed 17 June 2020).

61 Fan, L. and Miao, Z. (2012). Mitigating SSR using DFIG-based wind generation. *IEEE Transactions on Sustainable Energy* 3: 349–358.

62 Mohammadpour, H.A. and Santi, E. (2015). SSR damping controller design and optimal placement in rotor-side and grid-side converters of series-compensated DFIG-based wind farm. *IEEE Transactions on Sustainable Energy* 6: 388–399.

63 Varma, R.K. and Salehi, R. (2017). SSR mitigation with a new control of PV solar farm as STATCOM (PV-STATCOM). *IEEE Transactions on Sustainable Energy* 8: 1473–1483.

64 Varma, R.K. and Salehi, R. (2020). PV-STATCOM for mitigating subsynchronous resonance associated with inverter-based resources. Presented at the Panel Session, "Voltage Control and Low and High Voltage Ride-Through of Wind and Solar Photovoltaic Systems". In *Proc. 2020 IEEE PES General Meeting*.

65 Sharafdarkolaee, R. Salehi. (2019). Novel PV Solar Farm Control as STATCOM (PV-STATCOM) for SSR Mitigation in Synchronous Generators and Wind Farms. PhD Thesis. The University of Western Ontario, London, ON, Canada.

66 IEEE Subsynchronous Resonance Task Force of the Dynamic System Performance Working Group Power System Engineering Committee (1977). First benchmark model for computer simulation of subsynchronous resonance. IEEE Committee Report, *IEEE Transactions on Power Apparatus and Systems*, vol. 96, 1565–1572.

67 Mohammadpour, H.A., Ghaderi, A., and Santi, E. (2014). Analysis of sub-synchronous resonance in doubly-fed induction generator-based wind farms interfaced with gate – controlled series capacitor. *IET Generation Transmission & Distribution* 8: 1998–2011.

68 Ackermann, T. (2005). *Wind Power in Power Systems*. Wiley: Chichester.

69 Yazdani, A. and Iravani, R. (2010). *Voltage-Sourced Converters in Power Systems: Modeling, Control, and Applications*. Hoboken: IEEE Press and John Wiley and Sons Inc.

70 Leon, A.E. and Solsona, J.A. (2015). Sub-synchronous interaction damping control for DFIG wind turbines. *IEEE Transactions on Power Systems* 30: 419–428.

71 Faried, S.O., Unal, I., Rai, D., and Mahseredjian, J. (2013). Utilizing DFIG-based wind farms for damping subsynchronous resonance in nearby turbine-generators. *IEEE Transactions on Power Systems* 28: 452–459.

72 NERC (2018). BPS-Connected Inverter-Based Resource Performance. NERC, Atlanta, GA, USA. *Reliability Guideline.*

73 IEEE (2018). Impact of Inverter Based Generation on Bulk Power System Dynamics and Short-Circuit Performance, IEEE/NERC Task Force on Short-Circuit and System Performance Impact of Inverter Based Generation. New York, NY, USA. *IEEE PES Technical Report PES-TR68.*

74 CIGRE (1999). Coordination of controls of multiple FACTS/HVDC links in the same system. Technical Brochure No. 149.

9

EMERGING TRENDS WITH SMART SOLAR PV INVERTERS

The technology of smart photovoltaic (PV) inverters is undergoing a major evolution process. This chapter deals with some of the fast-emerging trends with smart solar PV inverters. Some application examples are presented of enhanced grid support capabilities enabled by integrating smart inverter functionalities with solar PV inverters, battery energy storage systems (BESSs), and electric vehicle (EV) chargers [1]. A new technology of "grid forming (GFM) inverters" that is presently being widely researched across the world is introduced.

Recent field demonstrations of advanced grid support functionalities that are not required by present standards are described. This chapter further presents some thoughts on potential financial compensation mechanisms to smart PV inverters for providing grid support functionalities that go beyond being just "good citizens" on the power transmission and distribution systems.

9.1 Combination of Smart PV Inverters with Battery Energy Storage Systems (BESS)

Integration of BESS with PV solar systems allows smart PV inverters to operate in all the four quadrants as depicted in Figure 2.1. This significantly enhances the capabilities of solar PV systems for providing different smart inverter functions as described in Chapter 2. Some application examples of these improved capabilities are presented below. An extensive coverage of modeling and performance of hybrid plants combining inverter-based renewable generation systems (solar PV and wind power systems) with BESS technology is presented in [2].

9.1.1 Increasing Hosting Capacity

A case study of PV hosting capacity increase in the Danish island of Bornholm through energy storage in combination with smart inverters having volt–var function is presented in [1, 3]. The study LV feeder system in the island of Bornholm, Denmark, is depicted in Figure 9.1 [3]. This feeder has 23 buses with 52 customers. It is assumed that all customers have PV systems of same rating and each is coupled with an energy storage system (ESS). Simulation studies are conducted for 50%, 75%, and 100% PV penetration scenarios. For the 100% penetration scenario, the PV inverter capacity is considered to be 5.2 kW for all customers, whereas for the 50% penetration case, the PV inverter capacity is considered to be 2.6 kW for all the customers. The maximum acceptable overvoltage is 5%. Furthermore, no PV inverter is allowed to absorb more than 48% of the nominal active power. Actual real power generation and system loading data are utilized in all the simulation studies.

Smart Solar PV Inverters with Advanced Grid Support Functionalities, First Edition. Rajiv K. Varma.
© 2022 The Institute of Electrical and Electronics Engineers, Inc. Published 2022 by John Wiley & Sons, Inc.

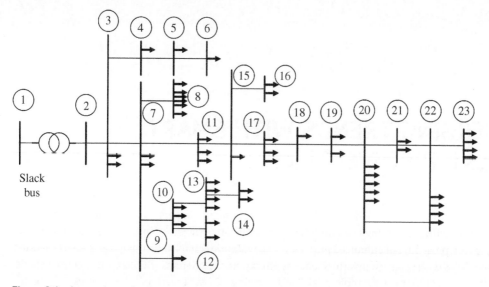

Figure 9.1 Low voltage feeder network in Danish island. *Source:* Hashemi and Østergaard [3].

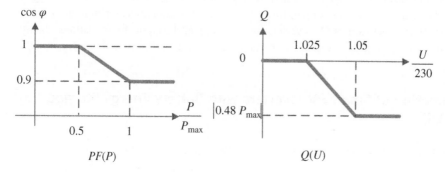

Figure 9.2 Smart inverter functions implemented on the PV inverters. *Source:* Hashemi and Østergaard [3].

Two smart inverter functions are considered on each PV solar system as shown in Figure 9.2 [3]. One is the PF (*P*) function wherein the reactive power absorption by the PV system is zero (power factor = 1) at 50% rated real power generation but increases to provide a power factor of 0.9 at the rated real power output. This implies that as the real power increases, the reactive power absorption also increases essentially to prevent voltage rise. The second smart inverter function is the *Q*(*U*) function wherein the reactive power absorption increases with the rise in voltage.

Figure 9.3 [3] shows the substantial increase in PV active power injection without causing any overvoltages for the condition of 75% PV penetration. The following inferences are made:

- Without any smart inverter control, only fixed power thresholds can be implemented. In this case, the PV active power needs to be curtailed at about 1.5 kW to prevent any overvoltage.
- With the proposed method and the application of PF (*P*) smart inverter control, the PV injected power can be increased to 2.5 kW without any overvoltage even during high sunny periods. In this smart inverter control, the maximum reactive power absorption occurs simultaneously with the maximum active power generation by the PV systems.

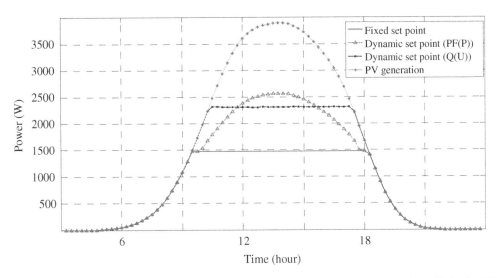

Figure 9.3 Real power that can be injected by PV systems without violating overvoltage limits for 75% PV penetration. *Source:* Hashemi and Østergaard [3].

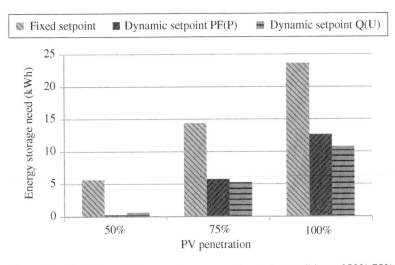

Figure 9.4 The ESS need for overvoltage prevention in the conditions of 50%, 75%, and 100% PV penetration. *Source:* Hashemi and Østergaard [3].

- When the $Q(U)$ smart inverter control is used, each PV inverter absorbs varying levels of reactive power in accordance with the voltage at their point of common coupling (PCC) resulting in a higher level of overall PV power injection.

The energy storage need for overvoltage prevention for different levels of PV penetration is illustrated in Figure 9.4 [3]. The following observations are made:

- For 50% PV penetration, an ESS of size 5.5 kWh per customer is required to prevent overvoltages. However, this need is reduced to about 1 kWh if dynamic set points resulting from use of smart inverter functions are utilized.

- For 75% PV penetration, the ESS requirement is reduced substantially from 14 to 5 kWh per customer if dynamic set points are utilized. A reduction is ESS rating from 23.5 to 12.5 kWh is seen for 100% PV penetration.

It is concluded that using 5-kWh ESS on customer sites can increase the PV hosting capacity to 50%. The hosting capacity can be further increased to 75% if smart inverter functions are utilized. This study demonstrates that:

1) A high rating of ESS is needed for overvoltage prevention for all PV penetration levels in the absence of smart inverter functions; and
2) The $Q(U)$ or the volt–var smart inverter function is more effective in limiting overvoltages at higher PV penetration levels.

This study thus makes the case that smart inverter functions can reduce the need and thereby the investment for ESS in limiting voltage-related PQ issues.

9.1.2 Capacity Firming

The power output of wind and solar power systems is quite intermittent due to varying winds and cloud coverage. ESS can help maintain the power output from such renewable energy systems at a firm (committed) level over a period of time. The ESS can also assist in rapidly alleviating power swings and voltage variations in the grid caused by intermittent power generation.

A 402 kW/282 kWh sodium nickel chloride BESS has been installed in Rankin Avenue substation in Mount Holly, NC in 2012 [4, 5]. This BESS provides centralized solar-induced power swing mitigation. It senses the real power loading on the substation and utilizes the BESS power to smoothen the ramp rates caused by solar intermittency due to cloud passages. It also alleviates the power swings from several solar plants on the distribution feeder, including that from a large 1.2 MW solar farm 3 miles away [4].

9.1.3 Preventing Curtailment of Wind/Solar Plant Outputs and Managing Ramp Rates

The power output from intermittent wind and solar power plants need to be curtailed to avoid unacceptable voltage and frequency variations on the grid. Grid codes restrict the rate at which wind or solar power can be ramped up or down, also to prevent voltage and frequency variations on the networks. Ramping capability is the rate at which generation levels can change in time, typically expressed in megawatts per minute. ESS can exchange active power with the wind/solar plants to ensure that the ramp rates specified by utility operating grid codes are implemented.

A ramp rate control strategy in solar PV farms is proposed in [6]. It is shown that an ESS installed for other reasons in a solar farm (e.g. storing excess energy) can be additionally utilized for ramp rate control. In this strategy, the ESS power exchange is controlled at the inverse of the ramp rate of the solar PV farm to offset the resulting voltage fluctuations at any instant of time.

A case study of a low-voltage distribution feeder in New South Wales in Australia is presented to demonstrate this ramp rate control [5, 6]. This feeder has several households (HHs) with rooftop solar PV systems each coupled to BESSs, as depicted in Figure 9.5 [6]. A 12 kWh regulated lead-acid battery is considered for this study. The voltage at HH 28 with and without the proposed ramp rate control is illustrated in Figure 9.6 [6]. If a decline in PCC voltage occurs due to a sudden decrease in solar power output, the BESS discharges real power to counteract the negative ramp rate of the solar power. On the other hand, if the voltage rises due to a sudden increase in the solar radiation (after a

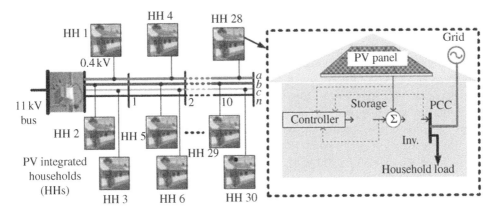

Figure 9.5 Test feeder with PV-storage integrated systems in Australia. *Source:* Alam et al. [6].

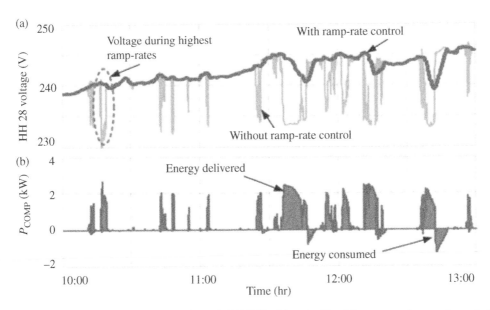

Figure 9.6 (a) Voltage fluctuation at household HH28 with and without the proposed ramp rate control of the BESS. (b) energy exchange from the BESS. *Source:* Alam et al. [6].

cloud has passed), the BESS absorbs real power to alleviate the positive power ramp rate of the solar plant. With this BESS control strategy, the ramp rate of the PCC voltage at HH28 is controlled to within 0.1 mV/s which is within the utility specified norm.

9.2 Combination of Smart PV Inverters with Electric Vehicle Charging Systems

EV charging systems and solar PV systems have several similarities. Both connect to the grid in a dispersed manner and at similar voltage level distribution systems. Both operate with power electronic converters. While EV charging systems involve storage and are bidirectional in nature, i.e.

can function both as a generator and a load, PV systems primarily function as generators. For the above reasons, EV charging systems can provide similar grid support functions as solar PV systems. A coordinated control of EV charging systems with solar PV distributed generators (DGs) and wind DGs can therefore help increase the hosting capacity of DGs and alleviate the need for expensive network reinforcements.

The potential of incorporating voltage support functions from EVs in a low-voltage distribution network in Denmark with high penetration of PV installations in is described in [1, 7].

The study system comprises a Danish low-voltage network connected to a 10 kV medium-voltage grid having three-phase short circuit power of 20 MVA, as illustrated in Figure 9.7 [7]. The MV/LV transformer is rated 400 kVA 10.5/0.42 kV. The low-voltage feeder line bifurcates into three radial sections coinciding with physical streets where the HHs are located. The line comprises 14 nodes and 13 line segments with total length of 681 m. Part A represents 17 houses located in Hormarken Street while part B represents 26 HHs located in Græsmarken Street. In addition, there is a street light connected to the grid in Græsmarken Street at node 608.

The solar PV systems are equipped with reactive power capability (RPC) as depicted in Figure 9.8 [7]. The EV chargers have a similar reactive power control characteristic except that their reactive power exchange (injection/absorption) is just opposite to that of the PV inverters.

This section reports results for cases when both PV systems and EV chargers have their RPC activated. These results correspond to winter weeks when the loading is higher.

The minimum voltages before and after EV reactive power control activation are presented in Table 9.1 [7]. It is seen that RPC from EV charging systems successfully increases the overall voltages and brings them within the utility specified limits of ±10%.

This study demonstrates the complementary voltage support offered by reactive power control of EVs, thereby helping integrate more PV systems in presence of EVs.

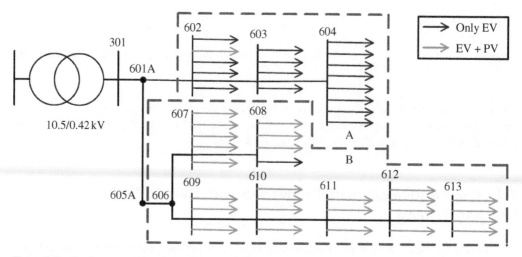

Figure 9.7 Study system. *Source:* Knezović et al. [7].

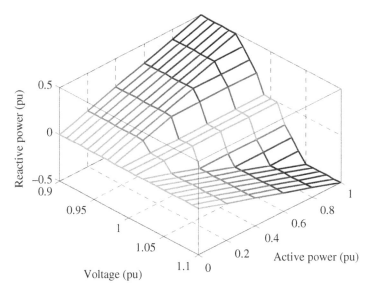

Figure 9.8 Reactive power control capability of PV inverters. *Source:* Knezović et al. [7].

Table 9.1 Minimum voltages at selected junction points before and after EV RPC activation – winter week.

Node	Minimum voltage without RPC (pu)	Minimum voltage with RPC (pu)	Relative voltage increase (%)
301	0.9898	0.9947	0.50
601A	0.9475	0.9659	1.94
602	0.9410	0.9607	2.09
604	0.9293	0.9518	2.42
606	0.9219	0.9433	2.32
607	0.9196	0.9411	2.34
608	0.9186	0.9403	2.36
609	0.9142	0.9366	2.45
613	0.8978	0.9205	2.53

Source: Knezović et al. [7].

9.3 Combination of Smart PV Inverters with Battery Energy Storage Systems (BESS) and EV Charging Systems

The PV systems can potentially provide reactive power with their entire inverter rating at night i.e., as PV-STATCOM, as described in Chapter 4. However, this reactive power control capability gets reduced during daytime if active power is not curtailed. Meanwhile, the active power exchange of BESS and EV systems is much less than their inverter ratings during several periods during daytime thus making available a substantial amount of reactive power. This provides a motivation to combine the reactive power capabilities offered by solar PV, BESS, and EV stations over a 24/7 period for STATCOM operation to optimally control voltages in utility networks or to provide any other smart inverter functions. Such a capability is described below.

The active power output (P) and reactive power capability (Q) of a PV system over 24 hours for a sunny day, derived from California Independent System Operator (CAISO) data [8] are illustrated in Figure 9.9 [9]. The available reactive power for PV-STATCOM operation is calculated as,

$$Q = \sqrt{S^2 - P^2} \tag{9.1}$$

where S is the MVA rating of the inverter.

During nighttime, full inverter capacity is available for reactive power control since active power production is zero. During noontime, when PV is generating maximum active power, the inverter capacity available for reactive power control is assumed to be zero.

The active power output P of a BESS derived from CAISO data [8] is portrayed in Figure 9.10 [9]. The charging of BESS is considered as positive P, whereas discharging of BESS is considered as negative P. The remaining inverter capacity available for STATCOM operation calculated according to (9.1) is also illustrated in Figure 9.10 [9]. The active power output of an EV charging station over 24 hours, estimated based on the number of charging stations in California, and the average power exchange data from a typical charging station [10] is depicted in Figure 9.11 [9]. The Q capability is also shown in Figure 9.11. The EV charging occurs mostly during either nighttime, early morning, or late evening. Thus, EV has room for reactive power support during noontime, when PV is not available for voltage regulation. The discharging/charging of BESS is distributed throughout the day, and thus it can complement the reactive power capabilities of PV and EV. The active power output and reactive power capability of the combination of PV, BESS, and EV, in the 40 : 30 : 30 ratio, are depicted in Figure 9.12 [9].

Since the distributed energy resources (DERs) complement each other in their capability for reactive exchange, their combination can provide significant reactive power support over a 24 hours period. This capability of the combination of DERs as smart inverters to provide reactive power support on 24/7 basis opens new opportunities for their combined use for providing several grid support functions

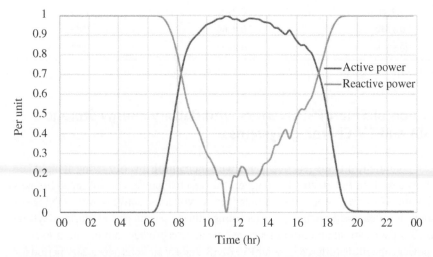

Figure 9.9 Active power output and reactive power capability of a solar PV farm for a sunny day. *Source:* Reprinted with permission from CIGRE [9] © 2020.

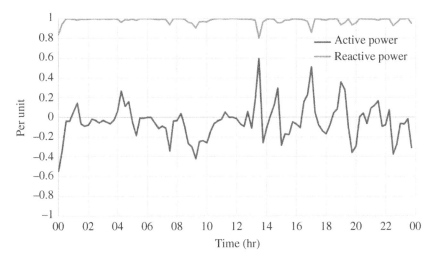

Figure 9.10 Active power output and reactive power capability of a BESS for the sunny day. *Source:* Reprinted with permission from CIGRE [9] © 2020.

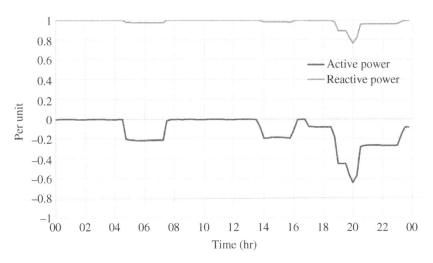

Figure 9.11 Active power output and reactive power capability of an EV for a weekday. *Source:* Reprinted with permission from CIGRE [9] © 2020.

Energy Savings Through CVR and Line Loss Reduction

It has been demonstrated in literature and various utility implementations of conservation voltage reduction (CVR) that a reduction of voltage by 1% can reduce the load by 1–2% [11–14]. This is especially significant during peak load conditions. Traditionally, CVR is implemented by adjusting the voltage set points of rotating generators, transformer taps, voltage regulators, and capacitor switching. However, these devices have response time of several seconds or few minutes. The application of volt–var control of smart inverter for CVR has been studied on Hawaiian Electric Company (HECO) distribution systems and Pacific Gas and Electric Company (PG&E) distribution systems [11]. With centralized voltage optimization (VO) without smart PV inverters, the annual energy consumption is reduced by 1.51% in HECO system. This energy saving increased by 1.37% with smart inverter control of PV inverters [11].

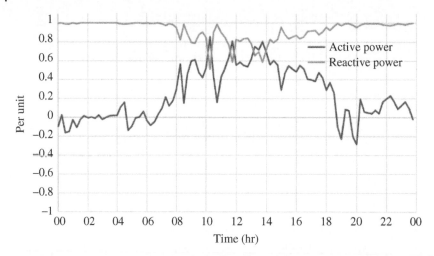

Figure 9.12 Active power output and reactive power capability of combination of PV, BESS, and EV in ratio of 40 : 30 : 30. *Source:* Reprinted with permission from CIGRE [9] © 2020.

Line losses are typically 4–5% in North American transmission and distribution systems. Considering an average Ontario demand of 18 000 MW during daytime, the line losses are 720 MW. This line loss, if saved, can easily supply the power needs of almost 400 000 people. Hence, even a fractional reduction in line losses can have substantial benefits. The reduction in line losses by application of smart PV inverter as PV-STATCOM is described in Chapter 5. A combination of line loss reduction and CVR in the IEEE 14 bus system using smart inverters with volt–var control both during night and day is shown to reduce active power demand by 2.11% leading to substantial energy savings [15].

A new technique of achieving energy savings in utility networks by simultaneous reduction of line losses and CVR on a 24/7 basis is presented in [9]. This is achieved by a patented smart inverter control of PV systems, BESSs, and EV charging stations as STATCOM.

The proposed novel STATCOM control on the combination of PV, BESS, and EV for line loss reduction and CVR in the IEEE 14 bus system results in energy saving which is sufficient to power 4354 U.S. homes, annually. An implementation of this technique on a larger provincial or regional grid would potentially result in substantially larger savings. With the rapid growth of PV, BESS, and EV systems in utility networks worldwide, the proposed STATCOM control on these DERs can provide substantial energy savings for utilities both during night and day.

9.4 Grid Forming Inverter Technology

With increasing efforts for decarbonizing electric grids and trend toward 100% renewables largely with inverter-based resources (IBRs), unique challenges are being experienced worldwide in maintaining a stable and secure power system [16]. These challenges include low system inertia, faster dynamics, and extensive control requirements. To meet such challenges, significant research and development activity is presently taking place to develop new smart inverter technologies known as "Grid Forming (GFM) inverters" [16–23]. These inverters are conceptually different than the "Grid Following (GFL) inverters" that are described in this book. The basic features of both these smart inverters are described below.

9.4.1   Grid Following Inverters

These inverters are based on voltage source converters (VSCs) and behave like virtual current sources due to their precise control of output current. These inverters are dependent on the electric power grid for synchronizing their phase-locked loops (PLLs) and thus "follow" the grid behavior by responding to electrical quantities measured from the grid [22]. The grid is assumed to have a high inertia and high short circuit strength. The GFL inverters cannot operate in a standalone mode and hence a potentially 100% IBR-based grid cannot be constituted by GFL inverters alone.

9.4.2   Grid Forming Inverters

These inverters are also based on VSCs but operate as a voltage source. They can control their terminal voltage and frequency independent from the grid. Instead of relying on the grid voltage for their synchronization they can "form" their own synchronism with the grid and control the grid voltage and phase through their own internal controllers. Hence these inverters are termed "grid forming" inverters. These inverters can maintain constant voltage and frequency as long as their current limits are not violated, and provide the functionality of a synchronous generator to a large extent. GFM inverters can provide blackstart capability as they are not dependent on synchronization from the grid. They need to be equipped with sufficient energy storage in form of BESS or supercapacitors, while providing blackstart services. These inverters are essential for creating future grids with very high, potentially 100%, IBR penetration [17].

9.4.3   Considerations in the Application of Grid Forming Inverters

GFM technology is in R&D stage at present (in 2020) [16, 18]. The exact design and controller requirements are being actively researched. The impacts of large implementation of GFM inverters in electric power grids in terms of control interaction with synchronous generators and other GFL inverters; system protection and relaying, are still being determined. Also, the required percentage amount of GFM inverters in power grids with a high IBR penetration scenario is presently under investigation. The GFM capability of inverters is restricted by their physical constraints such as inverter current limits. Additional energy buffer (e.g. BESS, supercapacitors) is needed to alleviate these power limit concerns. Converting a "grid following inverter" to "grid forming inverter" requires physical changes in equipment and is not achieved by simple retrofits. Also, there are tradeoffs related to design of GFM inverters as higher current ratings will increase the cost of inverters [18]. Market incentives also need to be developed to support the growth GFM technology [16].

It is clear from several studies [17], that the goal of very high penetration of IBRs and deep decarbonization of electric power grids can only be achieved with the use of GFM inverters. However, such a system cannot be totally based on GFM inverters. A future bulk power system (BPS) is expected to have a combination of both GFM inverters and GFL inverters [22].

9.5   Field Demonstrations of Smart Solar PV Inverters   ·

This section describes in a chronological order, some important worldwide utility demonstrations of smart PV inverters, which also have the potential of creating new revenue streams for solar PV systems.

9.5.1 Reactive Power Control by Solar PV Plant in China

The reactive power capability of a 30 MW solar PV plant was demonstrated in China [24] around 2014. The PV plant had 30 units of 1 MW, with each unit having two inverters. The individual units were connected to a 10 kV collector bus which was subsequently connected to a 110 kV power system. The PV plant capabilities of reactive power control, voltage control, and power factor control were demonstrated.

Reactive power capability was tested at an active power output of 8 MW. A step change in reactive power output from approximately 5.0 Mvar (capacitive) to 3 Mvar (inductive) was achieved in 0.4 s. This performance complied with the requirements of the Chinese National Standards GB/T 19964–2012 and GB/T 29319-2012. Voltage control tests further showed that the voltage can be controlled to a specified level of 1 pu with a response time of about 2 s. The power factor was then changed in different steps between 0.95 to −0.95. It was shown that the power factor could be successfully controlled to the desired setpoint in less than 1.6 s.

9.5.2 Active Power Controls by a PV Plant in Puerto Rico Island Grid, USA

In 2015, the National Renewable Energy Laboratory (NREL), AES, and the Puerto Rico Electric Power Authority (PREPA) conducted a demonstration project on a 20 MW utility-scale PV plant [25]. The objective was to demonstrate the capability of such a large PV plant to provide several ancillary services based on active power control including participation in automatic generation control (AGC), provision of droop response, and fast frequency response.

The PREPA island system is characterized by low inertia. During noon hours when the PV production is high, the displaced conventional generation makes the inertia even lower. This allows an opportunity for solar PV plants to participate in frequency support through active power controls.

At the time of conducting the tests, the total installed generation capacity in Puerto Rico was about 6 GW, largely contributed by petroleum and coal, with increasing participation by natural gas generators. The system also included 173 MW of wind and solar PV generation.

The AES's 20-MW Ilumina PV power plant is located in Guayama, Puerto Rico. The plant comprises 20 integrated arrangements of 1 MW each, utilizing a total of 40 inverters rated 500 kWAC each. The inverters were supplied by Green Power Technologies Corporation (GPTech). The available PV power at any time of the day is estimated based on pyranometers measurements of solar irradiance and solar radiation flux density. There are a total of 20 pyranometers installed to cover the entire solar plant.

Fast Frequency Response

The fast frequency response (FFR) tests were conducted with 10% curtailment of the PV plant output at three different power production levels: high, medium, and low. The PV plant was expected to deploy the curtailed reserve as fast as possible (within 500 ms) in response to the underfrequency event. The FFR test results conducted at high-power production are illustrated in Figure 9.13 [25]. An external frequency trigger signal simulated a step change in frequency from 60 to 59 Hz. The PV plant FFR got triggered within a short delay of 100-ms. The plant successfully deployed its curtailed reserve within another 100 ms. Once the FFR deployment was completed, the peak power tracking (PPT) algorithm got activated and increased the PV plant output to its maximum available power. Similar FFR tests at different levels of PV power production successfully demonstrated the capability of the PV plant to provide FFR service.

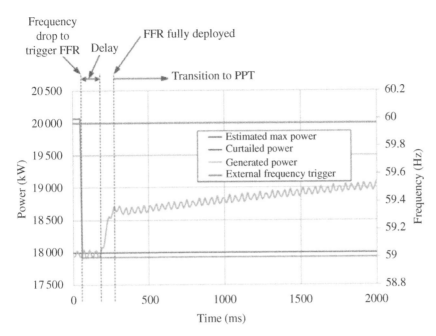

Figure 9.13 FFR test during high PV power production with 10% curtailment. *Source:* Reprinted with permission from the National Renewable Energy Laboratory [25].

9.5.3 Reliability Services by a 300 MW Solar PV Plant in California, USA

In August 2016, CAISO, First Solar, and the NREL conducted a demonstration project on a 300 MW PV power plant owned by First Solar in California [26]. The objective was to demonstrate that the large utility-scale PV plant had the capability of providing different essential ancillary services to the electric grid. These capabilities included functions based on active power control such as AGC and frequency regulation, droop response, and reactive power-based controls such as voltage control and power factor control. Reactive power control was also demonstrated during periods of very low solar irradiance.

In the First Solar's 300 MW PV plant, ten PV inverters rated 4 MVA 480 V each, are connected through pad-mounted transformers to constitute a 40 MVA block. Eight such PV inverter blocks are then integrated with the 230 kV bus through two 34.5 kV/230 kV 170 MVA transformers. Two switched capacitors are connected on the 34.5 kV bus to meet FERC's power factor requirements for large generators interconnections [26].

9.5.3.1 Automatic Generation Control (AGC) Test

This test was performed to demonstrate the ability of the PV plant to follow the active power signals issued by CAISO's AGC system. CAISO's AGC normally sends a direct active power set point signal to all participating generating units every four seconds. For this test, the active power output of the PV plant was reduced by 30 MW to create a headroom within which the plant can both increase and decrease its power output based on CAISO's active power set points. Also, all ramp-rate settings in the PV power plant controller were set to a very high level of 600 MW/min (10 MW/s).

The morning AGC test was conducted around 10 a.m., whereas the afternoon test was performed around 1 p.m. on 24 August 2016. The PV plant curtailed its output by 30 MW in each case and started responding to the CAISO's AGC signal. Figure 9.14 [26] and Figure 9.15 [26] illustrate

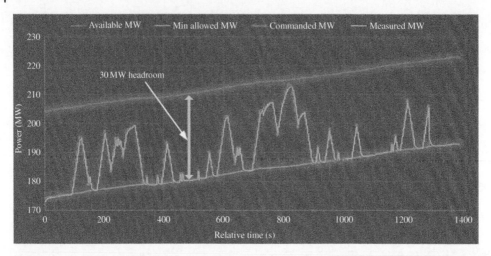

Figure 9.14 Morning AGC test results. *Source:* Reprinted with permission from the National Renewable Energy Laboratory [26].

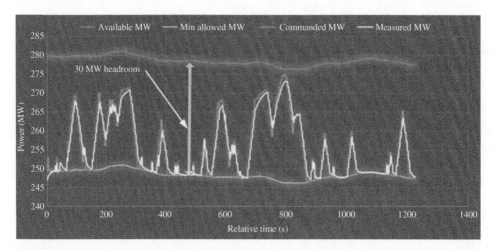

Figure 9.15 Midday AGC test results. *Source:* Reprinted with permission from the National Renewable Energy Laboratory [26].

the PV plant's response which was very close to the AGC signals from CAISO thus confirming successful AGC performance. The plant demonstrated good AGC performance even during cloudy periods. The regulation accuracy by the PV plant is seen to be significantly better than fast-ramping gas turbine technologies.

A recommendation was made that a relatively small and short-term energy storage system may be used to reduce the AGC error to essentially 0% by taking care of small control inaccuracies due to cloud impact and uncertainties of peak power calculation methods [26].

9.5.3.2 Droop Test During Underfrequency Event

Underfrequency droop tests were conducted on the 300 MW PV plant with both 3% and 5% droop settings [26]. The power plant control (PPC) curtailed the power of the PV plant by 10% for these tests. The minimum power level for downregulation was set to 20% lower than the available peak power for the different droop tests. This was done to minimize the revenue loss of the PV plant.

The results are depicted in Figure 9.16 [26]. The plant increased its power output during the initial period of frequency decline. Subsequently, the plant started decreasing its power output as

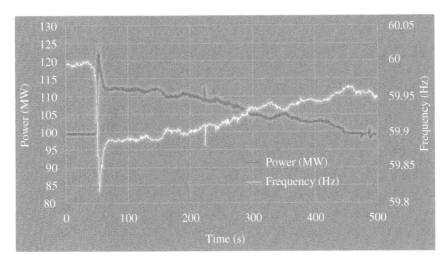

Figure 9.16 Droop response of PV plant during an underfrequency event. *Source:* Reprinted with permission from the National Renewable Energy Laboratory [26].

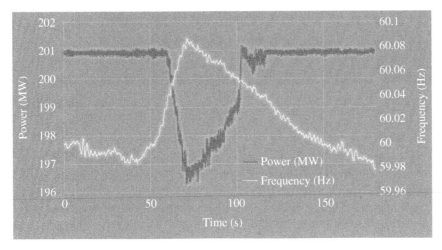

Figure 9.17 Droop response of PV plant during an over frequency event. *Source:* Reprinted with permission from the National Renewable Energy Laboratory [26].

frequency started approaching its nominal value. The power output showed a linear relationship with the frequency outside the deadband. Minor nonlinearity was observed which was attributed to decreasing solar irradiance and increased power output variability due to cloud conditions during the afternoon [26]. It was noted that finetuning of PPC controls would obviate this nonlinearity in future. Nevertheless, the overall test demonstrated the successful capability of the PV plant in providing frequency regulation in accordance with the droop setting.

9.5.3.3 Droop Test During Overfrequency Event
Droop test was also conducted on the 300 MW PV plant during an overfrequency event. The results of 5% droop test conducted in the morning hours are depicted in Figure 9.17 [26].

The PV plant decreased its power output as the system frequency increased. The plant subsequently increased its power output as the frequency started to reduce and achieve its predisturbance normal value. The PV power output displayed a linear relationship with frequency

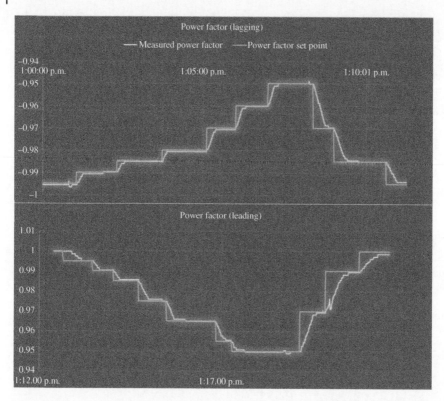

Figure 9.18 Power factor control tests in both leading and lagging mode. *Source:* Illustration from First Solar. Reprinted from [26].

after the deadband region was crossed. This was in accordance with the droop curve which the PV plant was expected to follow. Different overfrequency tests were performed with different droops (3% and 5%). Consistent and accurate downregulation performance of the PV plant was observed in all the tests conducted. These field tests successfully demonstrated the frequency regulation capability of solar farms during overfrequency events.

9.5.3.4 Power Factor Control Test

In this test, reactive power ramp rates and power factor limits were set at ±100 Mvar/min and ±0.95, respectively. Tests were performed for power factor control in both leading and lagging mode. The PV plant was producing almost its rated power. Figure 9.18 [26] shows the results of these tests. The PV plant achieved its power factor set points with the specified ramp rates without creating any adverse impact on system stability.

Reactive power set point control test was also performed to show the plant's capability to provide both inductive and capacitive reactive power at the 230 kV bus as depicted in Figure 9.19 [26]. Subsequently, the power output of the PV plant was curtailed to zero and the capability of the PV plant to inject or absorb 100 Mvar reactive power was demonstrated as shown in Figure 9.20 [26].

NOTE: This field demonstration however did not demonstrate STATCOM capability at night. This is because special control schemes are needed for grid-tied inverters to operate as STATCOM when a PV array is fully de-energized, and a certain amount of active power needs to be drawn from the grid to compensate for inverter losses [26].

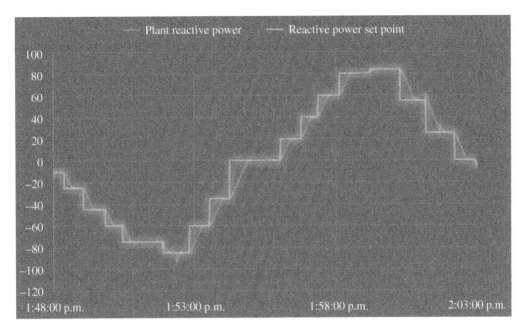

Figure 9.19 Reactive power control test results. *Source:* Illustration from First Solar. Reprinted from [26].

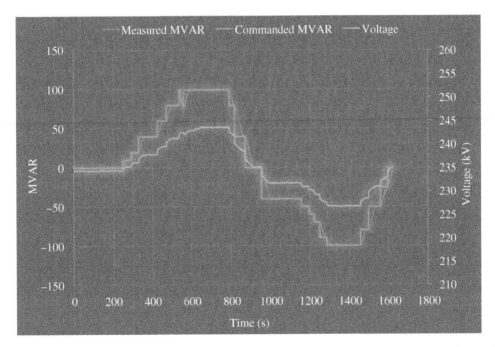

Figure 9.20 Reactive power production test at no active power (P ≈ 0 MW). *Source:* Reprinted with permission from the National Renewable Energy Laboratory [26].

9.5.4 Night and Day PV-STATCOM Operation by a Solar PV Plant in Ontario, Canada

On 13 December 2016, the first in Canada (and perhaps first in the world) utility demonstration was performed of a novel nighttime and daytime technology of utilizing solar PV farm as STATCOM, named PV-STATCOM, developed by this book's author [27]. The field demonstration was conducted on a 10 kW solar PV farm installed in the utility network of Bluewater Power Distribution Corporation, in Sarnia, Ontario. It was demonstrated that the solar farm autonomously transformed into a STATCOM and ensured continuous stable operation of a critical induction motor during large disturbances which would have otherwise destabilized the motor. The PV-STATCOM was thus shown to be a new FACTS Controller which could provide a 24/7 dynamic voltage control functionality as a STATCOM.

Shutdown of critical induction motors used in industries such as petrochemicals, process control, mining, automotive, medicines, etc. can result in significant financial loss to the industries. This first time utility implementation of PV-STATCOM technology demonstrated the:

- effectiveness of PV-STATCOM control to stabilize a critical induction motor during nighttime when conventional solar farms are idle, and also during daytime; and
- capability of PV-STATCOM to provide dynamic reactive power control at the same response speed (one to two cycles) as a conventional STATCOM.

9.5.4.1 Study System

The field demonstration of PV-STATCOM technology was conducted on the 10-kW grid-connected solar PV farm located on Confederation Street in the utility system of Bluewater Power Distribution Corporation, Sarnia, Ontario, Canada. The single line diagram of the study system and the photographs of the demonstration site are shown in Figures 9.21 and 9.22, respectively. The photograph of the demonstration site for day and nighttime operation is shown in Figure 9.22a,b, respectively. The solar PV farm is connected to Pole 325 using the switch S_1. Pole 325, in turn, is connected through a 150 kVA 600 V/4.16 kV transformer to the 4.16 kV distribution feeder 14F1 of Bluewater Power distribution system, although not shown in Figure 9.21. The voltage and current of the PV panels are 280 V and 35.7 A, respectively, for maximum power point (MPP) operation. Solar power is fed to the grid using a commercial utility inverter rated 10 kW, 600 V AC, and 475 V DC, operating at unity power factor. Switches S_2 and S_4 are used to isolate the existing PV inverter from the circuit. A 3-hp induction motor M_2 is used to operate the tracking system of the PV panels. This motor is connected to the 208 V terminal of transformer T_1 (Bus 1) using switch S_3.

The utility inverter is controlled by its proprietary control provided by the manufacturer and is not amenable to any modifications. For this reason, a separate three-phase two-level 10 kVA inverter to be controlled as PV-STATCOM was built in this book-author's laboratory at University of Western Ontario, and used for the field demonstration. The PV-STATCOM controller is implemented on the dSPACE controller board which generates appropriate firing pulses for the IGBTs of the PV-STATCOM inverter based on the selected control objectives and operation mode.

In this field demonstration, the 10 kVA PV-STATCOM is used to stabilize a 5 hp induction motor M_1 during a major system disturbance. This large system disturbance is initiated by switching a 10 kvar inductive load. In this demonstration, a variable line inductor L_1 rated 5.1 mH in combination with the system short circuit reactance L_S at Pole #325 is used to vary the effective short circuit inductance of the grid, L_g, as seen by the PV-STATCOM. Switch S_7 is used to isolate the PV-STATCOM and load from the grid.

For the nighttime operation of PV-STATCOM, the PV panels are disconnected from PV-STATCOM inverter using the switch S_5.

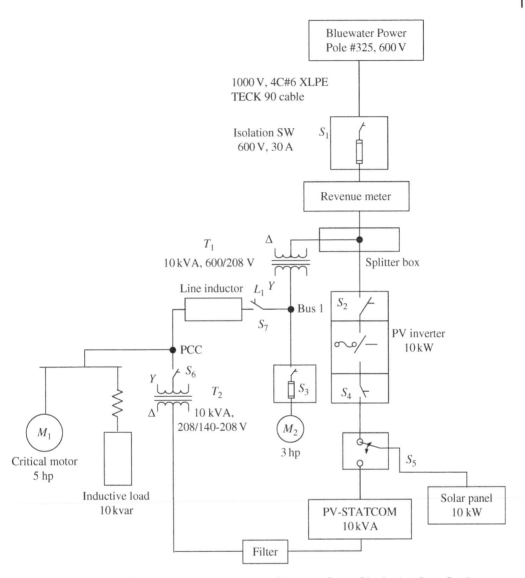

Figure 9.21 Schematic diagram of the study system at Bluewater Power Distribution Corp., Sarnia. *Source:* Varma et. al. [27].

9.5.4.2 PV-STATCOM Controller

The PV-STATCOM operating modes are depicted in Section 4.7.3.1. The detailed PV-STATCOM controller is described in [27]. The main modifications for PV-STATCOM operation are described below.

In a conventional PV system, the DC-link reference voltage ($V_{DC, ref}$) is provided by the maximum power point tracking (MPPT) module. In the proposed PV-STATCOM operation, the controller curtails active power generation during a disturbance and uses the entire inverter capacity to exchange (inject/absorb) reactive power. This objective is achieved by setting $V_{DC, ref}$ to open-circuit voltage V_{oc} of the solar panels at which voltage the power output of the PV panels reduces to zero. Once, the system returns to stable operation, the PV-STATCOM controller ramps up active power to the maximum available power according to the solar irradiance. The power is ramped up by varying $V_{DC, ref}$ from the open-circuit voltage V_{oc} to the voltage corresponding to the MPP V_{MPP}.

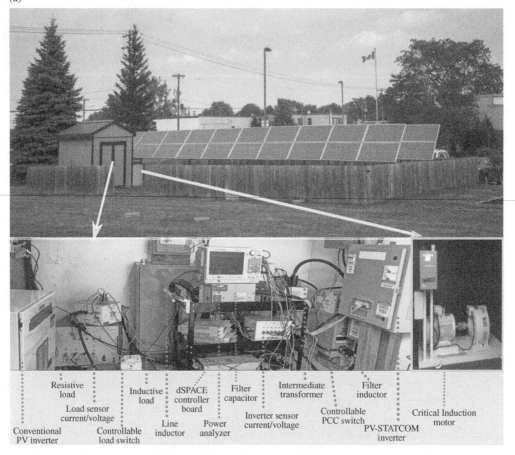

(a)

Resistive load · Inductive load · dSPACE controller board · Filter capacitor · Intermediate transformer · Filter inductor · Load sensor current/voltage · Line inductor · Power analyzer · Inverter sensor current/voltage · Controllable PCC switch · Critical Induction motor · Conventional PV inverter · Controllable load switch · PV-STATCOM inverter

(b)

Figure 9.22 (a) Field implementation of PV-STATCOM at Bluewater Power Distribution Corporation, Sarnia, ON, Canada. (b) PV-STATCOM field demonstration site during nighttime. *Source:* Varma et. al. [27].

It is noted that during daytime, when the sun is available, the DC voltage control uses solar power to compensate the inverter switching losses, and thereby keep the DC-link capacitor charged to the desired voltage reference level. Since at night the DC input power from solar panels to the PV inverter is zero, the DC-link capacitor is charged from the grid using a "Nighttime Charging circuit" utilizing the diodes across the IGBT switches of the inverter. Once the DC-link capacitor is charged, the charging circuit is disabled to prevent losses in the charging circuit. Thereafter, the DC voltage reference is switched to its rated value $V_{DCrated}$ for STATCOM operation. The required active power for compensating switching losses (1–3%) and for maintaining DC voltage at the reference value $V_{DCrated}$ is provided from the grid, as in STATCOMs [28, 29].

9.5.4.3 Response of Conventional PV Inverter to a Large Disturbance During Daytime

A large system disturbance is simulated by switching a 10 kvar inductive load at PCC during daytime for a period of 1.2 s, without the PV-STATCOM controller activated. The responses of solar PV farm in Full PV mode (conventional unity power factor mode) and the induction motor for this scenario are depicted in Figure 9.23.

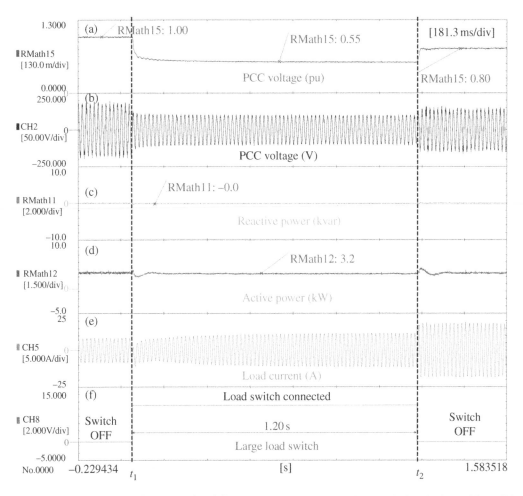

Figure 9.23 Response of the conventional PV inverter for large load switching during daytime without PV-STATCOM controller. (a) PCC rms voltage (pu), (b) PCC instantaneous voltage (V), (c) Reactive power output of inverter (kvar), (d) Active power output of inverter (kW), (e) Motor current (A), and (f) Status of the large load switch. *Source*: Varma et. al. [27].

Figure 9.23a–f illustrates the PCC rms voltage (pu), PCC instantaneous voltage (V), reactive power output of the inverter (kvar), active power output of the inverter (kW), motor current (A), and status of the large load switch, respectively.

$t < t_1$: The PCC voltage is 1 pu during this time interval. The inverter is operating at unity power factor by injecting the maximum available active power of 3.2 kW to the grid and keeping reactive power at zero. The motor is operating at steady state by providing the required load torque by consuming 8 A.

$t = t_1$: A disturbance is created by connecting 10 kvar inductor to the PCC. Due to this large load switching, the PCC voltage drops to a low value of 0.55 pu. Since the inverter is operating at unity power factor, the inverter continues injecting 3.2 kW active power to the grid while maintaining its reactive power output at zero. Due to the voltage drop, the motor starts consuming a large reactive power and the motor current increases to 12 A.

$t = t_2$: The large load is disconnected and the voltage starts recovering. However, due to inadequate reactive power support, the motor continues consuming a higher current of 16 A. Due to the large motor current, the PCC voltage recovers to 0.8 pu only. The motor eventually stalls.

This study shows that the induction motor will become unstable and stall during a large disturbance in the absence of any dynamic reactive power support from the solar farm.

9.5.4.4 Response of PV-STATCOM to a Large Disturbance During Daytime

The PV-STATCOM control is now enabled on the solar farm. The PV-STATCOM inverter and induction motor response for the same large load switching at PCC for a period of 1.2 s during daytime with PV-STATCOM controller are shown in Figure 9.24. Figure 9.24a–g illustrates the PCC rms voltage (pu), PCC instantaneous voltage (V), reactive power output of the PV-STATCOM inverter (kvar), active power output of the PV-STATCOM (kW), PV-STATCOM inverter current (i_{PV}) (A), motor current (A) and status of the large load switch, respectively.

$t < t_1$: The PV-STATCOM is operating at unity power factor by generating 4.1 kW power.

$t = t_1$: The large load is connected and consequently, the system voltage drops to 0.65 pu. The PV-STATCOM detects the drop in the voltage and autonomously switches to Full STATCOM mode. In this mode, the active power is curtailed, and full inverter capacity is utilized for injecting reactive power. The PV-STATCOM inverter reduces its active power from 4.1 kW to 0 and injects 9.4 kvar reactive power within 0.024 s (approximately 1.4 cycle). The fast reactive power injection is able to bring the PCC voltage to 0.99 pu and thus prevent the motor from stalling.

$t = t_2$: The large load is disconnected at $t = t_2$. The controller detects it and switches to Full PV mode. The reactive power output is reduced to zero and active power is ramped up to 4.1 kW. The motor continues stable operation and consumes its pre-disturbance current of 8 A.

This study demonstrates that the PV-STATCOM is able to inject almost 1 pu reactive current within 1.4 cycles. This response time matches that of commercial STATCOMs [28]. The rapid reactive power support by PV-STATCOM successfully prevents motor instability during a large disturbance. It is further demonstrated that the PV-STATCOM control is able to ramp the active power to its pre-disturbance level within three cycles after the large load is disconnected.

9.5.4.5 Response of Conventional Inverter to a Large Disturbance During Nighttime

Studies are now performed during nighttime. The responses of the solar PV farm and the induction motor for the switching of 10 kvar inductor at PCC during nighttime for a period of 1.2 s, without PV-STATCOM controller are shown in Figure 9.25. Figure 9.25a–f illustrate the PCC rms voltage (pu), PCC instantaneous voltage (V), reactive power output of the inverter (kvar), active power output of the inverter (kW), motor current (A), and status of the large load switch, respectively.

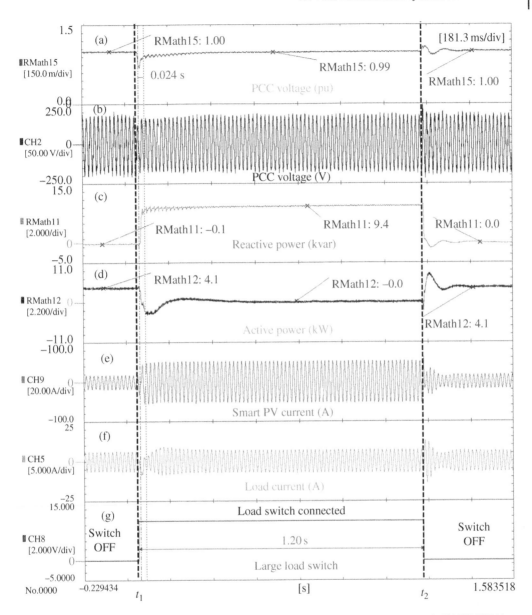

Figure 9.24 Response of the PV-STATCOM for large load switching during daytime with PV-STATCOM controller. (a) PCC rms voltage (pu), (b) PCC instantaneous voltage (V), (c) Reactive power output of PV-STATCOM (kvar), (d) Active power output of PV-STATCOM (kW), (e) PV-STATCOM current (A), (f) Motor current (A), and (g) Status of the large load switch. *Source:* Varma et. al. [27].

$t < t_1$: The inverter is idle as solar power is zero at night. The motor is operating at steady state by providing the required load torque, by drawing power from the grid.

$t = t_1$: A large disturbance is created by connecting 10 kvar inductor to the PCC. Due to this large load switching, the PCC voltage drops to 0.5 pu. Due to this large voltage drop, the motor starts consuming a high amount of reactive power and the motor current increases to 12 A.

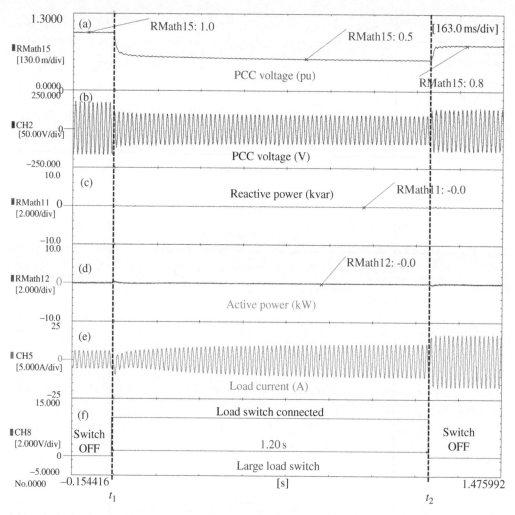

Figure 9.25 Response of the conventional PV inverter for large load switching during nighttime. (a) PCC rms voltage (pu), (b) PCC instantaneous voltage (V), (c) Reactive power output of inverter (kvar), (d) Active power output of inverter (kW), (e) Motor current (A), and (f) Status of the large load switch. *Source:* Varma et. al. [27].

$t = t_2$: The large load is disconnected, due to which the voltage starts recovering but only up to 0.8 pu. Since no reactive power support is available the motor consumes a larger current of 16 A and eventually stalls.

This study demonstrates that the induction motor will become unstable and stall during a large disturbance at night when no reactive power support is available.

9.5.4.6 Response of PV-STATCOM to a Large Disturbance During Nighttime
Studies are conducted during nighttime with the PV-STATCOM control activated. The PV-STATCOM inverter and induction motor responses for the large load switching at PCC during nighttime, for a period of 1.2 s, with PV-STATCOM controller are shown in Figure 9.26. Figure 9.26a–g illustrates the PCC rms voltage (pu), PCC instantaneous voltage (V), reactive power

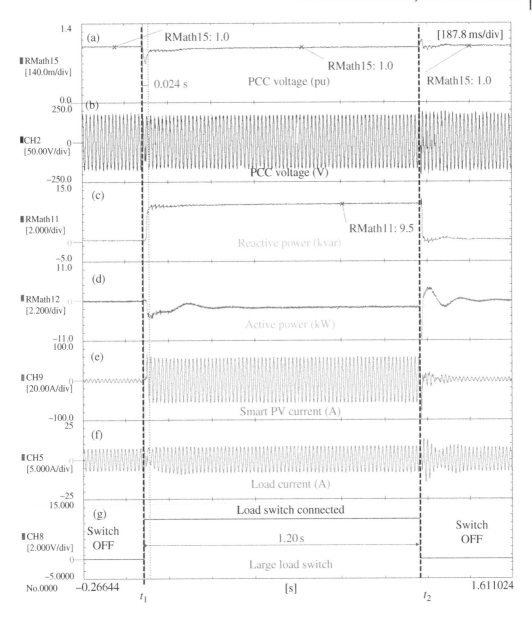

Figure 9.26 Response of the PV-STATCOM for large load switching during nighttime. (a) PCC rms voltage (pu), (b) PCC instantaneous voltage (V), (c) Reactive power output of PV-STATCOM (kvar), (d) Active power output of PV-STATCOM (kW), (e) PV-STATCOM current (A), (f) Motor current (A), and (g) Status of the large load switch. *Source:* Varma et. al. [27].

output of the PV-STATCOM inverter (kvar), active power output of the PV-STATCOM (kW), PV-STATCOM inverter current (i_{pv}) (A), motor current (A) and status of the large load switch, respectively.

$t < t_1$: The system is operating in a steady state condition with a PCC voltage of 1 pu. The PV-STATCOM is idle. The motor is providing the load torque by consuming 8 A from the grid.

$t = t_1$: The large load is connected and as a consequence, the bus voltage drops to 0.65 pu. The PV-STATCOM detects the drop in the voltage and switches to Full STATCOM operation. In this mode, the full inverter capacity is utilized for injecting reactive power. The PV-STATCOM inverter injects 9.4 kvar reactive power. The speed of response for the controller is seen to be 1.4 cycle. This fast reactive power injection is able to bring the PCC voltage to 1.0 pu and thus prevent the motor from stalling.

$t = t_2$: The large load is disconnected at $t = t_2$. The controller detects it and ceases reactive power injection. The motor continues to operate in a stable manner with its pre-disturbance current level of 8 A.

9.5.4.7 Summary

This study demonstrates that dynamic reactive power support by solar PV farm with the proposed PV-STATCOM control can prevent motor instability during large disturbance at nighttime, when conventional solar PV farms are usually dormant, as well as during daytime. Following are the conclusions of the field studies:

1) The switching of large inductive load causes a drop in the PCC voltage to about 0.5 pu for about 1.2 s both during night and day. The motor becomes unstable and stalls in both cases.
2) The PV-STATCOM control on the solar farm successfully ensures stable operation of the motor. The motor continues to operate in a stable manner despite the switching of the large load both during night and day.
3) The PV-STATCOM regulates the voltage to 1 pu in about 1.4 cycles, both during night and day. This speed of response is identical to actual STATCOMs
4) During daytime, the PV-STATCOM further restores the active power generation of the solar farm from zero to its pre-disturbance level in less than three cycles after the large load is switched off.

With the unprecedented growth of PV solar farms world-wide, it is quite likely that solar farms may find themselves installed in the vicinity of large critical induction motors. This novel functionality of solar farms as PV-STATCOM can enable these solar farms to provide a dedicated stabilization service to the critical motors on a 24/7 basis at a far lower cost than STATCOMs and SVCs that are normally used to stabilize such large critical motors. The proposed PV-STATCOM functionality can therefore open up a new revenue stream for solar farms in addition to the sale of active power [27].

9.5.5 Nighttime Reactive Power Support from a Solar PV Plant in UK

On 4 November 2019, a field demonstration of nighttime reactive power service from a 4 MW solar PV plant in East Sussex, UK, was performed [30]. This PV plant is owned by Lightsource BP. This was the first time in the United Kingdom that a solar PV plant with reconfigured inverters provided reactive power support at night for voltage control on the distribution network of UK Power Networks which is managed by National Grid Electricity System Operator (ESO). It was also demonstrated that this reactive power service at night did not have any adverse impact on customers.

The test was performed as a part of the "Power Potential" project led by National Grid ESO and UK Power Networks and aims to create a new reactive power market for DERs in the southeast of the United Kingdom.

9.5.6 Commercial Ancillary Grid Services by Solar PV Plant in Chile

In August 2020, the First Solar's 141 MW Luz del Norte PV power plant in Chile was announced to be the world's first known utility-scale solar facility licensed to commercially deliver a range of ancillary grid services including AGC [31]. This plant has been added to the portfolio of large-scale power generators that are approved for providing ancillary grid services by Chile's independent system operator (ISO), Coordinador Eléctrico Nacional [31].

9.5.7 Nighttime Reactive Power Control by a Solar PV Plant in India

An overvoltage problem on the 400 kV grid has been experienced during nighttime near the location of 2050 MW ultra-mega solar plant in the Southern Region of India [32]. This overvoltage is caused due to light loading conditions, which gets exacerbated during monsoon periods when loads are low and the nearby buses are also experiencing high voltages. The high voltages occur despite installed bus reactors. Typically, EHV lines need to be opened to alleviate the overvoltage, which impacts the reliability and lifespan of circuit breakers.

As a potential solution to the overvoltage problem, nighttime reactive power absorption tests were conducted in February 2021 on 1750 MW inverter blocks (out of 2050 MW) which had the capability for providing reactive power in the night [32]. A total of 36 Mvar (1 Mvar absorption per 50 MW block) was implemented on 16 February, and 180 Mvar absorption (5 Mvar absorption per 50 MW block) was implemented on 17 February 2021. A grid voltage reduction of 3 kV was successfully achieved with 128 Mvar reactive power absorption, without occurrence of any technical issues.

Subsequently, a two-month pilot project involving operation of solar PV inverters in the night mode (voltage control mode) was started on 1 August 2021. Initial studies have shown that a reactive power absorption of 350 Mvar can be achieved, which can be extended to 500 Mvar. This has reduced the need for opening multiple 400 kV lines for obviating the overvoltage issue [S.R. Narasimhan, email communication on Indian power systems internet group INPOWERG, 20 August 2021].

9.6 Potential of New Revenue Making Opportunities for Smart Solar PV Inverters

Although smart PV inverters conform to the requirements of various Standards and Grid Codes, they offer unique capability of providing several grid support functions and ancillary services that go beyond the present Standards but are needed by power grids for ensuring stability and reliability. These grid support functions can be classified into four categories:

9.6.1 Providing Ancillary Services Without Curtailing PV Power Output

There are many jurisdictions worldwide where financial compensation is provided by the system operator for providing reactive power and voltage control [33], such as in Australia, UK, USA, Canada, etc.

Smart PV inverters have demonstrated capability of providing such reactive power and voltage control services which they can provide even without curtailing their active power output. Hence, they can easily compete in the ancillary services market and earn substantial revenues.

9.6.2 Providing Ancillary Services by Curtailing PV Power Output

9.6.2.1 Fast Frequency Response and Frequency Regulation Services

The field demonstrations of grid support functionalities by solar PV plants in 2015 in Puerto Rico [25], in 2016 in California [26], and commercial ancillary grid support services in Chile in 2020 [31] have clearly demonstrated the capability of smart solar PV inverters to provide fast frequency response and other frequency regulation services such as participation in AGC, which are much needed by the grid.

Solar PV plants can provide very rapid frequency response and primary frequency response. If the primary frequency response were to come from conventional synchronous generators, they will need to operate at lower output levels at reduced efficiency. This option is more costly than that from solar PV plants.

It is however understood that solar PV farms need to curtail their revenue-making power output to provide such services. Hence, market mechanisms need to be developed to incentivize solar PV systems such that the financial compensation for providing FFR exceeds the loss of revenue from sale of energy [34].

9.6.2.2 Flexible Solar Operation

The smart inverter capability of rapidly varying active power output from solar PV plants allows solar plants to provide grid balancing services [34]. Solar PV systems can be operated in a fully flexible manner, i.e. can be dispatched both downwards and upwards based on grid requirement. The system operator keeps reserve capacity "headroom" available in conventional thermal generators in case of demand becoming more than the forecasted value. Similarly, the system operator sets a reserve capacity "footroom" in conventional thermal generators to allow the power to be dispatched down if there is load that goes below the forecasted amount. With increasing penetration of solar PV, they can provide the needed headroom and footroom for the system operator. This provision of balancing services from solar plants allows thermal generators to operate more efficiently by reducing the need for cycling and load-following services, thereby resulting in reduced consumption of fuel [34]. Furthermore, this obviates the need to commit (operate) inefficient thermal generation plants, which decreases the amount of solar curtailment especially during periods of high generation.

Even though this provision of balancing services from solar PV results in some loss of power production or "opportunity cost" for the solar farm, this loss of power is still lower than curtailment of solar power under normal circumstances. On the other hand, this balancing service helps reduce overall operational costs and CO_2 emissions in the power system. Electricity markets, therefore, need to develop mechanisms for compensating solar PV systems for becoming active participants in grid balancing and providing such ancillary services to the grid.

9.6.3 Providing Reactive Power Support at Night

In the 2017 EPRI Report [35], a guidance is provided to utilities that joint industry efforts should be initiated to discuss the advantages of advanced inverters performing grid support functions (such as volt–amp reactive (VAR) support) continuously i.e. during nighttime.

In the 2018 IEEE PES Report [36], it is recommended that smart inverters may be enhanced in future to provide functions to help mitigate grid issues. These include reactive power support and voltage support during nighttime.

In the 2018 NERC Report [23], different benefits are described from dynamic reactive power support from PV inverters during periods of no active power output, i.e. during nighttime. The report further states that reactive power support during these times may be able to lower the transmission investments that would otherwise be needed in absence of such dynamic reactive power support capabilities [23].

In the 2019 NERC Reliability Guideline [37], one of the recommendations is that transmission operators may require inverter-based resources to exchange reactive power with the BPS (to provide voltage control) when no active power is generated, i.e. during night for PV inverters.

During 2009–2012, a new technology of utilizing solar PV farms as STATCOM (termed PV-STATCOM) was developed by this book's author to provide several STATCOM functionalities at night with full inverter capacity and during daytime with inverter capacity remaining after active power generation [38–41].

In 2016, an enhanced night and day PV-STATCOM operation utilizing full inverter capacity was field demonstrated on a utility network to stabilize a critical induction motor [27].

In 2019, a PV plant control for providing reactive power support at night was field demonstrated in the United Kingdom for mitigating overvoltage issues typically arising from lightly loaded transmission lines and cable networks [31].

In 2020, reactive power support at night was implemented at a large solar plant in India to alleviate overvoltages caused by low loading conditions in the grid [32].

Clearly, these nighttime advanced grid support functionalities can bring several benefits and financial savings for utilities. Grid regulations and standards need to be developed for facilitating such functionalities. Market mechanisms also need to be developed to financially incentivize solar PV systems for providing such grid support functions.

The IEEE Standard P2800, presently under development in 2021, is expected to mandate several advanced grid support functions from solar PV plants and other inverter based resources [42].

9.6.4 Providing STATCOM Functionalities

There are several grid support needs where utilities require installation of FACTS Controllers including [28, 29]:

- Dynamic reactive power compensation
- Dynamic voltage control
- Increasing power transfer capacity of transmission lines
- Improvement of system transient stability limit
- Power oscillation damping
- Alleviation of voltage instability
- Mitigation of fault-induced delayed voltage recovery (FIDVR)
- Stabilization of critical motors
- Mitigation of subsynchronous resonance (SSR)
- Improvement of high-voltage DC (HVDC) converter terminal performance
- Load compensation
- Grid integration of wind power generation systems, and in some cases solar PV power systems

In 2017, an NREL report on field demonstration of reliability services by a 300 MW solar PV plant in California [26] concluded that fast response by PV inverters coupled with plant-level controls make it possible to develop other advanced controls, such as STATCOM functionality, power oscillation damping controls, subsynchronous controls oscillations damping and mitigation, etc. Furthermore, one of the future plans of the project team includes "...demonstrating true PV STATCOM functionality during nighttime hours...".

In 2018, a NERC Report [23] stated that appropriately configured PV inverters can function like a STATCOM at night. Provision of such reactive power support may be able to decrease the transmission investments that would otherwise be required in absence of such dynamic reactive power support capabilities [23]. According to inverter manufacturers, the incremental cost to enable this capability at the solar PV plant is significantly lower than a transmission-connected dynamic reactive power resource. However, there will be some minor costs associated with supplying the additional heating losses in the inverter during zero power output conditions, added operations and maintenance costs, in addition to some degradation in lifespan of the inverter components [23].

Starting from 2009, a suite of novel patented technologies for utilizing solar PV systems as STATCOM, termed PV-STATCOM were developed by this book's author for providing various advanced grid support functionalities as described in Chapters 4–6. The PV-STATCOM technology can provide several STATCOM functionalities both during nighttime and during any period of system need during daytime, with full inverter capacity. These grid support functionalities include:

- Dynamic voltage control during night and day [38, 43]
- Increasing connectivity of neighboring wind farms [41]
- Enhancement of solar farm connectivity [44]
- Stabilization of a remotely located critical motor [45]
- Reduction of line losses [9, 15, 46]
- Increasing power transmission capacity [39, 41, 47]
- Power oscillation damping [48]
- Mitigation of SSR [49]
- Mitigation of FIDVR [50]
- Simultaneous fast frequency control and power oscillation damping [51]

Several of these applications could only be accomplished by the installation of a physical STATCOM or Static Var Compensator (SVC). The implementation of PV-STATCOM controls on an existing solar PV plant in the same network provided a similar grid support functionality and performance increase as a STATCOM [52]. In some applications, the PV-STATCOM provided a dedicated 24/7 service as a STATCOM. In other words, PV-STATCOM reduced or in some cases totally eliminated the need for installing a physical STATCOM.

It is explained in Chapter 4 that transforming an existing solar PV plant into a PV-STATCOM is more than 10 times cheaper than installing a new equivalent size STATCOM or SVC. Also, the time required to transform an existing solar PV system into a PV-STATCOM is significantly shorter than that needed for constructing afresh an equivalent size STATCOM, including the time required to obtain all regulatory and building permissions. Obviating the need for fresh construction of STATCOM further prevents deforestation (hurting the environment) and eliminates associated greenhouse gas emissions (no vehicle usage for moving people, equipment, etc.)

Even if a large solar plant has to be derated (considered to be of smaller rating) for providing the short term overcurrent capability as a STATCOM, and a financial discount is incorporated to cover the decrease in inverter lifespan due to minor additional heating associated with nighttime reactive power flow (refer Chapter 4), the cost of PV-STATCOM will still be substantially lower than installing a new STATCOM.

Solar PV farms are not only becoming larger (largest being 2200 MW in 2020) but a substantial number of solar plants are becoming comparable to the size of transmission level STATCOMs and SVCs. Considering the unprecedented growth of solar PV systems worldwide, it is quite likely that solar PV plants will get installed on networks where advanced grid support functions as above are needed by the concerned utility.

If a utility or customer has determined that it must invest in a FACTS Controller to solve any of its transmission or distribution system problem, and if an equivalent size solar PV plant is available nearby, the solar plant with PV-STATCOM technology can provide the same service on a 24/7 basis to the utility/customer at about 10 times lower cost. This brings a substantial cost saving for the utility/customer.

The solar PV farms can, therefore, seek a share in these savings for providing the required FACTS functionality to the utility both during night and day. This can constitute a new revenue-making opportunity for solar PV farms in addition to the sale of active power during the day.

Such an arrangement however requires development of (i) new standards, (ii) appropriate agreements among the different stakeholders, i.e. the utility, system regulator, solar PV farm owner, inverter manufacturer, benefiting customer, etc., and (iii) suitable mechanisms for financial compensation to the smart solar PV farms.

9.7   Conclusions

The widespread deployment of energy storage systems and applications of EVs offer new capabilities to solar PV farms when integrated with these technologies in delivering smart inverter functionalities. Although only a few applications are described, the potential applications are many and growing.

This chapter briefly introduced the concept of a new smart inverter technology of "grid forming inverters," which is in an extensive research and development stage. It has been concluded that very high penetration levels of IBRs can only be possible if GFM inverters are installed. However, even in that scenario, a larger fraction of inverters will still be "grid following inverters," which is the subject matter of this book.

The main focus of this chapter has been to describe the field demonstrations of several novel smart solar PV inverter functions which can provide significant cost savings and benefits to power transmission and distribution systems. These functionalities include fast frequency response, frequency regulation, flexible solar operation, reactive power support at night, and night and day PV-STATCOM technology for providing several FACTS functionalities.

This chapter makes a case that smart solar PV inverters offer many grid support functionalities which can significantly help in enhancing the overall grid stability and reliability. PV-STATCOM is one of several such smart solar PV inverter technologies. Smart inverter functionalities continue to be developed by researchers and inverter manufacturers around the globe. These novel advanced grid support functionalities need to be recognized and facilitated by standards, worldwide. Furthermore, financial markets need to be developed to incentivize such technologies for the benefit of power systems.

References

1 CEATI International Inc. (2020). Enhancing Connectivity of DGs by Control Coordination of Smart Inverters. CEATI International Inc., Montreal, QC, Canada, *Report No. T184700 #5178.*

2 NERC (2021). Performance, Modeling, and Simulations of BPS-Connected Battery Energy Storage Systems and Hybrid Power Plants. NERC, Atlanta, GA, USA, *Reliability Guideline.*

3 Hashemi, S. and Østergaard, J. (2018). Efficient control of energy storage for increasing the PV hosting capacity of LV grids. *IEEE Transactions on Smart Grid* 9: 2295–2303.

4 CEG (2013). Duke energy's storage projects. http://www.cleanegroup.org/webinar/duke-energys-energy-storage-projects/ (accessed 10 April 2021).

5 CEATI International Inc. (2017). Electrical Energy Storage in Distribution Systems for Mitigation of Power Quality Issues. CEATI International Inc., Montreal, QC, Canada, Report No. T164700 #5173.

6 Alam, M.J.E., Muttaqi, K.M., and Sutanto, D. (2014). A novel approach for ramp-rate control of solar PV using energy storage to mitigate output fluctuations caused by cloud passing. *IEEE Transactions on Energy Conversion* 29: 507–518.

7 Knezović, K., Marinelli, M., Møller, R.J. et al. (2014). Analysis of voltage support by electric vehicles and photovoltaic in a real Danish low voltage network. In *Proc. 2014 49th International Universities Power Engineering Conference (UPEC)*, 1–6.

8 California ISO (2020). Todays outlook. http://www.caiso.com/TodaysOutlook/Pages/default.aspx (accessed 7 February 2020).

9 Varma, R.K., Hartley, T., Mohan, S. et al. (2020). Utility energy savings by novel smart inverter control of PV systems, BESS, and EV charging stations as STATCOMs. In *Proc. 2020 CIGRE Canada Conference.*

10 PlugShare (2020). EV charging station map – find a place to charge your car! https://www.plugshare.com/ (accessed 11 April 2020)

11 Ding, F., Nagarajan, A., Chakraborty, S., and Baggu, M. (2016). Photovoltaic Impact Assessment of Smart Inverter Volt–VAR Control on Distribution System Conservation Voltage Reduction and Power Quality. National Renewable Energy Laboratory, Golden, CO, USA, *Report No. NREL/TP-5D00-67296.*

12 Lefebvre, S., Gaba, G., Ba, A., D. Asber, A. Ricard, C. Perreault, et al. (2008). Measuring the efficiency of voltage reduction at Hydro-Québec distribution. In *Proc. 2008 IEEE Power & Energy Society General Meeting*, 1–7.

13 Wang, J., Raza, A., Hong, T. et al. (2018). Analysis of energy savings of CVR including refrigeration loads in distribution systems. *IEEE Transactions on Power Delivery* 33: 158–168.

14 Wang, Z. and Wang, J. (2014). Review on implementation and assessment of conservation voltage reduction. *IEEE Transactions on Power Systems* 29: 1306–1315.

15 Mahendru, A. and Varma, R.K. (2019). Reduction in system losses and power demand by combination of optimal power flow and conservation voltage reduction using smart PV inverters. In *Proc. 2019 IEEE Power & Energy Society General Meeting*, 1–5.

16 Matevosyan, J., Badrzadeh, B., Prevost, T. et al. (2019). Grid-forming inverters: are they the key for high renewable penetration? *IEEE Power and Energy Magazine* 17: 89–98.

17 TenneT TSO GmbH (2020). *The Massive InteGRATion of power Electronic devices (MIGRATE)*. TenneT TSO GmbH, Bayreuth, Germany, *Technical Brochure.*

18 GridLab and ESIG (2020). 10 Things You Should Know About Grid Forming Inverters. GridLab, Berkeley, CA, USA, *Report.*

19 Lasseter, R.H., Chen, Z., and Pattabiraman, D. (2020). Grid-forming inverters: a critical asset for the power grid. *IEEE Journal of Emerging and Selected Topics in Power Electronics* 8: 925–935.

20 NERC (2021). Grid Forming Technology - Definitions and Concerns to the Grid. NERC, Atlanta, GA, USA, *Draft White Paper.*

21 AEMO (2021). Application of Advanced Grid-scale Inverters in the NEM. Australian Energy Market Operator (AEMO), Melbourne, Australia, *White Paper.*

22 NERC (2018). BPS-Connected Inverter-Based Resource Performance. NERC, Atlanta, GA, USA, Reliability Guideline.

23 Lin, Y., Eto, J.H., Johnson, B.B. et al. (2020). Research Roadmap on Grid-Forming Inverters. National Renewable Energy Laboratory, Golden, CO, USA, *Report NREL/TP-5D00-73476.*

24 Huang J., Liu M., Zhang J., Dong W., and Chen Z, (2017). "Analysis and field test on reactive capability of photovoltaic power plants based on clusters of inverters", J. Mod. Power Syst. Clean Energy 5(2): 283–289.

25 Gevorgian, V. and O'Neill, B. (2017). Demonstration of active power controls by utility-scale PV power plant in an island grid. National Renewable Energy Laboratory, Golden, CO, USA, *Conference Paper NREL/CP-5D00-67255.* https://www.nrel.gov/docs/fy17osti/67255.pdf, (accessed November 16th, 2020).

26 Loutan, C., Morjaria, M., Gevorgian, V. et al. (2017). Demonstration of Essential Reliability Services by a 300-MW Solar Photovoltaic Power Plant. National Renewable Energy Laboratory, Golden, CO, USA, *Technical Report NREL/TP-5D00-67799.* https://www.nrel.gov/docs/fy17osti/67799.pdf, (accessed November 16th, 2020).

27 Varma, R.K., Siavashi, E.M., Mohan, S., and Vanderheide, T. (2019). First in Canada, night and day field demonstration of a new photovoltaic solar based Flexible AC Transmission System (FACTS) device PV-STATCOM for stabilizing critical induction motor. *IEEE Access* 7: 149479–149492.

28 Hingorani, N.G. and Gyugyi, L. (1999). *Understanding FACTS*. Piscataway, NJ: IEEE Press.

29 Mathur, R.M. and Varma, R.K. (2002). *Thyristor-Based FACTS Controllers for Electrical Transmission Systems.* New York: Wiley-IEEE Press.

30 Parnell, J. (2019). Lightsource BP makes solar pay at night. https://www.greentechmedia.com/articles/read/lightsource-bp-makes-solar-pay-at-night (accessed 25 November 2019).

31 First Solar (2020). First Solar Power Plant in Chile is World's First to Deliver Grid Services. https://investor.firstsolar.com/news/press-release-details/2020/First-Solar-Power-Plant-in-Chile-is-Worlds-First-to-Deliver-Grid-Services/default.aspx (accessed 4 January 2021).

32 Southern Regional Power Committee (SRPC), "Record notes of the special meeting on utilization of inverters at solar parks in night mode for controlling high voltages", Bengaluru, India, 27th July 2021, http://www.srpc.kar.nic.in/website/2021/meetings/special/rnsplnightmode-270721.pdf#page58 (accessed: 20th Aug. 2021).

33 Anaya, K.L. and Pollitt, M.G. (2018, 2020). Reactive power procurement: a review of current trends. *Applied Energy* 270: 1–16.

34 Energy and Environmental Economics, Inc. (2018). Investigating the Economic Value of Flexible Solar Power Plant Operation. Energy and Environmental Economics, Inc., San Francisco, CA, USA, *Report.*

35 EPRI (2017). Arizona Public Service Solar Partner Program – Advanced Inverter Demonstration Results. EPRI, Palo Alto, CA, USA, *Report 3002011316.*

36 IEEE (2018). Impact of IEEE 1547 Standard on Smart Inverters. IEEE Power & Energy Society, New York, NY, USA, *Technical Report PES-TR67.*

37 NERC (2019). Improvements to Interconnection Requirements for BPS-Connected Inverter-Based Resources. NERC, Atlanta, GA, USA, *Reliability Guideline.*

38 Varma, R.K., Khadkikar, V., and Seethapathy, R. (2009). Nighttime application of PV solar farm as STATCOM to regulate grid voltage. *IEEE Transactions on Energy Conversion (Letters)* 24: 983–985.

39 Varma, R.K., Rahman, S.A., and Seethapathy, R. (2010). Novel control of grid connected photovoltaic (PV) solar farm for improving transient stability and transmission limits both during night and day. In *Proc. 2010 World Energy Conference*, Montreal, Canada.

40 Varma, R.K., Das, B., Axente, I., and Vanderheide, T. (2011). Optimal 24-hr utilization of a PV solar system as STATCOM (PV-STATCOM) in a distribution network. In *Proc. 2011 IEEE Power & Energy Society General Meeting*, 1–8.

41 Varma, R.K., Rahman, S.A., Mahendra, A.C. et al. (2012). Novel nighttime application of PV solar farms as STATCOM (PV-STATCOM). In *Proc. 2012 IEEE Power & Energy Society General Meeting*, 1–8.

42 IEEE (2021), *Interconnection and Interoperability of Inverter-Based Resources Interconnecting with Associated Transmission Systems*, IEEE Draft Standard P2800.

43 Varma, R.K. and Siavashi, E.M. (2018). PV-STATCOM: A new smart inverter for voltage control in distribution systems. *IEEE Transactions on Sustainable Energy* 9: 1681–1691.

44 Varma, R.K. and Siavashi, E.M. (2019). Enhancement of solar farm connectivity with smart PV inverter PV-STATCOM. *IEEE Transactions on Sustainable Energy* 10: 1161–1171.

45 Varma, R.K., Mohan, S., and McMichael-Dennis, J. (2020). Multi-mode control of PV-STATCOM for stabilization of remote critical induction motor. *IEEE Journal of Photovoltaics* 10-6: 1872–1881.

46 Varma, R.K., Siavashi, E., Maleki, H. et al. 2016. PV-STATCOM: A novel smart inverter for transmission and distribution system applications. Poster Paper. In *Proc. 7th International Conference on Integration of Renewable and Distributed Energy Resources*, Niagara Falls, Canada.

47 Varma, R.K., Rahman, S.A., and Vanderheide, T. (2015). New control of PV solar farm as STATCOM (PV-STATCOM) for increasing grid power transmission limits during night and day. *IEEE Transactions on Power Delivery* 30: 755–763.

48 Varma, R.K. and Maleki, H. (2019). PV solar system control as STATCOM (PV-STATCOM) for power oscillation damping. *IEEE Transactions on Sustainable Energy* 10-4: 1793–1803.

49 Varma, R.K. and Salehi, R. (2017). SSR mitigation with a new control of PV solar farm as STATCOM (PV-STATCOM). *IEEE Transactions on Sustainable Energy* 8: 1473–1483.

50 Varma, R.K. and Mohan, S. (2020). Mitigation of fault induced delayed voltage recovery (FIDVR) by PV-STATCOM. *IEEE Transactions on Power Systems* 35-6: 4251–4262.

51 Varma, R.K. and Kelishadi, M.A. (2020). Simultaneous fast frequency control and power oscillation damping by utilizing PV solar system as PV-STATCOM. *IEEE Transactions on Sustainable Energy* 11-1: 415–425.

52 Varma R.K., Siavashi E.M., Mohan S., and McMichael-Dennis J., "Grid Support Benefits of Solar PV Systems as STATCOM (PV-STATCOM) Through Converter Control", *IEEE Electrification Magazine*, pp. 50–61, June 2021.

INDEX

Smart Solar PV Inverters with Advanced Grid Support Functionalities, First Edition. Rajiv K. Varma.
© 2022 The Institute of Electrical and Electronics Engineers, Inc. Published 2022 by John Wiley & Sons, Inc.

Printed and bound by CPI Group (UK) Ltd, Croydon, CR0 4YY